Wireless Communication Network Technology and Evolution

Wireless Communication Network Technology and Evolution

Shilin Wang
Nanjing Institute of Communication Engineering, China

Yunfei Cai
Shenzhen TCL New Technology Co., Ltd, China

Youyun Xu
Nanjing University of Posts & Telecommunications, China

Yuanyang Cai
Purdue University Northwest, USA

中国工信出版集团 電子工業出版社 PUBLISHING HOUSE OF ELECTRONICS INDUSTRY http://www.phei.com.cn World Scientific

Published by

World Scientific Publishing Co. Pte. Ltd.
5 Toh Tuck Link, Singapore 596224
USA office: 27 Warren Street, Suite 401-402, Hackensack, NJ 07601
UK office: 57 Shelton Street, Covent Garden, London WC2H 9HE

Library of Congress Cataloging-in-Publication Data
Names: Wang, Shilin, 1938– author. | Cai, Yunfei, author. | Xu, Y (Youyun), author. | Cai, Yuanyang, author.
Title: Wireless communication network technology and evolution / Shilin Wang, Nanjing Institute of Communication Engineering, China, Yunfei Cai, Shenzhen TCL New Technology Co., Ltd, China, Youyun Xu, Nanjing University of Posts & Telecommunications, China, Yuanyang Cai, Purdue University Northwest, USA.
Other titles: 880-01 Yi dong tong xin wang de yan jin. English
Description: Singapore ; Hackensack, NJ : World Scientific, [2022] | "Originally published in Chinese by Publishing House of Electronics Industry Co., Ltd." | Includes bibliographical references and index.
Identifiers: LCCN 2021059120 | ISBN 9789811245053 (hardcover) | ISBN 9789811245060 (ebook for institutions) | ISBN 9789811245077 (ebook for individuals)
Subjects: LCSH: Mobile communication systems. | 5G mobile communication systems.
Classification: LCC TK5103.2 .W3684 2022 | DDC 621.382/12--dc23/eng/20220204
LC record available at https://lccn.loc.gov/2021059120

British Library Cataloguing-in-Publication Data
A catalogue record for this book is available from the British Library.

移动通信网的演进
Originally published in Chinese by Publishing House of Electronics Industry Co., Ltd.

For any available supplementary material, please visit
https://www.worldscientific.com/worldscibooks/10.1142/12496#t=suppl

Desk Editors: Vishnu Mohan/Steven Patt

Typeset by Stallion Press
Email: enquiries@stallionpress.com

Foreword

by Pin Zhang

Wireless communication has become an indispensable basic method for the transmission and exchange of human language, text, pictures, images, data, symbols and other information. In the past 100 years, it has witnessed and promoted human progress and social development, and it will continue to play a role in information interaction in the process of human progress toward intelligent communication.

The invention and application of the telegraph in the mid-19th century fundamentally changed the way of mail delivery and remote information sharing for thousands of years. The continuous advancement of wireless communication technology in the 20th century had made radio broadcasting, fax and television widely popular. People sitting at home can know what major events are happening around the world, and they can listen and appreciate the songs and dances, customs and historical culture and arts from all parts of the world. Especially in the recent decades, with the integrated development of modern information and communication technologies (ICT) such as wireless communication and computer networks, the interconnection of people has achieved "global roaming, always online", which can be described as thousands of miles in a flash and horizons within easy reach.

Behind the provision of this convenient service is the huge investment in ICT infrastructure and the continuous development of wireless communication network technology for decades, including cellular mobile communication networks, wireless local area networks, wireless sensor networks and the Internet of things. The communication network exists in the form of a giant system and continues to explode. Therefore, whether as an individual or as a group, it is difficult for people to control this system alone or master all technologies by using one's own power. Then, how can a student or engineering technician in the ICT professional field with limited time and energy quickly understand the wireless communication network technology? In addition to the necessary professional basic training, it is essential for them to find a few comprehensive and professional technical books to read. This is exactly the purpose for which Professor Shilin Wang led the compilation of *Wireless Communication Network Technology and Evolution.*

Professor Wang and the other coauthors of this book have long been engaged in teaching and scientific research in related fields such as wireless communications and computer networks in universities. They also have rich experience in the design and product development of wireless mobile communications systems in the industry. Based on a large number of cutting-edge technical literature and academic books on wireless communication networks, they combined their own teachings and practical experiences, with full scanning and focused analysis, to explain the wireless communication network technologies' past and present, design ideas, key points, application methods, evolution processes and reasons. The contents of this book mainly focus on cellular mobile communication, wireless local area network (Wi-Fi), and wireless Internet of things (NB-IoT, eMTC).

The beginning of this book goes straight to the technical topic without any deviations. Each chapter begins with a brief summary, and the narrative style of this teaching plan is reflected in the subsequent chapters. This is one of the key features of this book.

The authors of this book closely follow the trends of the world and systematically introduce the technical principles and evolution of cellular communication networks, wireless local area networks and the Internet of Things. The three network technologies are integrated, and both common technologies and individual technologies are written, so that readers can clearly recognize the similarities and differences of the three networks. There are not only basic principles and formulas but also technical analysis, intuitive illustrations and interpretations, as well as experiences and cases throughout the book, thus explaining the profound things in a simple way. I believe this book can bring readers a wealth of knowledge and enjoyment as well as a distinctive reading experience. This book is very useful for engineering and technical personnel and university students.

Ping Zhang
Academician of Chinese Academy of Engineering
Professor of Beijing University of Posts and Telecommunications

Foreword

by Qihui Wu

When it comes to wireless communication, people naturally think of 4G or 5G. Indeed, these, which are the fourth- or fifth-generation wireless mobile communication systems, arc thc most succcssful representatives of the progress of wireless communication technology in the recent decades and one of the most powerful technological promoters of social innovation. However, wireless communication network technologies are not only limited to cellular mobile communications but also wireless local area networks, wireless sensor networks and the Internet of things.

In only 40 years, the mobile communication network has progressed from the first generation (1G) to the fifth generation (5G), enabling people's communication and information sharing to gradually break time and geographical limitations and realize the interconnection of people to people and people to things at any time and anywhere, and is now moving toward the Internet of Everything. With the evolution of mobile communication networks, wireless local area networks and the Internet of Things network have also developed. Although the application scenarios of these three networks are different from the other, under the perspective of wireless communication, many technologies and development trends are similar, such as transmission waveform (OFDM), high-order modulation (MPSK, QAM), channel coding (LDPC, HARQ), multi-antenna technology, frame structure and physical resource structure.

The authors of this book refer to these three networks as wireless communication networks and integrate these into one book. This may be the reason why the authors chose the title of the book *Wireless Communication Network Technology and Evolution*, which is one of the core focus points of this book. Chapters 1 and 2 mainly discuss the same technologies used in these three networks. Chapters 3–7 discuss the technologies used in different application scenarios of local area networks, Internet of Things and 5G, respectively. In the chapter on 5G technology, not only the relevant content of the R15 version but also the characteristics of the R16 version are discussed in detail. It also introduces the Starlink technology from the perspective of eliminating the digital divide, thus echoing the concept of air, space and ground integrated development of future wireless mobile communication networks (6G) to a certain extent.

The narratives of this book are all based on physical concepts. The authors combine their own experiences and selectively analyze several key technologies that restrict the performance of wireless communication networks. Although there are only limited mathematical problems involved, the authors always formulate profound interpretations close to the core concepts to enable more natural and smooth reading. This is another important characteristic of this book.

The authors of this book used to be my seniors and they have long been engaged in teaching and scientific research in professional directions, such as wireless communication and computer networks. Later, all of them worked in TCL Corporation and were engaged in the design, construction and maintenance of mobile phones, Wi-Fi and private networks. The academic research results and practical development experience accumulated by the authors are also reflected in this book, which is another key feature of this book.

I believe that the publication of this book will provide great help to related professional teachers, engineering and technical personnel, undergraduates and graduate students.

Qihui Wu
Professor of Nanjing University of Aeronautics and Astronautics
Yangtze River Scholar Professor of Chinese Ministry of Education

Preface

Wireless communication technologies, which are updated to a new generation in about every 10 years, are dedicated to obtaining faster network speeds and more functions for mobile devices not only in 1G, 2G (GSM), 2.5G (GPRS), 2.75G (EDGE), 3G (WCDMA), 3.5G (HSPA), 3.75G (HSPA+), 4G (LTE-A) and nowadays 5G (IMT-2020) but also in future 6G wireless communication network.

Remember beepers or the very first generation of mobile phones? Well, the author remembers them clearly. It was 1992 when the author constructed a wireless paging station in Guangdong, China. At that time, we took our mobile phones and climbed all over the hills of the Pearl River Delta (PRD) and installed the U-band wireless relay transceiver and the V-band radio frequency (RF) transmitter. We climbed towers of hundreds of meters to install the RF antennas and adjusted the paging system until the Beep Pager (BP) machine was able to communicate. At that time, the mobile phones that we used were actually called the 1G mobile phone, and the wireless paging system built by our paging station was a one-way system that can only transmit a few messages in one direction.

In 1999, which was also the first year that we abandoned the paging station and entered the TCL age, the author bought his first digital mobile phone which is regarded as the 2G mobile phone now. In addition to making phone calls, it could also send text messages. Until the year 2008, the author could not remember how many mobile phones he had changed because they were replaced in almost one to two years. There were brands such as Nokia, TCL, Ericsson, Sony Ericsson and others. After that, mobile phones could even connect to the Internet, even though the connection speed was too slow. Such phones are called the 2.5G (GPRS) mobile phones.

After 2008, smartphones have appeared and replaced the old mobile phones, which made online Internet access possible, but the connection speed was still not ideal. This generation of mobile phones is called the 3G mobile phone. At that time, we used the dual-mode mobile phone which supports GSM cards and WCDMA cards presented by the China Unicom. Such dual SIM cards and dual standby mobile phones were actually only transitional products.

After only a few years, a new standard mobile phone, which is the 4G (TD-LTE) mobile phone, had appeared around 2011. This kind of phone had greatly increased the network speed, which is nearly eight times faster than 3G, and is being widely used nowadays. The author first used the TCL mobile phone provided by the China Mobile for free (in fact, it is not free, it needs to be bundled with a one-year call fee) and later switched to iPhone and Huawei mobile phones.

With the rapid increase in the number of mobile phone users and connected devices, more and more 4G users started using a large amount of data networks at the same time, thus putting great pressure on the network. To solve this problem, 5G communication network came into being. As the next-generation wireless communication technology, 5G can run at GB-level speed, process GB-level data traffic and is 10 times faster than 4G. For example, downloading a full HD movie via 5G network can be done in just a few seconds or even faster. 5G will also become the basis of future instruments and equipment, thus making VR, autonomous driving, Internet of Things, smart home, cloud services and other application scenarios that we cannot imagine right now a reality.

But what exactly is 5G? 5G is the fifth-generation mobile communication technology. According to the requirements of the ITU, the 5G (IMT-2020) standard developed by 3GPP includes three scenarios: enhanced mobile broadband (eMBB), massive machine-type communication (mMTC) and ultra-reliable low-latency communication (URLLC). For example, 5G can support GB per second data rate, 3D video, ultra-HD display screen, cloud service, cloud entertainment, augmented reality (AR), industrial automation, business-critical applications, autonomous vehicles, smart homes, smart cities, smart voice and other application environments. At the same time, 5G can also improve the peak data rate, user-experienced data rate, mobility connection density, network energy efficiency, spectrum efficiency, and regional traffic capacity and so on.

Although the research of 5G is still in progress, we can already see several underlying technologies for 5G, such as millimeter wave, small cell, massive MIMO, beamforming, and full duplex.

With these technologies, the efficiency of 5G can be greatly improved. But to truly enter the 5G communication era, the continuous evolution of the current 4G LTE technology is required because in the early stages of 5G, both 5G and 4G should coexist. The advantage of such a coexistence is that, since the signal convergence of the 4G system is well constructed, 5G can use 4G as the base layer and function as the system enhancement layer which ensures a better user communication experience. But compared to the ultra-high bandwidth and speed of 5G, 4G LTE also needs to maintain a sufficient and powerful level to power up 5G, which introduces the concept of GB-level LTE by improving the speed of 4G into 1 Gbps. By supporting GB-level LTE mobile terminals, the user's connection experience can be significantly improved, especially in virtual reality, content streaming, cloud storage, and high-fidelity video and music caching. Taking virtual reality as an example, a GB-level LTE network can fully support immersive and fluent scene VR video streaming, allowing users to experience an unprecedented VR experience. Taking cloud storage as an example, GB-level LTE network allows opening files on the cloud as fast as opening them locally. As GB-level LTE network continues to develop in the future, it is expected that cloud storage will replace local storage. As an important milestone on the road to 5G, GB-level LTE will benefit the entire industry in many ways. For us, it also means opening the first door to the road toward 5G.

From the perspective of evolution, this book describes the development of radio networks with cellular radio networks (from 1G to 5G), wireless local area networks (WLAN) and narrowband Internet of Things (NB-IoT) as the main content and related technical measures to meet various application requirements. The top focus point of this book is 5G wireless communication.

To ensure a better reading experience, Chapter 1 introduces the basic principles of communication networks. Chapter 2 explains the evolution process, evolution theories and technical measures from 1G to 4G and also supplements some basic concepts which are involved in this book, including basic concepts of 4G and Wi-Fi networks. Chapter 3 describes the history, current situation and application examples of Wi-Fi development and looks forward to the future of Wi-Fi. Chapter 4 introduces the evolution history of the Internet of Things and several protocols, focusing on NB-IoT and its applications. Chapter 5 summarizes three application scenarios and eight major performance indicators and related technologies of 5G and discusses the main technical issues in the physical layer of the 5G access network. Chapter 6 introduces several core technologies of 5G. Chapter 7 compares the 5G research and application in China and the United States and finally the prospect of 6G.

This book is recommended for readers, including enterprise engineers, managers, and college-level, undergraduate, master's or other ICT students.

Chapters 1, 2, 5 and 7 are written by Shilin Wang. Chapters 3 (including 3.3 by Yuanyang Cai from Purdue University Northwest, USA) and Chapter 6 are written by Yunfei Cai, and Chapter 4 is jointly written by Shilin Wang and Yunfei Cai. Yuanyang Cai completed this book, taking care of the key processes including text entry, chart formula preparation, typesetting, and Youyun Xu reviewed the manuscript.

Acknowledgments

We would like to express our gratitude to all those who helped us during the writing of this book.

Yuhan Zhang (MASc) and Xin Ru (MASc) from Nanjing University of Posts and Telecommunications;

Dr. Jiasheng Xu from Shanghai Jiaotong University;

Dr. Shenghong Lin, Dr. Huawei Tong, Dr. Tianyou Li and Dr. Xinzhong Su from Nanjing University of Posts and Telecommunications;

Dr. Xiaochen Xia from Nanjing Institute of Communication Engineering;

Dr. Rui Jiang, Dr. Dapeng Li, Dr. Jian Chen, Dr. Xiaoming Wang, Dr. Li Wei and Dr. Haiming Li from Nanjing University of Posts and Telecommunications;

Xiaofei Zhao, a senior researcher at IoT think tank, Minjie Fan, a senior lecturer at Huawei Global Training Center and Liqing Feng, a senior lecturer at Huawei Global Training Center;

Professor Zhongmin Gan, Dr. Zhili Wang, Professor Qilin Fu and Dr. Jian Wang;

Professor Guyu Hu, Professor Yingfa Zhang, Professor Jingxia Liu, General Manager Wenbin Ling, Chairman Jian Kou, Professor Ming Zhang, Professor Ting Yang, Professor Chang Tian, Senior Engineer Xianghua PU, Senior Engineer Daqing Wang and Senior Engineer Jiayi Wang.

We would also like to thank Ms. Yingling Lin and Hong Lin (MA) for their concern and support.

Finally, we would like to express our sincere thanks to Academician Ping Zhang, Professor Qihui Wu, Chief Engineer Kailong Wang, Professor Guiyou Gao, Deputy Chief Engineer Hongwu Wang and Researcher Xiaochuan Dang for writing the preface and recommending this book amidst their busy schedules.

About the Authors

Shilin Wang is a Professor of Nanjing Institute of Communication Engineering and a technical consultant of TCL Industrial Research Institute and TCL mobile communication company. Since 1992, he has enjoyed the national special allowance for life. Due to outstanding achievements in scientific research, he has received second-class merit twice and also won the following awards for his works:

(1) four scientific and technological progress awards;
(2) state Council's A-class Product Award for bank data, voice and fax private network;
(3) the built-in antenna of mobile phone that won the national application patent;
(4) in 1987, his book *Modern Digital Modulation Technology* won the second prize of national natural science for excellent books;
(5) four papers were published in the International Conference on Communications (ICC), one in the *Journal of Electronic Science and Technology*, and more than 10 articles in *Telecommunication Science*;
(6) in 2015, he compiled modern digital modulation and its application in (i) constant envelope modulation and modern digital modulation and its application in (ii) high-order modulation, which has been highly praised by readers.

Yunfei Cai was graduated from Nanjing Institute of Communication Engineering, and is a Senior Engineer of Computer Science. With rich experience in ICT planning and construction, he has been engaged in IT systems and wireless data network construction for a long time; especially in TCL, he served as a senior manager focusing on the company's website, mailbox, intranet, WAN, WLAN, Data Center, audio and video conference system, VPN, desktop office virtualization, cloud, multimedia global intranet and other systems of planning, construction, operation and IT services. He participated in the construction of bank data, voice and fax private network, and won First Class Award of the State Council. From 1992 to 1998, he was responsible for the preparation, operation and maintenance of Dongjiang Paging Station in the Pearl River Delta. In 2015, he participated in the compilation of "modern digital modulation and its application": (i) constant envelope modulation and modern digital modulation and its application; (ii) high-order modulation.

Youyun Xu was graduated from Shanghai Jiao Tong University with a PhD degree in Information and Communication Engineering in 1999. He is currently a Professor at Nanjing University of Posts and Telecommunications and is serving as the Deputy Director of National Engineering Research Center of Communications and Networking. He has decades of experiences in mobile communication technology and R&D. Currently, he is in charge of a Sino-Spain international cooperation project under the National Key R&D Program. His current research interests include new generation wireless mobile communication systems (B5G/6G), new generation railway mobile communications, advanced channel coding and modulation techniques, multiuser information theory and radio resource management, wireless sensor networks, cognitive radio networks, and M2M communications. He is also the IEEE and a senior member of the Chinese Institute of Electronics.

Yuanyang Cai was a graduate student major in Electrical and Computer Engineering from Purdue University Northwest USA. He focuses on digital signal processing, deep learning network and automatic control. He obtained the bachelor's degree in Electrical Engineering and Intelligent Control from North University of China. He has experience in experiments and projects of communication theory, network security, network communication and big data. During his graduate study, he published several conference papers on fault signal detection, emotion recognition and so on. Now, under the guidance of Dr. Lizhe-Tan, he focuses on building and modifying the one-dimensional CNN network to reduce the computational complexity.

Contents

Chapter 1

Mobile Communication Basics

This chapter discusses the basics of mobile communication. To combine with practical application, we use the physical layer in the Long-Term Evolution (LTE) air interface (radio interface) as an example for discussion because the physical layer protocol almost covers the main basic knowledge in mobile communication.

The protocol architecture of the radio interface is shown in Figure 1.1 (see Figure 1.58 for details). The ellipse between the different layers is the Service Access Point (SAP). The physical layer is the bottom layer of the Open System Interconnection (OSI) reference model. It provides several physical channels for high-level services:

- Transmit various media data of the MAC sub-layer.
- Transmit various control signaling of the RRC sub-layer.
- Transmit the measured wireless channel state information to the RRC sub-layer.
- In addition to serving the higher layers, the physical layer also needs to transmit various reference signals generated by the physical layer itself.

These physical channels are all wireless channels. Now, we discuss the various problems that occur in the transmission of data in the wireless channel and the technical measures adopted to solve these problems.

It should be pointed out that the whole discussion process is based on physical concepts as the main and quantitative analysis as the supplement.

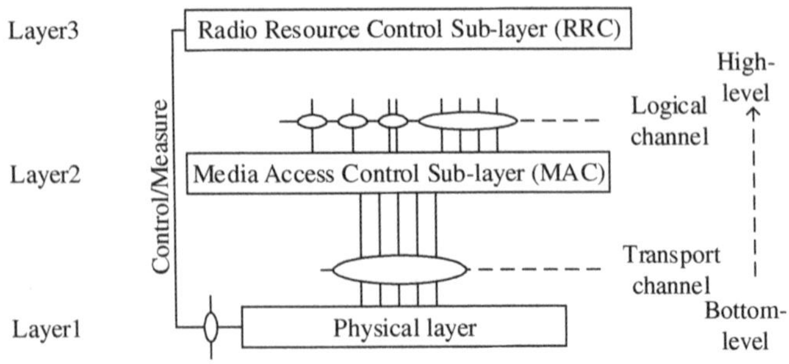

Figure 1.1 Radio interface protocol architecture.

Since the discussed questions have a strong causal relationship, the reading will be natural and smooth.

1.1 Methods to Improve Data Transmission Rate

With the development of mobile communication, the data rate through the MAC sub-layer is increasing. Therefore, the data rate of the physical channel in the physical layer is increasing, that is, the bandwidth of the transmitted data becomes wider. However, the physical channel bandwidth used to transmit data is limited, which is assigned by the National Radio Regulatory Agency (China's National Radio Regulatory Agency is the Ministry of Industry and Information Technology), and cannot be increased as we want. In this case, it happens that the data signal is larger than the channel bandwidth, making it unable to transmit the data signal from the physical channel. To solve this problem, four methods are discussed separately in the following sections.

1.1.1 *Adoption of high-order modulation*

When the data bandwidth BW_S is greater than the channel bandwidth BW_C, in order to ensure that the data can still be transmitted, we can narrow the bandwidth of the data output by the source and make it smaller than or equal to the channel bandwidth. Mathematically, this method is called mapping, and in the communication system, this method is called high-order modulation, as shown in Figure 1.2.

The function of mapping is to transform the input binary bit sequence into m-ary symbol sequence. A symbol in m-ary contains several binary

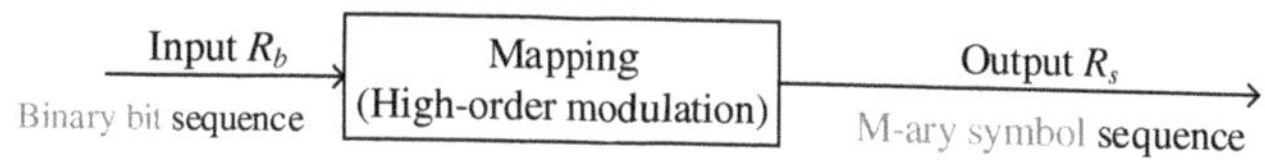

Figure 1.2 Mapping diagram.

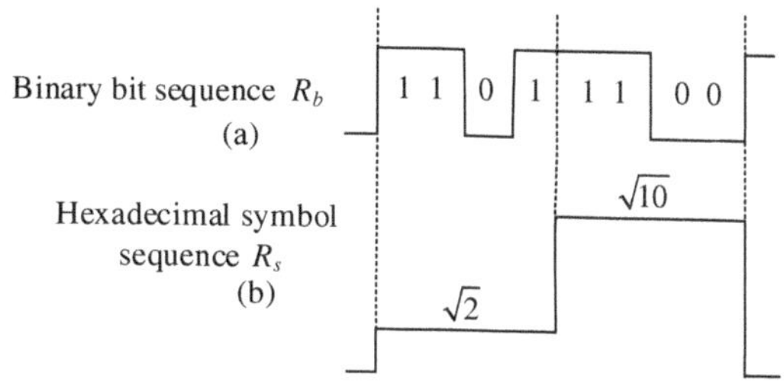

Figure 1.3 Binary mapping (2–16). In (b), $\sqrt{2}$ and $\sqrt{10}$ are the amplitudes of hexadecimal symbols (see Figure 1.9(a)).

bits. As a result, the output symbol rate *Rs* (symbols/sec) is less than the input bit rate R_b (bits/sec). In this way, the bandwidth of the symbol sequence, that is, the symbol signal, is narrowed.

Figure 1.3 shows an example of the conversion from binary to hexadecimal. It can be seen from the figure that when high-level modulation inputs four bits, it outputs one symbol, and one symbol contains four bits of information. There are totally 16 combinations (in hexadecimal system, it can be expressed as $(O–F)_{H}$). Therefore, high-order modulation correspondingly outputs 16 different symbols, and the modulator order M is 16. Obviously, the number of binary bits contained in a symbol output by high-order modulation is

$$n = \log_2 M \tag{1.1}$$

There are two high-order modulations used in mobile communications, which are discussed separately in the following sections.

1.1.1.1 *Phase Shift Keying (PSK)*

(i) *PSK constellation diagram*: Figure 1.4 shows the constellation diagrams of two-phase shift keying (BPSK), four-phase shift keying (QPSK) and eight-phase shift keying (8PSK). We use QPSK as an example to illustrate the meaning of constellations.

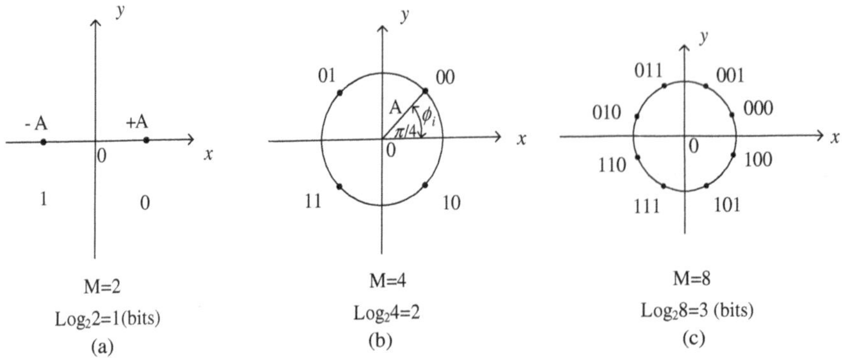

Figure 1.4 PSK constellation diagram: (a) BPSK, (b) QPSK and (c) 8PSK.

- The black dots · in the figure are called signal points. For QPSK, it represents the position of the four different 4-ary symbols output by the modulator in the Cartesian coordinate system. The number of signal points in the figure is equal to the modulation order *M*.
- The connection between signal points and the origin of the coordinate is called signal vector. The length of the vector is the amplitude *A* of the signal point (symbol). The angle between the vector and the *x*-axis is the initial phase of the vector ϕ_i (such as $\pi/4$ in the figure).
- The binary bit group near the signal point (for example, 00, 01, 11 and 01 in the figure) indicates that the signal point is generated by this group of binary bits after mapping. For example, the signal point $\phi_i = \pi/4$ in the figure is generated by bit 00. Therefore, all signal points in the figure have a fixed correspondence with the binary bit group.

(ii) *PSK signal expression table*: It can be seen from Figure 1.4 that all signal points of PSK modulation are uniformly (equal arc length) distributed on a circle with radius *A* (*A* is actually determined by the transmission power. The larger the power, the larger *A*, and vice versa). It can be seen that the amplitude in the PSK signal is the same, but the initial phase is different, which is determined by the binary bit group. Therefore, PSK signal uses phase to carry information, while the amplitude carries no information. The output PSK signal can be expressed as

$$S(t) = A\cos(2\pi f_0 t + \phi_i) \tag{1.2}$$

where f_0 is the carrier frequency.

$$\phi_i = \begin{cases} 0,\pi & \text{(BPSK)} \\ \pi/4, 3\pi/4, 5\pi/4, 7\pi/4 & \text{(QPSK)} \\ \pi/8, 3\pi/8, 5\pi/8, 7\pi/8, 9\pi/8, 11\pi/8, 13\pi/8, 15\pi/8 & \text{(8PSK)} \end{cases}$$

(iii) *PSK signal demodulator method*: The PSK coherent demodulator is shown in Figure 1.5. Following are the three main issues discussed here:

(1) *Principle of coherent demodulator*: It can be seen from the $S(t)$ of (1.2) that PSK demodulator first adjusts the early phase ϕ_i of $S(t)$ and then uses the constellation shown in Figure 1.4 to directly get the binary bit from ϕ_i. It can be seen that coherent demodulation only needs to solve the following two problems:

- *Get the initial phase ϕ_i of the input signal*: The multiplier and LPF in the figure are used to extract ϕ_i. The multiplier has two inputs: one inputs $\cos(2\pi f_0 t + \phi_i) + I(t)$, and the other inputs the local coherent carrier $\cos(2\pi f_0 t)$ provided by the receiver. When the interference $I(t)$ is not considered, it differs from the input signal $\cos(2\pi f_0 t + \phi_i)$ only by the factor of the initial phase ϕ_i. When the two signals are multiplied and the output

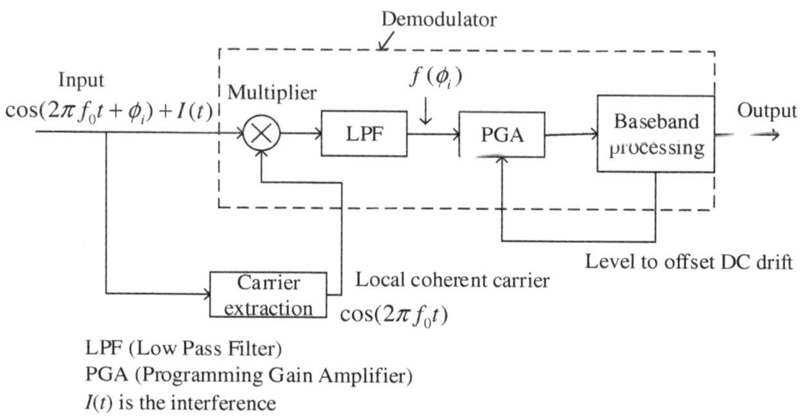

Figure 1.5 PSK coherent demodulator.

waveform after low-pass filtering is shown in Figure 1.6(a) (BPSK waveform is used as an example in this figure), it can be seen that when the "0" code is input, the half-wave output by the multiplier is positive, its DC component is > 0, and the output is positive after LPF filtering; when "1" code is input, the output half-wave of multiplier is negative, its DC component is < 0, and the output is negative after LPF filtering.

- *Use the value of $f(\phi_i)$ to solve the binary bit*: The baseband processor in Figure 1.5 is used to solve this problem. According to the characteristics of $f(\phi_i)$, it is not difficult to see that the baseband processor is actually a level judge, where the decision level is 0 volts, and the decision rule is as follows:

When $f(\phi_i) > 0$ volts, the decider outputs a "0" code.

When $f(\phi_i) < 0$ volts, the decider outputs a "1" code. (1.3)

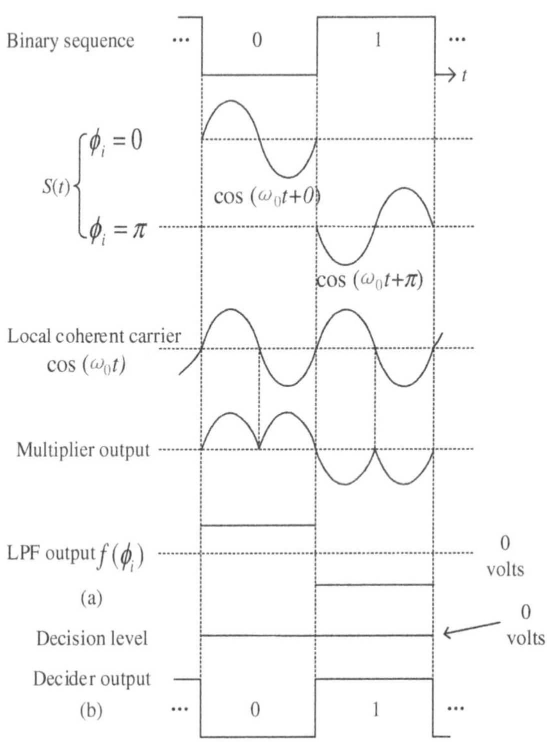

Figure 1.6 Principle of BPSK coherent demodulation.

The waveform is shown in Figure 1.6(b), and the figure shows the basic principle of the coherent demodulator.

(2) *How the local coherent carrier cos($\omega_0 t$) is obtained*: From the above analysis, we know that to achieve the coherent demodulation shown in Figure 1.5, the receiver must provide a local (receiver) coherent carrier, which can be obtained in the following two ways:

- *Extract the local coherent carrier directly from the input signal*: This is the method used in Figure 1.5. Since the input signal $S(t)$ = cos($\omega_0 t + \phi_i$) (I (t) is not considered), as long as ϕ_i is eliminated, the result is the local coherent carrier cos($\omega_0 t$). The way to eliminate ϕ_i is to undergo a nonlinear transformation. Take BPSK as an example, processing the double frequency operation on $S(t)$ can get cos 2($\omega_0 t + \phi_i$), so that $\phi_i = 0$ *or* π, and $2\phi_i = 0$ *or* 2π. So, cos 2($\omega_0 t + \phi_i$) = cos 2($\omega_0 t$). Then by dividing cos 2($\omega_0 t$) by two, we can obtain the local coherent carrier cos($\omega_0 t$). Similarly, for QPSK and 8PSK, by quadruple and octave processing, respectively, we can also obtain the local coherent carrier.
- *Send pilot signal*: In addition to transmitting the modulated signal, the transmitter also transmits a pilot signal. The configuration of the two in the frequency domain is shown in Figure 1.7.

Theoretically, it has been proved that modulated signal with symmetrical frequency spectrum can be produced after modulation with binary sequence such as 0 or 1. There is no discrete component of carrier (f_o) in its power spectral density. Therefore, an unmodulated carrier frequency can be inserted at the position of the carrier, which is called a pilot. The pilot signal is transmitted simultaneously with the modulated signal. It can be seen that the local coherent carrier cos($\omega_0 t$) can be obtained by

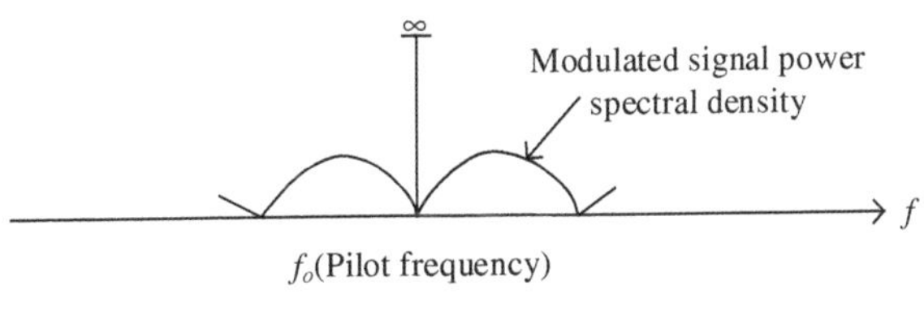

Figure 1.7 Diagram of inserting pilot frequency.

filtering with a narrowband filter with a center frequency of f_o at the receiver. The second-generation CDMA IS-95 protocol uses such a method of pilot insert.

(3) *DC drift and methods to offset DC drift* [7]: It can be seen from Figure 1.5 that the multiplier is actually a mixer. The two input signals of the mixer are of the same frequency. Therefore, the intermediate frequency output by the mixer is zero intermediate frequency, which is DC. In theory, the isolation between these two input terminals should be ∞. In fact, it is only a few decibels. This means that the two input signals leak each other. Especially the local coherent carrier leaks to the input signal of the multiplier due to the strong signal. Both local coherent carrier and signal leaking produce a zero-IF signal at the output of the multiplier. This mixing phenomenon is called self-mixing. Since the level of leakage is unstable, the zero-IF signal intensity generated by self-mixing is drifting, thus it is called the DC drift. Obviously, the signal level output by the LPF will also drift. This kind of drifting indefinite level is detrimental to the subsequent work of the arbiter. Therefore, we must try to eliminate the DC drift caused by self-mixing. The method used for this is to insert a PGA amplifier between the LPF and the baseband processor, and the baseband processor extracts a level that changes with the DC drift and controls the gain of PGA, thereby making the PGA output a more stable signal level.

1.1.1.2 *Quadrature Amplitude Modulation* (*QAM*)

(i) *Why do we need QAM?* It can be seen from the constellation in Figure 1.4 that the M signal points of PSK are evenly (equally spaced) distributed on a circle. In order to increase the data transmission rate, it can be seen from (1.1) that we can increase the modulation level M. For example, we can use 16PSK modulation with $M = 16$. At this time, 16 signal points will be evenly arranged on a circle, that is, another signal point will be added between any two adjacent signal points in Figure 1.4(c). When the radius of the circle does not change (that is, the transmit power does not change), the distance between two adjacent signal points of 16PSK is twice as small as that of 8PSK. Obviously, as M increases, the denser the signal points arranged on the circle, the smaller the distance between two adjacent signal points. The straight-line distance between signal points on the constellation diagram is called the Euclidean distance, and the influence of the decrease in Euclidean distance is illustrated in Figure 1.8. Figure 1.8(a) shows a certain signal point on the transmitting circle, which is an ideal signal point without any interference and noise. The corresponding signal vector has only one amplitude and one initial phase, so it is called an aggregated signal point. Figure 1.8(b) shows the receiver signal point, which is affected by various interference and noise. Since the values of interference and noise are random, the position of the signal point on the constellation diagram is also uncertain, and this signal point is a divergent signal point. Obviously, the magnitude and phase of the corresponding vector of the divergent signal point are also random. This means that each signal point at the receiver is a diverging signal point. When the Euclidean distance between two adjacent signal points is reduced, the two diverging signal points may be interwoven with each other. Therefore, the normal judgment of these two signal points can easily be mistaken. It can be seen that for PSK modulation, when the transmitter power is constant, the demodulation bit error rate P_b will increase as the modulation order M increases, or the effectiveness of system transmission will increase. However, the reliability of system transmission has decreased. Therefore, for PSK, the increase of M is limited. In practice, the maximum of M is 8, that is, 8PSK.

The fundamental reason for the limited increase in PSK modulation M is that PSK only uses the phase of the signal vector to carry information, while the amplitude of the signal vector is of constant amplitude and does not carry information. Therefore, M signal points can only be distributed on one circle. This method of arranging signal points makes it impossible to make full use of the signal plane formed by the x-axis and the y-axis.

To fully utilize the signal plane, both the amplitude and phase of the signal vector in QAM modulation carry information. In this way, M signal points are not distributed on a circle. Under such configuration, the

Figure 1.8 Signal points of the transmitter and receiver: (a) gathered signal points at the transmitter and (b) divergent signal point at the receiver.

Euclidean distance between adjacent signal points may not necessarily decrease as M increases, which is the biggest advantage of QAM. In order to highlight the difference between QAM and PSK, QAM is called quadrature amplitude modulation or amplitude and phase joint modulation.

(ii) *QAM constellation*: QAM modulation can also be written as MQAM, where the first M is the modulation order, which is called multiple quadrature amplitude modulation and $M = 2^n$. When n is an even number, the shape of the constellation is a rectangle (square). When n is an odd number, the shape of the constellation is a cross. Figure 1.9 shows two examples of constellation diagrams.

In practice, n is an even number, so the constellation diagrams are all rectangular. At present, 16QAM, 64QAM, 256QAM and 1024QAM are used in mobile communication networks and wireless local area networks.

(iii) *Rectangular constellation QAM signal demodulation method*: We take 16QAM in Figure 1.9(a) as an example to illustrate the QAM signal demodulation method. It can be seen from the figure that each signal point corresponds to four bits $a_1a_2b_1b_2$. Each signal point has a certain projection value (±1, ±3) on the y-axis and x-axis, respectively.

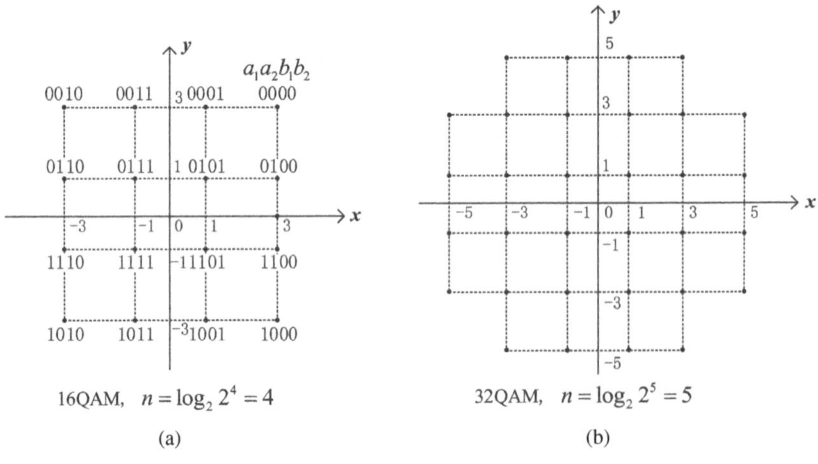

Figure 1.9 QAM constellation: (a) rectangle; (b) cross shape.

The corresponding relationship between these four bits and the projection value is shown in Table 1.1. It can be seen from the table that the correspondence between double bits a_1a_2 and the y-axis value is the same as the correspondence between double bits b_1b_2 and the x-axis value. The corresponding relationship is shown in Figure 1.10. The modulation corresponding to this constellation is called Pulse Amplitude Modulation (PAM). It can be seen that QAM modulation can be decomposed into two identical PAM modulations. Therefore, the demodulation of the QAM signal is transformed into the demodulation of the PAM signal.

The PAM demodulation method is shown in Figure 1.10. Set the three decision levels as 0 and ±2 between two adjacent projection values. The level decision rules are as follows.

Suppose the projection value of the received k-th quaternary symbol on the two axes is y_k and x_k (both values are divergent). Then, the following conditions apply:

Table 1.1 Correspondence between double bit and projection value.

Y	a_1a_2	X	b_1b_2
+3	00	+3	00
+1	01	+1	01
−1	11	−1	11
−3	10	−3	10
(a)		(b)	

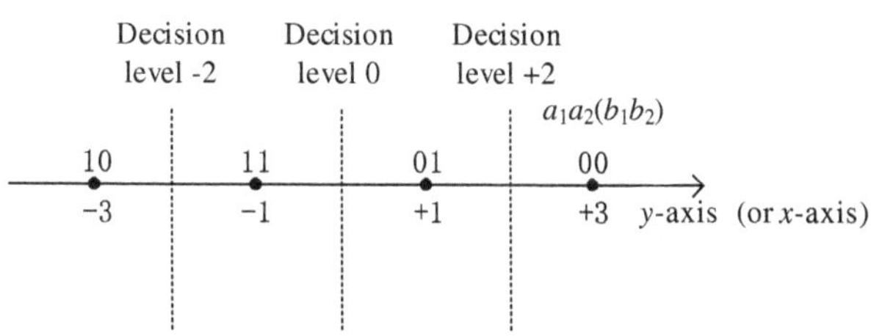

Figure 1.10 PAM modulation.

If y_k (or x_k) > +2, the PAM level is determined to be +3, and the decider outputs double-bit "00".

If $0 < y_k$ (*or* x_k) $\leq +2$, the PAM level is determined to be +1, and the decider outputs double-bit "01".

If $-2 < y_k$ (*or* x_k) ≤ 0, the PAM level is determined to be −1, and the decider outputs a double bit "11". (1.4)

If y_k (*or* x_k) < -2, the PAM level is determined to be −3, and the decider outputs a double bit "10".

Two kinds of high-order modulation MPSK and MQAM are discussed above. So far, we need to introduce a physical parameter, which is called channel frequency band utilization, which is defined as

$$\text{Channel band utilization} = \frac{\text{Data transfer rate}}{\text{Channel bandwidth}}\left(\text{b / s / Hz}\right) \tag{1.5}$$

It represents the data rate (b/s) that the channel can transmit for each 1 Hz bandwidth provided by the channel. If the transmitted data rate is higher, the channel frequency band utilization rate is higher, and vice versa. When the channel bandwidth is certain (for example, the channel bandwidth (MHz) specified in 4G LTE has six types, namely 1.4, 3, 5, 10, 15 and 20), and the above-mentioned high-order modulation is used, the higher the modulation order M, the higher the data rate of channel transmission, and the higher the channel bandwidth utilization rate. Mobile communication networks and wireless local area networks use this method to improve frequency band utilization. Both of them have undergone five generations of evolution. The former has gone through the process from QPSK, 8PSK to 16QAM, 64QAM and 256QAM (see Chapter 2). The latter has gone through the process from QPSK, 8PSK to 16QAM, 64QAM, 256QAM and 1024QAM (see Chapter 3).

1.1.2 *Adoption of MIMO technology* [14]

1.1.2.1 *What is MIMO?*

MIMO is referred to as multiple input and multiple output. It equips multiple antennas at the transmitter and receiver, and its structure is shown in Figure 1.11. So, MIMO technology is essentially a multi-antenna technology. The number of transmitting antennas is N_T, and the number of receiving antennas is N_R.

1.1.2.2 *Types of MIMO*

According to the number of receiving and transmitting antennas and its usage, MIMO can be divided into the following five categories:

(i) *When* $N_T = N_R = 1$, *it is called Single Input Single Output (SISO)*. It is the traditional communication method shown in Figure 1.12(a).

(ii) *Multiple receiving antennas*: As shown in Figure 1.12(b), one transmitting antenna and multiple receiving antennas are called Single Input Multiple Output (SIMO). In the figure, "+" means linear combination of N_R received signals. The N_R antennas at the receiver keep a certain distance between each other (the distance between the base station antennas is about 10 wavelengths of 10λ, and the distance between the terminal antennas is about half a wavelength of 0.5λ), so that the N_R signals received through different paths only have small cross-correlation, which means that there are both strong and weak N_R signals, and their fading is impossible at the same time. It is for this reason that as long as the N_R signals are combined linearly (i.e., the "+" box in the figure), the signal strength obtained is stronger than that of a single signal (under the same space environment). There are multiple methods for linear merging. The simplest method is equivalent to a switch, which is used to select the

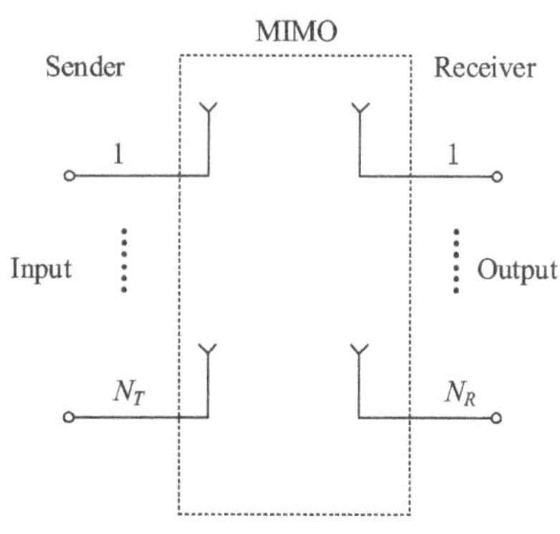

Figure 1.11 MIMO structure.

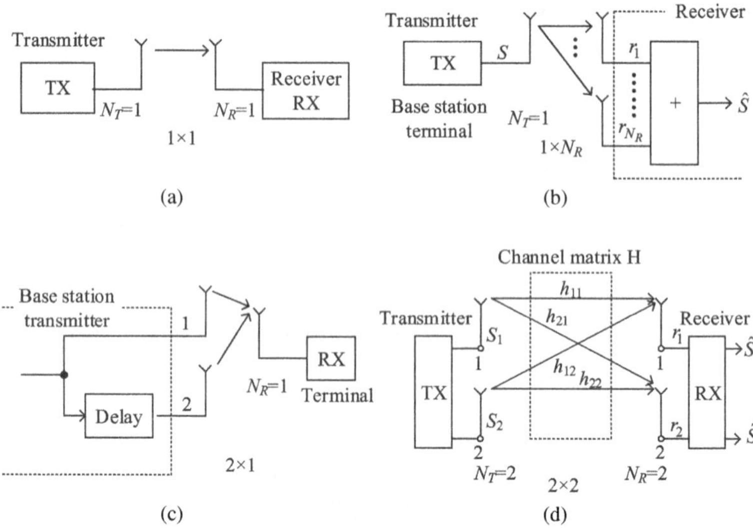

Figure 1.12 MIMO structure diagram: (a) SISO, (b) SIMO, (c) MISO and (d) MIMO.

strongest signal among the N_R signals and transmit it to the receiver for demodulation. As the signal-to-noise ratio is improved, the bit error rate is reduced. It can be seen that when the same bit error rate is guaranteed, the data transmission rate can be increased. Moreover, the channel bandwidth is not increased at this time, only space resources are used. Therefore, this method can also improve the channel frequency band utilization. SIMO technology is called receiving antenna diversity technology, which is a diversity method of space diversity.

(iii) *Multiple transmitting antennas*: As shown in Figure 1.12(c), two transmitting antennas and one receiving antenna are called Multiple Input Single Output (MISO). The transmitting two antennas emit the same signal, but one of the signals is delayed. Similar to multiple receiving antennas, MISO is the diversity of transmitting antennas. To be precise, it is a dual transmitting antenna delay diversity. It also belongs to the space diversity.

(iv) *Beamforming*: The multiple antennas at the transmitter or receiver are not used for spatial diversity (Figures 1.12(b) and (c)), nor are they used for spatial multiplexing (Figure 1.12(d)), but are used to form an antenna array (at this time, the antenna spacing on the base station side is $<10\lambda$). Each antenna in the array is equivalent to an antenna after the processing shown in Figure 5.1 in Chapter 5. Although each antenna in the array is an omnidirectional antenna (that is, a non-directional antenna), as shown in Figure 5.2(a) in Chapter 5. However, after equivalence, this antenna is a directional antenna, as shown in Figure 5.2(b) in Chapter 5. We have already known omnidirectional antennas that $G = 1(G_{dB} = 0)$, but the gain of directional antennas is $G > 1(G_{dB} > 0)$. Directional antennas make multiple non-directional antennas with $G_{dB} = 0$ into an equivalent antenna with $G_{dB} > 0$, which is called antenna beamforming. When the antenna gain is increased, the signal-to-noise ratio received by the receiver is increased, thereby reducing the bit error rate. When the same bit error rate is guaranteed, the data transmission rate can be increased, and therefore the channel frequency band utilization rate is also improved.

(v) *Space Division Multiplexing (SDM)*: Figures 1.12(b) and (c) are collectively referred to as space diversity. Space diversity simply uses multiple receiving antennas or multiple transmitting antennas. In spatial multiplexing, multiple antennas are used at both the transmitter and receiver to achieve true MIMO. The MIMO used in 4G LTE is 2×2, 4×4 and 8×8. Figure 1.12(d) shows a schematic diagram of 2×2 MIMO. MIMO is one of the important technologies applied in LTE, so it will be discussed in detail in the next section.

1.1.2.3 *Principle of spatial reuse*

We use 2×2 in Figure 1.12(d) as an example to discuss the principle of spatial reuse. The working principles of the transmitter and receiver are discussed in the following:

(i) *Transmitter*: The composition of the transmitter TX is shown in Figure 1.13. It can be seen from the figure that it splits (shunts) the input high-speed data into N_T ($N_T = 2$ in the figure) low-speed data streams, performs channel coding, and then performs hierarchical spatial coding. Then, each layer of data stream is modulated separately to obtain N_T output ports (two ports 0 and 1 in the figure). The N_T output ports are, respectively, connected to N_T transmitting antennas (antennas 1 and 2 in the figure). The advantage of this multi-antenna structure is SISO in

Figure 1.12 uses only one spatial channel to transmit data, while MIMO can use N_T independent spatial channels (when the antenna spacing is about 10λ) to transmit data. The sum of N_T data rates is the total data rate. Obviously, this method can increase the data transmission rate, but the channel bandwidth used is still the same as in SISO. Therefore, the utilization rate of the channel frequency band is also improved.

(ii) *Receiver:* Suppose the signals transmitted by the two antennas in Figure 1.12(d) are S_1 and S_2. The signals received by the two receiving antennas after transmission through the space channel are r_1 (with noise n_1) and r_2 (with noise n_2). We can obtain from the figure that r_1 and r_2 are

$$r_1 = h_{11}S_1 + h_{12}S_2 + n_1,\ r_2 = h_{21}S_1 + h_{22}S_1 + n_2 \tag{1.6}$$

The subscript i of h_{ij} in the formula represents the receiving antenna number (1 and 2), the subscript j represents the serial number of the transmitting antenna (1 or 2), and h_{ij} is the spatial channel transfer function formed between the jth transmitting antenna and the ith receiving antenna. Equation (1.6) can be expressed by the matrix as

$$\bar{r} = \begin{bmatrix} r_1 \\ r_2 \end{bmatrix} = \begin{bmatrix} h_{11} & h_{12} \\ h_{21} & h_{22} \end{bmatrix} \cdot \begin{bmatrix} S_1 \\ S_2 \end{bmatrix} + \begin{bmatrix} n_1 \\ n_2 \end{bmatrix}$$

$$\bar{r} = H \cdot \bar{S} + \bar{n} \tag{1.7}$$

where

$\bar{r} = \begin{bmatrix} r_1 \\ r_2 \end{bmatrix}$ is called the receiving vector (column matrix),

$\bar{S} = \begin{bmatrix} S_1 \\ S_2 \end{bmatrix}$ is called the emission vector,

$\bar{n} = \begin{bmatrix} n_1 \\ n_2 \end{bmatrix}$ is called the noise vector, and

$H = \begin{bmatrix} h_{11} & h_{12} \\ h_{21} & h_{22} \end{bmatrix}$ is called the channel matrix.

If the inverse matrix H^{-1} of the channel matrix H exists, then the two sides of equation (1.7) can be multiplied by H^{-1} to obtain

$$\begin{aligned} H^{-1}\overline{r} &= H^{-1} \cdot H \cdot \overline{S} + H^{-1} \cdot \overline{n} \\ &= \overline{S} + H^{-1} \cdot \overline{n} \end{aligned}$$

$$\begin{bmatrix} \hat{S}_1 \\ \hat{S}_2 \end{bmatrix} = \begin{bmatrix} S_1 \\ S_2 \end{bmatrix} + H^{-1} \cdot \overline{n} \tag{1.8}$$

It can be seen from equation (1.8) that as long as the channel matrix H can be obtained, the $\hat{S}_1$ and $\hat{S}_2$ corresponding to the transmitted signals S_1 and S_2 can be separated (demodulated) at the receiver. The only difference between the two is that $\hat{S}_1$ and $\hat{S}_2$ contains noise $H^{-1} \times \overline{n}$. And H^{-1} may increase the original noise $\overline{n}$. According to the demodulation principle of (1.8) and Figure 1.13, we can get the receiver structure, as shown in Figure 1.14. The channel estimation function block in the figure is used to extract the channel matrix H from the signals received by the two antennas. The extraction method and calculation method of (1.8) will be discussed later.

LTE Advanced (LTE-A) currently supports two spatial multiplexing. The first is called single-user MIMO, written as SU-MIMO. In the environment of the same time-frequency resource (see the discussion in Figure 5.4 in Chapter 5), all space division multiplexing (SDM) data streams are used for the same user UE. The second is called multi-user MIMO, written as MU-MIMO. In the same time-frequency resource environment, multiple SDM data streams are used for multiple user UE. The switching of the above two modes is controlled by eNodeB (enhanced base station).

1.1.3 *Adoption of carrier aggregation technology*

(i) *What is Carrier Aggregation (CA)?* The purpose of using CA is to enhance the channel bandwidth, which can increase the data transmission

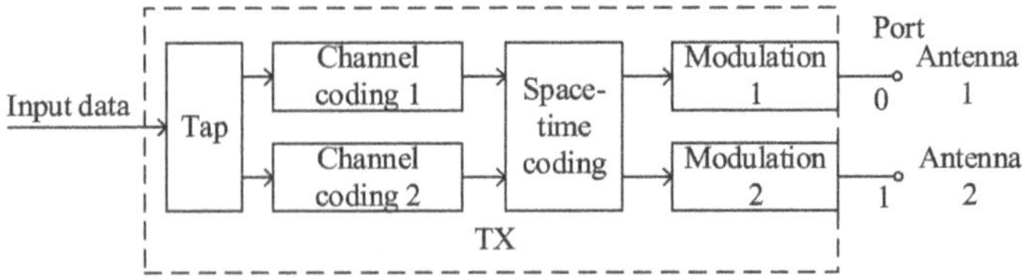

Figure 1.13 MIMO transmitter.

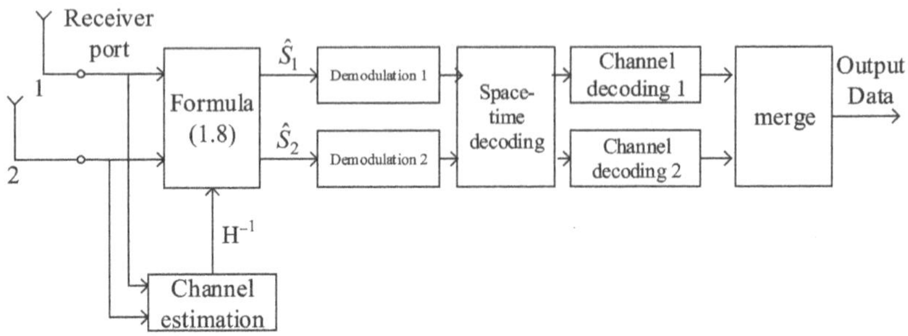

Figure 1.14 MIMO receiver.

rate. However, this method cannot improve channel frequency band utilization rate. The high-order modulation method discussed above can not only increase the data rate but also improve the utilization rate of the channel frequency band.

We know that the maximum bandwidth allowed by the system in LTE version 8 R8 is 20 MHz. However, LTE-A requires that the system bandwidth cannot be less than 40 MHz, and the maximum bandwidth is 100 MHz. In this case, it is impossible to require the National Radio Regulatory Agency to reallocate new frequency bands to meet this requirement. It can be seen that telecom operators can only solve this problem by using the spectrum resources they already have. To this end, 3GPP introduced CA technology in the R10 version. It manages to aggregate the existing frequency bands and idle frequency bands, thereby synthesizing them into a whole new band spectrum to increase the bandwidth.

(ii) *Carrier aggregation method*: There are three CA methods, as shown in Figure 1.15. The complexity of radio frequency equipment is different

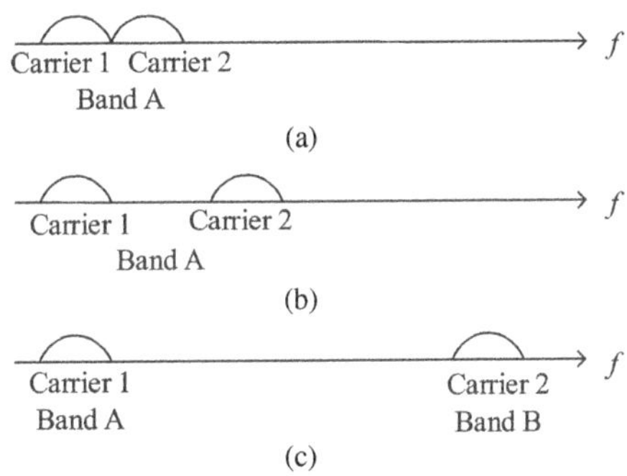

Figure 1.15 Carrier aggregation method: (a) in-band continuous carrier aggregation, (b) in-band discontinuous carrier aggregation and (c) inter-band non-contiguous carrier aggregation.

when different methods are implemented. The first type (Figure 1.15(a)) is the simplest. Related issues will be discussed in Chapter 2.

Each carrier participating in the aggregation is called a Component Carrier (CC). In LTE-A, two or more CCs are allowed to be used together. Up to five carriers are allowed, the maximum bandwidth of each carrier is 20 MHz, and the bandwidth after aggregation is 100 MHz.

In order to ensure the backward compatibility of LTE-A, it is required that each component carrier must be configured according to the R8/R9 version.

Generally, CA can be used for both uplink and downlink. Moreover, the number of aggregated component carriers and the bandwidth of each component carrier can be different, and the duplex mode can also be different. This is how it is applied in 5G. However, in 4G, only the downlink currently uses CA technology.

1.1.4 *Development of new frequency bands*

1G and 2G work frequency bands are 800–900 MHz. 3G mainly uses the 2 GHz frequency band. With the continuous development of 3G and 4G, new lower or higher frequency bands have been adopted. It has crossed the frequency range of 450 MHz–6 GHz. In addition to the sub-6 GHz frequency band used in 5G, the millimeter wave frequency band 24.25–100 GHz has also been developed. Higher frequency bands will also be used in 6G. For example, the terahertz (THz) (1 THz = 10^3 GHz) frequency band, which is a sub-millimeter wave frequency band with a frequency range of 100 GHz–10 THz. The purpose of developing higher frequency band is to increase the bandwidth of the carrier. For example, the maximum bandwidth of one carrier in millimeter wave can reach 400 MHz, while the maximum bandwidth of one carrier in 4G is only 20 MHz. When the carrier bandwidth increases, the system allows the system to transmit higher data rates.

1.2 Methods of Anti-Symbol Interference

In the previous section, high-order modulation was discussed. If the output m-ary symbol sequence is directly transmitted in the wireless channel, the multipath effect of the channel will cause interference between adjacent symbols, and this interference is called Intersymbol Interference (ISI). The existence of ISI increases the rate of demodulation error symbols. Especially in the case of high symbol rate, ISI will greatly reduce the reliability of the communication system, which is not allowed. It can be seen that the symbol sequence output by high-order modulation cannot be directly transmitted in the wireless channel. In order to reduce ISI, the symbol sequence output by the high-order modulation needs to be mapped again. This mapping is called Orthogonal Frequency Division Multiplexing (OFDM), as shown in Figure 1.16. The symbol sequence output by OFDM is directly transmitted in the wireless channel. It should be emphasized that the purpose of setting OFDM modulation is not to increase the data transmission rate, but to resist inter-symbol interference.

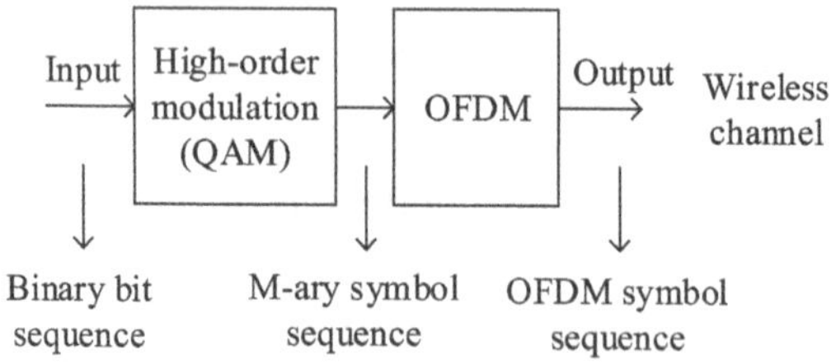

Figure 1.16 The purpose of OFDM.

The following section will first discuss the wireless channel transmission characteristics, then discuss the reasons for ISI, and finally discuss the principle of OFDM anti-ISI.

1.2.1 *Transmission characteristics of wireless channel*

The transmission characteristics of the wireless channel are very complicated. It is impossible and unnecessary for us to discuss it in depth here. But we will qualitatively discuss the following two characteristics according to the needs of practical applications.

1.2.1.1 *Multipath effect*

In wireless channels, radio waves are propagated by space waves. Space waves have many forms such as direct waves, reflected waves, scattered waves and diffracted waves. The space wave that directly connects the two communication parties is a direct wave, and the other types are space waves generated by obstacles encountered during the propagation of radio waves. When users use mobile communication networks outdoors to communicate, due to the large outdoor space, there are many types of obstacles encountered by the radio waves, for example, various buildings, woods, high slopes, rivers, moving vehicles, etc. When users use wireless local area network communication indoors, because the indoor space is small, there are fewer types of obstacles encountered by radio waves, for example, walls, ceilings, etc. It can be seen that there are multiple paths for both outdoor and indoor communications, as shown in Figure 1.17. Next, we discuss the influence of the multipath effect on the transmitted signal.

(i) *Multipath fading*: It can be seen from the figure that the path lengths taken by these space waves to reach the receiver are different. For example, the direct wave path is the shortest, and other space wave paths are longer than it. The signals arriving at the receiver through these different paths cannot be in phase. Therefore, the received signal vector is the sum of these different path signal vectors, as shown in Figure 1.18. Since the amplitude and phase of these space waves arriving at the receiver are random, the intensity of the received signal vector must also vary randomly over time. This phenomenon is called

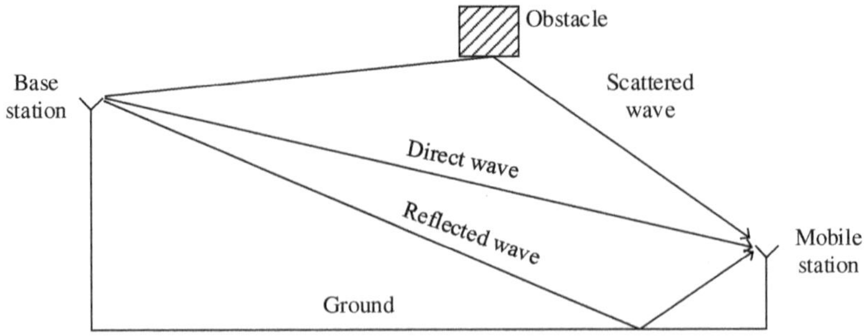

Figure 1.17 Multipath effect.

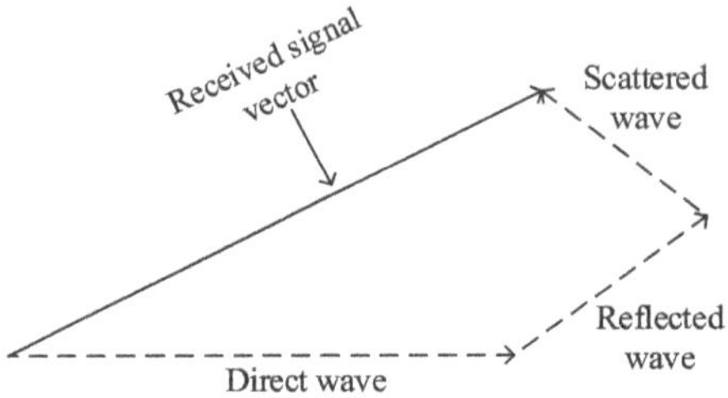

Figure 1.18 The vector sum of different space waves.

multipath fading, which means that the received signal strength produces fluctuations of different magnitudes due to multipath. Due to different propagation environments, multipath fading is divided into fast fading and slow fading. The fading rate of the former is fast and the fading rate of the latter is slow. In order to compensate for the fading of the received signal strength, there must be a system margin in the cell design. The size of the margin is determined by the communication reliability.

(ii) *Delay extension*: Due to the multipath effect, when the transmitter transmits a narrow pulse, the receiver will receive multiple narrow pulses, as shown in Figure 1.19. Moreover, the time delay for these pulses to reach the receiving end is different. The shortest path arrives at the receiving end first, and the longest path arrives at the receiving end last. The time difference between the two is called delay spread τ_m. Obviously, outdoor communication delay τ_m is larger than indoor communication delay. The influence of delay spread on the transmitted signal in the time domain is shown in Figure 1.20. When the transmitter transmits symbols

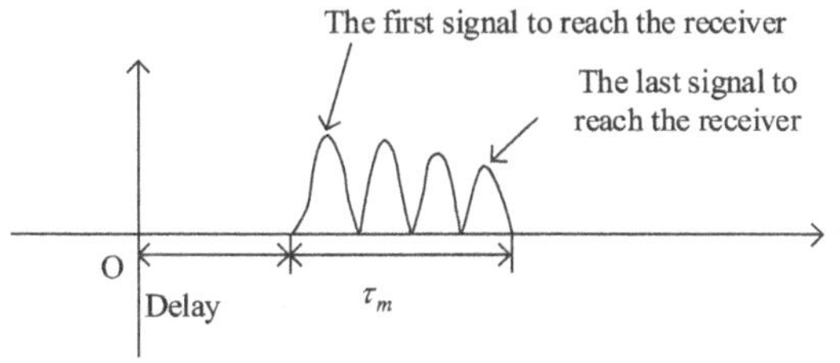

Figure 1.19 Delay spread.

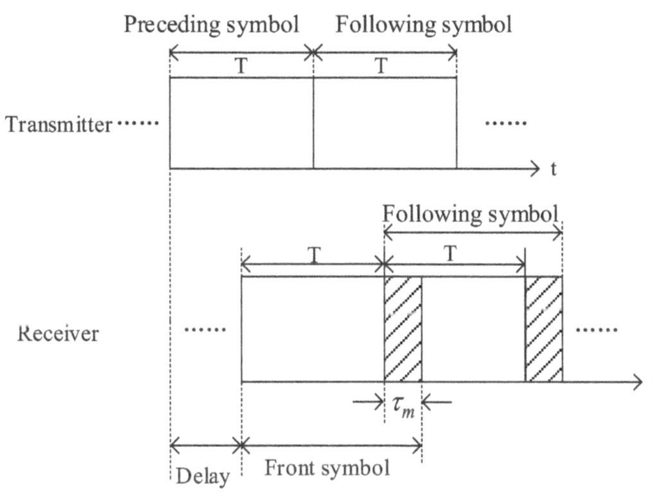

Figure 1.20 Intersymbol interference.

continuously, due to the delay τ_m spread, the received preceding symbols will be extended to the following symbols, resulting in ISI. It can be seen from the figure that in order to reduce the damage of ISI to the following symbols, T is required to be much larger than τ_m, which reduces the symbol transmission rate.

An important concept related to delay spread is coherent band width ΔB_c.

$$\Delta B_c \approx \frac{1}{\tau_m}(\text{Hz}) \tag{1.9}$$

The physical meaning of ΔB_c is that the frequency domain transfer function of the wireless channel is the same for all frequencies in the frequency band ΔB_c.

(iii) *Frequency selective fading*: Due to the multipath effect, the wireless channel has the characteristics of frequency selective fading. More precisely, the channel has selective fading in frequency bands with different coherence bandwidths ΔB_c. This means that the channel has different transfer functions in frequency bands with different coherent bandwidths ΔB_c. However, the transfer function of all frequencies within the same coherent bandwidth ΔB_c is the same. For this reason, we can divide the channel with a total bandwidth of B into several frequency bands or into several narrowband channels, and the bandwidth of each narrowband channel is smaller than the coherent bandwidth ΔB_c. As shown in Figure 1.21, the transfer function of each narrowband channel is the same, for example, the transfer function of the ith narrowband channel is H_i. Since it is a flat fading channel for each narrowband channel, the signal will produce linear distortion after passing through the frequency selective fading channel.

1.2.1.2 *Doppler shift*

(1) *What is Doppler shift?* As shown in Figure 1.22, when the mobile station communicates in a mobile state, the frequency of the received signal will change. The difference between this and the transmitted frequency is called the Doppler shift f_d:

$$f_d = \frac{V}{C} f_c \cos\theta = f_m \cos\theta \tag{1.10}$$

where

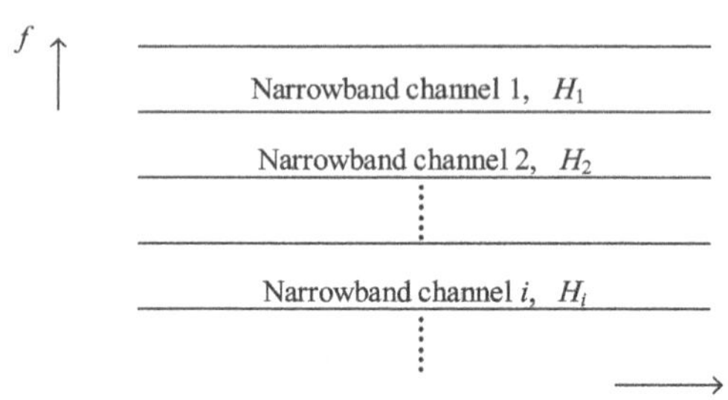

Figure 1.21 Frequency selective fading.

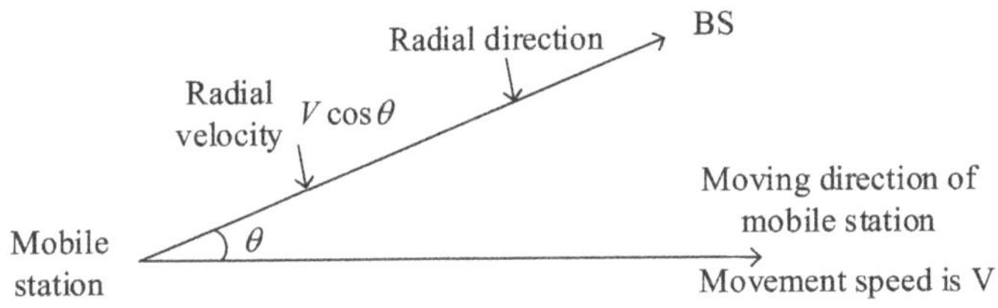

Figure 1.22 Doppler frequency shift.

$$f_m = \frac{V}{C} f_c \tag{1.11}$$

f_m is the maximum Doppler frequency shift, f_c is the emission frequency and C is the speed of light. It can be seen from equation (1.10) that as long as the radial velocity of the mobile station is not equal to zero when it moves, a Doppler shift will occur. As the movement speed and direction change, the magnitude and sign of f_d will also change. When the mobile station moves closer to the base station, $f_d > 0$ means that the received frequency is greater than the transmitted frequency. When the mobile station moves away from the base station, $f_d < 0$ means that the received frequency is less than the transmitted frequency.

(2) *The influence of Doppler frequency shift on the transmitted signal*:
(i) *Frequency domain*:
Doppler shift causes the frequency of the received signal to drift. For coherent demodulation (see Figure 1.5), the receiving end must eliminate the Doppler shift. Otherwise, the error performance of demodulation will deteriorate. It is not difficult to see that the communication quality of users on the high-speed rail is worse than in the urban area or indoors. This is because the fast-moving speed of the high-speed rail increases the Doppler frequency shift.

(ii) *Time domain*:
Another important concept related to Doppler shift is coherence time ΔT_c:

$$\Delta T_c \approx \frac{1}{f_d} \text{(sec)} \tag{1.12}$$

The physical meaning of ΔT_c is that the time domain transfer function of the wireless channel is the same in the time period ΔT_c.

In summary, it can be seen that the transmission function of the wireless channel is a two-dimensional function. It is both a function of frequency and a function of time and can be expressed by $H(f, t)$. But $H(, t)$ is a constant value in the coherence time period and the coherence bandwidth frequency band. Of course, $H(, t)$ is different in different time periods and in different frequency bands.

1.2.2 *OFDM working principle*

1.2.2.1 *OFDM modulator*

The principle of the OFDM modulator is shown in Figure 1.23, which details the principle of OFDM anti-ISI.

(1) *Serial/parallel conversion*: As seen in Figure 1.21, the wireless channel can be divided into a number of flat fading parallel narrowband channels (also called sub-channels). When the serial symbol sequence output by the high-order modulation QAM is passed through a 1: N serial/parallel converter, N parallel QAM symbol sequences can be the output. Then, they are allocated to N sub-channels with flat fading characteristics for transmission. The N pieces of data (d_1–d_N) given in the figure are data that are simultaneously transmitted within a time width T of an OFDM symbol.

(2) N *subcarriers*: N subcarriers must be configured in the N sub-channels as the carrier of the N sub-channels. N parallel QAM symbol

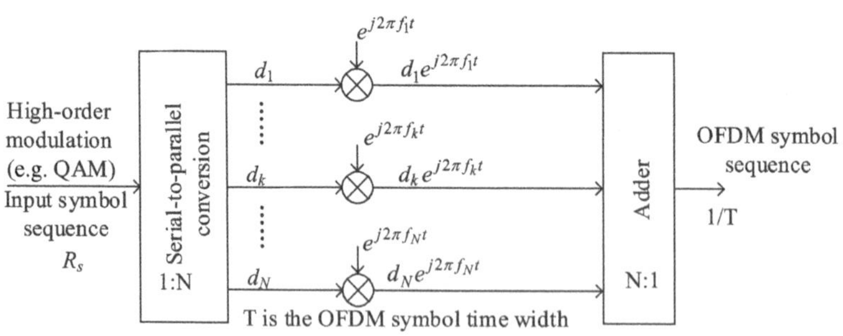

Figure 1.23 OFDM modulator.

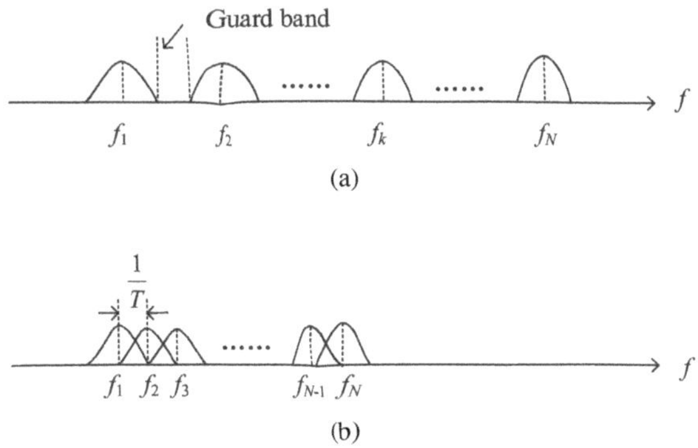

Figure 1.24 Two frequency division multiplexing schemes: (a) FDM, (b) OFDM.

sequences are, respectively, multiplied by N subcarriers. It can be seen that OFDM is a multi-carrier modulation method. There are two configurations of N subcarriers in the frequency domain, as shown in Figure 1.24. Figure 1.24(a) shows the traditional Frequency Division Multiplexing (FDM). The receiver only needs to filter the received signal through N band-pass filters with different center frequencies to demodulate the N symbol sequence. The frequency spectrum of each subcarrier of FDM cannot overlap, and guard bands are needed between adjacent channels. Therefore, the frequency band occupied after modulation is too wide, thereby reducing the channel frequency band utilization. While Figure 1.24(b) uses an Orthogonal Frequency Division Multiplexing (OFDM) scheme, which not only does not require a guard band, but adjacent spectrums can also overlap. Obviously, the frequency band after OFDM modulation is much narrower than FDM.

Some readers may ask how to demodulate in the case of overlapping spectrum? Will there be inter-channel interference (ICI) between adjacent sub-channels? We will discuss this issue in the following:

(i) *Subcarrier spacing* Δf:
If the adjacent subcarrier spacing $\Delta f = 1/T$, T is the OFDM symbol time width (IFFT operation time), then the value of each subcarrier is

$$f_k = k \cdot \Delta f = k/T, \quad k = 1,\ldots, N, \quad \Delta f = 1/T$$

$$f_1 = 1/T,\quad f_2 = 2/T,\quad f_3 = 3/T,\ \ldots,\quad f_N = N/T \qquad (1.13)$$

It can be seen that each subcarrier has an integer multiple relationship.

(ii) *N subcarriers are orthogonal in the time domain*:
When each subcarrier has an integer multiple relationship, then these subcarriers must be an integer multiple of the period within the OFDM symbol time width *T*. Figure 1.25 shows the waveforms of the four subcarriers. It can be seen that the number of sine waves in the *T* period is an integer.

The following uses subcarriers f_1 and f_2 as an example to prove that these two subcarriers are orthogonal in the *T* period. The proof process is shown in Figure 1.26.

Figure 1.26(c) is the waveform after two different subcarriers are multiplied. There are positive and negative waveforms. And the positive and negative waveforms are the same. Therefore, the average value of

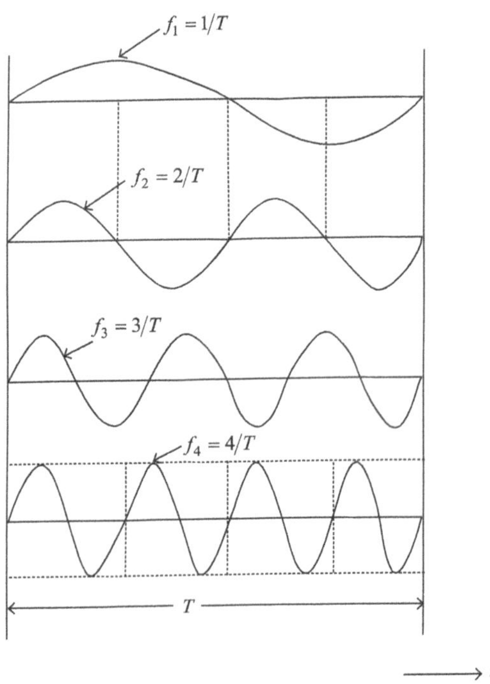

Figure 1.25 Example of four subcarriers.

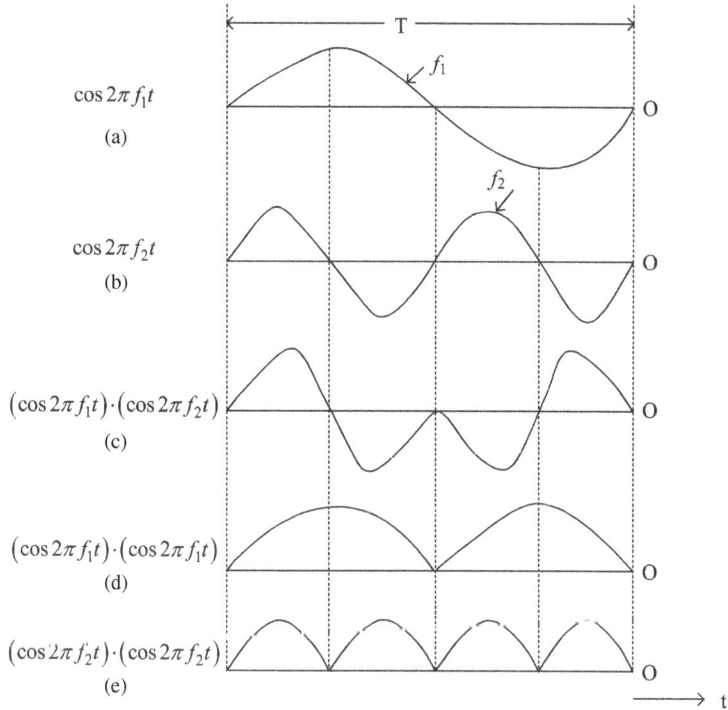

Figure 1.26 The orthogonality proof of two subcarriers.

these waveforms in the T period (that is, the integrated value of the waveform (c) in the T period) is zero.

Figures 1.26(d) and (e) are the waveforms of the multiplication of two identical subcarriers. The waveforms are all positive, so the average value of these waveforms in the T period is not zero.

In summary, the orthogonality of N subcarriers can be expressed as

$$\frac{1}{T}\int_0^T e^{j2\pi f_i t} \cdot e^{-j2\pi f_k t} dt = \begin{cases} 0 & i \neq k \\ \text{constant} & i = k \end{cases} \tag{1.14}$$

It can be seen from Figure 1.26 that the necessary and sufficient condition for N subcarriers to be orthogonal to each other is that within the OFDM symbol width T time, the period of N sine waves must be an integer. Otherwise, it is non-orthogonal.

Because of the above-mentioned orthogonality, the frequency spectrum of adjacent subcarriers can overlap, and different subcarriers can be separated during demodulation at the receiving end without ICI.

(3) *Adder*: It can be seen from Figure 1.23 that the output of N multipliers is N parallel signals. But there is only one wireless channel, so it is necessary to convert N parallel signals into a serial signal. This is the reason for the adder.

The function of the OFDM modulator in Figure 1.23 is to add up N sine waves during the T period of the OFDM symbol. And the amplitude of each sine wave is the QAM symbol. The time domain OFDM symbol is obtained after the adder. It can be seen that the function of the OFDM modulator is the same as the inverse Fourier transform. In actual engineering, the OFDM modulator is realized by the inverse Fourier transform method. In order to reduce the time delay of the inverse Fourier transform operation, the inverse fast Fourier transform (IFFT) in the discrete domain is used. The principle block diagram of OFDM in the discrete domain (digital domain) is different from the block diagram in the continuous domain (analog domain) in Figure 1.23, as shown in Figure 1.48(a) (without DFT in the figure).

1.2.2.2 *OFDM demodulator*

The OFDM demodulator is the inverse operation of the modulator, which realizes the function of Fourier transform. The principle block diagram in the continuous domain is shown in Figure 1.27. It uses the orthogonality of N subcarriers to separate N QAM symbols from the OFDM symbol through a multiplier and a low-pass filter and then undergoes parallel and serial conversion into a serial QAM symbol sequence. The demodulator in the discrete domain should be FFT, as shown in Figure 1.48(b) (without IDFT in the figure).

1.2.2.3 *Guard interval and cyclic prefix*

(1) *Guard interval*: The OFDM symbol sequence output by the modulator in Figure 1.23 is shown in Figure 1.28(a). Due to the above-mentioned delay extension (τ_m), the preceding symbols will still interfere with the following symbols. In order to eliminate the inter-symbol interference

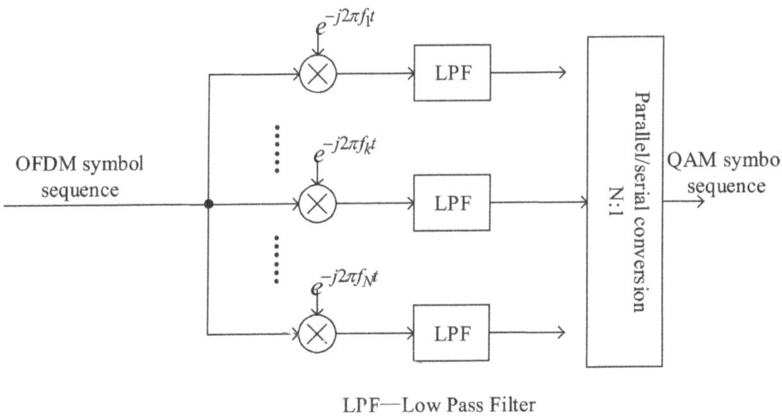

Figure 1.27 OFDM demodulator.

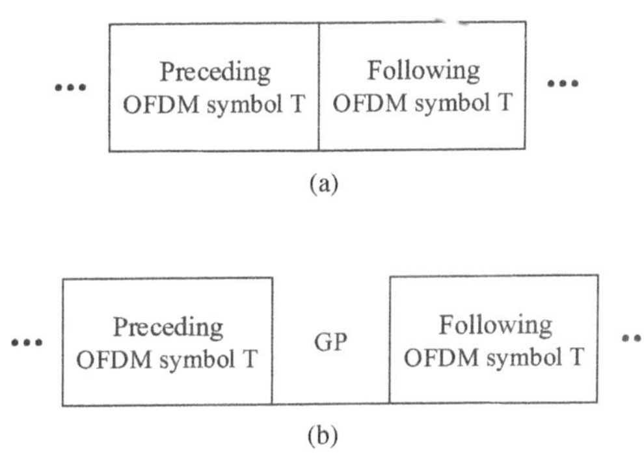

Figure 1.28 Guard interval.

(ISI), people first think of leaving a certain time interval between the preceding and following symbols, which we call the guard interval (*GP*). As shown in Figure 1.28(b), if $GP > \tau_m$ is selected, ISI can obviously be eliminated.

(2) *Cyclic prefix*: Please note that no signal is transmitted during the guard interval (*GP*). For this reason, ISI was eliminated when $GP > \tau_m$. But it also produces interference between sub-channels (ICI), which is illustrated with the example in Figure 1.29.

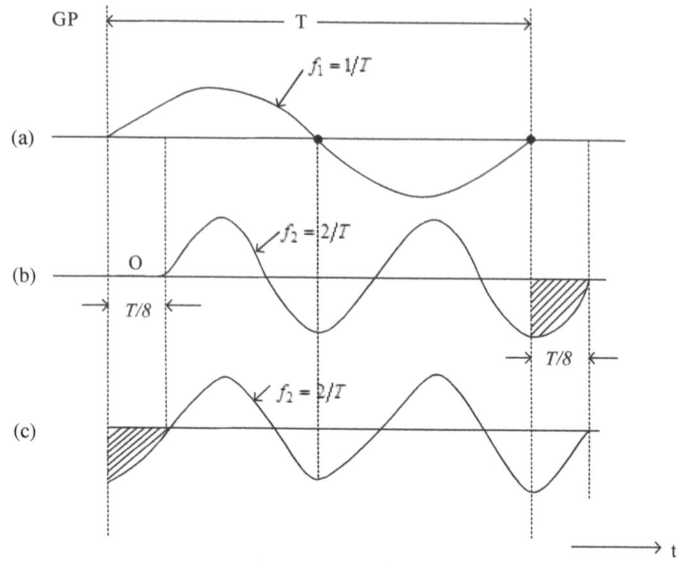

Figure 1.29 Cyclic prefix.

Assume that there are two subcarriers f_1 and f_2 within the OFDM symbol width T at the transmitter, as shown in Figures 1.26(a) and (b). f_1 sine wave has one period in T, and f_2 sine wave has two periods in T, so f_1 and f_2 are orthogonal to each other. They cause a delay in transmission when passing through the channel. If the two delays are different (because of the different frequencies), the received waveforms are, for example, as shown in Figures 1.29(a) and (b). Here we need to explain Figure 1.29(b). f_2 is $T/8$ time longer than the time delay of f_1, so that the zero level of $T/8$ in the *GP* is moved into T, which makes the level is zero in the previous $T/8$ time of f_2. But it also has the tail waveform of $T/8$ (the slashed waveform in Figure 1.29(b)) removed from T, so the waveform of f_2 is shown in Figure 1.29(b). It can be seen that although f_1 is still a cycle within T, f_2 is not an integer multiple of the cycle (7/4 cycles). At this time, f_1 and f_2 within T are no longer orthogonal to each other. This means that when the receiver demodulates the subcarrier f_1, f_2 interferes with the demodulation of f_1 and vice versa. This interference is called ICI.

It can be seen that in order to eliminate ICI, a signal must be transmitted in the *GP*. And the tail waveform in T should be cyclically moved to *GP*. As shown in Figure 1.30(a), the waveform moved to the front is

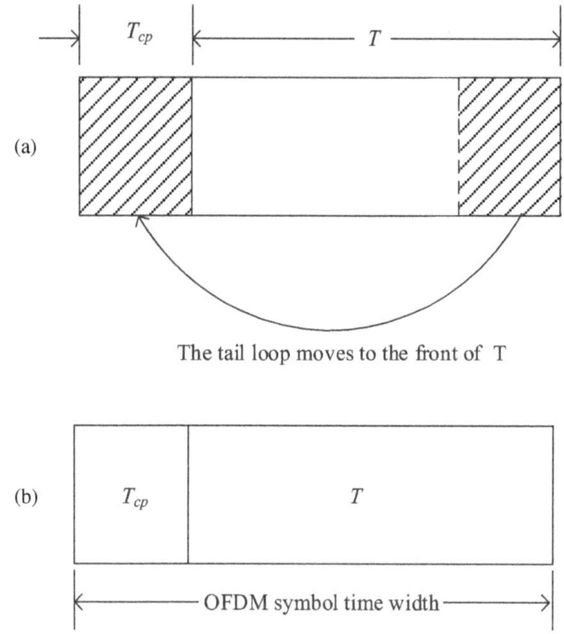

Figure 1.30 Cyclic prefix.

called cyclic prefix (*CP*). It can be seen that the time width of an OFDM symbol is composed of T and T_{cp}, where T is the operation time of IFFT, and T_{cp} is just a *CP*. It does not participate in the IFFT operation. Obviously in order to eliminate ISI, $T_{cp} \gg \tau_m$ is required. In order to improve power efficiency (because T_{cp} does not transmit information), $T > T_{cp}$ is required, usually:

$$T_{cp} = (2 \sim 4)\tau_m$$

$$T = (4 \sim 5)\,\tau_{cp} \tag{1.15}$$

1.2.2.4 *Channel estimation for OFDM system*

From the above analysis, the OFDM system is composed of three functional blocks, which are OFDM modulator, wireless channel and OFDM demodulator. The functions of the modulator and demodulator cancel each other (Fourier transform and inverse transform). Therefore, the

OFDM system can be simplified, as shown in Figure 1.31. From the figure, we can obtain

$$R_{kL} = d_{kL} H_{kL} + \mu_{kL} \tag{1.16}$$

where d_{kL} is the QAM symbol (frequency domain) transmitted by the Lth OFDM symbol on the kth subchannel, H_{kL} is the transmission function of the Lth OFDM symbol in the kth subchannel, μ_{kL} is the noise of the Lth OFDM symbol in the kth subchannel, and R_{kL} is the data corresponding to d_{kL} output by the FFT at the receiver.

If we have obtained the transfer function $\hat{H}_{kL}$ of the k-th sub-channel by estimating the channel, and the estimate is accurate, that is $H_{kL} = \hat{H}_{kL}$, then we can apply the following operations on (1.16):

$$\frac{R_{kL}}{\hat{H}_{kL}} = d_{kL}\frac{H_{kL}}{\hat{H}_{kL}} + \frac{\mu_{kL}}{\hat{H}_{kL}}$$

Since $H_k = \hat{H}_{kL}$, we have

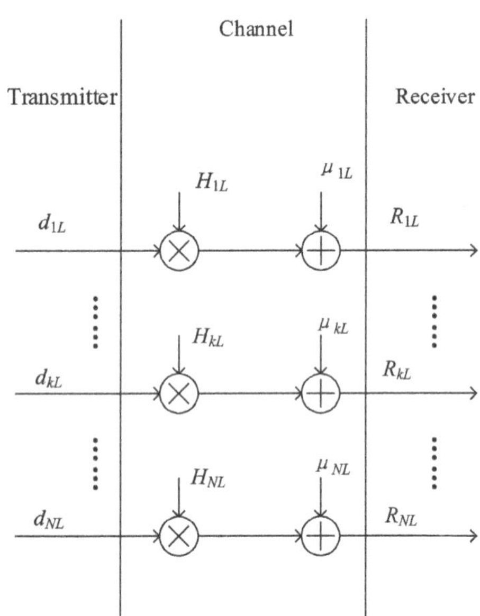

Figure 1.31 Simplified diagram of OFDM system.

$$\hat{d}_{kL} = \frac{R_{kL}}{\hat{H}_{kL}} = d_{kL} + \frac{\mu_{kL}}{\hat{H}_{kL}} \tag{1.17}$$

Take $R_{kL}/\hat{H}_{kL}$ from the above equation as the judgment variable. Since it is only disturbed by additive noise $\mu_{kL}/\hat{H}_{kL}$, it is reasonable to recover the data d_{kL} from it. This means that as long as we know $\hat{H}_{kL}$, we can demodulate the data. Now we discuss the methods to obtain $\hat{H}_{kL}$ by channel estimation. If the transmitted symbol d_{kL} is known, that is, the symbol has been predetermined by the transmitter and receiver in advance, we call it the reference symbol. In this way, we can apply the following operations on (1.16):

$$\hat{H}_{kL} = \frac{R_{kL}}{d_{kL}} = H_{kL} + \frac{\mu_{kL}}{d_{kL}} \tag{1.18}$$

The channel transfer function $\hat{H}_{kL}$ estimated by the method of transmitting known reference symbols differs from the real channel transfer function H_{kL} by only one additive noise term μ_{kL}/d_{kL}. Therefore, it is reasonable to regard $\hat{H}_{kL}$ as the channel transfer function.

Some readers may be confused about the usage of estimated $\hat{H}_{kL}$. We use Figure 1.32 to discuss. The ΔB_c, Δf, ΔT_c and T in this figure are coherence bandwidth, subcarrier spacing, coherence time and OFDM symbol time width, respectively. Since $\Delta B_c > \Delta f$ and $\Delta T_c > T$, there will be multiple subcarriers in ΔB_c and multiple OFDM symbols in ΔT_c. It can be seen that there are multiple QAM symbols in the resource grid formed by ΔB_c

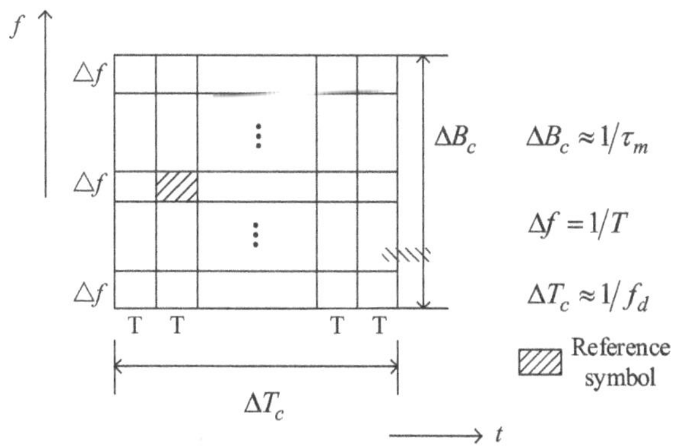

Figure 1.32 Usage of reference symbols.

and ΔT_c. Since these symbols are in the coherence bandwidth ΔB_c and coherence time ΔT_c, the transmission function of the channel is almost the same. In this way, we can simply configure the slashed symbol in the figure as reference symbol, and the transfer function of the symbol is obtained at the receiver. It is also the transfer function of all symbols in the aforementioned resource grid. Then, equation (1.17) should be used to find the criterion variable of each symbol. If necessary, several reference symbols can also be configured.

Of course, the channel transfer functions in different ΔB_c and ΔT_c ranges are different. Therefore, additional reference symbols need to be configured. In short, the density of reference symbols configured in the frequency domain or the time domain is determined by the correlation of the channel transfer functions in these domains. The smaller the correlation, the denser the configured reference symbols, and vice versa. The specific configuration of LTE is discussed later.

1.2.2.5 *OFDMA multiple access mode*

The multiple access modes in 2G, 3G and 4G are TDMA, CDMA and OFDMA, respectively. They are all orthogonal multiple access methods. However, it should be pointed out that in 4G LTE, only the downlink uses OFDMA. It stands for Orthogonal Frequency Division Multiple Access, and its principle is shown in Figure 1.33. Different users occupy different frequency bands, and each frequency band contains several subcarriers. The receiver uses the orthogonality of the subcarriers to separate different users.

1.2.2.6 *OFDM parameters*

(1) *Peak-to-Average Power Ratio (PAPR) average power*: The peak-to-average power ratio (PAPR) is the ratio of peak power to an OFDM symbol time width T, that is,

$$\text{PAPR} = \frac{\text{Peak power}}{\text{Average power}} \tag{1.19}$$

It can be seen from Figure 1.23 that the output waveform of the modulator is generated by the addition of N sine waves. If the peaks of

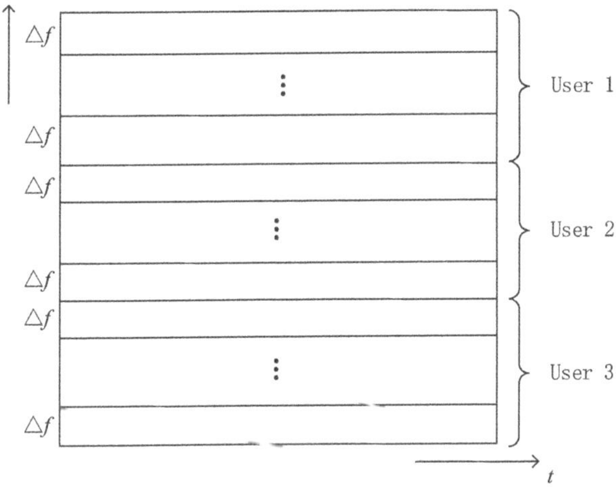

Figure 1.33 OFDMA.

all sine waves are aligned at a certain moment, then a large peak will appear when added up. Therefore, the PAPR of OFDM is relatively large. For a single carrier signal, we can obtain from equation (1.19) that its PAR is 3 dB (because average power is 1/2 of the peak power). But for multi-carrier OFDM waveforms, its PAPR is much greater than 3 dB.

The transmitting OFDM signal must be amplified by a high-frequency power amplifier. Since the OFDM signal envelope will have a large peak, it is required that the high-power amplifier should have a large linear range. Otherwise, when the signal peak enters the nonlinear region of the amplifier, the waveform will be distorted. As a result, intermodulation interference and out-of-band radiation between subcarriers are generated, which destroys the orthogonality between subcarriers, generates ICI and degrades system performance. To prevent signal peaks from entering the nonlinear region of the power amplifier, the following measures are adopted:

(i) *Base station*:
When the base station uses a class A power amplifier, it should work under high power compensation. High power refers to its large linear range. Power compensation refers to a reduction of, for example, 3 dB from the 1 dB power compression point to reduce the third-order

intermodulation distortion to a required value. The power amplifier range after reducing 3 dB can be regarded as the linear range, and the linear range is required to be very wide. In addition to low power efficiency (Class A power amplifier power efficiency is only 50%), such power amplifiers are also very expensive.

(ii) *Mobile terminal*:
Since the power source of the mobile terminal is a rechargeable battery, the requirements for power saving are particularly high. Therefore, class A power amplifiers with low power conversion efficiency cannot be used, but high-efficiency nonlinear power conversion such as class B, C, D, E, F, etc., should be used. However, in order to obtain a large linear range, linearization measures are required. This will increase the complexity of the power amplifier and greatly increase the cost of the power amplifier.

It can be seen from the above analysis that certain technical measures should be adopted in the OFDM system to reduce the impact of peak level on system performance. Two measures can be adopted for this. One is to try to reduce its peak level. For example, the LTE uplink does not use the OFDM waveform, but uses the DFTS-OFDM waveform (discussed later). The second is to try to reduce the probability of peaks. However, it should be pointed out that the problem has not been completely resolved so far, and it can be said that it is still a worldwide problem.

(2) *OFDM signal bandwidth*:
It can be seen from Figure 1.24(b) that the bandwidth of the OFDM signal is

$$B = N\Delta f = N/T\ (Hz) \tag{1.20}$$

where N is the number of subcarriers contained in an OFDM symbol, Δf is the subcarrier spacing, and T is the OFDM symbol time width, which is the operation time of IFFT/FFT.

(3) *Data transmission rate of OFDM system*:
The block diagram used to calculate the data transfer rate is shown in Figure 1.34. It can be obtained from the figure that

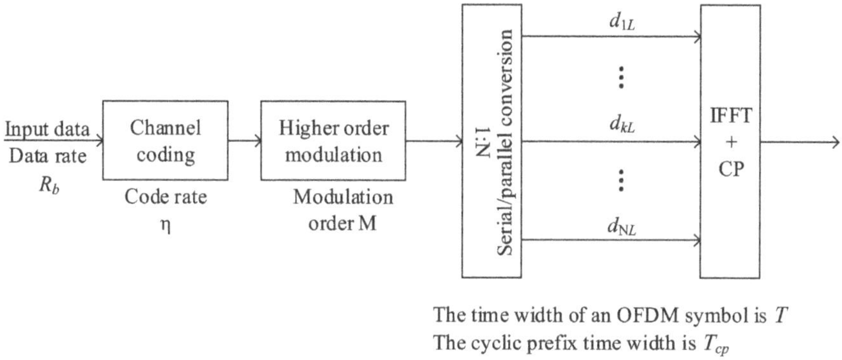

Figure 1.34 Block diagram used to calculate R_b.

$$
\begin{aligned}
R_b &= \frac{\text{The number of bits contained in an OFDM symbol}}{\text{Time spent on transmitting bits}} \\
&= \frac{\text{Number of bits contained in a subcarrier} \times \text{Number of subcarriers in an OFDM symbol}}{\text{IFFT operation time} + \text{Cyclic prefix time}} \\
&= \frac{(\eta \cdot \log_2 M) \cdot N}{T + T_{cp}}
\end{aligned}
\tag{1.21}
$$

The equation (1.21) is applicable to both mobile communication networks (such as LTE) and wireless local area networks (Wi-Fi).

1.3 Error Control Method

When data is transmitted in a multipath fading wireless channel, errors will always occur. Especially when transmitting high-speed data, the bit error rate at the receiver will increase significantly due to the reduction of bit energy. In order to ensure communication reliability, that is, to reduce the bit error rate to the specified requirements, two technical measures are required for the data at the receiver. First, the errors in the data must be corrected, which is called Forward Error Correction (FEC). But error correction is not enough, and FEC cannot correct errors beyond its error correction capabilities. Therefore, it is necessary to check the corrected data again to see if there are any errors that have not been corrected and notify the transmitter of the result. The signal fed back to the transmitter is called the confirmation signal. When there is no error, the confirmation signal is ACK. And when error occurs, the signal is NACK. If there is no error, the transmitter continues to transmit new data. If there is an error, the transmitter retransmits this data. Of course, whether to retransmit is determined by the upper layers of the physical layer, and the physical layer itself only checks if there is any error. We call this technology Automatic Repeat Request (ARQ).

It can be seen that the so-called error control refers to error correction and error detection of the data demodulated at the receiver. Combining FEC and ARQ, we get a solution called hybrid ARQ (HARQ). HARQ has three schemes, which are Type I, Type II and Type III. Among them, the initiator of the Type I ARQ scheme adds CRC supervision bits after the data block and then forms a new data block with the bit information and supervision and then performs FEC encoding on the new data block. The receiver first performs error correction on FEC decoding and then uses CRC decoding for error detection.

The principle block diagram of error control is shown in Figure 1.35. Next, we will discuss the principles of FEC, CRC and HARQ.

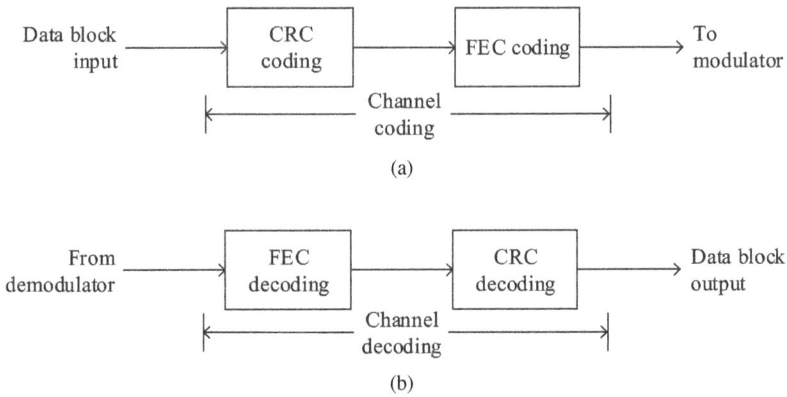

Figure 1.35 Error control principle diagram: (a) transmitter and (b) receiver.

1.3.1 *FEC code*

The FEC used in 3G and 4G is the turbo code proposed by the French in 1993. The data channel in the eMBB scenario in 5G uses Low-Density Parity Check (LDPC) code. The LDPC code was proposed by the Americans in 1963. Channel coding in Wi-Fi currently also uses this code. The signaling channel in the eMBB scenario in 5G uses the polar code proposed by the Turks in 2007. The three codes all use the maximum likelihood probability decoding method, which means that the data correct probability of this method is the largest. Please note that only the probability of correctness is the greatest, so there is still a chance that errors cannot be corrected. It should also be pointed out that because the amount of data transmitted on the data channel is larger than that of the signaling channel, HARQ is used on the data channel. At this time, HARQ confirmation information needs to be returned as feedback, and the information is used to tell the transmitter whether the data is successfully transmitted. HARQ is not used on the signaling channel. The channel coding principles of turbo, LDPC and polar will be discussed in Section 6.3.2.3 of Chapter 6.

1.3.2 *CRC code*

CRC code is called cyclic redundancy check code, which is a commonly used error detection code. It can be used to find a variety of errors in packet data, but it still cannot solve all errors.

1.3.2.1 *CRC encoding*

(1) *CRC code structure*: It is also a block code, which can be expressed as (n, k). n is the code group length after encoding (number of binary coded bits), and k is the information bits (the number of information bits) in the code group. We can see that the supervision bits in the code group is $r = n - k$ (number of binary coded bits). The relationship between the three is shown in Figure 1.36.

In the figure, the r supervision bits are produced by the CRC code generator polynomial $g(x)$ and k information bits. The encoded code group polynomial can be expressed as

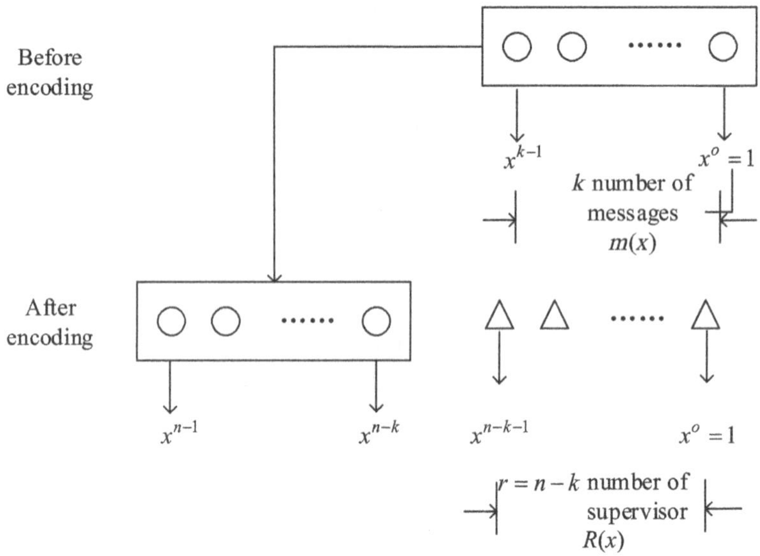

Figure 1.36 CRC code structure.

$$A(x) = m(x) \cdot x^r + R(x) \tag{1.22}$$

$$R(x) \equiv m(x) \cdot x^r \mod g(x) \tag{1.23}$$

where $A(x)$ is called the code group polynomial, $m(x)$ is called the information polynomial, and $R(x)$ is called the redundant polynomial.

Equation (1.23) shows the method of calculating $R(x)$: $R(x)$ is equal to the remainder after $m(x) \times x^r$ is divided by $g(x)$.The division can be performed in binary or hexadecimal domain. Here is an example.

The hexadecimal number of the information polynomial $m(x)$ is 4D6F746F0000. Each bit is a four-bit binary code. The relationship between the three bases is shown in Table 1.2. The hexadecimal number of the generator polynomial $g(x)$ is 11021. We can use long division in the hexadecimal field to find the hexadecimal value of $R(x)$.

Hexadecimal long division

Table 1.2 The relationship between the three bases.

Decimal	Binary	Hexadecimal
0	0000	0
1	0001	1
2	0010	2
3	0011	3
4	0100	4
5	0101	5
6	0110	6
7	0111	7
8	1000	8
9	1001	9
10	1010	A
11	1011	B
12	1100	C
13	1101	D
14	1110	E
15	1111	F

```
            48D88805
       _____________
11021 / 4D6F746F0000
        44084
      -------------
         96734
      -------------
         88108
      -------------
          E62C6
      -------------
          DD1AD
      -------------
            9119F
      -------------
            88108
      -------------
             90970
      -------------
             88108
      -------------
              88680
      -------------
              88108
      -------------
                57800
      -------------
                550A5
      -------------
remainder:   275B
```

In this way, the encoded hexadecimal code group can be obtained as 4D6F746F0000,275B.

The tables are all hexadecimal numbers, generating polynomials $g(x) = x^{16} + x^{12} + x^5 + 1$.

The above-mentioned long division calculation is a bit complex, and a fast algorithm of table look-up is given in the following. The generator polynomial $g(x) = x^{16} + x^{12} + x^5 + 1$ in 2G is 11021 in hexadecimal, while the highest power of $g(x)$ is 16. It can be seen from (1.23) that the highest power can only be 15, indicating that the supervision bit has only 16 bits, that is, 2 bytes. The shortest information bit is one byte, so the shortest code group length is three bytes. In this way, we can calculate the remainder of three bytes [*a* 0 0], as shown in Table 1.3.

The "*a*" byte in the table is divided into high half-byte (first four bits) and low half-byte (last four bits), which are represented by hexadecimal 0–F. The first column 0 to F in the leftmost column of the table is the value of the upper half-byte of "*a*". The 0 to F in the top first row of the table is the value of the lower half-byte of "*a*". The four-digit hexadecimal number where the upper half-byte and the lower half-byte intersect in the table are the two-byte supervision bits h_{a00} and L_{a00}. The first two digits are h_{a00}, and the last two digits are L_{a00}.

The following example illustrates how to use this table.

When the information bit is two bytes, we can use the look-up table method to find the two-byte supervision bit.

Two bytes of information bits are *a* *b*

00100000 10000001

Hexadecimal is 2 0 8 1

Step				
	a 0010 0000	*b* 1000 0001	R_1 0000 0000	R_2 0000 0000
(1) [*a* 0 0]	*a* 0010 0000	0 0000 0000	0 0000 0000	
(1) Look up the table		$h^1_{a00} = (24)_H$ 0010 0100	$L^1_{a00} = (62)_H$ 0110 0010	

(Continued)

(2) Modular two plus	h^1 1010 0101	L^1 0110 0010	
(3) Three bytes	h^1 1010 0101	L^1 0110 0010	R_2 0000 0000
(2) $[h^1 00]$	1010 0101	0000 0000	0000 0000
(1) Look up the table		$h^2_{[h^1 00]} = (E5)_H$ 1110 0101	$L^2_{[h^1 00]} = (4F)_H$ 0100 1111
		h^2 1000 0111	L^2 0100 1111

$$h^1 = b \text{ and } h^1_{a00} \text{ bitwise modulo two addition}$$
$$L^1 = R_1 \text{ and } L^1_{a00} \text{ bitwise modulo two addition}$$

$$h^2 = L^1 \text{ and } h^2_{[h^1 00]} \text{ bitwise modulo two addition}$$
$$L^2 = R_2 \text{ and } L^2_{[h^1 00]} \text{ bitwise modulo two addition}$$

The last two bytes of supervision bits are $h^2L^2 = (874F)_H$

$$R(x) = x^{15} + x^{10} + x^9 + x^8 + x^6 + x^3 + x^2 + x + 1$$

From the above calculation process, it can be seen that the fast calculation method of three bytes is used to calculate each byte in the information byte separately for three bytes and the look-up table is utilized to obtain the redundant bits of two bytes until all the information bytes are calculated, and the final two bytes are its supervision bits. Obviously, such processes can be implemented by programming.

Table 1.3 Three-byte [a 0 0] remainder table.

a	0	1	2	3	4	5	6	7	8	9	A	B	C	D	E	F
0	0000	1021	2042	3063	4084	50A5	60C6	70E7	8108	9129	A14A	B16B	C18C	D1AD	E1CE	F1EF
1	1231	0210	3273	2252	52B5	4294	72F7	62D6	9339	8318	B37B	A35A	D3BD	C39C	F3FF	E3DE
2	2462	3443	0420	1401	64E6	74C7	44A4	5485	A56A	B54B	8528	9509	E5EE	F5CF	C5AC	D58D
3	3653	2672	1611	0630	76D7	66F6	5695	46B4	B75B	A77A	9719	8738	F7DF	E7FE	D79D	C7BC
4	48C4	58E5	6886	78A7	0840	1861	2802	3823	C9CC	D9ED	E98E	F9AF	8948	9969	A90A	B92B
5	5AF5	4AD4	7AB7	6A96	1A71	0A50	3A33	2A12	DBFD	CBDC	FBBF	EB9E	9B79	8B58	BB3B	AB1A
6	6CA6	7C87	4CE4	5CC5	2C22	3C03	0C60	1C41	EDAE	FD8F	CDEC	DDCD	AD2A	BD0B	8D68	9D49
7	7E97	6EB6	5ED5	4EF4	3E13	2E32	1E51	0E70	FF9F	EFBE	DFDD	CFFC	BF1B	AF3A	9F59	8F78
8	9188	81A9	B1CA	A1EB	D10C	C12D	F14E	E16F	1080	00A1	30C2	20E3	5004	4025	7046	6067
9	83B9	9398	A3FB	B3DA	C33D	D31C	E37F	F35E	02B1	1290	22F3	32D2	4235	5214	6277	7256
A	B5EA	A5CB	95A8	8589	F56E	E54F	D52C	C50D	34E2	24C3	14A0	0481	7466	6447	5424	4405
B	A7DB	B7FA	8799	98B8	E75F	F77E	C71D	D73C	26D3	36F2	0691	16B0	6657	7676	4615	5634
C	D94C	C96D	F90E	E92F	99C8	89E9	B98A	A9AB	5844	4865	7806	6827	18C0	08E1	3882	28A3
D	CB7D	DB5C	EB3F	FB1E	8BF9	9BD8	ABBB	BB9A	4A75	5A54	6A37	7A16	0AF1	1AD0	2AB3	3A92
E	FD2E	ED0F	DD6C	CD4D	BDAA	AD8B	9DE8	8DC9	7C26	6C07	5C64	4C45	3CA2	2C83	1CE0	0CC1
F	EF1F	FF3E	CF5D	DF7C	AF9B	BFBA	8FD9	9FF8	6E17	7E36	4E55	5E74	2E93	3EB2	0ED1	1EF0

Note: The table occupies a total of 512 (2×16×16) byte units.

Binary long division calculation

$$m(x) = x^{13} + x^7 + 1$$

$$m(x) \cdot x^r = m(x) \cdot x^{16} = (x^{13} + x^7 + 1)\, x^{16} = x^{29} + x^{23} + x^{16}$$

$$g(x) = x^{16} + x^{12} + x^5 + 1$$

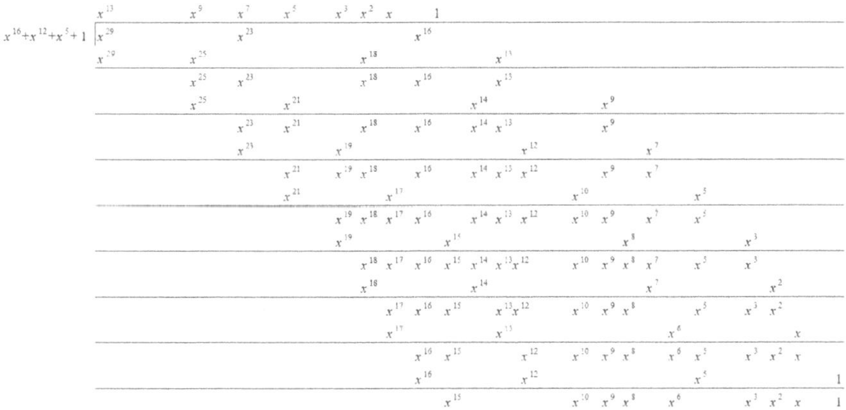

Then $R(x) = x^{15} + x^{10} + x^9 + x^8 + x^6 + x^3 + x^2 + x + 1$, it is the same as the result of the three-byte fast algorithm.

(2) *CRC code used in 4G LTE*: There are four CRC codes used in 4G LTE, which are called CRC24A, CRC24B, CRC16 and CRC8, and their generating polynomials are:

CRC24A $g(x) = x^{24} + x^{23} + x^{18} + x^{17} + x^{14} + x^{11} + x^{10} + x^7 + x^6 + x^5 + x^4 + x^3 + x + 1$

CRC24B $g(x) = x^{24} + x^{23} + x^6 + x^5 + x + 1$

CRC16 $g(x) = x^{16} + x^{12} + x^5 + 1$

CRC8 $g(x) = x^8 + x^7 + x^4 + x^3 + x + 1$

CRC24 is used to transmit data on the shared channel. CRC16 is used to transmit data on the downlink control channel and broadcast channel. CRC8 is used to transmit Control Quality Information (CQI) data.

It can be seen that the CRC24, CRC16 and CRC8 supervision bits are three bytes, two bytes and one byte, respectively. So, they can use four-byte, three-byte and two-byte fast algorithm, respectively.

1.3.2.2 *CRC decoding*

The decoding method of CRC code is as follows:

Calculate syndrome
$$s(x) \equiv B(x) \bmod g(x) = \begin{cases} 0, & B(x) \text{ has no error} \\ \neq 0, & B(x) \text{ has error} \end{cases} \tag{1.24}$$

where $B(x)$ is the received CRC code group, $g(x)$ is the CRC code generator polynomial and $s(x)$ is the remainder after dividing $B(x)$ by $g(x)$. If the syndrome $s(x) = 0$, it is considered that there is no error in $B(x)$, else there are errors in $B(x)$. Obviously, this calculation method can detect error.

The following proves why $s(x) = 0$ can be considered as there is no error in $B(x)$. Since there is no error in $B(x)$, $B(x)$ is the same as $A(x)$ in equation (1.22), that is,

$$B(x) = A(x) = m(x) \cdot x^r + R(x)$$

$$\text{and} \quad m(x) \cdot x^r + Q(x)\, g(x) + R(x) \tag{1.25}$$

where $Q(x)$ is the quotient formula obtained when calculating $R(x)$ (see equation (1.23)).

The equation (1.25) means that the divided formula $m(x).x^r$ should be equal to the quotient $Q(x)$ multiplied by the divisor $g(x)$ and add the remainder $R(x)$. Substitute (1.25) into (1.24), we arrive at:

$$s(x) = \frac{B(x)}{g(x)} = \frac{[Q(x)g(x) + R(x)] + R(x)}{g(x)} = \frac{Q(x)g(x)}{g(x)} = Q(x)$$

$$\therefore s(x) \equiv \frac{B(x)}{g(x)} = 0$$

(since the remainder in the above formula is zero)

1.3.3 *HARQ*

As mentioned earlier, there are three kinds of HARQ schemes, among which type I scheme is ARQ. The principles of type II and type III schemes are the same, and the difference lies in the soft merger method. Therefore, the HARQ principles and the soft merger method are discussed in the following sections.

1.3.3.1 *What is HARQ?*

Due to fading, interference and noise of the wireless channel, the demodulated and decoded data at the receiving end will always produce errors. Especially when transmitting high-speed data, this problem is particularly prominent. In order to reduce this kind of errors in mobile communication systems, the most commonly used method at present is to use HARQ, and its working principle is shown in Figure 1.35.

What is implemented in HARQ is the "stop-wait" protocol. In **one** process of the "stop-wait" protocol, the transmitter stops after transmitting a transmission block each time and waits for the receiver to confirm. This simple mechanism only needs one bit of feedback information. It is used to indicate whether the received transmission block is correct or incorrect. The correct feedback information is Acknowledgement (ACK), and the wrong feedback information is Negative Acknowledgement (NACK). If the feedback information received by the transmitter is ACK, a new transmission block is transmitted. If the received feedback information is NACK, the transmission block that has been transmitted needs to be retransmitted. Obviously, the shortcomings of this mechanism are obvious. That is, while the transmitter is waiting for the receiving end to feedback information, the transmitter cannot transmit data, which reduces the channel utilization rate and ultimately reduces the system throughput.

To solve this problem, **multiple** "stop-wait" processes are used concurrently in HARQ. And each process corresponds to a transmission block. In this way, when the transmitter is waiting for the feedback information from one HARQ process, it can transmit a new transmission block to another HARQ process. Therefore, the transmitter can continuously transmit data, thereby improving system throughput. Moreover, each HARQ process can independently generate feedback confirmation signals ACK/NACK. From the above analysis, we know that a HARQ process is actually a buffer that can store information. Moreover, the error correction and error detection of the transmission block are performed in the buffer, and an acknowledgement feedback signal ACK/NACK is generated according to the error detection result and transmitted to the transmitter. Figure 1.37 is an example of four parallel HARQ processes in the physical layer. In the figure, it is assumed that the first transmission block still has errors after error correction and error detection in process 0, until the first transmission block is retransmitted during the fifth transmission block period, and the error correction is performed again in process 0 and no errors are found after error detection. Only then can the first transmission block be transmitted to the MAC layer. The correct transmission blocks after error correction and error detection in all processes in the figure are transmitted to the MAC layer in parallel, and the buffer is cleared. The sequence numbers of the transmission blocks transmitted to the MAC layer may be out of order. In addition, 5GNR uplink/downlink uses asynchronous HARQ. The so-called asynchronous means that the retransmission can occur at any time. The LTE downlink also uses asynchronous HARQ. However, the uplink uses a synchronization protocol. The so-called synchronization protocol means that retransmission can only occur at a fixed time after the previous transmission. This fixed time is known to both transmitter and receiver. Asynchronous HARQ is more flexible in the time domain.

1.3.3.2 *What is HARQ for soft merge?* [14]

In the traditional HARQ operation, the receiver discards the transmission blocks that still have errors after error detection and requests retransmission. In fact, although the wrong transmission block cannot be decoded correctly, it still contains information. For this reason, we should not discard the wrong transmission block. Instead, it should be stored in the buffer and combined with the retransmitted transmission block to obtain a new single combined transmission block and then decode the combined transmission block. As we will find out in the following, the correct probability of decoding a combined transmission block is much higher than that of decoding a single transmission block.

The most commonly used (for example, 4G and 5G) soft merge method is called incremental redundancy (IR), and its basic principle is shown in Figure 1.38. IR punctures the coded bits output by the FEC

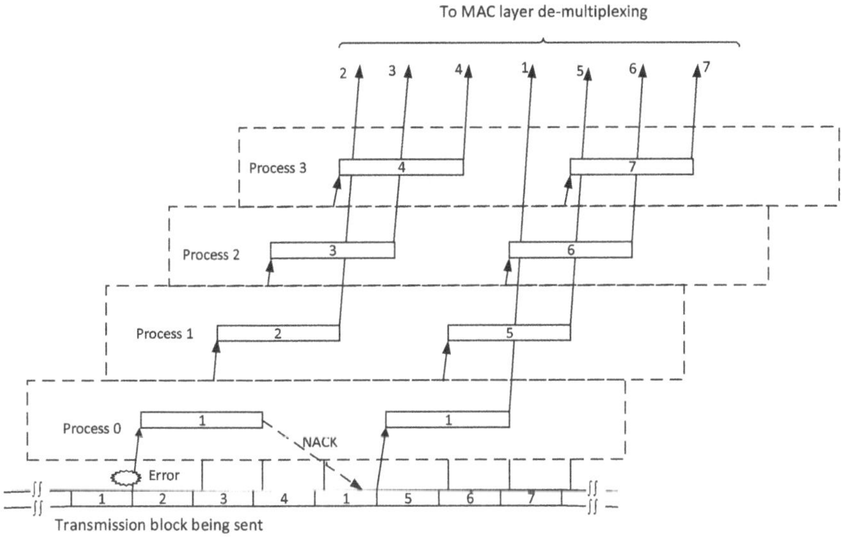

Figure 1.37 Examples of multiple HARQ processes.

encoder to obtain different redundancy versions. The information bits in these redundant versions are all the same, and the only difference is the redundant bits (check bits). There are a total of four coded bits in the redundancy version 1 of the first transmission (initial transmission) in Figure 1.38. There are three information bits and one redundant bit, so the equivalent coding rate is 3/4. The redundancy version 2 of the first retransmission is still four coded bits, of which three information bits are the same as in the initial transmission, but one redundant bit is different from the initial transmission. At this time, the received two redundant versions are soft combined. Obviously, the energy of three information bits is increased ($2E_b$), but the equivalent coding rate reduces to only 3/8. The redundancy version 3 of the second retransmission is still four bits, of which three information bits are the same as in the initial transmission, but one redundancy bit is different from the redundancy version 2. Then, the received redundancy versions 1, 2, and 3 are combined by software. Obviously, the energy of the three information bits increases ($3E_b$), but the equivalent coding rate drops again, which is only 3/12 = 1/4. In the third retransmission, the redundancy version is the same as the redundancy version 1. Therefore, the

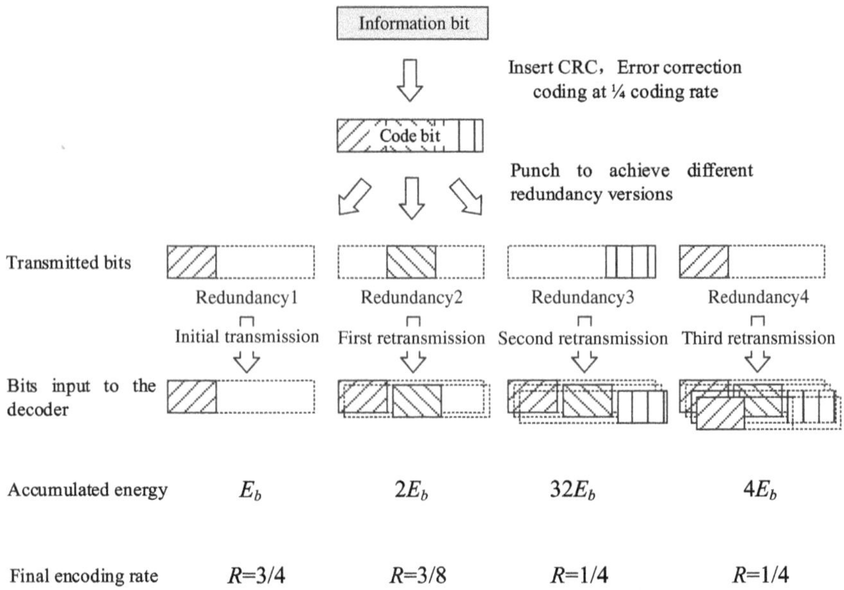

Figure 1.38 Example of incremental redundancy.

equivalent coding rate is still 3/12 = 1/4 except that the information bit energy increases to $4E_b$.

It can be seen from Figure 1.38 that as the number of retransmissions increases, there are two differences between the combined transmission block and the single transmission block.

- The accumulated energy of each information bit gradually increases.
- The equivalent coding rate gradually decreases.

The above two points will reduce the bit error rate after channel decoding, which is exactly what we need. This soft combination is called incremental redundancy (check bit) soft combination, also known as Chase combination (because it was first proposed by Dr. Chase). Its characteristic is that the information bits of the retransmitted transmission block and the initial transmission block are the same, but the check bits are different.

It should be noted that the length of the transmission block in Figure 1.38 is just an example, the actual transmission block length is much longer.

1.3.3.3 *Link rate adaptation and soft combination HARQ*

We know that the network side can adaptively change the modulation and coding scheme (i.e., MCS) according to the estimated instantaneous channel transmission characteristics, thus directly adjusting the coding rate. And the above HARQ with incremental redundancy soft combination can also indirectly adjust the coding rate. So, what is the difference between the two from the perspective of adjusting the bit rate? From the perspective of system throughput, indirect link adaptation only adds redundant bits (that is, increases the parity bit) when it cannot decode correctly to reduce system throughput. There is no need to add redundant bits when it can decode correctly. However, once the code rate drops in direct link adaptation, the system throughput directly decreases. Therefore, indirect link adaptation is better than direct link adaptation in terms of throughput. From the speed of system rate adjustment, direct link adaptation is faster than indirect link adaptation. Since the latter requires feedback to adjust the code rate indirectly, in terms of reducing system delay, direct link adaptation is better than indirect link adaptation.

1.3.3.4 *Comparison of HARQ and ARQ*

When the number of HARQ processes is one, it is ARQ. The following compares the multi-process HARQ with the single-process ARQ:

(i) *The order of output transmission blocks*: It can be seen from Figure 1.37 that the output transmission blocks after multi-process HARQ decoding may be out of order. For example, the output transmission blocks in the figure are in the order of 2, 3, 4, 1, 5, 6 and 7, rather than the input transmission block order 1, 2, 3, 4, 5, 6 and 7. For some data services, out-of-order transmission blocks are not allowed and should be changed into sequence. It can be seen from the following protocol stack in Figure 1.59 that when the lower layer (MAC layer) protocol uses HARQ, the transmission block order is rearranged at the upper layer (RLC layer) to make it a sequence.

For single-process HARQ, the output transmission blocks after decoding are still sequential.

(ii) *Effectiveness and reliability*: Multi-process HARQ can only retransmit once per round (each round means that all processes arrange a transmission block separately) because it adopts multiple processes. Therefore, the transmission effectiveness is high, but the transmission reliability is low. ARQ can perform multiple retransmissions each time, so the transmission effectiveness is lower than multi-process HARQ, but the transmission reliability is higher than multi-process HARQ. It can be seen that for a transmission system, multi-process HARQ and ARQ should be used in combination. This can give consideration to the effectiveness and reliability of the system. It can be seen from the protocol stack in Figure 1.59 that while the lower MAC layer uses multi-process HARQ, the upper RLC layer uses ARQ.

(iii) *Application*: The several radio networks discussed in this book usually use HARQ, but because radio network application scenarios are different, the number of processes used by HARQ is different.

- *Narrowband Internet of Things*:

The R13 version is single-process HARQ (that is, ARQ). The main purpose is to enhance the coverage of the cell and enable it to achieve one of the four major requirements of the IoT, that is, strong coverage.

The R14 version is a dual-process HARQ, so in addition to meeting strong coverage requirements, it can also take into account system throughput.

- *4G network*:

The highest number of processes in HARQ is eight.

- *5G network*:

The highest number of processes in HARQ is 16.

The purpose of choosing a larger number of processes for 4G and 5G is to improve the effectiveness of network transmission and enable it to achieve high data rate transmission in MBB and eMBB application scenarios.

By the way, HARQ technology is not currently used in wireless local area networks.

1.4 Diagram of Baseband Chip

The baseband chip (BB) is a complex-structured system on chip (SOC). BB needs to complete all the functions of the LTE protocol in the baseband part. It involves multiple technologies of the network layer (including RRC and NAS), the data link layer (including MAC, RLC and PDCP) and the physical layer in the Open Systems Interconnect (OSI) reference model. It consists of CPU, channel encoder/decoder, digital signal processor, modulator/demodulator and interface. Figure 1.39 shows only the schematic block diagram of physical layer data transmission. It contains the content of the previous Sections 1.1–1.3. BB chip is implemented in the discrete domain with a digital signal processing (DSP) method. Figure 1.39 shows the detailed diagram of the BB chip.

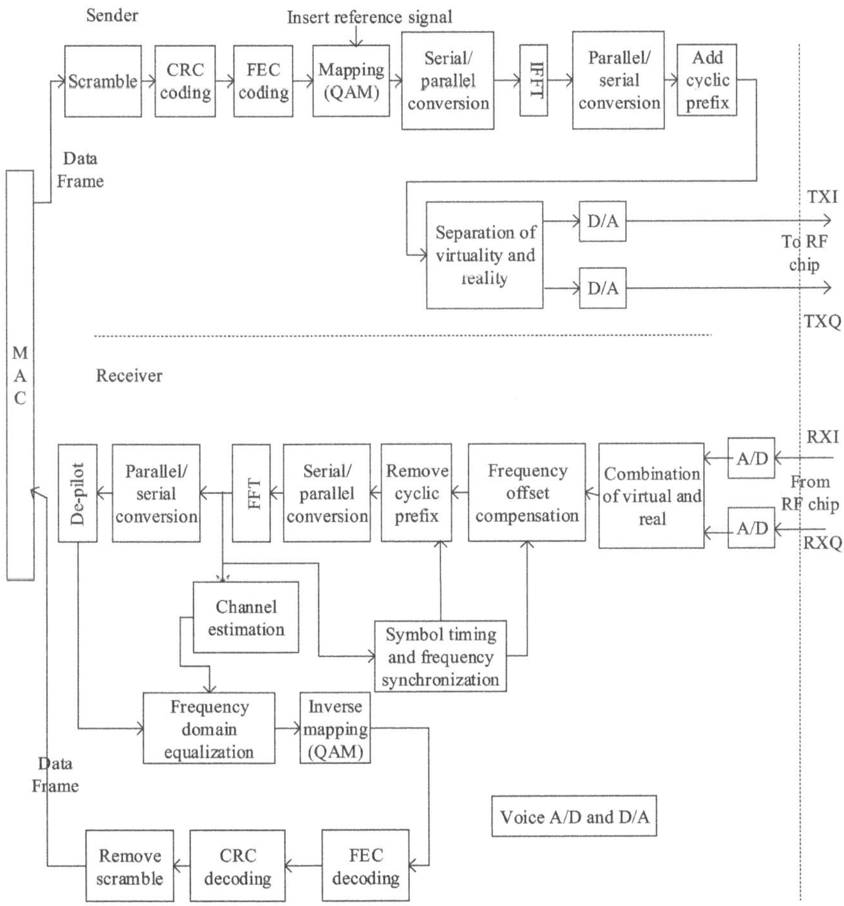

Figure 1.39 Diagram of baseband chip.

(1) *Transmitter*:

(i) *Scrambling*:
Continuous codes of "1" or "0" appears in the data, which is similar to DC and cannot be transmitted in the subsequent circuit blocking capacitor or transformer. The purpose of scrambling is to cut these continuous codes to make the probabilities of "1" and "0" codes the same. Therefore, scrambling is to transform the data into white noise.

(ii) *Separation of virtuality and reality*:
Figures 1.4 and 1.9 are both modulator constellation diagrams. The position of any signal point in the graph can be represented by a complex number "*a*"+ "*b*". "a" is the projection value of the signal point on the *x*-axis. All signal points on the constellation are projected on the *x*-axis to form pulse amplitude modulation (PAM) of the *x*-axis. "*b*" is the projection value of the signal point on the *y*-axis. All signal points on the constellation are projected on the *y*-axis to form PAM of the *y*-axis. It can be seen that the modulator is essentially composed of two PAMs orthogonal to each other. The effect of virtual and real separation is to divide the values on the *x*-axis and *y*-axis into two parallel paths for subsequent processing.

(iii) *D/A*:
The virtual and real separated outputs are two digital signals. They need to convert digital signals into analog signals through digital/analog (D/A) converters and transmit them to Radio Frequency (RF) chips. D/A is relatively a simple low-pass filter.

(iv) *Insert reference signal (symbol)*:
The inserted reference signal is set for purposes such as channel estimation at the receiver, and the channel transfer function is obtained from the channel estimation.

(2) *Receiver*: De-scrambling, combining virtual and real and A/D are the inverse operations of the transmitter. We focus on the following three issues:

(i) *Symbol timing and frequency synchronization*:
Symbol timing is to synchronize the timing of the receiver with the timing of the transmitter and then the CP in Figure 1.30 can be removed.

The transmitting N subcarriers are orthogonal to each other, but they will produce a frequency shift after they are synthesized by the wireless channel Doppler frequency shift and the radio frequency receiver frequency, which destroys the orthogonality of N subcarriers. The frequency synchronization and frequency offset compensation in the figure are used to eliminate this frequency shift, thereby restoring the orthogonality of the N subcarriers, so that ICI will not occur.

(ii) *Channel estimation and frequency domain equalization*:
Channel estimation refers to the calculation of equation (1.18) with known reference symbols to obtain the estimated value of the channel transfer function $\hat{H}_{kL}$.

The frequency domain equalization function is as follows.

It can be seen from the figure that the frequency domain data carried on the N subcarriers of the FFT output (after removing the pilot) is directly transmitted to the frequency domain equalizer. It can equalize the phase and amplitude of N frequency domain data according to the parameters obtained by channel estimation. Phase equalization means that it can eliminate the fixed phase error $\Delta\phi$ (see Figure 5.36(b) in Chapter 5) generated during frequency domain data transmission. Amplitude equalization means that it can eliminate the amplitude changes caused by frequency domain data during transmission to keep the amplitude constant. After frequency domain equalization, we use $\hat{H}_{kL}$ which has been obtained to perform the calculation of equation (1.17) to obtain the transmitting data d_{kL} (L is the Lth OFDM symbol, K: 1, 2,..., N, d_{kL} is the kth frequency domain data of the Lth OFDM symbol). Since the frequency domain data is processed by the frequency domain equalizer and then transmitted to the inverse modulation, the shape of the constellation diagram is not distorted except for the divergence of signal points (due to noise and interference) on the inverse modulation constellation.

(iii) *A/D*:
The two signals from the RF chip are analog signals. The BB chip works in the digital domain, so the input analog signal must be converted into a digital signal. So, two A/D converters are used in the figure. But it should be emphasized that the A/D in speech coding cannot be used here because the quantized signal-to-noise ratio of the A/D conversion is required to reach more than 100 dB, and the speech coding A/D cannot meet this requirement.

First, we should explain the meaning of quantization noise. When the continuous signal is sampled, we obtain a random sample value such as 1.05 volts. Then, the sampled value is quantified, which is rounded off into 1.0 volts. And finally, we encode the 1.0 volts. Obviously, the above rounding produces an error of 0.05 volts, and this is called the quantization error. Quantization error is also a kind of noise. In order to distinguish it from thermal noise, it is called quantization noise. Its power is called quantization noise power, and the ratio of signal power is called quantization signal-to-noise ratio (dB).

Modern Σ–Δ modulation can make the quantization signal-to-noise ratio reach more than 100 dB, and its working principle is shown in Figure 1.40. It is known from the time domain sampling theorem that if the highest frequency of the sampled signal is f_H, in order to restore the signal without distortion, the sampling frequency is required to be $f_s > 2f_H$, and when $f_s > 2f_H$, it is called Nyquist sampling. After analysis, it is known that the quantization noise power spectral density $N(f)$ is a uniform spectrum in the range of 0–f_s, that is, the spectral density is the same. The figure only draws the quantization noise power spectral density in the range of $0 - f_s = f_s/2$. At this time , the quantization noise power falling in the range of $0 - f_s$ is the area of in the figure. However, modern Σ – Δ modulation uses oversampling, that is, the sampling frequency is much greater than $f_s(= 2f_H)$. For example, if the sampling frequency $F_s = kf_s$, and k is a positive integer greater than 1, then since the quantization noise power is constant, the quantization noise power spectral density in the frequency range of 0–F_s is smaller than that during Nyquist sampling and is only $1/k$ of it. At this time, the quantization noise power falling in the range of 0–f_H is the area of in the figure, which is obviously smaller than that in Nyquist sampling. It can be seen that in order to reduce the power of quantization noise, the first measure adopted in modern modulation Σ–Δ is oversampling. In order to further reduce the power of quantization noise, the second measure used in modern modulation Σ–Δ is called noise shaping. It can be seen from the figure that it pushes the quantization noise power to the high-frequency band as much as possible. In this way, the power of the quantization noise falling in the range of 0–f_H is further reduced, as shown in the area of .

In summary, it is known that in modern Σ–Δ modulation, oversampling and noise shaping are used to reduce the quantization noise power and improve the quantization signal-to-noise ratio.

Voice A/D and D/A are also in the BB chip. There are three ways of speech coding, which are waveform coding, parametric coding and hybrid coding. Among them, waveform coding has good naturalness, but the coding rate is high. Parametric coding can only understand the waveform, but the coding rate is low. Hybrid coding is the combination of waveform coding and parametric coding, so its naturalness is better and the coding rate is relatively moderate. Hybrid coding is used in LTE. The language code is evaluated according to the MOS (Mean Opinion Score) five-point system. It is internationally stipulated that only voice codes with a score of 3.8 or higher are allowed to enter the public telephone network.

There is a dedicated chip for image encoding/decoding, which is not included in the BB chip. Voice signals, video signals and other multimedia signals are controlled through the MAC sub-layer and cannot directly enter the physical layer.

It should be pointed out that Figure 1.39 is only applicable to uplink/downlink transmission with OFDM symbols. But the LTE uplink transmits DFTS-FDMA symbols. Therefore, the BB chip used on the base station side and the terminal side are different, and the principle block diagram is also different from Figure 1.39, which is discussed in Section 1.9.

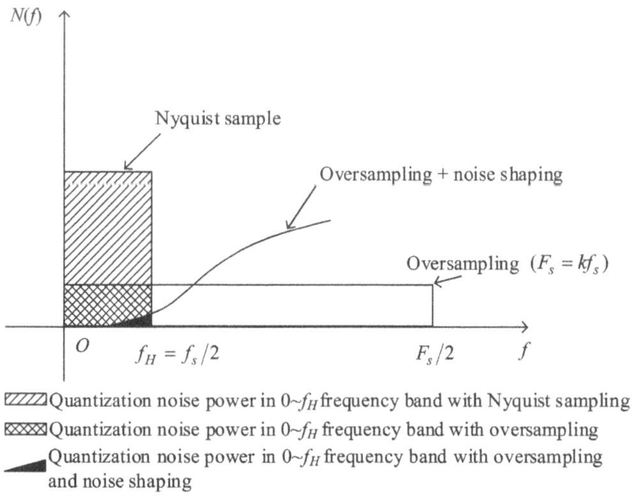

Figure 1.40 Modern $\Sigma - \Delta$ modulation working principle.

1.5 RF Chip Diagram

The principle diagram of the RF chip is shown in Figure 1.41. It is used to realize the function of the LTE protocol in the radio frequency part. The RF chip is explained as follows:

(1) *Transmitter*:
(i) *Quadrature amplitude modulation*:
The two parallel analog signals output by the transmitter in Figure 1.39 (BB chip) are directly transmitted to the quadrature amplitude modulated I channel and Q channel of the RF chip. The reason for using quadrature

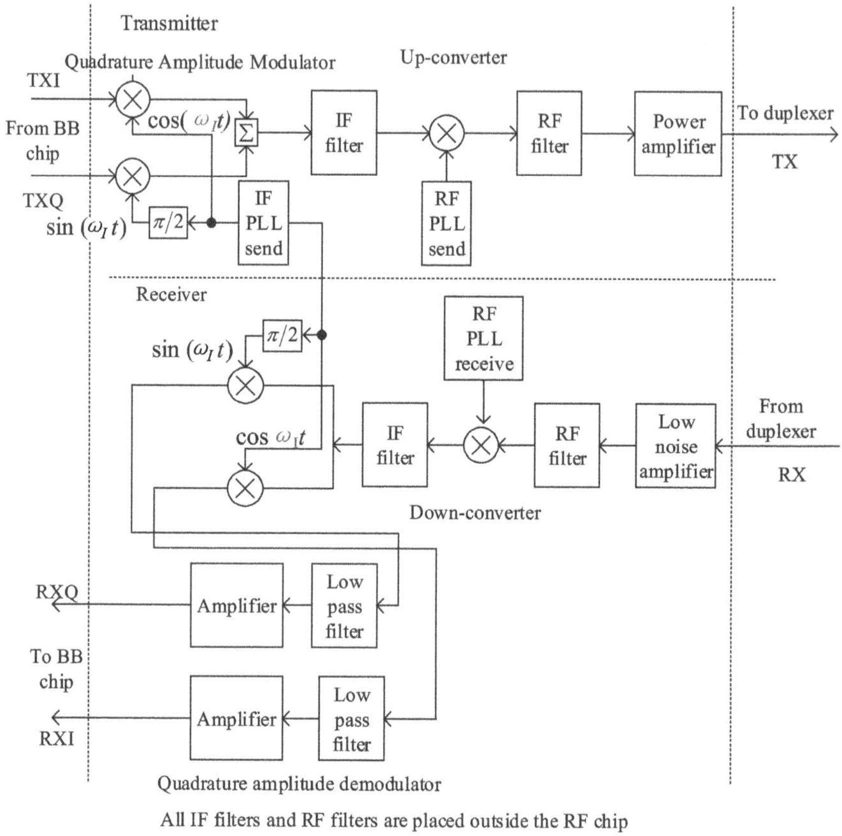

Figure 1.41 RF chip diagram.

amplitude modulation is as follows. These two signals represent the projection of the signal points on the constellation diagram of the x-axis and the y-axis, that is, the real part and the imaginary part. In order to distinguish the real and imaginary parts of the signal point, quadrature amplitude modulation is used. That is, the I channel signal modulates the amplitude of cos $\omega_I t$, and the Q channel signal modulates the amplitude of sin $\omega_I t$. The intermediate frequency carrier cos $\omega_I t$ and sin $\omega_I t$ are orthogonal, so this orthogonality can be used in the receiver to separate the real and imaginary parts. The intermediate frequency carrier is provided by the intermediate frequency synthesizer (IFPLL).

(ii) *Up-converter*:
The up-converter transforms the intermediate frequency ω_I into the required radio frequency ω_e through mixing, and the local oscillator frequency of the mixer is provided by the radio frequency synthesizer (RFPLL).

(2) *Receiver*:
(i) Down-converter and quadrature amplitude demodulator are the inverse operations of the corresponding parts of the transmitter.

(ii) *Low Noise Amplifier (LNA)*:
The magnitude of receiver thermal noise is mainly determined by LNA. The total noise power equivalent to the input of the LNA (under AWGN channel conditions) is

$$Ne = kTB(NF) \tag{1.26}$$

where $k = 1.38 \times 10^{-23}$ J/K is the Boltzmann's constant, $T = 273$+ambient temperature (°C) is the absolute temperature (K), B (Hz) is the receiver bandwidth and NF (dimensionless) is the LNA noise figure.

(iii) *Receiver reference sensitivity*:
Under the premise that the bit error rate (P_b) of the FEC decoder output at the receiver of BB chip reaches the specified value, the minimum signal power to be input at the LNA input is defined as the reference sensitivity (dBm) of the receiver, which is discussed in Section 4.4.6.1 of Chapter 4.

(3) *Frequency synthesizer block diagram*: The diagram of IF PLL and RF PLL is shown in Figure 1.42. When the loop is locked, $f_d = f_r$ and $f_s = Nf_d = Nf_r$. It can be seen that the frequency stability and accuracy of the f_s outputted by the frequency combiner are completely determined by the crystal oscillator.

(4) *Duplexer*: Figure 1.43(a) shows the frequency division duplex (FDD) device. The receiving and transmitting carrier frequencies are different. In order to transmit and receive at the same time without mutual interference, the two filters with center frequencies f_R and f_T shown in Figure 1.43(a) should be used. Moreover, there needs to be a sufficiently wide isolation band between the two filters. These two filters are called full duplex filters. However, some terminals (such as certain IoT terminals) may only need to support half–duplex, that is, the terminals do not need to work both upstream and downstream. Therefore, the full duplex filter can be omitted, which can simplify terminal equipment and reduce costs. In this case, the terminal is called a half–duplex FDD state. It should be pointed out that no matter whether the terminal is full duplex FDD or half–duplex FDD, the base station should work in the full duplex FDD state. Figure 1.43(b) shows a time division duplex (TDD) device with same transmitting and receiving carrier frequencies. But the time slots for receiving and transmitting are different, and the uplink and downlink will not interfere with each other. Therefore, a full duplex filter is not required, but a transceiver switch is required.

(5) *When the RF chip works in TDD, only one intermediate frequency filter and radio frequency filter of IF PLL and RF PLL in Figure 1.41 are required*: Mobile communication systems are all adaptive communication systems, which is based on changes in the characteristics of the transport channel and adaptively changing the system parameters to ensure that the system bit error rate requirements. Among them, power adaptation refers to the requirement of changing the bit energy E_b by changing the signal amplitude (changing the transmitter power) to ensure the system bit error rate under the condition of keeping the data transmission rate unchanged. The rate adaptation refers to the requirement of changing E_b by changing the data transmission rate under the condition of unchanged signal amplitude to ensure the system bit error rate. Changing the data transmission rate allows the selection of a different modulation and coding scheme (MCS). Since the modulation order is different when the modulation

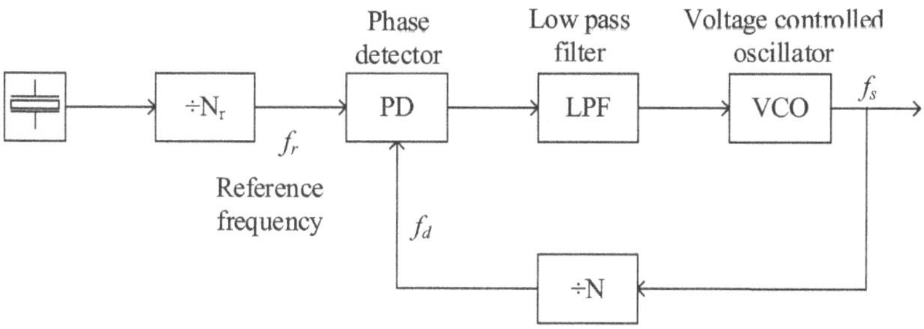

Figure 1.42 Diagram of frequency synthesizer.

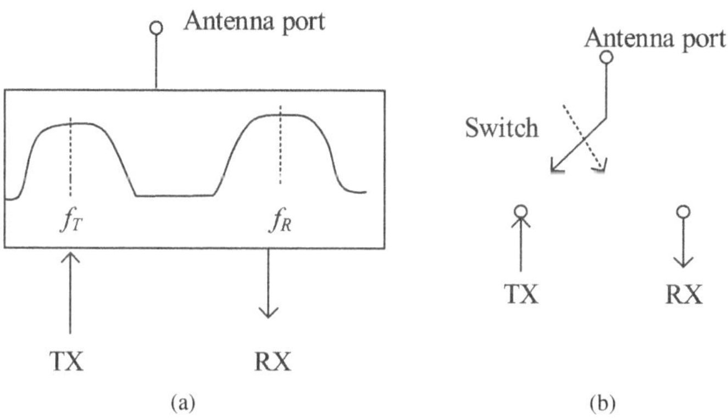

Figure 1.43 Duplexer: (a) FDD; (b) TDD.

mode is different, the coding rate is different when the channel coding mode is different, and the modulation order and code rate will change the data transmission rate.

1.6 BB Chip and RF Chip Interface

The following main interfaces between the chips are shown in Figure 1.44:

(1) *Three-wire interface*: When enable is activated, the data outputted by BB enters the latch in RF in a synchronous manner (clock). The data in the latch can control the following parameters of RF:

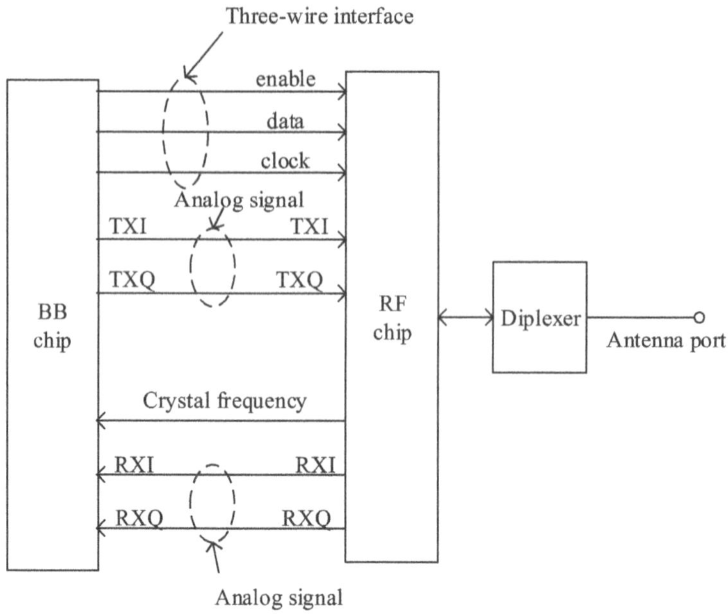

Figure 1.44 Chip interface.

(i) Transmit power, (ii) Transmit frequency and (iii) Frequency divider *N* value in RF PLL.

It can be seen that the working state of the RF chip is controlled by the BB chip through a three-wire interface.

(2) *Analog signal*: The two sets of analog signals are the real and imaginary signals of BB→RF and RF→BB.
(3) *Crystal frequency*: The various timings and frequencies required in the BB chip are all generated by the crystal oscillator in the RF chip.
(4) When working in time division duplex (TDD), the antenna switch control signal (see Figure 1.43(b)) and the transmitter and receiver switches in the RF chip (not shown in Figure 1.41) are all provided by the BB chip.

1.7 LTE Frame Structure and Resource Configuration

It can be seen from Figure 1.39 that the LTE uplink/downlink is used to transmit data in the form of a frame structure. To be precise, the downlink transmits OFDM symbols, and the uplink transmits DFTS-OFDM symbols. The transmitted OFDM symbols and DFTS-OFDM symbols are collectively referred to as radio resources. The radio frame in LTE is composed of several time slots, and these resources are configured in the time slots for transmission.

The LTE radio frame structure and resource configuration vary with the duplex mode of the communication system. The uplink/downlink is also different under the same duplex mode. So, it needs to be discussed separately.

1.7.1 *Frequency Division Duplex (FDD)*

1.7.1.1 *Downlink*

(1) *Frame structure*: The structure of the downlink frame is shown in Figure 1.45. This frame structure is called a type one frame structure, and it has the following characteristics in the time domain:

(i) The duration of a radio frame is 10 ms.
(ii) A radio frame contains 10 sub-frames, and each sub-frame has a duration of 1 ms.
(iii) A sub-frame contains two time slots, and each time slot has a duration of 0.5 ms.
(iv) A slot contains several OFDM symbols, and the number of configurations is different according to different situations. Shown in the figure are seven OFDM symbols, which refer to the case of standard cyclic prefix.
(v) The duration of an OFDM symbol is equal to the operation duration T of the cyclic prefix T_{cp} + IFFT. T_{cp} or T may also be different in OFDM with different serial numbers.

The following three-time units in the above frame structure are also used in other frame structures:

(1) Radio frame duration $T_f = 10$ ms.
(2) Time slot duration $T_{\text{slot}} = 0.5$ ms.
(3) Sampling interval $T_s = \frac{1}{30720}$ ms.

Figure 1.45 explains how the value of T_s is obtained.

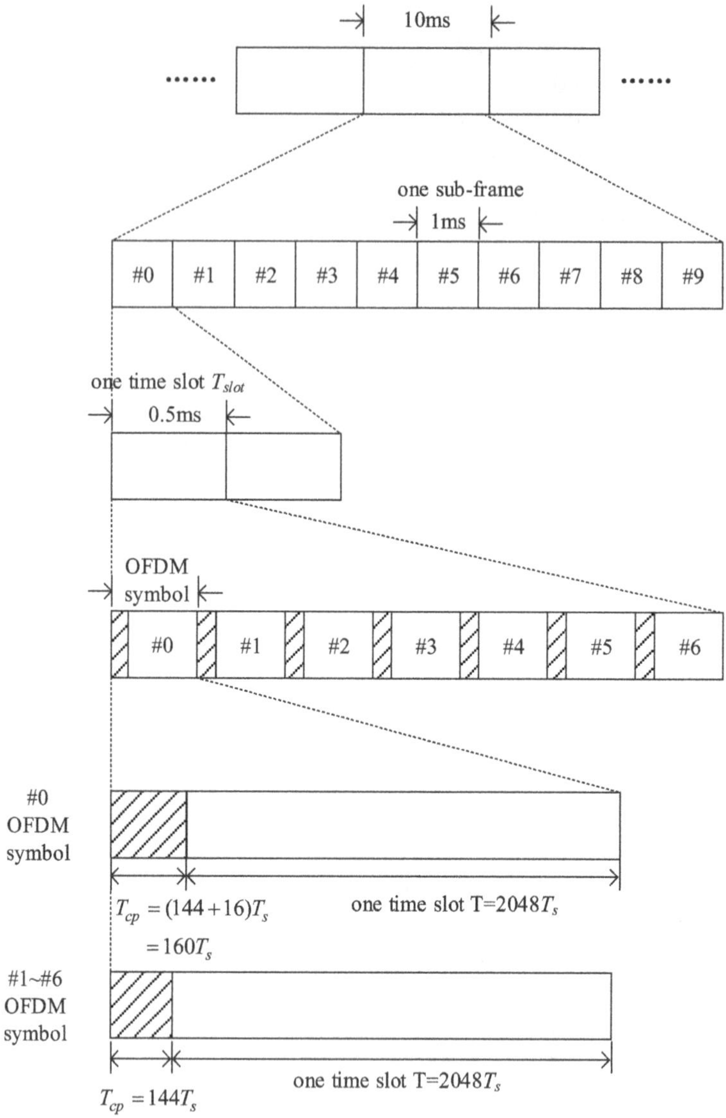

$$T_s = \frac{1/15}{2048} = \frac{1}{15 \times 2048} = \frac{1}{30720}(\text{ms})$$

Figure 1.45 Downlink frame structure.

In LTE, the subcarrier spacing has been determined as $\Delta f = 15$ KHz, which can be obtained from the equation (1.13) that $T = 1/15$ ms. When sampling is performed in the T period and the number of samples is 2048, the sampling interval is obtained.

$$T_s = T/2048 = \frac{1/15}{2048} = \frac{1}{15 \times 2048} = \frac{1}{30720}\text{ms}$$

(2) *Resource grid*: It has been pointed out earlier that the OFDM symbol of the radio resource is transmitted in the radio frame slot. The OFDM symbol is a two-dimensional resource composed of time domain and frequency domain. Therefore, the OFDM symbols transmitted in the time slot can be represented by Figure 1.46. Obviously, due to two-dimensional reasons, the graph becomes a grid, which is called a resource grid. The resource grid can be described with the following parameters:

(i) *Resource unit (RE)*:
RE is composed of an OFDM symbol (time domain) and a subcarrier spacing (frequency domain). Obviously, the resource grid is composed of several REs, so RE is the smallest unit in the resource grid.

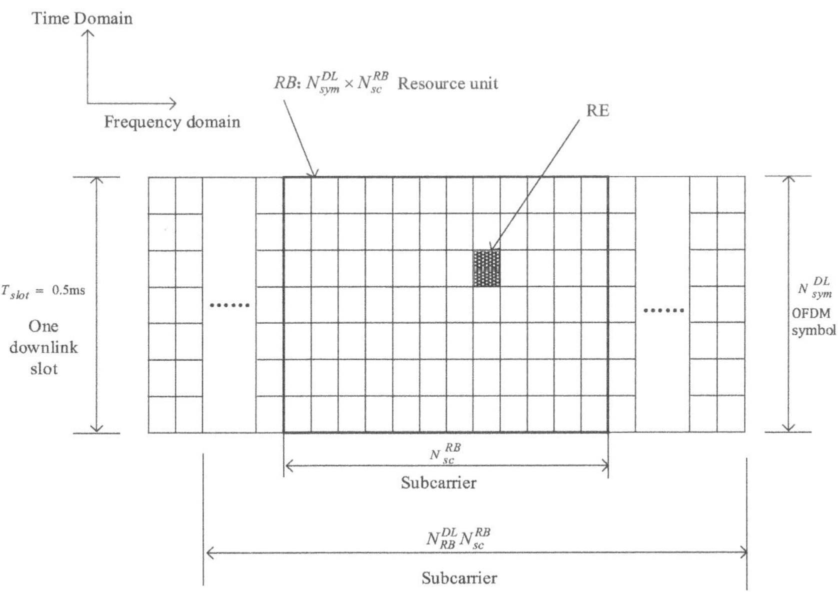

Figure 1.46 Downlink resource grid.

(ii) N_{sym}^{DL}:
It means that the downlink contains several OFDM symbols (time domain) in a time slot, and the number is different in different situations. The figure shows seven and it refers to the standard cyclic prefix.

(iii) N_{sc}^{RB}:
It means that a resource block in the downlink contains several subcarriers (frequency domain), and the number is different in different situations. The figure shows 12, which refers to the standard cyclic prefix.

(iv) *Resource Block (RB)*:
It is composed of N_{sym}^{DL} consecutive symbols in the time domain and N_{sc}^{RB} consecutive subcarriers in the frequency domain, so an RB contains $N_{sym}^{DL} \times N_{sc}^{RB}$ REs.

(v) N_{RB}^{DL}:
It represents the number of resource blocks used by a radio frequency carrier in the downlink, which is specified in LTE as

$$6 \le N_{RB}^{DL} \le 110 \tag{1.27}$$

Obviously, the bandwidth occupied by the downlink is

$$B = N_{RB}^{DL} \times N_{sc}^{RB} \times \Delta f\,(\text{Hz}) \tag{1.28}$$

(vi) *Representation method of the position of resource unit RE in the resource grid*:
The position of RE should be defined in two dimensions of time and frequency, that is, expressed by (K, L). K and L represent the values of RE in frequency domain and time domain and $K = 0,1,\ldots,\left(N_{RB}^{DL} \times N_{sc}^{RB} - 1\right)$, $L = 0,1,\ldots,(N_{sym}^{DL} - 1)$. The value of RE itself is represented by a_{KL}. When a_{KL} is not used for physical channel or physical signal transmission, then $a_{KL} = 0$.

(3) *Resource allocation*: The resource configuration in a time slot in the downlink is shown in Table 1.4. There are three different configurations, which are discussed separately in the following:

Table 1.4 Downlink time slot resource configuration.

Configuration		N_{sc}^{RB}	N_{sym}^{DL}
Standard cyclic prefix	$\Delta f = 15$kHz	12	7
Extended cyclic prefix	$\Delta f = 15$kHz	12	6
Extended cyclic prefix	$\Delta f = 7.5$kHz	24	3

(i) *Standard cyclic prefix*:

#0 OFDM symbol: $T_{cp} = (144 + 16)T_{s,}$ $T = 2048T_s$.

#1–#6 OFDM symbol: $T_{cp} = 144T_{s,}$ $T = 2048T_s$

$$7 \text{ OFDM symbles occupied time} = \left[(144+2048)\times 7+16\right]T_s = 15360T_s$$
$$= 15360 \times \frac{1}{30720} = 0.5(\text{ms})$$

Bandwidth occupied by one resource block = 12×15 = 180(KHz)

(ii) *Extended cyclic prefix* ($\Delta f = 15$ KHz):

#0–#5 OFDM symbol: $T_{cp} = 512T_{s,}$ $T = 2048T_s$

6 OFDM symbles occupied time = $(512 + 2048) \times 6T_s$
= $15360T_s = 0.5$(ms)

Bandwidth occupied by one resource block = 12×15 = 180(KHz)

(iii) *Extended cyclic prefix* (($\Delta f = 7.5$ KHz):

#0–#2 OFDM symbol: $T_{cp} = 1024T_s$, $T = 4096T_s$

3 OFDM symbols occupied time = $(1024 + 4096) \times 3T_s$
= $15360T_s = 0.5$(ms)

Bandwidth occupied by one resource block = 24 × 7.5 = 180(KHz).

It can be seen from the above data that the T_{cp} in the extended case is longer than that in the standard case. It is used in the following situations: when the cell coverage space is large, its delay spread τ_m increases. In order to resist ISI and ICI, a longer T_{cp} is needed.

(4) *Downlink reference signal*: Downlink reference symbols have many uses, such as downlink quality measurement, channel estimation and cell search. In the case of the standard cyclic prefix, the position occupied by the reference symbol in the resource grid is shown in Figure 1.47. Figure 1.47(a) shows the position occupied by reference symbols in a pair of resource blocks when there is only one antenna port in the downlink. Figure 1.47(b) shows the position occupied by reference symbols in a pair of resource blocks when there are two antenna ports in the downlink. Figure 1.47(c) shows the position occupied by reference symbols in a pair of resource blocks when there are four antenna ports in the downlink. It can be seen from the figure that there are five subcarriers between two adjacent reference symbols (which are the same) in the frequency domain. In the time domain, two adjacent reference symbols (which are different) are separated by three OFDM symbols.

Since the reference signal configuration positions of different antenna ports are different, the terminal must know which antenna port of the base station that the currently received signal comes from, so as to know the position of the reference signal and its purpose. The information about the antenna port can be obtained by the terminal from the downlink signaling.

1.7.1.2 *Uplink*

(1) *Uplink transmission waveform*: The OFDM waveform has a fatal weakness, that is, the peak-to-average power ratio of the waveform is large. In order to reduce the nonlinear distortion, the power amplifier is required to be in a linear working state in a large range. The power conversion efficiency of such a successful amplifier is low and the cost is high. It is still acceptable for a limited number of base stations powered by AC. Therefore, the downlink transmission waveform of the base station is still OFDM. But this weakness is unacceptable for the terminal, since terminals (especially mobile phones) are powered by batteries, and power saving is very important. In addition, the number of terminals is very large and involves thousands of households, so the

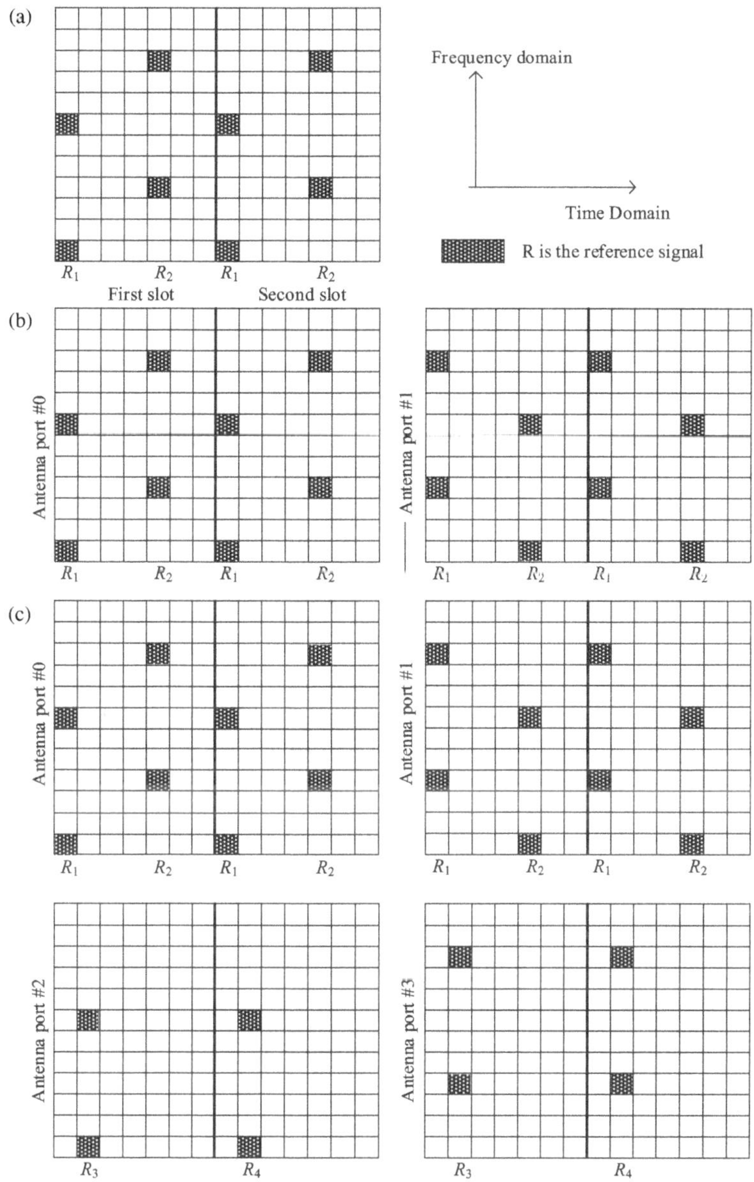

Figure 1.47 Configuration of reference symbols.

cost must be low. For these reasons, the uplink transmission waveform of the terminal is not OFDM, but the DFTS-OFDM (DFT Spread OFDM) waveform, which has a power peak-to-average ratio smaller than that of the OFDM waveform. The principle diagram is shown in Figure 1.48.

In the figure, DFT is the discrete Fourier transform, and IDFT is the inverse of DFT. The difference between the DFTS-OFDM waveform diagram (a) and the OFDM waveform diagram 1.39 is only the addition of the DFT function block. Now we need to discuss why it can reduce the power peak-to-average ratio.

In Figure 1.39, QAM symbols from high-order modulation are serial/parallel converted, and their output is a parallel N-channel QAM symbol sequence. Each QAM symbol sequence is transmitted by a narrowband (15 KHz in LTE) subcarrier, as shown in Figure 1.49(a). Therefore, it is a multi-carrier transmission scheme. Each subcarrier is narrowband. In Figure 1.48(a), after serial/parallel conversion of high-order modulated QAM symbols, the output parallel N-channel QAM symbol sequence is first transformed (spread spectrum) into frequency domain symbols through DFT. Each frequency domain symbol is transmitted by a narrowband subcarrier, so that the bandwidth occupied by any QAM symbol sequence is $N\Delta f$, as shown in Figure 1.49(b). It occupies a bandwidth of 60 KHz, which shows that it is not a multi-carrier transmission scheme at this time, but a normal single-carrier transmission scheme. Therefore, the power peak-to-average ratio is smaller than the OFDM waveform. In fact, it can be considered from Figure 1.48(a) that the DFT (FFT) transform offsets the effect of the IFFT, making the multi-carrier transmission

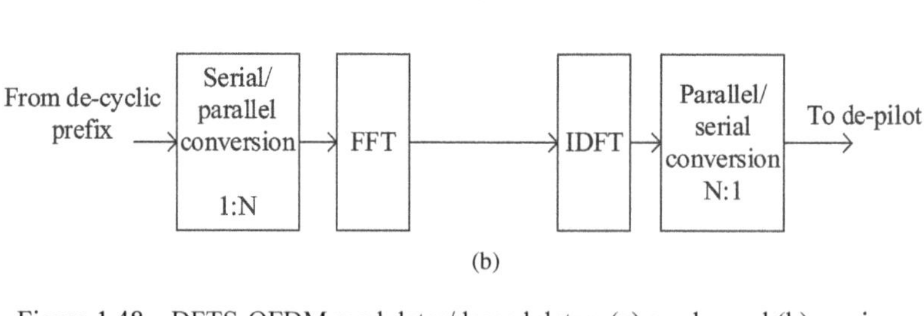

Figure 1.48 DFTS-OFDM modulator/demodulator: (a) sender and (b) receiver.

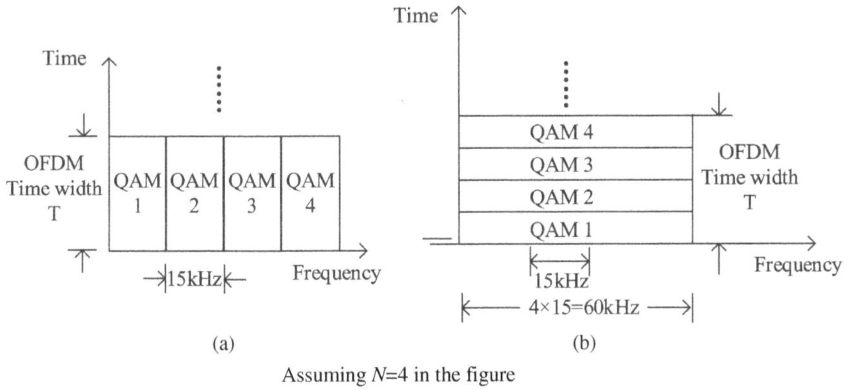

Figure 1.49 Comparison of DFTS-OFDM and OFDM.

scheme convert to a single carrier transmission scheme. It is precisely because the uplink is a single-carrier modulation method, so different terminals should use single-carrier frequency division multiple access (SC-FDMA) for uplink multiple access.

Figure 1.48(b) is the inverse transform of Figure 1.48(a), so it is not explained here.

Since the LTE downlink/uplink transmission waveforms are OFDM and DFTS-OFDM, the BB chip on the base station side and the terminal side are different. The base station side should change the corresponding part of the receiver in Figure 1.39 to Figure 1.48(b). The terminal side should change the transmitter in Figure 1.39 to Figure 1.48(a).

(2) *Frame structure*: The structure of the uplink frame is the same as that of the downlink frame in Figure 1.45, and the radio of frame duration T_f, time slot duration T_{slot} and sampling interval T_s is also the same as the downlink frame.

(3) *Resource grid*: The uplink resource grid is also the same as the downlink resource grid in Figure 1.46, but the following parameters in Figure 1.46 should be changed to uplink parameters:

$$\text{One downlink time slot} \longrightarrow \text{One uplink time slot}$$

$$N_{sym}^{DL} \longrightarrow N_{sym}^{UL}$$

$$N_{RB}^{DL} \longrightarrow N_{RB}^{UL}$$

$$\text{OFDM symbol} \longrightarrow \text{DTFS-OFDM symbol}$$

(i) *Resource unit RE*:
The uplink RE is composed of a DFTS-OFDM symbol (time domain) and a subcarrier spacing (frequency domain).

(ii) N_{sym}^{UL}:
It means that the uplink contains several DFTS-OFDM symbols (time domain) in one slot.

(iii) N_{sc}^{RB}:
It means that a resource block in the uplink contains several subcarriers (frequency domain).

(iv) *Resource Block (RB)*:
It is composed of N_{sym}^{UL} in the uplink time domain and N_{sc}^{RB} in the frequency domain, so one RB contains $N_{sym}^{UL} \times N_{sc}^{RB}$ RE.

(v) N_{RB}^{UL}:
It represents the number of resource blocks used by a radio frequency carrier in the uplink and is specified in LTE that

$$6 \le N_{RB}^{UL} \le 110 \tag{1.29}$$

The occupied bandwidth is

$$B = N_{RB}^{UL} \times N_{sc}^{RB} \times \Delta f\,(\text{Hz}) \tag{1.30}$$

Generally, the bandwidth occupied by the terminal is less than or equal to the bandwidth of the base station.

(vi) The representation method of the resource unit RE position in the resource grid is the same as the downlink, that is, it is represented by (*K*, *L*), where

$$K = 0, 1, \ldots, (N_{RB}^{UL} \times N_{sc}^{RB} - 1)$$

$$L = 0, 1, \ldots, (N_{sym}^{UL} - 1)$$

(4) *Resource allocation*: As shown in Table 1.5, it can be seen from the comparison with Table 1.4 that there is no $\Delta f = 7.5$ KHz configuration in

Table 1.5 Uplink time slot resource configuration.

Configuration		N_{sc}^{RB}	N_{sym}^{UL}
Standard cyclic prefix	$\Delta f = 15\text{kHz}$	12	7
Extended cyclic prefix	$\Delta f = 15\text{kHz}$	12	6

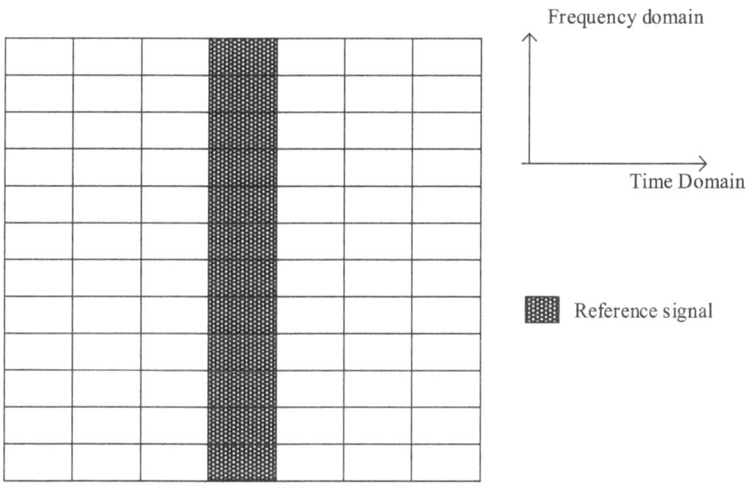

Figure 1.50 Uplink reference symbol configuration.

the uplink. The other two configurations are the same. The configuration of the downlink $\Delta f = 7.5$ KHz is used for Multimedia Broadcast Multicast Services (MBMS). But there is no such service on the uplink.

(5) *Uplink reference signal*: In the case of the standard cyclic prefix, the position of the uplink reference signal in a time slot is shown in Figure 1.50, which is different from the downlink configuration method. The reference signal and other uplink transmission signals are time-division multiplexed.

1.7.2 *Time Division Duplex*

1.7.2.1 *Frame structure*

Time Division Duplex (TDD) has two frame structures, the type 1 frame structure is the same as Figure 1.45. The type 2 frame structure is shown in Figure 1.51. A radio frame is 10 ms and is divided into 2 half-frames;

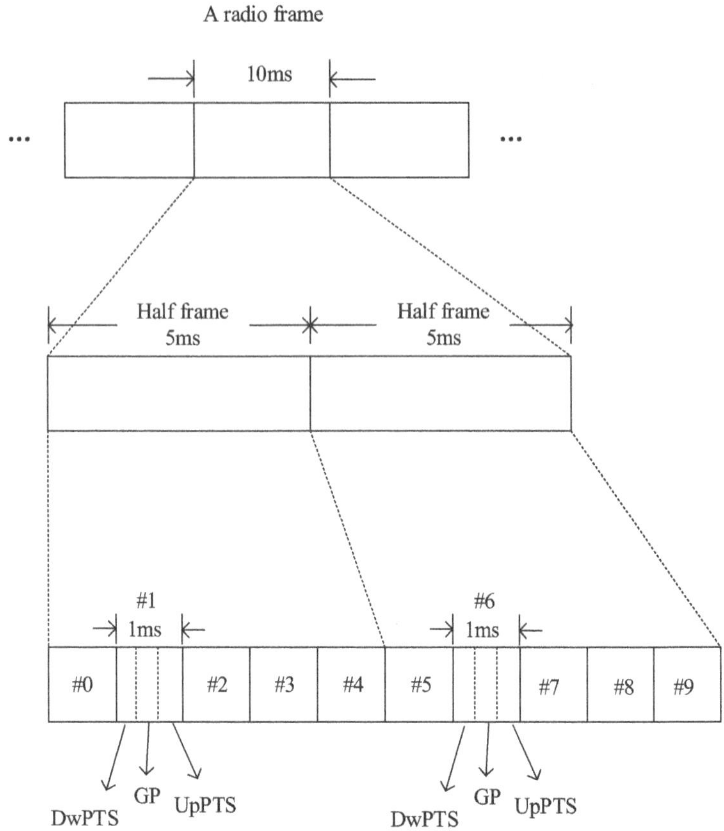

Figure 1.51 TDD Type 2 frame structure.

each half-frame is 5 ms, and the half-frame is composed of 5 sub-frames, and each sub-frame is 1 ms. There are two types of sub-frames: one is called a normal sub-frame, which contains two 0.5 ms time slots, and the other is called a special sub-frame, which is composed of three time slots: Downlink Pilot Time Slot, downlink synchronization interval (DwPTS), Guard Period, guard interval (GP) and Uplink Pilot Time Slot, uplink synchronization interval (UpPTS).

(1) *Uplink/downlink sub-frame configuration*: The configuration of the uplink/downlink sub-frame is shown in Table 1.6. It can be seen from the table that different sub-frames are allocated for transmission of uplink or downlink data. *S* (switching) in the table is the above-mentioned special

Table 1.6 Uplink/downlink sub-frame configuration.

Configuration number	Switching cycle	Sub-frame number (U: Uplink, D: Downlink, S: Switch)									
		0	1	2	3	4	5	6	7	8	9
0	5 ms	D	S	U	U	U	D	S	U	U	U
1	5 ms	D	S	U	U	D	D	S	U	U	D
2	5 ms	D	S	U	D	D	D	S	U	D	D
3	10 ms	D	S	U	U	U	D	D	D	D	D
4	10 ms	D	S	U	U	D	D	D	D	D	D
5	10 ms	D	S	U	D	D	D	D	D	D	D
6	5 ms	D	S	U	U	U	D	S	U	U	D

sub-frame. The *S* sub-frame enables the system to switch from downlink transmission to uplink transmission. This configuration is called semi-static configuration. It can be seen from the table that the numbers #0 sub-frame and #5 sub-frame are all allocated to the downlink. The number #2 sub-frames are all allocated to the uplink. The number #1 sub-frames are all special sub-frames. In 10 sub-frames, there are 2 *S* sub-frames, which means the uplink/downlink switching period is 5 ms, and only 1 *S* sub-frame means the uplink/downlink switching period is 10 ms. The so-called semi-static configuration means that there are multiple (seven) rather than only one up/down configuration schemes within 10 ms. If there is only one, then it is a fully static configuration. The role of special sub-frames is discussed in the following.

(2) *The role of special sub-frames*:

(i) *DwPTS*:

It is used for downlink synchronization, so that the terminal timing can be synchronized with the base station timing that arrives at the terminal. Figure 1.52 is the base station frame signal arriving at the terminal. There are zero power zones on the left and right sides of the DwPTS time slot. In this way, after the terminal has searched for these two zero-power regions, it can find the downlink synchronization code transmitted in the DwPTS time slot, thereby realizing downlink synchronization. Downlink synchronization is also required in the case of FDD.

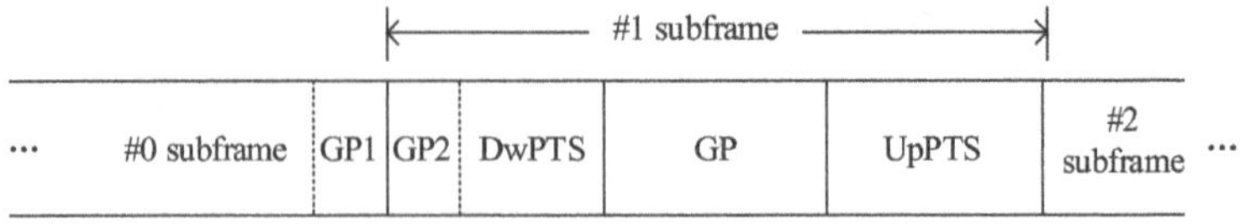

Figure 1.52 Downlink synchronization.

(ii) *UpPTS*:
UpPTS is similar to DwPTS, so that the base station timing can be synchronized with the terminal timing that arrives at the base station. For TDD, the uplink must be synchronized, that is, the signals of different terminals must be seated when they arrive at the base station to avoid mutual interference. Figure 1.53 illustrates how the uplink is synchronized. Figure 1.53(a) is the eNodeB frame signal on the base station side, which reaches the terminal through path transmission. The frame signal generated by the terminal after downlink synchronization is shown in Figure 1.53(b), and its timing is Δt later than that of the base station side due to path delay. At this time, if we want the signal transmitted by the terminal in the UpPTS time slot to accurately fall within the UpPTS time slot in Figure 1.53(a), the terminal must transmit it ahead of time $2\Delta t$ than the UpPTS in Figure 1.53(b) because the signal needs to go $2\Delta t$ from the base station to the terminal and back from the terminal to the base station (assuming that the terminal is not moving). In fact, users in the cell are at different distances from the base station and may be in motion. In this way, different users' need to transmit in advance time is different and also keep changing. For these reasons, if we want to accurately achieve uplink synchronization, the time that the user needs to transmit in advance is determined by the base station through a closed-loop operation and then notified to the terminal. The time for uplink transmission in advance just falls within the guard interval GP. It can be seen that the choice of GP size is related to the size of the cell coverage.

1.7.2.2 *Resource grid*

(1) The downlink resource grid is the same as Figure 1.46.
(2) *Uplink resource grid*: For a Type 1 frame, the uplink resource grid structure is the same as the uplink resource grid in FDD. For Type 2 frame, the uplink resource grid structure is shown in Figure 1.54,

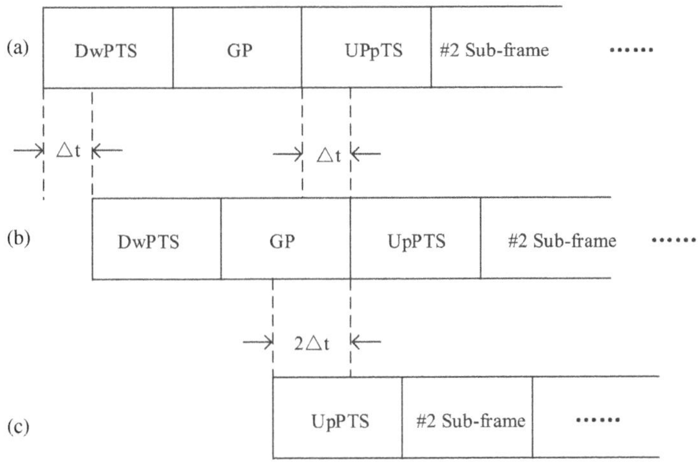

Figure 1.53 Principle of uplink synchronization (a) eNodeB timing, (b) UE timing and (c) UE advanced sending time (2Δt).

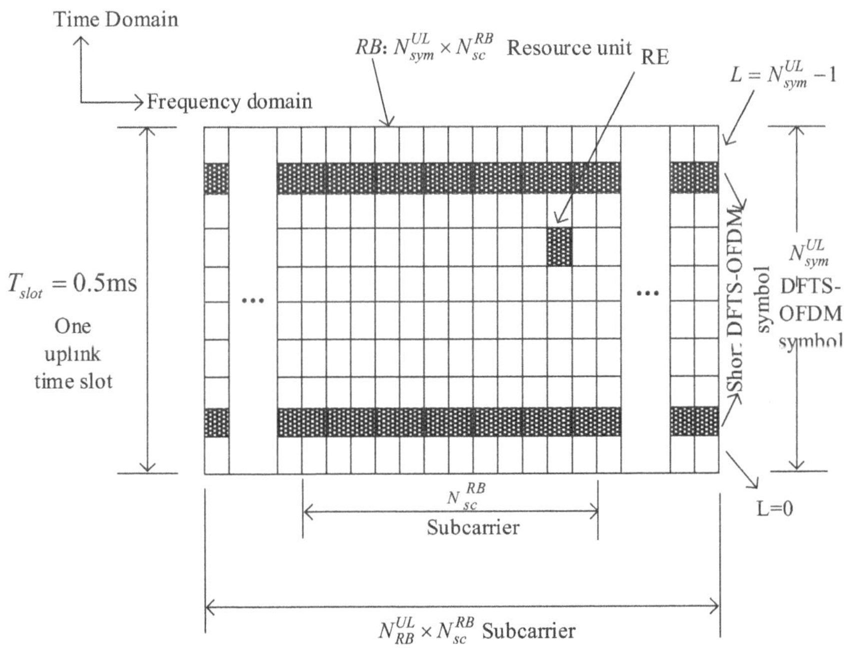

Figure 1.54 Uplink resource grid.

Table 1.7 Resource allocation.

(a) Downlink

Configuration		N_{sc}^{RB}	N_{sym}^{UL}
Standard cyclic prefix	$\Delta f = 15$kHz	12	9
Extended cyclic prefix	$\Delta f = 15$kHz	12	8
	$\Delta f = 7.5$kHz	24	4

(b) Uplink

Configuration		N_{sc}^{RB}	N_{sym}^{UL}
Standard cyclic prefix	$\Delta f = 15$kHz	12	10
Extended cyclic prefix	$\Delta f = 15$kHz	12	9

where $L = 1$ and $(N_{sym}^{UL}-2)$ are short DFTS-OFDM symbols, that is, its time length is shorter than other DFTS-OFDM symbols in the grid. They are used to transmit uplink modulation reference signals.

(3) *Resource allocation*: It can be seen from the above analysis that OFDM signals have two outstanding advantages (Table 1.7):

(i) It can resist ISI caused by multipath channel (also known as time dispersion channel).
(ii) OFDM is a two-dimensional signal composed of frequency domain and time domain. Therefore, the number of subcarriers and the number of OFDM symbols in a time slot can be configured according to the requirements of different channels (such as required bandwidth) and signals (such as different number of bytes). And this configuration is very convenient.

For the above reasons, OFDM waveforms are used in both 4G and 5G.

1.8 Terminal Category

Terminals (such as smartphones and IoT terminals) that can support the peak rate of uplink/downlink data in the radio access network vary with baseband chip performance, MIMO scale, radio frequency bandwidth and

Table 1.8 Terminal category.

Terminal equipment	Down/Up peak (Mbps)	Maximum number of downlink MIMO	3GPP release
Cat 0	1/1	1	R12
Cat 1	10/5	1	R8
Cat 2	50/25	2	R8
Cat 3	100/50	2	R8
Cat 4	150/50	2	R8
Cat 5	300/75	4	R8
Cat 6	300/50	2/4	R10
Cat 7	300/100	2/4	R10
Cat 8	3000/1500	8	R10
Cat 9	450/50	2/4	R11
Cat 10	450/100	2/4	R11
Cat 11	600/50	2/4	R12
Cat 12	600/100	2/4	R12
Cat 13	390/50	2/4	R12
Cat 14	390/100	2/4	R12
Cat 15	3900/1500	8	R12

Note: The rates in the table are all rounded values.

carrier aggregation number. The rate can divide the terminal into different levels. For 4G terminal equipment, it is called LTE Cat (LTE UE Category), that is, the LTE user equipment level. Such a level given by 3GPP is shown in Table 1.8.

1.9 LTE RF Frequency Band and Bandwidth

1.9.1 *LTE frequency band*

LTE supports both FDD and TDD duplex modes at the same time. FDD requires paired frequency bands, while TDD is an unpaired frequency band. For this reason, 3GPP has defined two frequency bands for LTE.

(1) *Paired frequency bands*: As shown in Table 1.9, there are 27 frequency bands.

Table 1.9 Paired frequency bands.

Band number	Uplink range (MHz)	Downlink range (MHz)	Main area
1	1920–1980	2110–2170	Europe, Asia
2	1850–1910	1930–1990	America (Asia)
3	1710–1785	1805–1880	Europe, Asia (America)
4	1710–1755	2110–2155	America
5	824–849	869–894	America (Asia)
6	830–840	875–885	Japan (UTRA only)
7	2500–2570	2620–2690	Europe, Asia
8	880–915	925–960	Europe, Asia
9	1749.9–1784.9	1844.9–1879.9	Japan
10	1710–1770	2110–2170	America
11	1427.9–1447.9	1475.9–1495.9	Japan
12	698–716	728–746	America
13	777–787	746–756	America
14	788–798	758–768	America
17	704–716	734–746	America
18	815–830	860–875	Japan
19	830–845	875–890	Japan
20	832–862	791–821	Europe
21	1447.9–1462.9	1495.9–1510.9	Japan
22	3410–3490	3510–3590	Europe
23	2000–2020	2180–2200	America
24	1626.5–1660.5	1525–1559	America
25	1850–1915	1930–1995	America
26	814–849	859–894	America
27	807–824	852–869	America
28	703–748	758–803	Asia
29	—	717–728	America

Note: UTRA, Universal Terrestrial Radio Access.

(2) *Unpaired frequency band*: As shown in Table 1.10, there are 12 frequency bands.

Please refer to the Appendix of this book for more details about the frequency bands used by LTE.

Table 1.10 Unpaired frequency bands.

Band number	Frequency range (MHz)	Main area
33	1900–1920	Europe, Asia (except Japan)
34	2010–2025	Europe, Asia
35	1850–1910	America
36	1930–1990	America
37	1910–1930	
38	2570–2620	Europe
39	1880–1920	China
40	2300–2400	Europe, Asia
41	2496–2690	America
42	3400–3600	Europe
43	3600–3800	Europe
44	703–803	Asia

1.9.2 *LTE-specified channel bandwidth*

It can be seen from (1.28) and (1.30) that the channel occupied bandwidth is determined by the number of occupied resource blocks (N_{RB}^{UL} or N_{RB}^{DL}) and the number of subcarriers contained in a resource block (N_{sc}^{RB}) since it is known that

$$\Delta f = 15\text{kHz}, \quad N_{sc}^{RB} = 12 \quad 6 \leq N_{RB} \leq 110$$

According to this, the channel bandwidth specified by LTE is shown in Table 1.11, and there are six usable bandwidths.

1.9.3 *Location of RF carrier frequency*

It can be seen from Table 1.11 that LTE stipulates a total of six radio frequency bandwidths. Their center frequency, that is, the radio frequency carrier frequency is the same. As shown in Figure 1.55, the carrier frequency is configured on the OFDM DC subcarrier position. The advantage of this is in the previous Section 1.1.1.1 (Figure 1.5), it was said that

Table 1.11 Channel bandwidths.

Channel bandwidth (MHz)	Number of resource blocks
1.4	6
3	15
5	25
10	50
15	75
20	100

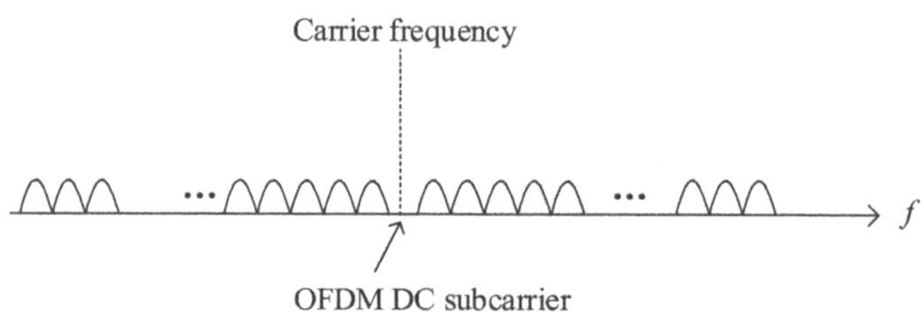

Figure 1.55 The position of the carrier frequency in the frequency domain.

due to self-mixing, the DC (zero frequency) subcarrier will have level drift and cause interference. However, no information is provided about the DC subcarrier in Figure 1.55, so this interference is avoided.

1.10 How to Choose the Modulation Order of MQAM

M is the modulation order of QAM. It can be seen from equation (1.1) that when *M* is larger, the number of information bits n contained in a QAM symbol is also larger. This can increase the data transmission rate of the system. Therefore, increasing the modulation order is one of the effective methods to increase the data transmission rate, that is, increasing the modulation order is one of the methods to enhance the effectiveness of system transmission. It can be seen from Figure 1.9(a) that when the MQAM constellation is rectangular, the position (distance from the origin of the coordinate) of all signal points in the figure in the coordinate is determined by the power of the transmitter. When the power is larger, the

Euclidean distance between adjacent signal points is larger, and vice versa. We know that the number of signal points on the constellation diagram will increase when M increases. When the transmitter power remains unchanged, the Euclidean distance between adjacent signal points in the constellation diagram inevitably decreases due to the above reasons, thereby increasing the bit error rate of the system, which makes the transmission reliability of the system decline. It can be seen that from the choice of modulation order that the effectiveness and reliability of system transmission always constitute a contradiction. That is, the reliability will decrease if the effectiveness increases, and the effectiveness will decrease if the reliability increases. According to the above analysis point of view, we will discuss how to choose the value of M in several wireless communication networks.

(1) *Wireless Local Area Network (WLAN)*: WLAN has the following characteristics:

- The wireless coverage is small. Especially for home Wi-Fi, the distance between the terminal and the router is short. Therefore, the signal received by the terminal and router is strong.
- The terminal moves slowly, so the transmitted data is called the nomadic data. Since the moving speed is nomadic, the transmission characteristics of the wireless channel are better.
- The number of users in the network is small, so the interference between users is small.

The first feature described above makes the received signal strong. The second and third features make the receiver interference small. Therefore, the signal-to-noise ratio at the input of the receiver is high. For this reason, even when the transmitter power does not increase, when there are many signals on the constellation diagram, each signal point on the constellation diagram at the receiver still has a high signal-to-noise ratio, which can ensure that the bit error rate at the receiving end is still relatively low. It can be seen that a larger value of M can be used in wireless local area networks. Known from Chapter 3, the maximum M value is 1024, that is, $n = 10$.

(2) *Cellular mobile communication network (2G–5G)*: In the cellular mobile communication network working in the low- and medium-frequency

bands, because a carrier is allowed to occupy a narrow bandwidth, increasing the modulation order M is one of the effective means to increase the system data transmission rate. Known from Chapter 2, 2G/3G/4G/5G modulation methods are 8PSK/16QAM/64QAM/256QAM, respectively. It can be seen that as the data rate continues to increase, its modulation order is gradually increasing. Some readers may ask, why not increase the value of M to 1024? This is because when the value of M increases, the bit error rate of the system will deteriorate. Compared with the local area network, the cellular mobile network operating in the low- and medium-frequency band has the following characteristics:

(i) The wireless coverage is large, so the signal received by the terminal and the network side is weak.
(ii) The terminal moves fast, so the transmission characteristics of the wireless channel are poor.
(iii) There are many users in the cell, so multiple access interference is high.
(iv) When M increases, if we still want the Euclidean distance between adjacent signal points on the constellation diagram to be large, the only way is to increase the transmitter power. But this solution is not advisable for terminals (especially mobile phones) because after increasing the transmission power, the power consumption and cost of the terminal will increase. For battery-powered terminals, increased power consumption means shorter battery life. Increasing costs will also influence thousands of users, so the plan to increase the transmission power is not feasible.

Due to the above reasons, the maximum M value in cellular mobile networks so far is smaller than that in wireless local area networks.

(3) *5G working in high-frequency band*: Working in a high-frequency band, such as 5G in the millimeter wave frequency band, the bandwidth that a carrier is allowed to occupy is much wider than in the low- and medium-frequency bands. The maximum bandwidth of the medium- and low-frequency bands is 20 MHz, but it is 400 MHz in the high-frequency band. Therefore, in this case, the main way to increase the data transmission rate is to increase the bandwidth instead of increasing the modulation order M. Therefore, the maximum M working in the high-frequency band for 5G is still 256.

(4) *HDTV Network*: At present, the Chinese cable television network transmits 2K TV signals. The standard set-top box and TV are both 2K. The network downlink uses 64QAM modulation. On June 6, 2019, China's Ministry of Industry and Information Technology simultaneously issued 5G commercial licenses to the three major operators and China Radio and Television. China Radio and Television can use the 700 MHz golden frequency band it has mastered to form a high-definition wireless TV network. Due to the advantages of low path loss and strong penetration in the 700 MHz frequency band, the network can transmit 4K or even 8K TV, and its standard configuration is 4K. It can be seen that the downlink data rate of this network (for 2K, 4K and 8K corresponding data rate, see Chapter 5 in Section 5.1.1.2) is definitely higher than the cable TV network. Since 700 MHz is a mid-low frequency band, which allows a narrow bandwidth, increasing the modulation order M is one of the most important means to increase the data rate. It can be predicted that the downlink M value in the future network to be built must be greater than 64 and even 1024. The reasons are as follows:

(i) The 700 MHz channel has good transmission characteristics, so it is called the golden band.
(ii) The TV network is a broadcast network, single-point transmission and multi-point reception. Therefore, it is only necessary to increase the single-point (similar to a base station) transmit power to increase the signal power of the receiving point, thereby ensuring a low bit error rate at the receiving end.

(5) *Narrowband Internet of Things (NB-IoT)*: As the name implies, narrowband means a lower transmission rate. The system does not need to use high-order modulation, so there is no problem of M value selection. In fact, we know from Chapter 4 that all low-speed data are transmitted in NB-IoT. Therefore, it is enough to use low-level modulation, which can achieve the technical requirements of strong coverage in the Internet of Things.

1.11 4G Network Architecture [14]

The network architectures of both 5G and NB-IoT evolved (enhanced) based on the 4G network architecture. From the perspective of network architecture, 4G network architecture is a basic technology of cellular mobile communication networks, so we introduce it in Chapter 1. And for the different application scenarios of 5G and NB-IoT network architectures, we only introduce the differences from 4G.

1.11.1 *Network composition*

The 4G network consists of a core network and an access network. The former is called Evolved Packet Core Network (EPC), and the latter is called LTE Radio Access Network (LTE RAN, hereinafter written as RAN). EPC and LTE RAN together are called Evolved Packet System (EPS).

For the entire 4G network, the functions of the above two networks are different. The access network RAN is responsible for all functions related to radio access in the 4G network, for example, scheduling, radio resource management (referring to the allocation of time-frequency resources in the grid of Figure 1.46 to terminals that have access to the network, and different terminals occupy different resources in the grid), retransmission protocol (HARQ), modulation and coding (MCS) and multiple antenna schemes. It can be seen that they are all functions closely related to wireless communication. Without these functions, the connection between the terminal and the network cannot be realized. The core network EPC is responsible for all functions that are not related to radio access in the 4G network. And these functions are essential for realizing the integrity of the entire 4G network. They include authentication , billing functions and end-to-end connections. In terms of communication technology, these functions really have nothing to do with whether the terminal and the network can be connected. But it is necessary for the entire network, for example, for the following:

Certification: By authenticating users, it can ensure that only legitimate users can access the network, while illegal users are denied access. Legal users refer to users who have registered on the network and do not owe fees.

End-to-end connection: Only with this feature can users stay online.

Billing function: It refers to charging rules and charging users according to the rules.

There are two reasons for separating the entire network function instead of concentrating it in one network:

(1) The separation of functions allows the same core network to connect to different access networks. For example, the 4G core network EPC

can connect to the 4G access network (LTE) and 5G access network (NR) at the same time. This is the 5G non-independent networking solution (see Figure 5.11 in Chapter 5). Of course, the 5G core network 5GCN can also be connected to 4G and 5G access networks at the same time (see Figure 5.12 in Chapter 5).

(2) The core network and the access network can evolve independently. For example, the 5G network has a different application scenario from 4G, which makes the 5G core network and 4G different, but the two access networks are not much different. Therefore, the 4G core network should independently evolve into a 5G core network. There are similar problems for NB-IoT.

The following section first briefly introduces the functions of the core network EPC and then focuses on the protocol stack of the radio access network RAN.

1.11.2 *Core network*

The 4G core network architecture is shown in Figure 1.56. The EPC only supports access to the packet switched (PS) domain. Therefore, it can access the Internet, but not the circuit switched (CS) domain. Its function can be divided into two planes, namely control plane and user plane. The former is used to transmit the signaling between the core network and the terminal. This type of signaling is called non-access stratum (NAS) signaling. The latter is used to transmit business data. Each plane contains several logical nodes, and the core network functions are completed by these nodes. The functions of these nodes are briefly introduced in the following sections.

(1) *MME (Mobility Management Entity)*: It is a node in the control plane and has the following main functions:

- *Bearer connection/release to the terminal*:

After the terminal is authenticated (the NAS signaling is required here) and confirmed as a legitimate user, the terminal should be connected to the network (bearer connection). When the service data transmission is completed, the terminal connection should be released.

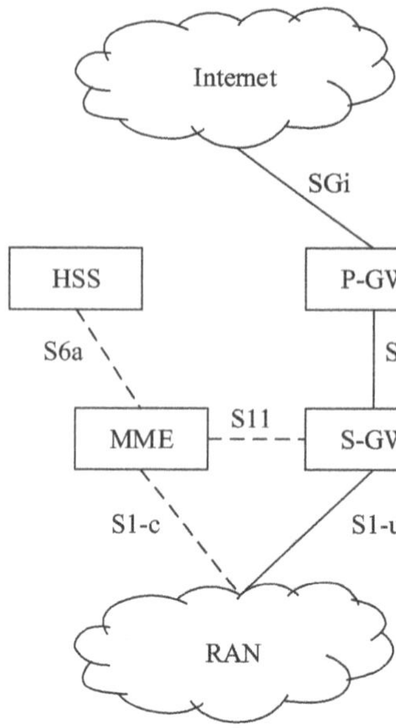

Figure 1.56 Core network architecture.

• *The control terminal switches from the idle state to the active state*:
A terminal in an idle state will wake up periodically to receive paging messages (if any) from the MME. If the terminal is included in the paging message, the terminal will switch the state to the active state through random access (discussed later).

• *Security key management*:
When the terminal transmits data (signaling data and service data) through the network, the data needs to be encrypted (see the PDCP layer in Figure 1.59). The key is managed by the MME. It should also be pointed out here that when the MME authenticates the user, another node HSS in the control plane must be used, called the home subscriber server. In fact, it is a database that stores information about registered (subscribed) users.

(2) *S-GW (Serving Gateway)*:
It is a node in the user plane and has the following main functions:

- It is the mobility anchor point when the terminal moves between eNodeB (abbreviated as eNB). This means that when the terminal moves

within the registration area (discussed later) allocated by the network side, although the eNB has been handed over (that is, the eNB connected to the terminal is different), the terminal can still reside in the same MME/S-GW. This can reduce the number of NAS signaling interactions and the EPC signaling load. If the moving range of the terminal exceeds the registration area allocated by the network side, in this case, the terminal needs to initiate a new registration update process. At this time, it will reside on the new MME/S-GW. It is also the mobility anchor point of the terminal in other 3GPP technologies (GSM/GPRS and HSPA) (see Figure 6.16 in Chapter 6).
- It is responsible for processing information collection and statistics required in billing.

(3) *P-GW (PDN gateway, packet data gateway)*:
It is a node in the user plane and has the following functions:

- Connects the EPC to the Internet.
- Assigns an IP address to the terminal (because it is connected to the Internet, it must be assigned an IP address).
- Quality of Service (QoS) enhancements are carried out according to the policies controlled by Policy and Charging Rules Function (PCRF).
- It is also the mobility anchor point for EPC to connect to non-3GPP radio access technologies (such as CDMA2000).

1.11.3 *Radio access network*

1.11.3.1 *RAN architecture*

The LTE radio access network uses a flat architecture, in which there is only one logical node eNB, as shown in Figure 1.57. The figure is explained as follows:

(1) *eNB has two implementation methods*:
- It is a three-sector base station, which handles the transmission of three cells.
- A baseband processing unit is connected to multiple remote radio units to cover a large number of indoor cells or several cells along highways.

(2) *eNB connects to EPC through S1 interface*: The user plane part of the S1 interface is connected to the S-GW node in the EPC. The S1 interface control plane is partially connected to the MME node in the EPC. It should be noted that one eNB can also be connected to multiple MMEs/S-GWs through the S1 interface. And one MME/S-GW can also be connected to multiple eNBs through the S1 interface. The benefits of this many-to-many connection have already been mentioned, that is, when the active terminal moves between different eNBs, it can reduce the number of NAS signaling interactions and the EPC signaling load.

(3) *eNBs are connected to each other through the X2 interface*: When the terminal is in the active state and moves between different eNBs, in order to realize the handover of different eNBs (that is, the terminal can be connected to different eNBs), different eNBs must be connected through the X2 interface. The X2 interface is also divided into two planes, namely the control plane X2-c and the user plane X2-u.

1.11.3.2 *RAN protocol stack*

The RAN protocol stack is used for the radio interface between the terminal UE and the network. The interface is divided into three layers and two sides. Layer three refers to the physical layer (L1), data link layer (L2) and network layer (L3) used by the protocol stack. The two sides are divided into user plane and control plane by function. The RAN protocol stack is shown in Figure 1.58, which is explained as follows:

- In the above figure,
  PHY is the physical layer.
  The three sub-layers of MAC, RLC and PDCP are the data link layer (L2).
  RRC and NAS are the network layer (L3).
- In the control plane

The non-access stratum NAS is used to transmit signaling between the core network and the terminal. The RRC layer is used to transmit signaling between the access network and the terminal. In order to distinguish it from NAS, this layer is called the access layer (AS).

It can be seen that there are two types of signaling between the terminal and the network, namely NAS signaling and AS signaling. In the process of terminal access to the network, how these two types of signaling are dealt with will have a great impact on network delay and terminal power consumption. As we will see later in the discussion, the processing methods of 5G and NB-IoT are different. In 5G, in order to reduce the transmission delay (a requirement in the uRLLC application scenario), a new RRC inactive state is added. When NB-IoT works in small data packets (the number of bytes contained in the transmission block is small), in order to reduce the power consumption of the terminal, the small data packets will be transmitted together in the NAS signaling. In addition, a new RRC suspend/resume process has been added.

- In the figure, the four protocol entities PDCP, RLC, MAC and PHY are common to the two planes, and they are also the user plane protocol stack.

The functions of the above four protocol entities are shown in Figure 1.59. It is a function of the downlink LTE protocol stack, and the uplink is similar. In order to illustrate the changes of the downstream data flow when passing through the four protocol entities, Figure 1.60 shows an example. There are three IP data packets in the figure, and the protocol stack has two radio bearers. One radio bearer has two data packets, and the other radio bearer has one data packet. It can be seen from Figure 1.60 that the impact of these four protocol entities on the data flow has a common feature, that is, the data flow before the header is called Service Data Unit (SDU), and the data flow after the header is called Protocol Data (PDU). It can be concluded that for any protocol layer (such as RLC sub-layer), the data entity from/to the higher protocol layer (such as PDCP sub-layer) is called the SDU of that layer (such as RLC SDU). The corresponding data entity from/to the lower protocol layer (such as the MAC sub-layer) is called the PDU (such as RLC PDU) of this layer.

The following section discusses the functions of the four general protocol entities, which are actually the functions of the user plane protocol stack in the access network.

1.11.3.3 *User plane protocol stack*

(1) *Packet Data Convergence Protocol (PDCP)*:
It has the following functions:

- Performs IP header compression to reduce the number of bits transmitted on the radio interface.
- Responsible for encryption to prevent eavesdropping.

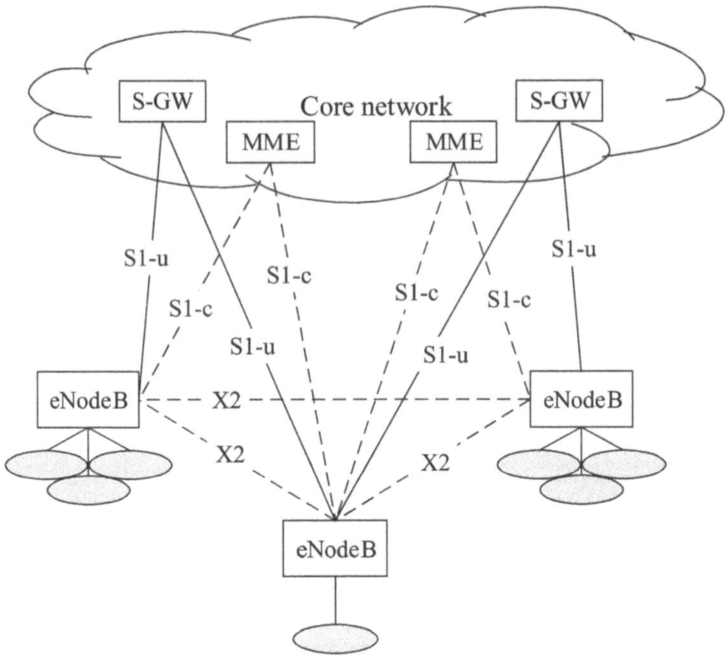

Figure 1.57 LTE radio access network architecture.

Performs corresponding decompression and decryption at the receiving end.

- Deletes the upstream duplicate data packets.

This problem needs to be explained: When a terminal that has been connected to the network moves between different eNBs as in Figure 1.57, if an eNB handover occurs, the terminal may be connected to two adjacent eNBs at the same time. This is also a dual connection. The HARQ buffer is emptied during handover. The PDCP in the terminal will retransmit all uplink data packets that have not yet been delivered to the eNB. Since the terminal is in a dual connection state, some uplink data packets may be received repeatedly. At this time, the network is required to test PDCP to delete duplicate data packets. By the way, during handover, PDCP should forward undelivered downlink data packets from the eNB before the handover to the eNB after the handover.

- PDCP is also responsible for integrity protection for the control plane.

Integrity protection refers to preventing the intrusion of fake base station signals.

(2) *Radio Link Control (RLC)*:
It has the following functions:

- Responsible for the segmentation and concatenation of PDUs (also RLC SDUs) from PDCP to form RLC PDUs of appropriate size. The principle of segmentation and cascade is shown in Figure 1.61.

It should be pointed out that for the uplink, the concatenation of RLC SDUs must be performed after the terminal receives the uplink scheduling authorization. This will increase network latency. Therefore, in 5G, in order to reduce the time delay, the cascading function is removed and only segmentation is performed without cascading because in this way, the RLC PDU can be pre-assembled before the uplink scheduling authorization is received.

- Retransmission ARQ mechanism
  MAC has HARQ, and RLC has ARQ. The reason for this combination has been discussed in Section 1.3.3.4.
- Responsible for reordering to ensure that SDUs are delivered to the higher-level PDCP in order.

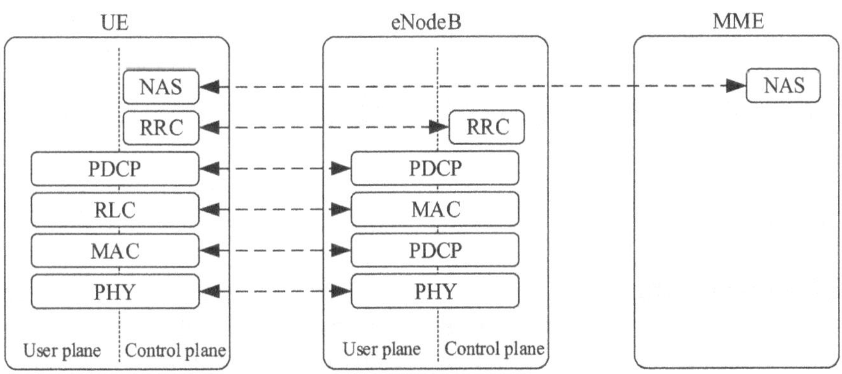

Figure 1.58 RAN protocol stack.

The principle has been discussed in the previous Section 1.3.3.4. It should be pointed out that it takes time to reorder SDUs, which also increases network delay. Therefore, in order to reduce the delay in 5G, the function of sequential delivery is removed.

(3) *Media Access Control (MAC):*
It has the following functions:

- HARQ
- When carrier aggregation is used, it is responsible for the multiplexing (transmitter) and de-multiplexing (receiver) of the transmission data of each component carrier. The transmitting end can be seen in Figure 2.6(a) in Chapter 2.
- Processes the multiplexing of logical channels and maps them to transport channels.
- Responsible for uplink/downlink scheduling.

The latter two functions are more complicated and will be discussed separately in the following:

- Logical channel multiplexing: It can be seen from Figure 1.59 that the MAC layer provides services for the RLC layer through logical channels. The PHY layer provides services for the MAC layer through the transport channel. This means that the MAC layer maps logical channels to transport channels. The method to achieve this mapping is to use the multiplexing of logical channels. Figures 1.62 and 1.63 are the mapping block diagrams of the downlink/uplink, respectively. The two types of channels are different by mapping, because the transport channel is generated by multiplexing the logical channel. The two types of channels are discussed in the following:

(1) *Logical channel*: Logical channels are classified according to the type of information it carries. Since there are control plane and user plane information in the system, it can be divided into two categories. One type is the control channel, which is used to transmit the control information needed during system operation and the parameters that need to be configured after scheduling. The other is the traffic channel, which is used to

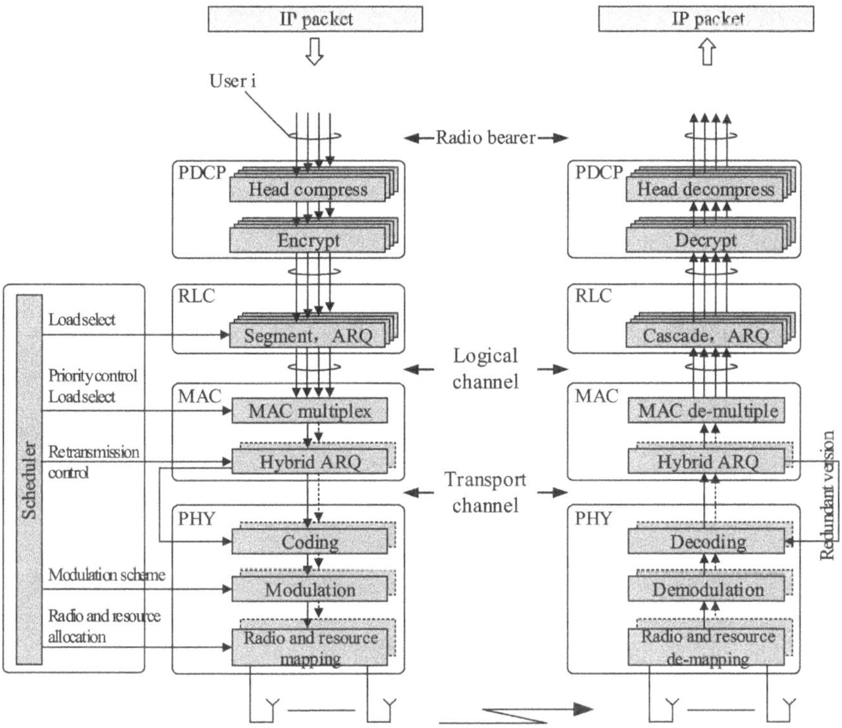

Figure 1.59 LTE Layer 4 protocol function (downlink).

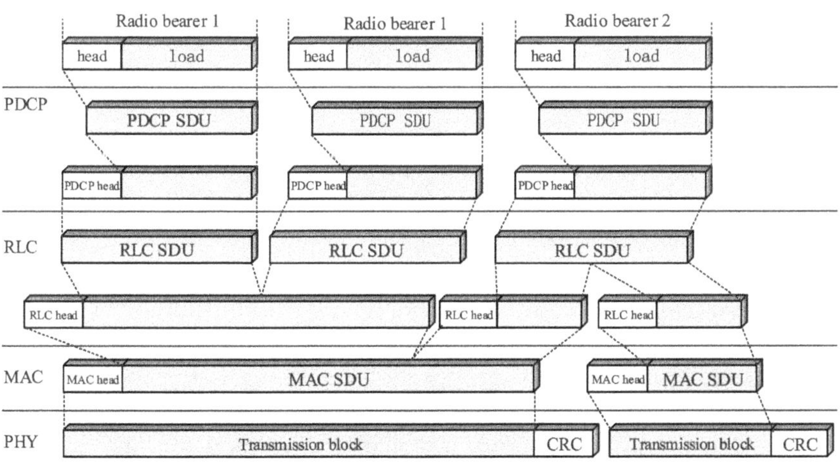

Figure 1.60 Example of LTE data flow.

transmit user data. The types of logical channels in LTE are given as follows:

- *BCCH (Broadcast Control Channel)*:

Used to broadcast system information, so the terminal must obtain this system information to be able to access the network.

- *PCCH (Paging Control Channel)*:

It is used for paging terminals when the network does not know which cell they are located in, so paging messages need to be transmitted in multiple cells.

- *CCCH (Common Control Channel)*:

Used to transmit control information related to random access.

- *DCCH (Dedicated Control Channel)*:

It is used to transmit control information to/from the connected network terminal. The control information configured on the DCCH of different terminals is different. The above-mentioned CCCH is the control information of the transmission terminal during connection establishment and is different from the DCCH.

- *MCCH (Multicast Control Channel)*:

It is used to transmit and receive control information required for MTCH (MBMS traffic channel). There is no MBMS in R13 of 5G and NB.

- *DTCH (Dedicated Traffic Channel)*:

Used to transmit user data from/to the terminal (of course the terminal that has been connected to the network).

- *MTCH (Multicast Traffic Channel)*:

It is used to transmit the downlink data of MBMS service.

Among the above seven types of logical channels, only DTCH and MTCH are traffic channels, and the remaining five are control channels.

(2) *Transport channel*:

Because the transport channel is generated by logical channel multiplexing, it is impossible to classify the transport channel according to the type of information it carries. Instead, it is classified according to the characteristics of information transmitted on the radio interface, and these characteristics refer to:

(a) The data is arranged as transmission blocks. In each transmission time interval (TTI), when space division multiplexing is not used, at most

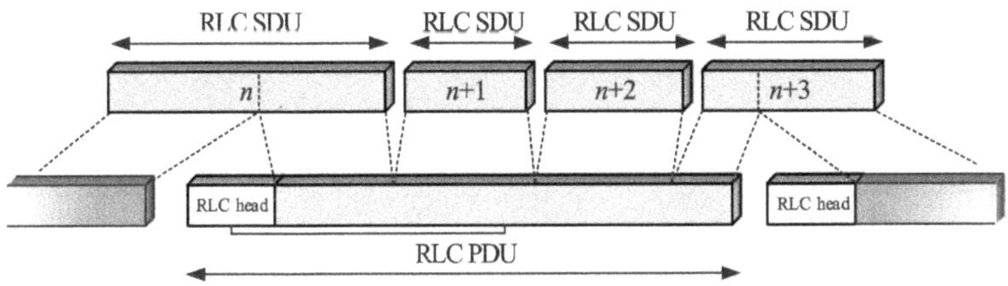

Figure 1.61 RLC segmentation and cascade schematic diagram.

one transmission block of dynamic size is transmitted on the radio interface. When using space division multiplexing, at most two transmission blocks can be transmitted in each TTI.

(b) Each transmission block has a transmission format (TF). TF refers to the size of the transmission block (the number of bytes contained, which is dynamic), the modulation and coding scheme (MCS), and the antenna mapping scheme (related to the multi-antenna scheme).
(c) By changing the transmission format, the MAC layer can change the data transmission rate.

According to the above classification method, LTE defines the following transport channels:

- *BCH (Broadcast Channel)*:

It has a fixed format. It can be seen from Figure 1.62 that it is used to transmit part of the BCCH information.

- *PCH (Paging Channel)*:

It can be seen from Figure 1.62 that it is used to transmit paging information from PCCH. PCH supports DRX (discontinuous reception). This allows the terminal to wake up only at a predetermined time to receive the PCH. Obviously, it saves power than the terminal continuous receiving mode, and the longer the DRX interval, the more power the terminal saves.

- *DL-SCH (Downlink Shared Channel)*:

It is the main transport channel for downlink data transmission and has the following functions:

(a) Supports system data transmission rate adaptation (see Figure 1.59).
(b) Supports time-frequency domain channel-related scheduling (see Figure 1.59).

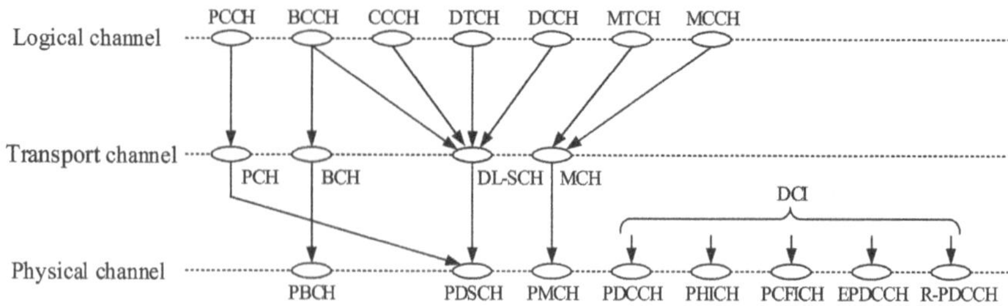

Figure 1.62 Downlink channel mapping.

(c) HARQ with soft merger (see Figure 1.59).
(d) Supports space division multiplexing.
(e) Supports DRX.
(f) The transmittable part is not mapped to BCH and BCCH information (see Figure 1.62).
(g) There are multiple DL-SCHs in a cell, and each scheduled terminal uses one DL-SCH in the TTI. For carrier aggregation, the terminal can receive multiple DL-SDHs, and receive one carrier for each component carrier.
(h) In some sub-frames, a DL-SCH is allowed to carry system information.

• *MCH (Multicast Channel)*:
Used to transmit MBMS information.
• *UL-SCH (Uplink Shared Channel)*:
It is the uplink transport channel corresponding to DL-SCH, which is used to transmit uplink data.
• *RACH (Random Access Channel)*:
It is used to transmit the uplink information required when the terminal accesses the network, but this information does not form the above-mentioned transmission block.

○ Up/down link scheduling: All users in the LTE access network share the time-frequency resources shown in Figure 1.46. The time-frequency resources occupied by each user are uniformly allocated by the scheduler in the MAC (see Figure 1.59, it seems that the scheduler is a separate entity, in fact it is a part of the MAC layer). Since the wireless channel is a multipath channel, when the terminal is in different locations in the cell, its time domain and frequency domain characteristics are different. Therefore, the scheduler must allocate time-frequency resources and configure parameters required in wireless communication according to the current characteristics of the wireless channel.

The uplink/downlink scheduler is located in the MAC layer of the eNB. But its scheduling decision (determined by the characteristics of the uplink/downlink wireless channel) can be independent of each other. The scheduler allocates resources in units of resource block pairs. The resource block pair means 1 ms in the time domain (two LTE time slots, 14 OFDM symbols in total) and 180 KHz in the frequency domain (12 subcarrier spacing, 12×15 KHz = 180 KHz). This definition is exactly the same as the pair of resource blocks given in Figure 1.47. This means that the eNB makes a scheduling decision every 1 ms interval and transmits the scheduling information to the selected terminal. The uplink/downlink scheduler is shown in Figure 1.64. They have the following functions:

(1) *Downlink scheduler*:

- Responsible for (dynamically) controlling which terminals are transmitted (of course, selecting terminals with good transmission characteristics).
- Controls which resource blocks in the grid DL-SCH of each terminal is transmitted on.
- Selects the downlink transmission format (TF refers to the transmission block size, modulation and coding scheme MCS and antenna mapping).
- The multiplexing of logical channels is controlled by the eNB scheduler.
- Since the scheduler can control the data transmission rate, RLC segmentation and MAC multiplexing will also be affected by scheduling decisions, as shown in Figure 1.59.

The downlink wireless channel state information is reported by the terminal to the eNB.

(2) *Uplink scheduler*:

- Controls which terminals can transmit on their UL-SCH (of course, select terminals with good transmission characteristics).

- Controls which resource blocks the UL-SCH of each terminal is transmitted on in the grid (including component carriers during carrier aggregation).
- Selects the transmission format of the uplink.
- The multiplexing of logical channels is not controlled by the eNB scheduler, but by the terminal itself. The parameters can be configured by the eNB scheduler. This is different from the downside.

The uplink wireless channel state information is obtained from the sounding reference signal (SRS) transmitted by each terminal to the eNB.

As we will see later in the discussion, in order to reduce the delay, 5G adopts the following two measures in the scheduling:

- The time slot is shortened to reduce scheduling time. The time slot in LTE is 0.5 ms, while in 5G it is only 0.125 ms.
- The scheduling time in LTE must be at the time slot boundary (that is, the start or end time). However, the scheduling time in 5G can be at a certain time in the time slot. In order to distinguish it from LTE, this time slot is called mini-slot. The mini-slot method can reduce the scheduling time (see Figure 5.33 in Chapter 5).

(iv) *Physical layer (PHY)*:
The physical layer has the following functions:

- Coding.
- HARQ processing in the physical layer.
- Modulation.
- Multi-antenna processing.
- Map signals to appropriate physical time-frequency resources.
- Mapping transport channels to physical channels.

It can be seen from Figures 1.62 and 1.63 that physical channels can be divided into two categories. One type is mapped from the transport channel, that is, the physical channel corresponding to the transport channel. The other type is physical channels that do not correspond to transport channels, and they are called L1/L2 control channels. Among them, the downlink control information (DC1) is the information necessary for the terminal to correctly receive and decode the downlink data. The uplink

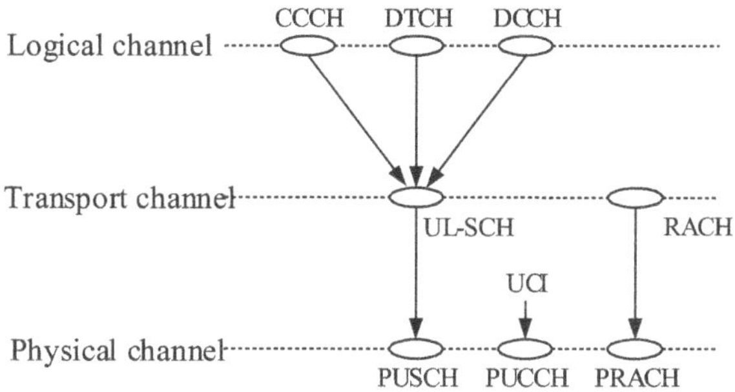

Figure 1.63 Uplink channel mapping.

control information (UCI) is used to provide information about the state of the terminal for the scheduler and the HARQ protocol.

Eleven types of physical channels are defined in LTE, of which five are mapped from transport channels, and the other six are L1/L2 control channels.

- *Mapped from the transport channel*:

  - PDSCH (Physical Downlink Shared Channel): It is the main physical channel used for the transmission of unicast data and also for the transmission of paging information.
  - PBCH (Physical Broadcast Channel): Part of the system information required by the bearer terminal to access the network (the other part of the system information is transmitted in the PDSCH).
  - PMCH (Physical Multicast Channel).
  - PUSCH (Physical Uplink Shared Channel): It is the uplink channel corresponding to the PDSCH, and each terminal has at most one PUSCH on each uplink component carrier.
  - PRACH (Physical Random-Access Channel): It is used for terminal random access (which will be discussed later).

- *L1/L2 control channel*:

  - PDCCH (Physical Downlink Control Channel): For the transmission of downlink control information, it mainly includes the scheduling decision needed to receive the PDSCH and the scheduling grant that triggers the PUSCH transmission.

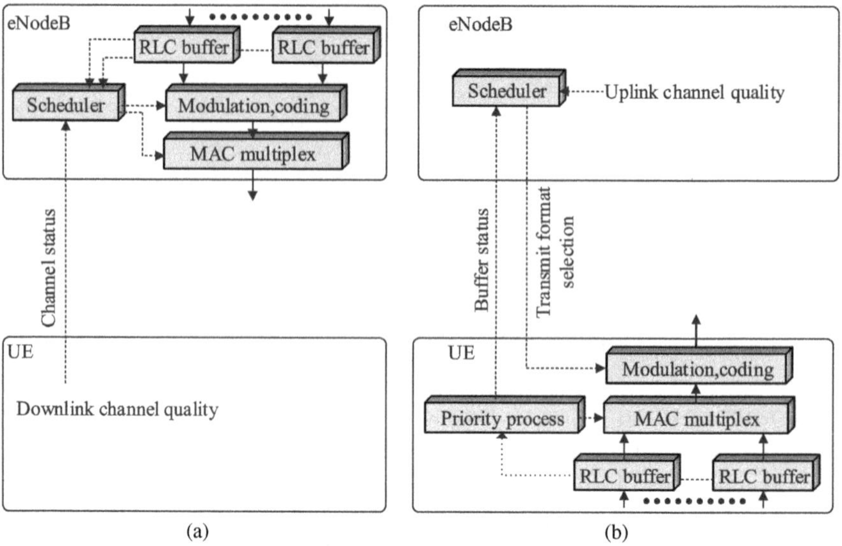

Figure 1.64 Scheduler block diagram. (a) downlink and (b) uplink.

- ePDCCH (enhanced Physical Downlink Control Channel, enhanced PDCCH): Its function is the same as PDCCH, but the transmission mode is more flexible (enhanced).
- R-PDCCH (Relay Physical Downlink Control Channel, Relay PDCCH): It is mainly used to carry L1/L2 control signaling on the link from the master eNB to the relay.
- PHICH (Physical HARQ Indicator Channel): It is used to carry HARQ confirmation to indicate whether a certain transmission block of the terminal needs to be retransmitted.
- PCFICH (Physical Control Format Indicator Channel): It is a channel that provides the information needed to decode the PDCCH for the terminal, and each component carrier has only one PCFICH.
- PUCCH (Physical Uplink Control Channel): It is used for the terminal to transmit HARQ confirmation, to transmit channel state report, and to request time-frequency resources required for uplink transmission. Each terminal has at most one PUCCH.

For the physical channel mapped from the transport channel, the following two points need to be pointed out:

(1) Except for PBCH and PRCH, the formats of the other three physical channels all change dynamically with scheduling. Among them, PCH (multiplexed on PDSCH) cannot use HARQ (downlink). This is because the uplink has not been established when the terminal is paged.
(2) PBCH must be a fixed format. This is because the terminal must obtain the system information from the broadcast channel before it is possible to access the network, that is, the terminal obtaining the system information is only the first step for the terminal to access the network. Therefore, the format of PBCH must be fixed. In this way, the terminal can know this format in advance and use the format to demodulate the broadcast system information.

1.11.3.4 *Control plane protocol stack*

Figure 1.58 shows that there are two control planes in the 4G network. The NAS control plane is located between the MME and the terminal in the core network. The main functions include authentication and security management. Through paging, the terminal can be converted from an idle state to an active state, and it is also responsible for assigning IP addresses to the terminal. These issues have been discussed in Section 1.11.2. The other is the AS control plane, which is located between the RRC (radio resource control layer) and the terminal in the eNB. Its protocol stack is composed of the RRC protocol entity and the four general protocol entities mentioned above. The four common protocols have been discussed, so only the RRC protocol needs to be discussed here.

RRC is responsible for handling RAN-related control procedures and has the following functions:

- *System information broadcast*: *The terminal must know the system information to access the network*: Send a paging message from the MME (located in the core network) to notify the terminal of the connection request. It can be seen that when the terminal has not accessed a certain cell (the terminal is in an idle state), the paging message can be used.
- *Connection management, including connection bearer and mobility*: The following concepts are involved in the connection management process. After the terminal has accessed a certain cell, the terminal and the network can communicate at this time. In order to ensure the

communication quality, the scheduler should make scheduling decisions based on the wireless channel state and configure all the parameters required for the communication between the terminal and the radio access network through related signaling. This process is called "establishing the RRC context" between the terminal and the network. The concept of saving RRC context is used in the following discussion. Obviously, it means that both the terminal side and the network side save all the parameters needed for communication. When the terminal moves, the RRC context changes. Therefore, the saved RRC context also needs to be continuously updated.
- *Mobility management, such as cell reselection*: This means that when the terminal has not yet connected to the network (the terminal is in a spatial state), which cell the terminal accesses is determined by the terminal itself. This situation is called terminal-controlled mobility. After the terminal has access to the network, cell switching may occur when the terminal moves between eNBs. However, whether the terminal switches to a new cell is determined by the network. This situation is called network-controlled mobility. It is different from terminal-controlled mobility.
- *Measurement configuration and report*: It mainly measures and reports the status of uplink/downlink wireless channels.
- *Processing of terminal capabilities*: When a connection is established, the terminal will inform the network what capabilities it has because not all terminals can support all the functions described in the specification.

The above functions of RRC are all realized through the transmission of RRC messages (AS signaling). The transmission process is similar to the user plane in Figure 1.59. The difference is that the radio bearer (actually it is a data radio bearer DRB) is changed to a signaling radio bearer (SRB). The RRC message accesses PDCP through SRB and use the same four-layer (PDCP, RLC, MAC and PHY) protocol as the user plane. As mentioned earlier, since logical channels are classified according to the type of information transmitted, various messages in RRC should be mapped to the following logical channels in Figures 1.62 and 1.63:

- Broadcast system information is mapped to BCCH.
- Paging information is mapped to PCCH.

- Other types of information arc mapped to CCCH during terminal establishment and will be mapped to DCCH once the connection is established.

Control plane and user plane data can be multiplexed in the MAC layer (see Figures 1.62 and 1.63) and transmitted within the same TTI.

The following discuss four issues related to the control plane:

(i) *4G RRC state machine*: As mentioned earlier, there are two control planes in the 4G network. The NAS control plane is responsible for handling the signaling between the core network and the terminal. The AS control plane is responsible for handling the signaling between the access network and the terminal. Obviously, the working state of the terminal in the network is determined by the above two types of signaling processing conditions. According to this, the terminal can be in two working states, as shown in Figure 1.65.

RRC-IDLE (idle state): This state has the following characteristics:

- There is no RRC context in the radio access network (that is, "no RRC connection" in the figure), and the terminal does not belong to any cell at this time.

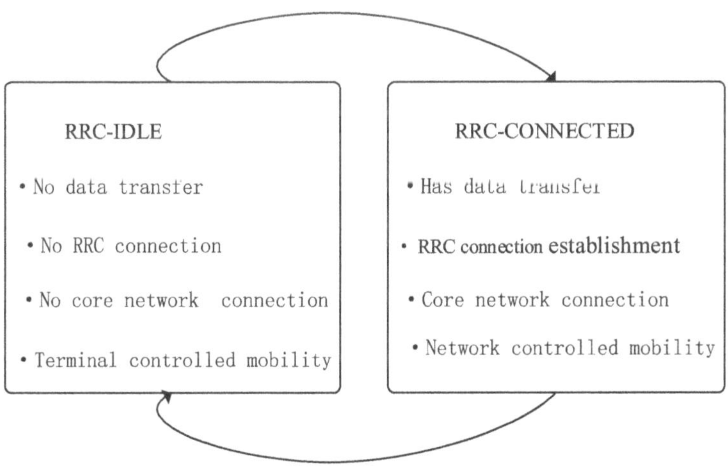

Figure 1.65 4G RRC status.

- From the perspective of the core network, the terminal is now in CN-IDLE (that is, “no core network connection” in the figure).
- Terminal-controlled mobility.
- Cannot transfer data.
- In order to save power, the terminal stays dormant most of the time, but periodically wakes up to receive paging messages (if any) from the network.
- The terminal can be converted to the connected state by random access, and converting to the connected state using the paging mode is called the called mode.

RRC-CONNECTED (connection status): This state has the following characteristics:

- The RRC context of the access network and the terminal has been established (i.e., “RRC connection establishment” in the figure). At this time, the cell to which the terminal belongs is known, and the terminal is configured with a C-RNTI (Cell-Radio Network Temporary Identifier).
- From the perspective of the core network, the terminal is already in CN-CONNECTED (i.e., “core network connection establishment” in the figure).
- Network-controlled mobility.
- There is data transmission.
- In order to reduce terminal power consumption, discontinuous reception of DRX can be configured. Since both the eNB and the terminal have established the RRC context in the connected state, it is relatively quick to leave the DRX and start receiving or transmitting data because there is no need to establish a connection through related signaling at this time.

The above two states can be converted to each other, as shown in Figure 1.65.

(ii) *5G RRC state machine*: In order to reduce network latency, 5G terminals have added a new state, RRC-INACTIVE (inactive state), which has the following characteristics:

- No data transfer.
- The RRC connection is established.

- The core network connection is established.
- Terminal-controlled mobility.

This shows that from the perspective of the core network, the terminal is still CN-CONNECTED. In this way, the speed of data transmission from the inactive state to the connected state is very fast. As the core network signaling is not required to participate, it is sufficient to directly use the RRC context for transmission.

(iii) *NB-IoT RRC state machine*: In order to reduce terminal power consumption, NB-IoT terminals have added a new state and a new process.

(1) Power Save Mode (PSM) state.
(2) RRC Suspend/Resume process.

The above two issues will be discussed in Chapter 4.

(iv) *Random access*: Random access means that the terminal initiates a request to establish a connection (this is the primary call of the terminal, and paging is when the terminal is called). In LTE, random access is used in the following situations (Figure 1.66):

- Transition from RRC-IDLE state to RRC-CONNECTED state.
- When the terminal needs to maintain uplink synchronization with the cell, for example, when a cell is switched, the terminal needs

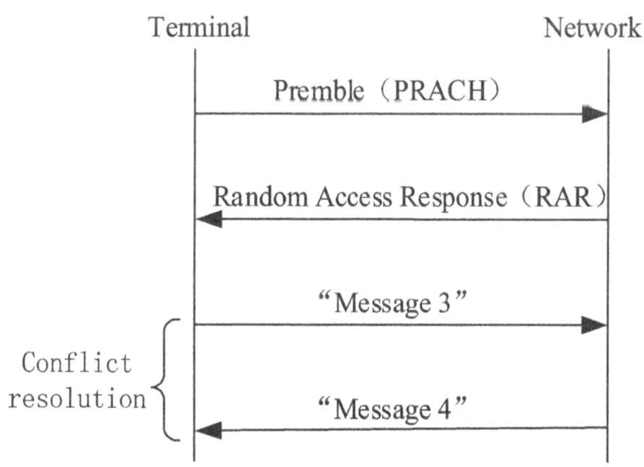

Figure 1.66 Random access process.

to establish uplink synchronization with the new cell after the handover.

The random access process is divided into four steps, as shown in Figure 1.66:

Step 1: The terminal transmits a preamble, also known as the physical random access channel (PRACH) (see Figure 1.63).
Step 2: The network transmits a random access response (RAP), indicating that the preamble has been received. And according to the delay of the received preamble, a time advance command is transmitted to adjust the transmitting time of the terminal so as to achieve uplink synchronization (see Figure 1.53).
Steps 3 and 4 (Messages 3 and 4): These are the terminal and the network exchange messages. The purpose is to resolve conflicts caused by multiple terminals in a cell that may simultaneously transmit the same preamble. If successful, message 4 will also move the terminal to the connected state.

The aforementioned random access procedure is also applicable to 5G.

Chapter 2

Evolution from 1G to 4G

2.1 Introduction

In order to improve people's material and cultural living standards and tackle the explosive growth of the number of Internet users, the application scenarios and technical performance of cellular mobile communication networks have continuously increased and improved. We have experienced four generations of development in the past 40 years. At present, the fifth generation (5G) is being put into use, and the sixth generation is under development in many countries. The evolution process is briefly described in Figure 2.1.

At the end of 1978, Bell Labs in the United States created the world's first mobile cellular phone system, which is called the Advanced Mobile Phone System (AMPS). In the 1980s, the United States established a cellular mobile phone network; after this, many countries established similar networks, which are called the first-generation (1G) mobile communication networks. The 1G mobile communication system uses analog communication and frequency division multiple access (FDMA) technology, which is only a voice cellular phone system.

However, it should be pointed out that due to the use of analog communication and FDMA technology in 1G, there were inherent drawbacks in performance, including small network capacity, poor security and poor anti-interference ability. In addition, there was no ITU to formulate international standards like in the present, so different countries could only formulate national standards according to their own national conditions, which made international roaming a very serious problem.

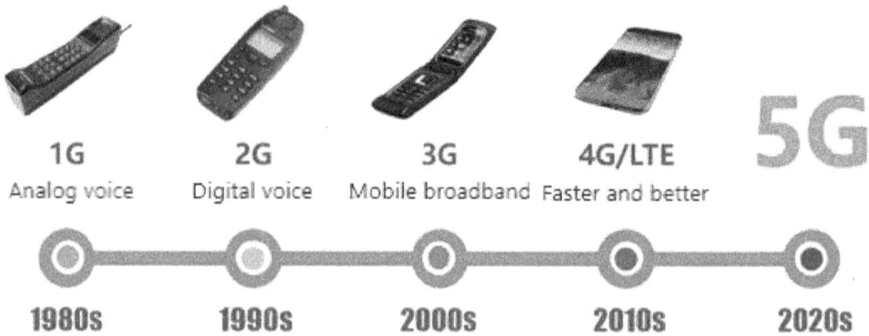

Figure 2.1 Evolution of mobile communications.

In order to solve the problems in the 1G system, the ITU began to use the second-generation (2G) technology and formulate the international standards. This system utilized the digital communication technology and was put into use in the 1990s.

From 1G analog communication to 2G digital communication, the performance of the network has improved qualitatively. The fundamental reason for this is the waveform characteristics of the digital signal itself; that is, its values are discrete in time and amplitude. It is precisely because the digital signal is transmitted in the network that the network performance will be improved in the following three aspects.

First of all, because the digital signal is easy to encrypt and the receiving end can adopt the detection method of threshold judgment for the signal, the security and anti-interference ability of the network are greatly enhanced.

Second, due to the characteristics of digital signals, the number of multiple access methods available in the network has increased. In addition to FDMA, time division multiple access (TDMA), code division multiple access (CDMA) and orthogonal frequency division multiple access (OFDMA) can also be used. All these are orthogonal multiple access methods. In the case of limited spectrum, FDMA restricts the number of users that can access the network. Therefore, for TDMA, CDMA and OFDMA, multiple access methods are used in 2G, 3G and 4G, respectively.

Finally, since it is a digital signal that is transmitted in the network, it lays the foundation for the development of multimedia transmission services in the network.

It should be pointed out that voice signals are mainly transmitted in 1G and 2G networks (but there are also low traffic and low-rate GPRS data in 2G). We call this network service the single-media (i.e., voice) service.

With the improvement of people's material and cultural living standards, in addition to providing the network to conduct voice calls, many other services have been realized such as the transmission of video (movies, TV, pictures), text and other information, and services such as online games. We call this kind of network service the multimedia business.

In these media, except for the voice, which is a digital signal, the rest are data signals. In the binary domain, both the digital signal and the data signal are composed of "1" and "0" codes. They are the same in form, but are completely different in essence, mainly in the following three points.

First, the data rate of multimedia signals is much higher than that of voice signals. Therefore, multimedia signals are usually called broadband signals.

Second, the data signal requires that the bit error rate (P_b) after demodulation at the receiving end is much lower than that of the digital signal (data $P_b < 10^{-5}$, digital $P_b < 10^{-3}$).

Third, the data signal interaction method is packet switching (PS), while the digital signal interaction method is circuit switching (CS).

In order to realize the transmission of multimedia services, the third-generation (3G) and fourth-generation (4G) networks that were put into use at the end of the last century and the beginning of this century all adopted data communication technology.

From a data communication point of view, 4G networks are only a transition from 3G networks. Because the 4G network only transmits a data rate and spectrum efficiency higher than 3G, there is no essential difference between the two. Therefore, the multimedia signals transmitted by 4G networks are usually called large-flow broadband signals.

For this reason, we will directly discuss 4G network instead of 3G.

The transition from 2G digital communication through 3G to 4G data communication makes a qualitative leap in network design thinking; that is, the entire network is designed according to the data communication mode without considering voice transmission. The 4G network designed under such guidance allows users to smoothly enjoy high-definition (HD) videos and online games, and the most important thing is that the 4G network can be successfully connected to the Internet through a gateway to form the mobile Internet. This solves the problem of mobile broadband terminals such as smartphones surfing the Internet in cellular networks. The service transmitted in the mobile Internet is called OTT (Over the Top) service. From Chapter 1, we know that the user's smartphone is called an Internet phone.

It is necessary to briefly describe the problem of smartphones. Two chips are used in smartphones to transmit data (instead of digital), namely, baseband chip (BB) and radio frequency chip (RF). These two chips can be collectively referred to as a data card. Among them, BB uses digital signal processing (DSP) methods to complete the calculation of various data signals in the 4G protocol (air interface LTE and LTE-A) network interface layer (including the physical layer and data link layer) and various data signal operations in the network layer. And, RF uses analog circuits to complete the calculation of various analog signals in the physical layer. BB controls the working status and parameters of RF through a three-wire interface. Since BB works in the digital domain and RF works in the analog domain, the two must be connected through D/A and A/D converters. Among them, D/A is relatively simple; it is just a low-pass filter. But, A/D is very complicated. It cannot use the A/D converter in voice coding, but must use a new A/D converter with a quantized signal-to-noise ratio of more than 100 dB, such as modern Σ-Δ modulated A/D converters composed of over-sampling, noise shaping and digital extraction techniques.

The chips of Chinese mobile phone manufacturers are mainly imported from abroad, and the most complex baseband chip comes from Qualcomm. China's Huawei announced on July 28, 2018 that its independently developed 4G baseband chip KIRIN 980 has succeeded after years of research and development, and its performance can reach LTE cat 21. The full name of LTE cat x is LTE UE-category x. LTE refers to 4G networks, UE refers to user equipment, such as mobile phones, category means the level and x is the number of levels, which refers to the uplink and downlink peak (maximum) data rate level of the 4G LTE network of the mobile phone after using a certain baseband chip. Huawei has achieved a level of 21, and its downlink peak rate is 1.4 Gbps = 1.4/8G Bps $\approx$ 0.175 GBps. Finally, the baseband chip for the base station developed by Huawei is named "Tiangang".

As mentioned above, the 4G network is designed according to the data communication mode and does not specifically consider the issue of voice transmission. Thus, there are two solutions for users in the 4G network to conduct voice calls.

First, every time a user makes a voice call, the 4G network automatically falls back to the 2G network, and initiates or receives a voice call in the circuit-switched manner. The communication protocol LTE requires that the system must be backward compatible, so that the 4G has the functions of a 2G. But, at this time, the 4G data card is unused. It is impossible to make calls while surfing the Internet, and this is a defect of the solution.

Second, the voice signal is regarded as a data signal, so it can still be transmitted by a data card. The advantage of this method is that we can make calls while surfing the Internet. In order to reduce voice delay and ensure voice quality, a fixed channel is set up for both parties in the network. Therefore, this method is essentially a circuit switching method. This kind of call is called VoLTE (Voice over LTE). Currently, the three major telecom operators in China have adopted the VoLTE solution, and it will be discussed in Section 2.5.12.

In summary, it can be seen that the cellular mobile communication network from 1G to 4G has been developed with the continuous increase of transmission services. It also requires the continuous improvement of network speed and spectrum efficiency, and the continuous reduction of network delay. Although these generations of network technologies are different in performance, their main purpose is to solve the problem of information interaction between people. Borrowing the name of Internet of Things, we can call these networks Internet of People. From the perspective of network applications, 4G networks can be classified as mobile broadband (MBB) application scenarios.

The 5G mobile communication network has been put into commercial use around 2020. It is not difficult to imagine that 5G networks will support more application scenarios such as the Internet of Things in addition to the Internet of People. The 5G network and the Internet of Things will be discussed in Chapters 4 and 5.

Although radio network has a history of more than forty years, it is still evolving and keeping pace with the times. From the dial-up era to the adoption of broadband Internet, to smart phones and to the early days of the Internet of Things, radio networks have been constantly developing new technologies that can adapt to markets. If history can give us some enlightenment, then wireless communication networks will certainly continue to provide us with more convenience in the coming years.

The Internet of Things is an interconnection network of everything, which solves the problem of information interaction between things and

Table 2.1 Comparison of mobile communication systems.

Generation	System name	Multiple access	Data service type	Modulation	Maximum channel bandwidth	Peak data rate up/down (bps)	Maximum spectral efficiency up/down (bps/Hz)	One-way wireless delay (ms)
2G	GSM	FDMA + TDMA	SMS	GMSK	200 KHz	9.6K/14.4K	0.05/0.07	600
	GPRS(2.5G)			GMSK		9.6K/115K	0.05/0.58	
	EDGE(2.75G)			8PSK		384K/384K	1.92/1.92	
3G	WCDMA	FDMA + CDMA	SMS Internet access	QPSK, 16QAM	5 MHz	64K/7.2M	0.01/1.44	75
	HSPA(3.5G)					384K/14.4M	0.08/2.88	
	HSPA + (3.75G)					5.76M/21.6M	1.15/4.32	50
4G	TD-LTE FDD-LTE	上行 SC-FDMA/ 下行OFDMA	SMS Internet access video	64QAM (MIMO + OFDM)	20 MHz	50M/100M	2.5/5.0	10
	LTE-A			64QAM (CA + MIMO + OFDM)	100 MHz	500M/1G	5.0/10	
5G	IMT-2020	See Chapter 5 5G						1

Notes:
(1) 3G WCDMA HSPA, only SISO is used and the downlink peak rate is 7.2 Mbps.
(2) 3G WCDMA HSPA+, supports 2 × 2 MIMO and the downlink peak rate is 42 Mbps. The 21.6 Mbps in the table refers to SISO.
(3) 4G 3GPP LTE, supports SISO, 2 × 2 MIMO, 4 × 4 MIMO, when 4 × 4 MIMO is supported and the downlink peak rate is 100 Mbps.
(4) 4G 3GPP LTE-A, supports up to 8 × 8 MIMO and the downlink peak rate is 1 Gbps.

people. Therefore, we can say that 4G changes lives and 5G changes society.

The main performance of different generations of mobile communication systems can be summarized in Table 2.1. We can see that with the evolution of each generation, the average network delay is required to be smaller and smaller, and 5G requires a smaller delay due to changes in the application scenarios.

It can be seen from Table 2.1 that with the increase in the types of data transmission services and the rapid increase in the number of mobile users, the data transmission rate and spectrum efficiency (i.e., channel frequency band utilization) of the mobile communication system must also be greatly improved. These two performance indicators are determined by the size of the channel capacity.

2.2 Channel Capacity — Shannon Formula

In order to clarify the theoretical basis of radio network evolution, it is necessary to review the concept of channel capacity first.

In 1948, C.E. Shannon gave the channel capacity of disturbed continuous channels as

$$C = W \log_2\left(1+\frac{S}{N}\right)(bps) \tag{2.1}$$

Thus, the channel spectrum efficiency can be obtained as

$$C/W = \log_2\left(1+\frac{S}{N}\right)(bps/Hz) \tag{2.2}$$

where W is the channel bandwidth and S/N is the power signal-to-noise ratio in (2.1).

2.2.1 *Application of Shannon's formula in mobile communication*

In Figure 2.2, both the transmitter and receiver are multi-antenna systems (2T2R in the figure). When multiple antennas on both sides are located geographically, the signals between the antennas are uncorrelated (because only uncorrelated signals can be separated at the receiving end). We use such systems to discuss the application of Shannon's formula.

There are four paths (channels) in the figure. When they are all Additive White Gaussian Noise (AWGN) channels, the channel capacity of any path can be calculated using (2.1). In (2.1), the signal power S should be the power on the input of the transmitting antenna on the path. N is the total equivalent white noise power on the output of the receiving antenna on the path. And, $N = n_0 W$, where n_0 is the white noise unilateral power spectral density (Watts/Hz), and W(Hz) is the bandwidth occupied by the path. We should point out the following:

(1) *The meaning of channel capacity* C: It is the maximum data rate that can be transmitted during point (transmitting point) to point (receiving point) communication; that is, C is an upper bound value. The physical meaning is that as long as the actual data rate R transmitted on the path is lower than C, there must be a channel coding scheme that can make the received bit error rate P_b arbitrarily small.
(2) *The actual path in mobile communication is not an additive Gaussian noise channel, but a multipath fading channel.* Therefore, the channel capacity between any two points in the system is not equal to (2.1). However, (2.1) can still be used as an estimate of channel capacity when qualitatively analyzing the system.

2.2.2 *Strategies to improve system data transmission rate*

The following two strategies are in practice to improve the system data transmission rate:

(1) Under the condition that the actual transmission data rate of the system is lower than the channel capacity, every means should be tried to find various technical methods to make the actual transmission data rate as close as possible to the channel capacity. For example, high-order modulation is a very effective technical method.
(2) Improve the channel capacity of the system, so that the system has the potential to transmit high-rate data.

The method to increase the system channel capacity will be discussed here. From equation (2.1) and Figure 2.2, the following methods can be used:

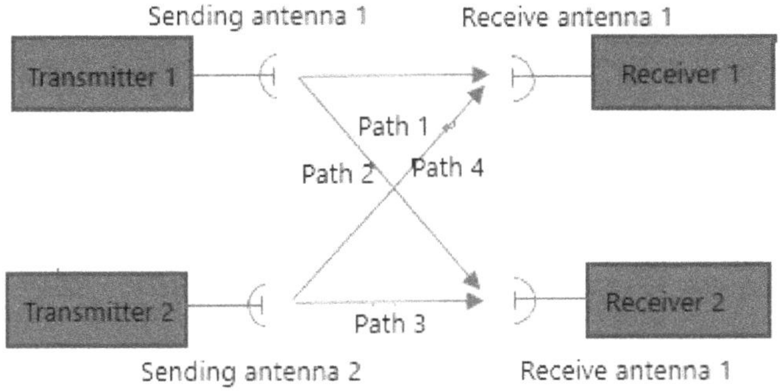

Figure 2.2 Multi-antenna system (2T2R).

(1) **Adopt multi-antenna technology (improve spectrum efficiency):** It can be seen from Figure 2.2 that when the number of antennas increases, the number of data transmission paths also increases. The total number of paths is equal to the product of the number of transmitting antennas and receiving antennas. The number of data streams that the transmitter can simultaneously transmit is equal to the number of transmit antennas. The total data rate transmitted by the transmitter is equal to the sum of the rates of each data stream. Therefore, the system data rate will increase with the increase of the number of transmitting antennas, which is called spatial multiplexing. Of course, the data rate transmitted on each path still cannot exceed the channel capacity for point-to-point communication. The spatial diversity multiple of each data stream received by the receiver is equal to the number of receiving antennas. Therefore, the spatial diversity gain of the receiver increases as the number of receiving antennas increases. Under the condition of ensuring the same bit error rate at the receiver, the increased spatial diversity gain at the receiver can be used to increase the rate of the data stream, thereby increasing the system data rate. The MIMO technology used in the mobile communication system is proposed based on this principle. It is a multi-antenna system, in which any pair of transmitting and receiving antennas will occupy different transmission spaces. Therefore, MIMO is a technology that makes full use of space resources. The detailed analysis of multi-antenna technology is carried out in Section 5.6.11.

(2) **Increase the channel bandwidth W:** It can be seen from equation (2.1) that increasing W can increase the channel capacity. However, it cannot increase the channel capacity indefinitely. This is because when it is

the additive white Gaussian noise, the noise power $N = n_0 W$ increases as W increases. In extreme cases,

$$\lim_{W\to\infty} C = \lim_{W\to\infty} W \log_2\left(1+\frac{S}{n_0 \mathrm{W}}\right)$$
$$= \frac{S}{n_0}\lim_{W\to\infty}\frac{n_0 W}{S}\log_2\left(1+\frac{S}{n_0 \mathrm{W}}\right) = \frac{S}{n_0}\log_2 \mathrm{e} = 1.44\frac{S}{n_0} \quad (2.3)$$

It can be seen that even if the channel bandwidth increases infinitely, the channel capacity is still limited. Nevertheless, it is still a very practical method to increase the data rate of system transmission by increasing the channel bandwidth. The Carrier Aggregation technology used in the mobile communication system and the development of new frequency bands can both be used to increase the channel bandwidth.

2.3 Technical Measures Adopted in the Evolution

From the viewpoint of improving the data transmission rate and spectrum efficiency, we discuss the technical measures adopted in various generations of mobile communication systems in the evolution process. Through the analysis of Table 2.1, we can get the following results.

2.3.1 *2G technical measures*

2G mobile communications were put into use in the 1990s, among which GSM and NCDMA (Narrow CDMA) are two representative systems. The replacement of 1G by the 2G system completed the leap from analog communication to digital communication. But, the main service of 2G is still voice call. Therefore, it only provides digital voice services and low-rate data services. Among them, GPRS (General Packet Radio Service) is used to provide low-rate data services. EDGE (Enhanced Data rate for GSM Evolution) is an upgraded version of GPRS, which is a transition technology from GSM to 3G. Compared with GSM, it uses high-order modulation 8PSK, which improves the data transmission rate. It has been pointed out earlier that only when the actual data rate of the system is lower than the channel capacity can the use of high-order modulation to increase the system data transmission rate be an effective method.

Because the bandwidth of the 2G digital mobile communication system is relatively narrow (only 200 KHz), which limits the application of data services, it is impossible to realize high-rate data services and mobile multimedia services. It is necessary to briefly explain the concept of media here. Media refers to the materialization of information and expressing it in a form that is easily recognized by humans and machines. Therefore, media is a method of propagating and expressing information. Multimedia includes three elements, that is, video, audio and text. Hypermedia is composed of five elements, that is, video, audio, text, image and graphics. ITU's definition of multimedia services is that multimedia is a service with integration, interactivity and synchronization. Integration refers to the ability to store, transmit, process, and display multimedia and hypermedia information. Interactivity refers to the mutual control ability between people and business systems. Synchronization means that the images, sounds and texts displayed on the multimedia service terminal work in a synchronized manner. For example, a user wants to retrieve a segment of a certain historical event. The moving (or fixed) image of the event is stored in an image database, and the text narrative and voice description are stored in other databases. The multimedia service terminal extracts the required information from different databases through different transport channels, and synchronizes the sound, image and text to form a whole and transmits it to the user.

2.3.2 *3G technical measures*

3G was put into use at the end of the last century. It evolved from 2G, so it must be compatible with 2G functions; that is, it can transmit voice and SMS. This compatibility is called backward compatibility. At the same time, 3G is mainly based on data transmission services and needs to access the Internet. Therefore, 3G can transmit mobile multimedia services with rich content such as voice, data and video. It can be seen that 2G has completed the leap from analog communication to digital communication, and 3G has completed the leap from digital communication to data communication. Since the beginning of 3G (including 3G), all generations have been communication systems based on data transmission services.

The mainstream international standards of 3G include WCDMA, CDMA2000 and TD-SCDMA. Among them, the former two use frequency division duplex (FDD), and the latter uses time division duplex (TDD). HSPA (High Speed Packet Access) and its enhanced version HSPA+ are high-speed downlink/uplink packet access technologies adopted by WCDMA/TD-SCDMA. They are an upgraded version of the transition to 4G, and they greatly enhance the Internet access ability of 3G.

In order to increase the data transmission rate, 3G has adopted two measures:

(1) The channel bandwidth has been increased from 0.2 MHz in 2G to 5 MHz in 3G.
(2) High-order modulation QPSK, 8PSK (TD-SCDMA) and 16QAM are used.

2.3.3 *4G technical measures*

At the beginning of this century, 4G was put into use. In order to adapt to the rapid growth of mobile users and the development needs of multimedia services, mobile communications are required to provide higher data transmission rates. The 3G system could no longer meet the above requirements. Therefore, 3G evolved to 4G, and 4G is backward compatible with 3G. Currently, 4G has two international standards, namely, LTE (3.9G) and LTE-A (LTE-Advanced, 4.0G). LTE is divided into time division duplex (TD-LTE) and frequency division duplex (FDD-LTE). LTE-A is an upgraded version of LTE.

In order to increase the data transmission rate, LTE has adopted three measures:

(1) The channel bandwidth has been increased from 5 MHz for 3G to 20 MHz for 4G.
(2) High-order modulation 64QAM is used.
(3) MIMO technology is used.

The LTE base station is a multi-antenna system called an evolved Node B. It was obtained after enhancement and evolution on the basis of 3G base station Node B. Compared with the existing Node B of 3G, it integrates part of the RNC (Radio Network Controller) function, reducing the level of communication protocol.

LTE-A is an upgraded version of LTE, so LTE-A must be backward compatible with LTE. For this reason, the three technical measures it adopts are the same as LTE. However, in order to further increase the data transmission rate, the channel bandwidth is increased from LTE's 20 MHz to LTE-A's 100 MHz. However, it should be emphasized that the method of increasing the channel bandwidth in the previous generations of mobile communications that has been discussed is very simple. It is directly approved by the National Radio Regulatory Agency to widen the bandwidth occupied by the system, which is to allocate frequency resources to users through administrative means. But, the method of increasing the channel bandwidth in LTE-A is not like this. It is obtained through a technical means called carrier aggregation (CA), which will be discussed specifically in the next section.

Finally, the following two points should be pointed out about 4G:

(1) It can be seen from Table 2.1 that the data transmission rate of the 4G system is relatively high. At this time, the bit width is relatively narrow, so the inter-symbol interference becomes a very serious problem. In order to resist inter-symbol interference, OFDM technology is used in 4G, so that the multiple access method is different from 3G. In 4G, orthogonal frequency division multiple access (OFDMA) is used in the downlink, and single-carrier frequency division multiple access (SC-FDMA) is used in the uplink, which has a peak-to-average power ratio (PAR) smaller than OFDMA (see Section 1.7.1.2).

(2) Globally, most countries and regions use FDD-LTE. The TD-LTE, which is mainly promoted by China, is only used in China and some small telecom operators in India and Japan. South Korea completed the world's first LTE-A network in June 2013. So far, many LTE-A networks have been put into commercial use worldwide, which are distributed in most countries and regions.

2.4 Working Principle of Carrier Aggregation (CA) [1]–[3]

2.4.1 *The background of proposing CA*

From the above discussion, we know that in order to increase the data transmission rate, three measures are adopted in LTE. However, if we want to further increase the data transmission rate based on them, these measures seem to have the following technical bottlenecks:

(1) It is difficult to further increase the modulation order. Under certain transmitter power, increasing the modulation order will shorten the Euclidean distance between the signal points on the constellation diagram, thereby deteriorating the bit error rate performance. It can be seen that only 64QAM can still be used in LTE-A.
(2) It is difficult to increase the number of antennas in MIMO systems. In the LTE standard, a base station can be configured with multiple antennas, while user equipment (UE, such as mobile phone) can only be configured with one antenna currently. Since the 2–3 GHz radio frequency band is being used, the wavelength is 10–15 cm and the size of UE is small, so it is impossible to install multiple antennas. Thus, it is impossible to increase the data transmission rate by increasing the number of antennas in the currently adopted frequency band.
(3) It is also impossible to apply a wider bandwidth by administrative means. This is because the current spectrum resources below 4 GHz are very crowded. In this case, increasing the bandwidth is almost impossible. Thus, only the channel bandwidth allocated in LTE can be used in LTE-A.

Based on the above reasons, in order to increase the system bandwidth, two international standardization organizations, 3GPP (Third Generation Partnership Project) and IEEE 802.16 m WG (Work Group), have proposed CA. The basic principle of CA is to aggregate different carriers of multiple existing LTEs to transmit data to one or several users in LTE-A. This is equivalent to increasing the system bandwidth and increasing the data rate.

2.4.2 *Carrier aggregation scheme*

There are two carrier aggregation schemes, as shown in Figure 2.3. Scheme (a) is called continuous CA, which means that multiple available component carriers are adjacent to each other and in the same frequency band. Therefore, it is also called Intra-band continuous CA. Scheme (b) is called non-continuous CA, which means that multiple available component carriers are separated from each other on the frequency axis. When the component carriers are in the same frequency band, it is called

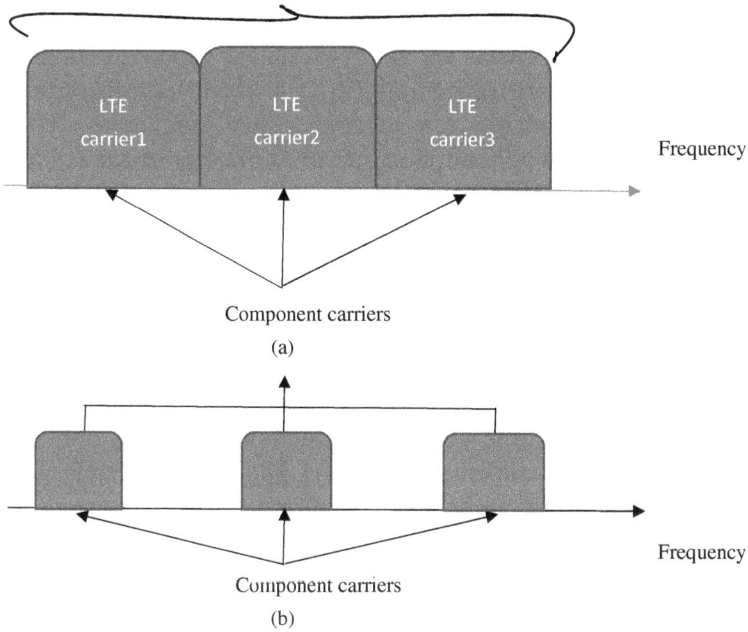

Figure 2.3 Carrier aggregation scheme.

Intra-band discontinuous CA. When component carriers are in different frequency bands, it is called Inter-band discontinuous CA. It should be noted that the channels in the same frequency band are flat fading channels, but the fading characteristics of different frequency bands are different, which one must pay attention to in CA design. The original LTE carriers participating in CA are called component carriers. Regardless of the scheme, all component carriers must be aggregated to serve one or several users in LTE-A. The two schemes are compared as follows.

(1) Continuous CA has the following characteristics:

(i) The UE in the original LTE system can become the UE of LTE-A without much change. This is because the frequency range spanned by continuous CA is relatively narrow (up to 100 MHz). Thus, it is possible to use an RF module and an FFT module to realize the LTE-A function.
(ii) The management and configuration of radio resources in the system are similar to the original LTE, so the algorithm is relatively simple.

(2) Discontinuous CA has the following characteristics: Since the spectrum resources below 4 GHz are very crowded, a variety of communication systems will be arranged in this frequency band, so it is difficult for mobile communication network to configure a continuous 100 MHz bandwidth. In this case, discontinuous CA is a more practical and feasible solution. It can make use of all the spectrums in the existing mobile communication network, such as the spectrums occupied by 2G, 3G and LTE, and even use some unused scattered frequency bands. Since the spectrum of discontinuous CA is pieced together using the above method, it has the following characteristics:

(i) The LTE-A UE of discontinuous CA should have multiple RF receiving units and baseband units (while in continuous CA, only one RF unit and one baseband unit are needed). The reason is that the component carriers of discontinuous CA will span a wide frequency range, and the channel transmission characteristics in different frequency bands are different. Figure 2.4 [1] shows the relationship between

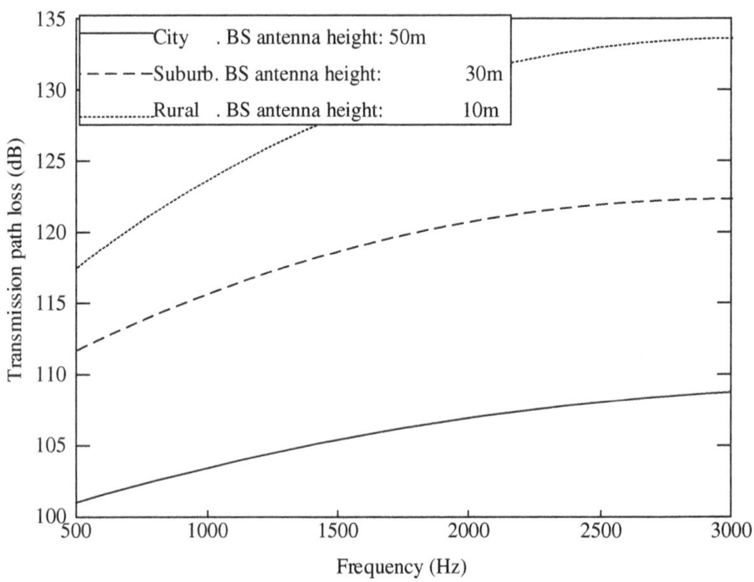

Figure 2.4 The relationship between propagation path loss and different frequencies and communication environments. The wireless channel model is the Okumura Hata model. The communication distance is 1 km. The UE antenna height is 1.5 m.

propagation path loss and frequency. It can be seen from the figure that the propagation path losses are different in different frequency bands. For this reason, when different antennas in the LTE-A eNB transmit the same power, in order to better cover the cell, the modulation and coding schemes (MCS) in different component carrier systems are different. Generally, high-order modulation should be used in low-frequency component carrier systems (because the low-frequency road loss is small). Therefore, there must be multiple RF and baseband units in the UE.

(ii) The UE in LTE-A with discontinuous CA should have multiple FFT modules (while in continuous CA, there is only one FFT module). The reason is that the component carriers of discontinuous CA will span a wide frequency range, and the Doppler shifts in different frequency bands are different, as shown in Figure 2.5 [1]. It can be seen from the figure that the Doppler shifts of different component carriers are different, and the corresponding channel coherence time is also different, resulting in different OFDM design parameters. Therefore, there must be multiple FFT modules in the UE.

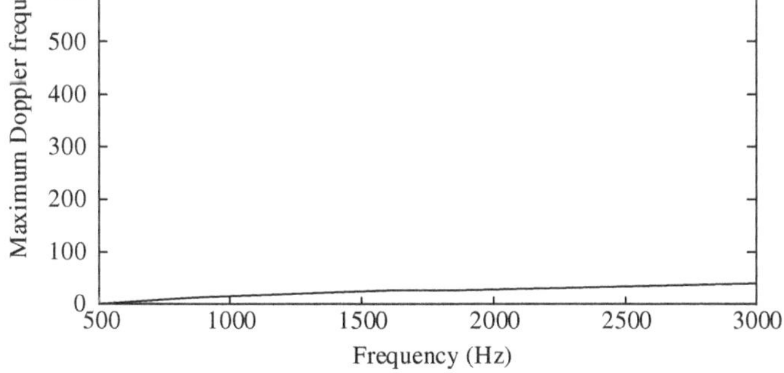

Figure 2.5 The relationship between maximum Doppler shift and different frequencies. The moving speed is 2 km/h

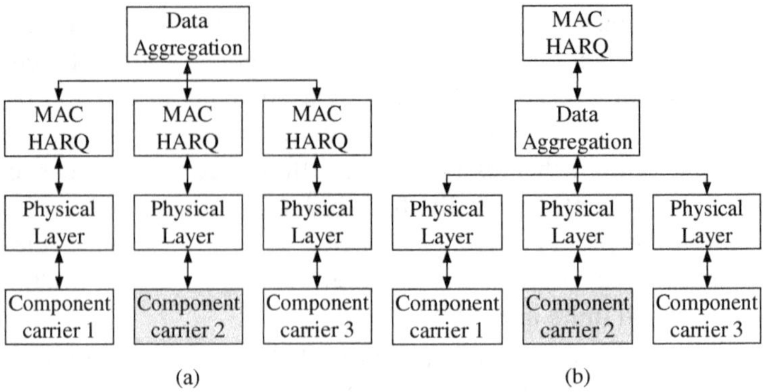

Figure 2.6 Data aggregation scheme.

2.4.3 *Data aggregation scheme*

It is known from the working principle of carrier aggregation that when the original data are at different levels of the communication protocol, different LTE-A systems will be obtained, as shown in the two schemes in Figure 2.6.

The data in Figure 2.6(a) are aggregated in the MAC layer, and the data need to be split into branches in the MAC layer. The number of branches is equal to the number of component carriers used by CA. Therefore, this scheme has the following characteristics:

(1) There are multiple independent hybrid automatic repeat requests (HARQ), and the number of HARQ is equal to the number of component carriers.
(2) The parameters (such as transmit power and MCS) configured for each component carrier in the physical layer are independent, and must be configured for different carriers.

It can be seen that this solution is equivalent to several LTE systems working in parallel at the same time. Therefore, the parameters configured in the physical layer and MAC layer of each LTE system can be used in LTE-A. Of course, LTE-A must be backward compatible with LTE.

In Figure 2.6(b), data are aggregated in the physical layer, and data need to be split into branches in the physical layer. The number of branches is equal to the number of component carriers used in CA, so this scheme has the following characteristics:

(1) Only one HARQ is needed at the MAC layer.
(2) Since there is only one HARQ, the parameters must be configured for the entire aggregated bandwidth.

It can be seen that the design of this scheme is not as flexible as the scheme in Figure 2.6(a). It cannot be that the design of each branch in the scheme of Figure 2.6(a) can be carried out independent of each other. In addition, its backward compatibility with LTE is more complicated than the scheme in Figure 2.6(a). Therefore, the plan (a) is preferred in practice.

The following points need to be pointed out about CA:

(1) The base stations of the above two schemes must have a central baseband processing unit, which is used to coordinate all antennas participating in carrier aggregation to transmit data at the same time (called time synchronization).
(2) MIMO and CA are both multi-antenna systems, but there is a difference between the two. In MIMO, each LTE antenna uses the same carrier frequency and bandwidth to transmit data, realizing the function of data multiplexing. But, each LTE antenna in CA uses different carrier frequencies and bandwidths (some bandwidths may also be the same) to transmit data. In addition to realizing the data multiplexing function, it can also increase the bandwidth of the entire system. Thus, different antennas in MIMO are located in the same cell, while different antennas in CA are located in different cells.

2.4.4 *Bit error rate performance of CA*

Simulation shows the following:

(1) In a broadband and high-speed mobile environment, as long as each component carrier in CA is synchronized in time, the impact of the Doppler frequency shift on the data bit error rate is limited, as shown in Figure 2.7 [1], which shows the comparison between each component carrier with and without a protection band.
(2) FA-Frequency Aliasing between adjacent component carriers will greatly deteriorate the bit error rate performance of the downlink data, as shown in Figure 2.8 [1]. In order to avoid frequency aliasing,

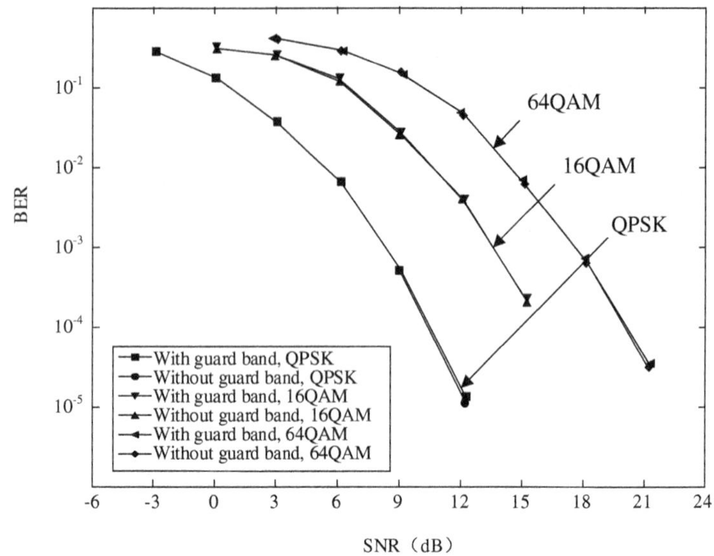

Figure 2.7 The influence of Doppler frequency shift on bit error rate with or without protection band.

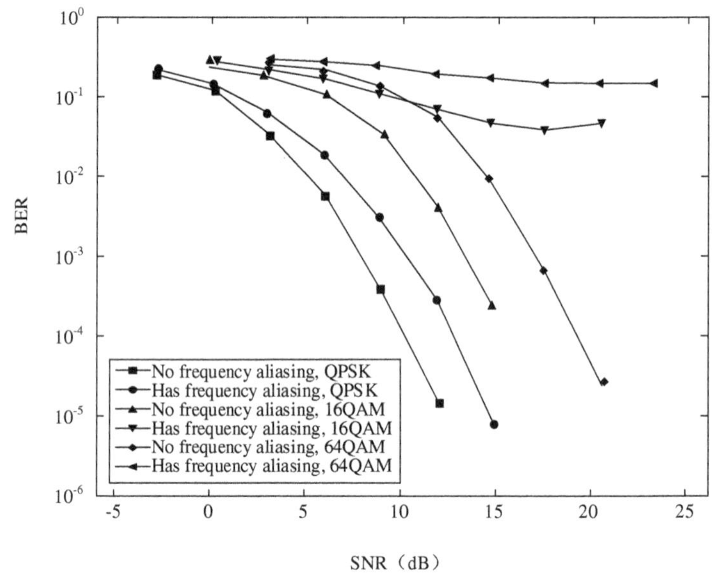

Figure 2.8 The influence of frequency aliasing on bit error rate performance.

whether it is a continuous CA scheme or a discontinuous CA scheme, a certain frequency protection band needs to be set between the frequency spectrums of adjacent carriers.

The following section will discuss several useful concepts in networking. The author originally planned to write a separate chapter. But due to space limitation, the author has to put it in Chapter 2. However, the content of this following section is not closely related to the previous content. Please understand this arrangement of the author.

2.5 Several Basic Concepts

The following sections provide brief introductions to several basic concepts commonly used in the 4G mobile communication network and WLAN (Wi-Fi).

2.5.1 *Diversity and multiplexing*

The MIMO system has two methods which are spatial diversity and spatial multiplexing.

(1) **Space diversity:** It refers to the use of multiple transmitting antennas to transmit the same information signal through different spatial paths. The receiver can receive multiple independent fading signals of the same signal with one antenna (the number of received signals is the same as the number of transmitting antennas). Diversity gain can be obtained after signal processing, thereby improving the reliability of reception. Currently, a commonly used technology is Space-Time Block Coding (STBC), which is an important form of coding based on transmit space diversity. The most basic of such forms is the Alamouti scheme designed for two transmit antennas (see Figure 1.12(c)). In addition, there is also a beam forming (BF) technology that can achieve the abovementioned space diversity of transmit antenna. It uses multiple antennas in the base station to form an antenna array, and uses digital signal processing technology to shape the antenna beam (equivalent to the pattern of a directional antenna) and scan it. In this way, directional transmission and reception can be realized, thereby enhancing the signal-to-noise ratio and improving the reliability of the receiver.

(2) **Spatial reuse:** It is a technology that divides the transmitted data into several parallel data streams (also called several hierarchical data streams), and then transmits them on different antennas through mapping, thereby increasing the data transmission rate. The current commonly used spatial multiplexing technology is the Vertical Layered Space-Time (VLST) scheme (see Figure 1.13).

2.5.2 *Transmitting data rate*

(1) The amount of data transmitted by the physical layer of the transmitter per unit time is called the transmitter data rate.
(2) Both mobile communication networks and wireless local area networks are rate adaptive systems. The data rate transmitted is high when the channel characteristics are good, and vice versa. The largest data rate is called the peak data rate. The uplink and downlink data rates usually refer to the peak data rate.
(3) Data transmission is based on bits. The data rate unit is b/s or bps (bit per second). Larger units include 1 Kb = 1000 b, 1 Mb = 1000 Kb, 1 Gb = 1000 Mb, 1 Tb = 1000 Gb and so on.
(4) K always means 1000.

The standard of International Units clearly formulates that the amount of 1KB is 1000B, where K represents thousands in decimal, and K is Kilo.

But, for a binary number, 1KiB = 1024B, where Ki represents thousands in binary (Kilo binary). It can be seen that 1KB = 1024B is an incorrect use of the International Units standard prefix. But, since it is now widely being used, it is difficult to correct.

According to the International Electrotechnical Commission (IEC), the International Units standard prefix has only a decimal meaning, and no longer uses the binary meaning. If so, there will be no confusion between binary Ki = 1024 and decimal K = 1000, which is very clear.

In reality, we can only judge what K represents based on the context.

2.5.3 *Throughput of receiving end data*

The amount of data (bps) that the receiving physical layer can correctly receive in a unit of time is called the receiving data throughput. It can be seen that the size of the throughput is not only related to the transmitting data rate and system bit error rate but also related to the network speed of the transmitted data.

2.5.4 *Streaming media*

Various audio and video files transmitted on the network can be called multimedia (see the definition of multimedia in Section 2.3.1). There are two methods for multimedia transmission on the network. The first is the traditional download method, which is suitable for downloading by offline users. The second is the streaming method, which is characterized as streaming and broadcasting. It is suitable for online users (such as mobile Internet access). In order to emphasize the difference from the first method, people call the multimedia transmitted by the second method streaming media. It can be seen that streaming media is still a kind of multimedia, and only the transmission method is different compared with general multimedia. These two methods are discussed separately as follows:

(1) Traditional download method: When the user selects the file (multimedia) to be downloaded, the network downloads the entire content of this file to the terminal and stores it. After that, users can open it for reading or playing. Since multimedia files are usually relatively large, this method has the following two disadvantages:

The user terminal must prepare a large storage space.

Due to the limitation of network bandwidth, the network speed is relatively slow, so it takes a long time to download all the file, which makes the waiting time too long for users.

(2) Streaming download method: In order to overcome the disadvantages of the traditional methods, people proposed the streaming method in which multimedia files are transmitted in a way that they can be viewed while downloading. This method is the so-called streaming transmission. The following briefly describes the working process of this method.

The streaming media provider compresses and packs the data of the continuous file according to the streaming protocol, and stores it in the server. When the server transmits the compressed package to the user terminal, the user does not have to wait until the entire file is downloaded. Instead, when a part of the compressed package is downloaded, it starts to decompress and play. When user terminal plays the received content, it is simultaneously receiving subsequent data packets in the background. Therefore, the user terminal can play the selected content continuously after a short wait at the beginning (pre-store a piece of data as a buffer before playing, so a short delay will occur at the beginning). Obviously, the realization of the streaming method depends on whether the network has a download speed which is sufficient to support the amount of data required for smooth playback, and the user terminal has a multitasking operating system. This transmission method can also be called cloud download.

In order to achieve streaming, various related organizations and enterprises have formulated a series of protocols. At the application layer, the RTSP (Real Time Streaming Protocol) is currently the most used. The biggest feature of this protocol is that it can interact between the terminal and the server during streaming; that is, the user terminal can replay the content that has just been watched (listened) or skip the current content to the new content later in the play process. It should be noted that after streaming a media file, the user terminal does not store the data of the file. If we want to watch the same file again, we must stream it again.

It can be seen from the above that compared with traditional methods, streaming has the following advantages:

The memory capacity required by the user terminal can be greatly reduced.

The time that users wait is greatly reduced.

It has improved network online interaction capabilities.

2.5.5 *Data traffic*

It should be emphasized that the data flow mentioned here is only defined from the following narrow perspective. That is, when users use the network to transmit data (note that it is not a phone call), what standard should the network operator apply to charge for? The following is a brief process description of downloading an A/V file from the Internet. First, the user sends an application to the network, requesting what kind of file to download from the network, such as a movie, a song or a certain webpage. This request needs to be represented by several data bytes, but the operator does not charge for the uploaded request. When the network receives the upload request, it finds the required A/V file on the corresponding server, and transmits all the data bytes contained in the file to the user one by one through the network. The operators charge for the transmission of the downlink data. The amount of charge should be determined by the size of the file, that is, by the number of bytes contained in the file. Therefore, from a charging perspective, the so-called data traffic of a file refers to the number of bytes occupied by the file stored in the server. In this way, when users download files of different sizes, the fees they need to pay are also different. It can be seen that the purpose of defining data flow is to find a basis for operators to charge users, and of course this basis is reasonable. Similarly, when the user terminal uploads data files (such as videos, emails and files) through the network operator, in principle, it is also required to pay according to the number of bytes transmitted. According to this narrow definition, the concept of data traffic is also applicable to the upload process.

Finally, it is worth mentioning that data traffic, also known as network traffic, was originally proposed as a measure of network performance. It refers to the total amount of data being transmitted on the network at a given point in time. Usually, network data appear in a packaged manner. They are the so-called load on the network. Network traffic is used to characterize the important content of a network transmission and flow control capability. Reasonable network traffic organization is an important factor to ensure the quality of a network service.

Data storage is based on byte (B), such as 1B = 8b, 1KB = 1000B, 1MB = 1000KB, 1GB = 1000MB and 1TB = 1000GB.

2.5.6 *Network speed*

When a user connects to the Internet through a 4G mobile communication network or a wireless local area network, it is called surfing the Internet. At this time, data exchange is necessary between the user and the Internet. The data transmitted by users to the Internet are called upload data, and the speed of such transmissions is called upload speed. The data transmitted to the user by the Internet are called download data, and the speed of such transmissions is called download speed. Since 4G mobile networks, wireless local area networks and the Internet all use packet switching (PS) to transmit data, the data transmission speed of the network (referred to as network speed) should refer to the number of data bytes transmitted by the network in a unit of time. That is, the number of data bytes received by the receiver in a unit time, and its unit, is Bps (Byte per second). The actual internet speed is related to the following factors:

(1) Via 4G mobile network: The 4G mobile network consists of a core network and a radio access network. The core network is connected to the Internet through a gateway, and the user terminal is connected to the core network through a radio access network. It can be seen that the network speed at this time is related to the following factors:

a. The working conditions of the Internet.
b. The working status of the core network.
c. The channel transmission characteristics of the radio access network (the radio access network is a rate adaptive system).
d. Number of users in the sector covered by the radio access network.

There are two ways to transmit data in the communication network, which are circuit switching (CS) and packet switching (PS). Circuit switching refers to the need for the network to establish a fixed link for both communicating parties to ensure the data flow between the two parts. At this time, all the communication resources that make up the fixed link will be exclusively shared by the parties. Even during the idle period when there is no data transmission between the two parts, these communication resources cannot be used by other users. When packet switching is used, the network does not establish a fixed communication link for both communicating parties. Instead, all user data are packaged in the format required by packet switching. Then, these data packets are stored and forwarded through different nodes in the network until they reach the destination. It can be seen that at this time, all communication resources in the network are not exclusive to any user, but shared by all users. The number of users in the network is constantly changing, and the number of data packets transmitted in the network is also a variable. Therefore, the data transmitted by each user, and even the route (path) taken by each data packet, may be different. Obviously, their delays are also different. The more users there are, the longer the network transmission delay will be.

(2) Via wireless LAN: The network structure of surfing the Internet via Wi-Fi is shown in Figure 2.9. In view of the fact that the Chinese wired broadband network has entered various users in the community, the specific configuration in Figure 2.9 is as follows.

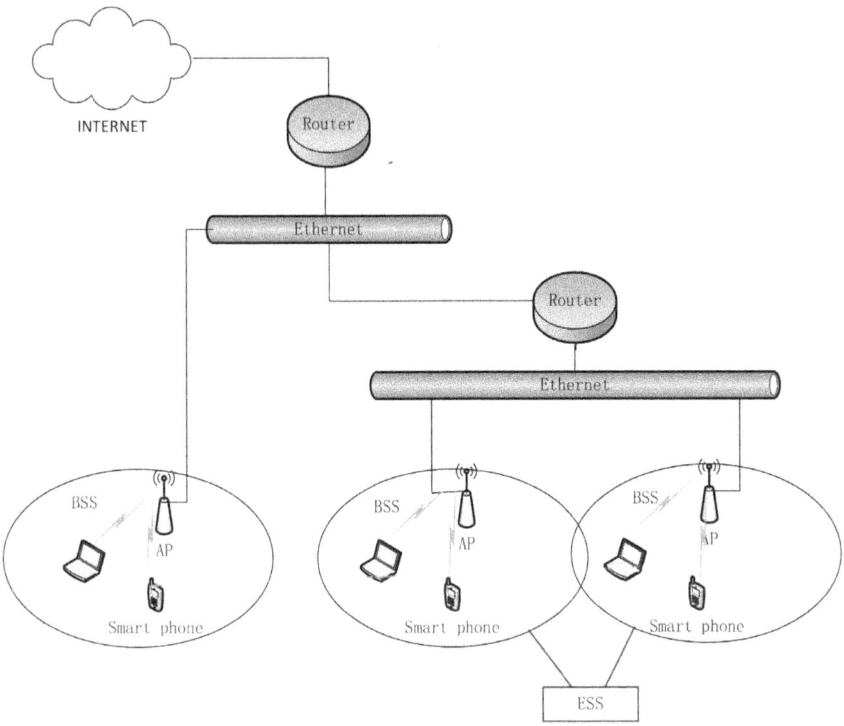

Figure 2.9 WLAN network architecture.

The access point (AP) is a Wi-Fi router. It forms a radio access network with terminals equipped with Wi-Fi functions, such as mobile phones, tablets and laptops. These devices are collectively referred to as STA (Station). They are connected to the AP with an optical modem with a network cable to realize photoelectric conversion, modulation and demodulation. Then, the optical signal is transmitted on the broadband optical fiber, and connected to the distribution system (DS) through the optical modem. The most commonly used DS is the Ethernet. The Ethernet is connected to the Internet through a gate bridge (or portal). In the above configuration, except that the router is purchased by the user, the rest are provided by the network operator. The following briefly introduces the role of the main components in the network:

- **Station STA:** STA is a user terminal device of the Wi-Fi radio access network. In the past, user terminals needed to be connected to a wired broadband network through a network cable to access the Internet, which limited user mobility. But, after using the Wi-Fi radio access network, users can realize data transmission under slow moving conditions. This kind of slow movement is also called nomadic movement, and the transmitted data are called nomad data.

- **Access point (AP):** It is equivalent to the base station of BSS. The network can realize the addressing of the station through AP. When the network in Figure 2.9 is connected to other networks, AP is also called a hotspot. These other networks include the Internet or non-802.11 series 802.X wireless LANs, such as 802.15 (Bluetooth) and 802.16 (Broadband radio access WiMAX).
- **Basic service set (BSS):** There are two types of BSSs, which are called infrastructure BSS and independent BSS. The latter is also called Ad Hoc Network. The former consists of an access point and several stations. In this structure, any site within the BSS range can access the Internet through AP, and can communicate with each other through AP. The BSS in Figure 2.9 is an infrastructure-based basic service set. The latter has no access point and consists of several sites (at least two sites). For example, Bluetooth is such a structure.
- **Service Set Identifier (SSID):** SSID is the identifier (account) of the BSS. It is stored in the AP and broadcasted by the AP in the BSS range. Users can determine which SSID AP to connect to by setting different SSIDs by themselves.
- **Distribution System (DS):** The role of DS is to address different APs, and then APs can address different stations. DS can connect multiple BSSs to form ESS through AP. This allows roaming over a longer distance.
- **Extended Service Set (ESS):** Because ESS has roaming characteristics, it can be applied in the following situations. For example, some large enterprises install a BSS on each floor. When they have roaming functions, they can ensure a seamless connection when users move between different floors. Another example is this ESS-type installation along the bustling streets of cities, which enables users on foot or in a car (because the moving speed must be slow on bustling streets) to achieve seamless communication or Internet access.
- **Portal:** Its role is equivalent to a network bridge. ESS is connected to the Internet through a bridge (or portal). In addition, the ESS bridge can also be connected to other non-802.11 series LANs.

From the above introduction, we can understand that when a user accesses the Internet through a Wi-Fi network, the network speed is related to the following factors:

a. The working conditions of the Internet.
b. The performance of the wired broadband network (such as the bandwidth of the optical fiber) and working conditions.
c. The channel transmission characteristics of the radio access network BSS (it is also a rate adaptive system).
d. The number of users in the BSS of the radio access network.

From the perspective of Wi-Fi evolution, the following two factors need to be particularly emphasized for network speed:

a. The protocol the AP adopts, and whether it supports MIMO, MU-MIMO and OFDMA. For MU-MIMO, the number of channels of MIMO that are supported.
b. Terminal type. In actual situations, the real bottleneck of network speed is not the AP or router, but the terminal (see Table 1.8). No matter how high the specifications of APs or routers are, if the terminal performance is not enough, the network speed will be slow. In theory, high-end APs or routers do provide higher speeds and can connect more terminals; however, this is based on the fact that all terminals must support MU-MIMO or OFDMA. When a terminal accesses the Internet via a 4G network, its network speed is also related to the terminal level.

In summary, network speed and network delay are actually physical quantities that change over time. Therefore, the usual network speed and delay should be the average speed and average delay of the network in a certain period of time. The average network delay, download speed and upload speed of the network shown in Figure 2.9 can all be measured with a mobile phone.

The above discussion involves the concepts of gateways, routers and bridges (gate bridges or portals). It is necessary to make a brief explanation here.

- **Gateway:** The gateway is also called the network connector and the protocol converter. It implements network interconnection above the network layer in the Open System Interconnection Reference Model (OSI/RM). Therefore, it is a complex network interconnection device, which is only used for the interconnection of two networks with different high-level protocols.
- **Router:** Routers are used to connect multiple logically separated networks. To access the WAN, a local area network needs a router to provide conversion services. The BSS in Figure 2.9 is connected to the wired broadband network through a router. It realizes network interconnection on the network layer of OSI/RM (simpler than gateway). Routers are divided into local routers and remote routers. The former is used to connect network transmission media, such as optical fiber, coaxial cable and twisted pair. The local router used in Figure 2.9 is used to connect the DS medium (optical fiber).
- **Bridge (gate bridge or portal):** The bridge is also called a bridge connector, which is used to connect wireless LAN, wired LAN or other networks. The most common way to extend the LAN range is to use a bridge. It implements network interconnection on the data link layer of OSI/RM (and simpler than routers).

2.5.7 *Billing methods of network operators*

The three major Chinese network operators (China Mobile, China Unicom and China Telecom) need the following investments when building communication networks.

First, they need equipment investment, such as building base stations and core networks.

Second, they need frequency investment. The frequency spectrum used by the mobile communication network is controlled and allocated by the National Radio Regulatory Agency. Any spectrum managed by the National Radio Regulatory Agency is called licensed spectrum. Network operators need to pay the cost of occupying national resources when using a licensed spectrum. Here, we want to say that operators in other countries (such as the United States) obtain frequency resources through auction, that is to say, operators in these countries have to pay a one-time fee for the frequency resources obtained through auction.

The ISM (Industrial Scientific and Medical) spectrum is used in Wi-Fi. As long as these frequency bands meet the requirements of the National Radio Regulatory Agency (such as restrictions on the transmission power), then there is no charge, and any individual or group can use it. This type of spectrum is called unlicensed spectrum.

The third is the cost of network maintenance, management and technical personnel.

It can be seen that users should pay network operators when they use the communication network. In order to charge reasonably, operators should follow the following two principles from a technical perspective:

- When the communication resources occupied by the user are exclusive, the user should be charged according to the length of the communication time.
- When the communication resources occupied are shared by all users, it is reasonable to charge according to the amount of data (bytes) contained in the transmitted file, that is, according to the data flow rate.

According to these two principles, we discuss how to charge 4G mobile communication networks and wireless local area networks as follows:

(1) 4G mobile network

a. *Phone*

There are two ways to transmit voice over 4G network.

> First, when the 4G network transmits voice, it automatically falls back to the 2G network (since the 4G network is backward compatible). The 2G network uses circuit switching to transmit voice signals. It can be seen that voice and data cannot work at the same time in this case. So, it is impossible to make a phone call while surfing the Internet.
>
> Second, the voice data are still packaged in the format required by the packet switching mode, so that the voice and data can be transmitted in the same 4G network. We call this kind of voice LTE (VoLTE). But, in order to ensure a short voice connection time and good transmission quality, the 4G network still uses a dedicated channel to transmit voice data. So, VoLTE is still a high-definition phone, so that data and voice can be simultaneously transmitted in a 4G network. Therefore, the function of making calls while surfing the Internet can be realized. It can be seen that VoLTE phones work similarly to circuit switching.

In summary, users in 4G networks should be charged according to the time of the call.

b. *Data*

In 4G networks, packet switching is used to transmit data, so operators should charge based on data traffic. Under this principle, operators have also adopted some other strategies. The charging standards are divided into two types, which are inside and outside the package. The former is cheaper and the latter is more expensive. For example, China Telecom's RMB 79 per month package can cover 200 minutes of telephone calls and 700 MB of data per month. If it exceeds that, it will be charged according to the standard outside the package.

(2) Wi-Fi network

a. *Home*

Although a Wi-Fi network uses packet switching to transmit data, its communication resources at home are still exclusively shared by the family (because we can set passwords so that others cannot access it). So, it should be charged according to the length of communication. That is, users need to pay to the broadband network operator once a month.

b. *Public*

The Wi-Fi network provided in public places is a kind of welfare for users, so it is free to use.

2.5.8 *Application of Wi-Fi in Smart Home*

The Smart Home system uses technologies such as the Internet to integrate and control electronic products or systems in the home, such as computers, lighting systems, kitchen equipment, security systems, air conditioning equipment, communication systems, robots, and video and sound systems.

Since mobile phones have successfully connected to the Internet, in order to ensure that family members use their mobile phones to control these smart devices at home (short-range) and outside (remote), the key issue is that these smart devices must also be able to access the Internet.

Since the current wired broadband network has entered the home, smart homes can access the Internet in both wired and wireless methods.

But, the former must be connected to a lot of network cables, which is not only inconvenient to use but also sabotages the home layout. Thus, the wireless method is better. The short-distance home wireless technologies include radio frequency technology, infrared technology, Bluetooth, ZigBee (wireless personal area network) and Wi-Fi.

No matter which technology is adopted, these smart homes must use these technologies to form a wireless local area network to integrate all the devices in the smart home. In addition to Wi-Fi, there is a problem of communication protocol conversion when using other wireless technologies for networking. That is, a gateway must be set up in the local area network to convert the communication protocol used by the corresponding technology to Wi-Fi protocol. Then, the local area network can be connected to the wired broadband network. This not only increases the technical difficulty but also increases the cost. But, if Wi-Fi technology is used for networking, there will be no such problem. At this time, as long as all the smart devices in the smart home become a site in the BSS, we can conveniently use mobile phones to perform short-range or remote control of all the devices.

2.5.9 *Icon*

Since 4G networks are backward compatible, the following icons will appear on the cell phone signal bar:

G: Indicates that the mobile phone did not find 4G and 3G networks, and currently uses 2G networks.
E: Indicates that the mobile phone uses the 2.75G network.
H: Indicates that the mobile phone uses the 3.5G HSPA network.
H+: Indicates that the mobile phone uses the 3.75G HSPA+ network.
4G LTE: Indicates that the mobile phone uses a 4G LTE network.
4G+: Indicates that the mobile phone uses a 4G+ network (its performance is between 4G LTE and 4G LTE-A).

Refer to Table 2.1 for the meaning of the above symbols.

2.5.10 *Terminal category*

See Table 1.8 of Chapter 1 for details on terminal categories and 3GPP release levels. The table also provides the different peak rates of uplink/

downlink data in the radio access network that vary with baseband chip performance, MIMO scale, radio frequency bandwidth, and carrier aggregation number.

2.5.11 *User surfing process* [9]

We take 4G as an example to illustrate how users connect to the Internet through 4G networks. The following first discusses the architecture of the 4G network, and then discusses the process of surfing the Internet.

2.5.11.1 *4G network architecture*

From Figure 6.16, the LTE architecture can be obtained as shown in Figure 2.10. It is an all-IP network. Both data transmission and voice transmission adopt packet switching (PS) technology.

The LTE architecture is composed of three major parts, namely, user equipment (UE), evolved radio access network (E-UTRAN) and evolved packet core network (EPC). It can be seen from Figure 2.10 that the user plane and control plane of EPC are clearly divided. The upper half of the EPC in the figure is the control plane, and the lower half is the user plane. The transmission of signaling is represented by a dashed line, and the transmission of user data is represented by a solid line in the figure.

In the control plane, the base station eNB connects the MME and the HSS, which is responsible for processing signaling messages such as registration, handover, calling and paging with the UE.

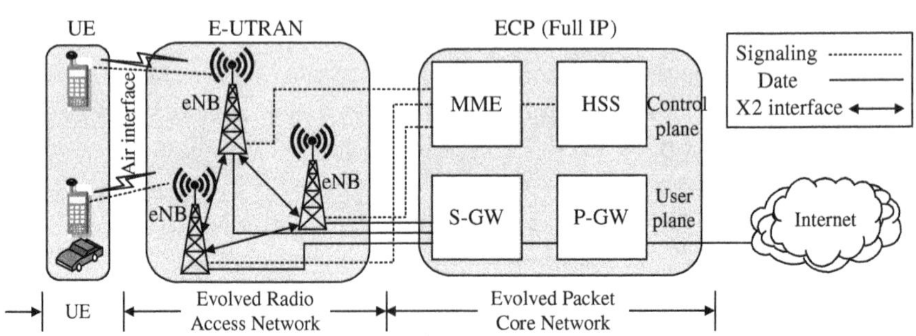

Figure 2.10 Simplified LTE architecture.

In the data plane, the base station eNB transmits IP datagrams between the user equipment UE and the core network.

Packet Data Network Gateway is a gateway router or border router from the core network to the Internet. It is a smooth evolution and upgradation from GGSN (Gateway GPRS Support Node) (see Figure 6.15). It is the interface between the core network and 3GPP or non-3GPP external networks. In real networks, 2G/3G GGSN and 4G P-GW are a converged device (see Figure 6.16). The P-GW is also the anchor point of the core network to the outside world. The P-GW is responsible for assigning IP addresses to all UEs and ensuring the implementation of QoS (Quality of Service). The datagram of UE is encapsulated in the GTP-U tunnel at the base station eNB (will be discussed later). It first moves from eNB to S-GW, and then to P-GW through the all-IP core network EPC. This tunnel mode also has the effect of ensuring the QoS. For example, the LTE network can guarantee that the voice packet delay between the UE and the P-GW does not exceed 100 ms, and the voice packet loss rate is less than 1%. This ensures that the quality of voice communication is still good when transmitting on the all-IP network. The reason that circuit switching is no longer retained in the LTE network is that the mainstream of mobile communication traffic is now data communication (such as browsing websites, reading message and using WeChat for audio or video communication). Retaining circuit-switched components for a small number of mobile phone communication services will make the network more complex and will greatly increase the construction and operating costs of networks. Adopting an all-IP network is a major change in the LTE network structure.

The Serving Gateway is a gateway router between the radio access network and the core network, evolved from the Serving GPRS Support Node (SGSN) (see Figure 6.15). In a real network, 2G/3G SGSN and 4G S-GW are a converged device (see Figure 6.16). The S-GW is responsible for the forwarding and routing of data packets at the user level and functions as a router. The S-GW is also responsible for tunnel management from eNB to S-GW and S-GW to P-GW. S-GW is the anchor point of mobility in the data plane. During the communication process, UE may switch between a different eNB of the LTE system or even roam to different 3GPP access systems (such as 2G GSM or 3G UMTS). If these eNB and 2G/3G base stations are all connected to a certain S-GW, then the S-GW remains unchanged. All the data flow out from the same S-GW, and then forward to the P-GW.

SGW and PGW can be implemented on the same physical node or different physical nodes.

Home Subscriber Server (HSS) is a central database that contains basic user data stored by network operators.

The mobility management entity (MME) is a signaling entity responsible for all signaling exchanges between the base station and the core network as well as between the user and the core network. A large core network requires multiple MMEs to handle a large amount of signaling exchange. When a user accesses the LTE network for the first time, the base station eNB must communicate with MME so that MME and the user can exchange authentication information. MME must obtain a user's relevant information from HSS.

In Figure 2.10, some components are omitted, such as the PCRF (Policy and Charging Rules Function) unit, which are not introduced here.

LTE must be backward compatible with 3G and 2G. Therefore, many mobile phones are marked with 4G/3G/2G functions. This means that if the location of the mobile phone is not covered by the 4G network, the mobile phone can still use the functions of the 3G or 2G network.

Initially, LTE transferred telephone communication services to the original 3G UMTS or 2G GSM circuit-switched network for processing to ensure the quality of communication. This is called CSFB (Circuit Switched Fallback), which means returning to the 3G/2G circuit-switched network to process communication services. But, this treatment method is transitional. In 2012, IP-based VoLTE (Voice over LTE) came out. VoLTE can provide high-quality communication, but the operation of VoLTE depends on the IP Multimedia Subsystem (IMS) connected to the P-GW. IMS does not belong to LTE, but belongs to the scope of IP services. It is another packet-switched network system besides LTE.

2.5.11.2 *Data transmission process in LTE network*

The following discusses the data transmission process in the LTE network, which requires the concept of the LTE protocol stack.

When the UE (such as a mobile phone) is turned on, it registers with the LTE network in order to use network resources to transmit IP data services. The data path in the LTE network consists of two parts, which are the air interface radio link (UE→eNB) and the tunnel in the

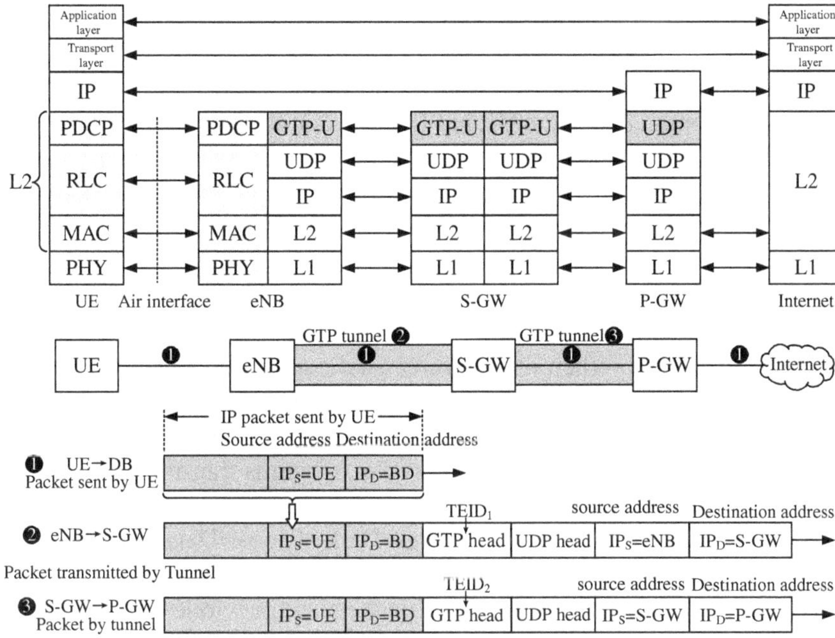

Figure 2.11 LTE protocol stack (user plane).

core network (eNB→S-GW→P-GW). The tunnel will be explained in detail here. Figure 2.11 shows the LTE protocol stack (user level). The communication of data in the tunnel mentioned above uses the GPRS tunneling protocol (GTP). The letter U at the end of GTP-U indicates that the user plane data are transmitted. LTE also uses another control plane tunnel protocol GTP-C to transmit related signaling (limited to space, this part of the content is omitted). When the context is clear, GTP-U can also be abbreviated as GTP. As long as the UE does not exceed the coverage of the P-GW when it moves, the IP address assigned to the UE by the P-GW does not change.

After UE registration is completed, if there is no data service for a period of time (for example, 10 to 30 seconds), then in order to save precious radio spectrum resources, the air interface link between the UE and the eNB will be released quickly. The GTP-U tunnel between the eNB and the S-GW will also be released subsequently, so that the UE air interface enters an idle state. However, the GTP-U tunnel between the S-GW and the P-GW and the IP address of the UE are still maintained. The P-GW believes that the IP address is currently used by the UE temporarily, and retains information about the UE.

When the UE is in the idle state of the air interface, there are two different situations. That is, the UE has IP packets transmitted to the Internet, or the Internet has IP packets transmitted to the UE, which are discussed separately in the following (the process of making a phone call between UEs is not discussed here):

(1) Assume that the UE wants to visit Baidu (www.baidu.com): First, the UE should transmit a connection request to the base station eNB of the cell where it is located. When eNB receives the connection request, it establishes a GTP-U tunnel between the air interface link and the eNB → S-GW. Note that the original GTP-U tunnel between S-GW → P-GW still exists. Then, the UE transmits an IP packet, which is transmitted from the IP layer to the lower layer 2 (now L2 has three sub-layers, see Figure 1.58).

The uppermost layer of L2 is the PDCP (Packet Data Convergence Protocol) sub-layer. The main function of the PDCP sub-layer is to support more efficient transmission of IP packets on the wireless link, including compression/decompression of the IP header. When the UE transmits a series of IP packets, each IP packet has a 20-byte IP header. Most parts of these headers are repeated. If an appropriate compression algorithm is used to compress the header, a large amount of wireless channel resources can be saved. After receiving the compressed header IP packet, the PDCP sub-layer of the base station eNB decompresses it.

The next sub-layer is the Radio Link Control (RLC) sub-layer. The RLC sub-layer can provide three operating modes with different reliability levels. For example, for the Acknowledged Mode (AM), when transmitting data, the PDCP protocol data unit passed down from the PDCP sub-layer is segmented or spliced to make the length suitable for wireless channel transmission. When receiving data, the protocol data unit is reorganized and then uploaded to the PDCP sub-layer. The RLC sub-layer also has the functions of packet reordering, repeated data detection and data retransmission using the error detection protocol ARQ.

The bottom of L2 is the media access control MAC sub-layer, which performs multiplexing and de-multiplexing between the logical channel of the RLC sub-layer and the transport channel of the physical layer below. In an environment with poor wireless channel quality, the MAC sub-layer adopts HARQ. The principle of HARQ is discussed in Section 5.6.13. In addition, the MAC sub-layer also dynamically allocates radio resources to UE in accordance with the eNB scheduling program, thereby ensuring the QoS.

When transmitting, the physical layer encodes and modulates the data transmitted from the MAC sub-layer, inserts bits into the appropriate time slots in each frame and transmits them out. When receiving, the physical layer needs to demodulate and decode, and upload the received bits to the MAC sub-layer. The physical layer also uses an adaptive modulation and coding AMC (Adaptive Modulation and Coding) technology. The base station eNB dynamically adjusts the modulation method (QPSK or 16QAM or 64QAM) and coding rate adopted by the physical layer according to the channel conditions fed back by the user terminal. When the quality of the wireless channel is poor, the transmission rate of the physical channel may be much lower than its peak rate to ensure the transmission reliability of the wireless link.

In Figure 2.11, the destination addresses of the IP packet ❶ transmitted by the UE are IPD = BD (BD stands for www.baidu.com) and the source address of the IP packet ❶ is IPS = UE (UE stands for the IP address of the UE).

When the PDCP of the base station eNB decapsulates the received data, it is encapsulated with the protocol GTP-U, and a GTP tunnel endpoint identifier (TEID) is written into the GTP header, as shown in the figure TEID1. At this time, the IP packet transmitted by the UE has been encapsulated in a new IP packet ❷ and transmitted in the tunnel (eNB→S-GW). The destination addresses of the IP packet ❷ are IPD = S-GW and source address is IPS = eNB.

After the S-GW receives the IP packet, it uses the same method to decapsulate it, then encapsulate it into another new IP packet ❸ transmitted in the GTP-U tunnel (S-GW → P-GW) and put another GTP tunnel. The endpoint identifier TEID2 is written into the GTP header. The destination addresses of IP packet ❸ are IPD = P-GW and the source address is IPS = S-GW.

We know that between the eNB and the S-GW as well as between the S-GW and the P-GW, many IP datagrams are encapsulated and transmitted in respective GTP tunnels. Therefore, in order to identify different tunnels, a GTP tunnel endpoint identifier (TEID) must be allocated for each tunnel transmitted by the UE to a certain destination address. This means that each UE is allocated a different tunnel. Then, all IP datagrams transmitted by the UE to the same destination address are encapsulated in this tunnel for transmission. The IP datagrams transmitted by different

UEs are assigned to different GTP tunnel endpoint identifiers TEID, and thus are transmitted in different tunnels.

Finally, the P-GW decapsulates the IP packets received from the GTP-U tunnel, obtains the IP packets transmitted by the UE and forwards them to the target website on the Internet.

Why does LTE use the protocol GTP to re-encapsulate the UE's IP packets into the GTP tunnel for transmission? This is because in the LTE cellular mobile communication system, the base station eNB associated with the user equipment UE often changes when the UE is roaming. This means that the same UE may use a different eNB or even different S-GWs at different times. This makes the P-GW/GGSN and S-GW/SGSN in the packet switching domain of the core network unable to forward IP packets to the UE based on the UE's IP address and using the traditional routing protocol. But, we know that in an LTE network, the geographic locations of all eNB, S-GWs and P-GWs are fixed, so that the core network EPC can only be responsible for routing within the core network. Due to the adoption of the GTP tunnel mode, the IP packet transmitted by the UE to the Internet is encapsulated by the core network as a new IP packet and transmitted to the P-GW in the tunnel. Later, the P-GW will forward it to other routers in the Internet. In the same way, the IP packets transmitted from the Internet to the UE are always forwarded to the P-GW first, and the P-GW is responsible for determining which tunnel to forward to the S-GW and eNB, and finally from the eNB to the destination UE.

It is not difficult to see that the round-trip IP packet transmitted from a UE to the target server passes through a total of four tunnels, two uplink tunnels and two downlink tunnels. A total of four different GTP tunnel endpoint identifiers TEID are used in the GTP header.

(2) The target server transmits data to user equipment UE: The target server does not know the air interface status of the UE, but only the IP address of the UE. The target server uses the IP address of the UE as the destination address to form an IP packet and transmit it out. The router on the Internet can find the P-GW where the UE resides according to the destination address of the IP packet (because the IP address of the UE is allocated by the P-GW, the router on the Internet can find the P-GW according to the IP address of the UE).

The P-GW can pass the corresponding GTP-U tunnel through the IP address of the UE, encapsulate the IP packet into a GTP-U packet and forward it to the S-GW in the tunnel. There will be two situations in the following process:

- **The GTP-U tunnel between S-GW and eNB exists:** At this time, the S-GW transmits the GTP-U packet to the eNB through the tunnel. The eNB decapsulates the GTP-U packet, adopts PDCP/RLC/MAC/PHY layer encapsulation on the air interface link and transmits the data to the UE.
- **The GTP-U tunnel between S-GW and eNB does not exist:** Now, the UE is in an air interface idle state. This situation is more complicated, because all base stations eNB in LTE do not know where the UE is. Therefore, the S-GW has to temporarily buffer the received IP packets and trigger the mobility management entity MME to page the UE (sometimes called the wake-up UE). So, how can MME page to UE?

MME can broadcast and page UEs in the entire LTE network, but the cost is too high. When network operators build LTE networks, they divide the entire coverage area into many tracking areas (TA). The network operator assigns each TA with a tracking area identity (TAI), which is the only such TA in the world. Identification includes country code, network operator code and TA code here. TA is the basic unit of location update and paging in the LTE system. One TA can cover multiple cells. When the UE is in the standby state, it must listen to the broadcast of the neighboring eNB in order to know which TA it is in. The UE must periodically report its TAI to the MME of the core network so that the MME can page itself. In order to avoid the core network signaling load being overloaded when the UE frequently switches between TA areas, the MME writes a group (1–16) of TAs into a tracking area list (TAL) and transmits it to the UE. When the UE roams across TAs within this TAL range, it does not need to transmit a TA update message to the MME. If the MME needs to page the UE at this time, it only needs to page within a small area of a TAL. Figure 2.12 illustrates the process of updating the TAL of the UE. We see that different TAs can contain different numbers of cells. For example, TA1 contains only one cell, but TA4 contains five cells. Assume that the UE is initially at location ❶, belonging to TA1. The UE reports this location information to the MME, and then the MME transmits a TAL1 to the UE. Figure 2.12 indicates that TAL1 contains three tracking areas, TA1, TA2 and TA3. When the UE roams to locations ❷ and ❸, its

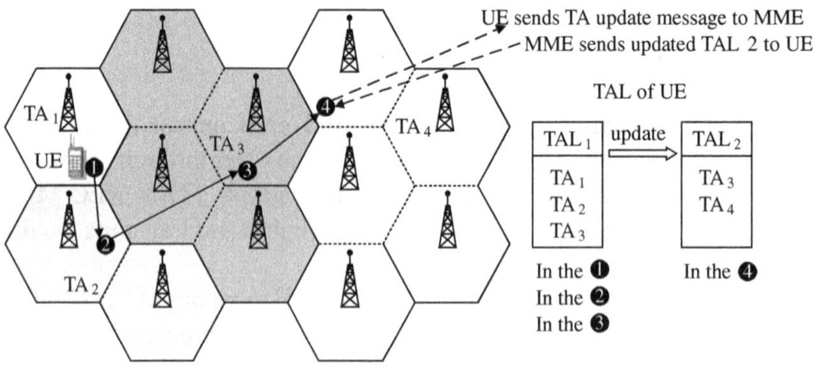

Figure 2.12 Update of UE tracking area list.

tracking area list remains unchanged, so the UE does not transmit the TA update messages to the MME. But, when the UE roams to location ❹, it finds that the newly arrived TA4 is not in its tracking list TAL1, so it transmits a TA update message to the MME. The MME then transmits the updated TAL2 to the UE. At this time, as long as the UE roams in TAL2 (that is, within the range of TA3 and TA4), it does not need to transmit TA update messages to the MME, thus reducing the signaling pressure on the core network. Note that one TA can belong to multiple TALs; for example, TA3 is in both TAL1 and TAL2.

Therefore, when the MME is paging the UE, it does not need to broadcast in the entire LTE network, but only needs to transmit a paging message to a small number of base stations eNB in the TAL where the UE is located. When a certain base station eNB pages the UE, the UE responds to the paging in the cell and triggers the eNB to establish a GTP-U tunnel with the S-GW. After that, the S-GW forwards the buffered IP packet to the eNB and then to the UE.

Due to the burstiness of packet-switched traffic, and to save radio air interface resources at the same time, the UE is often in an idle state, so frequent paging may occur.

Furthermore, when the UE is roaming under conditions when the air interface link is established, the UE must continuously measure the strength of the cell pilot signal during roaming, and report the measurement result to the base station eNB. If the eNB finds a more suitable cell, it will trigger the UE to switch, establish an air interface link in the new cell and release the air interface link of the old cell, and also remove the GTP-U tunnel between the eNB and the S-GW from the old cell and switch to the new cell. Due to the large coverage of the S-GW, usually the GTP-U tunnel between the S-GW and the P-GW will not be reestablished. During the switching process, data communication will not be interrupted.

2.5.12 *VoLTE*

VoLTE (Voice over LTE) is called long-term evolution voice bearer, and it is a high-speed wireless communication standard for mobile phones and data terminals. It is based on the IP Multimedia Subsystem (IMS) network, which acts as a special configuration file (defined by the GSM Association in PRD IR.92) for the Control Plane and the Media Plane of the voice service on LTE. The voice service (control and media plane) is transmitted as a data stream in the LTE data bearer network, instead of relying on the traditional circuit switched voice network. The VoLTE voice and data capacity is more than three times that of 3G UMTS and more than six times that of 2G GSM.

The most direct experience that VoLTE technology brings to 4G users is that the connection waiting time is shorter, and it offers a higher-quality and more natural voice and video call effect. VoLTE is essentially different from 2G and 3G voice calls. It is an end-to-end voice solution built on 4G networks under all-IP conditions. Compared with 2G and 3G voice calls, VoLTE can improve its voice quality by about 40% because it uses high-resolution codec technology. VoLTE brings users a lower access delay (waiting time after dialing), which is 50% less than 3G, which is about 2 seconds, while 2G is (6–7) seconds. In addition, phone drops occur from time to time in 2G and 3G, but VoLTE has almost zero dropped calls.

It can be seen that VoLTE essentially uses IP to carry voice services. Voice over IP is a more macro concept. VoLTE is only a means to implement VoIP. VoLTE does not continuously occupy air interface resources. Therefore, it can accommodate more users than traditional circuit domains. The resource utilization is higher. The 5G voice service is also carried by IP, imitating the name of 4G, and it can be called Vo5GNR, or abbreviated to VoNR.

Chapter 3

Evolution of Wireless Local Area Network

Wireless Local Area Network (WLAN) is commonly known as the Wi-Fi. It cannot be seen, touched or heard, but it has a great impact on the world today and will continue its influence. High-speed wireless connections without relying on cables can be seen everywhere, from our home to offices or public places, which fundamentally changed our way of life. It is no exaggeration to say that due to the available Wi-Fi connection at anytime and anywhere, we can have a better life and be able to use our laptops, tablets and smartphones in a more direct, simple and high mobility way, while no longer needing to worry about having a bunch of messy wires around.

Cellular mobile communication started in the late 1970s and has been around for 40 years now. As shown in Figure 3.1, Wi-Fi has been around for 20 years since the IEEE passed the 802.11 standard.

3.1 The Past and Present of Wi-Fi [16]

3.1.1 *The origin of Wi-Fi*

The so-called Wi-Fi is based on the IEEE 802.11 standard and is used for Wireless Local Area Network (WLAN) communication technology.

Let's take a look at the history of Wi-Fi and see how it overcame many technical challenges and evolved into the ultra-fast, highly convenient Wi-Fi standard that we know today.

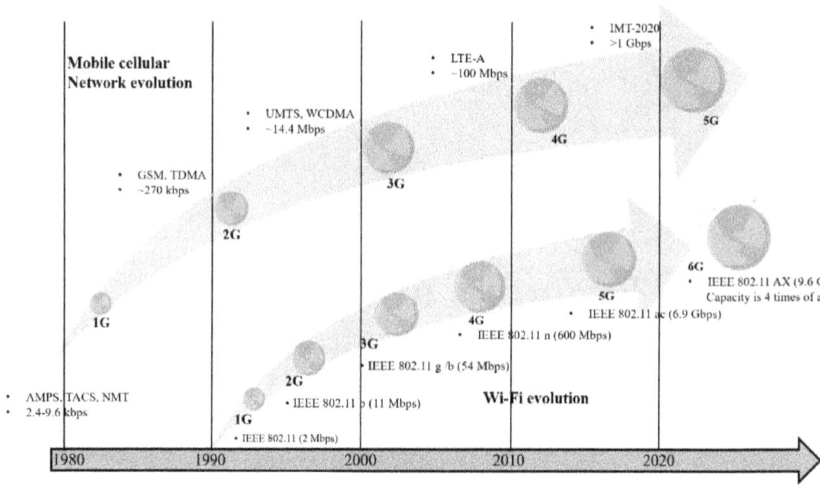

Figure 3.1 Evolution of cellular mobile network and WLAN (Wi-Fi).

Although we now think that 802.11 wireless technology is the main way to connect personal devices and smartphones to the Internet, in fact it was originally invented for connecting cash registers. In the late 1980s, retail tool and cash register (POS) computer system manufacturer NCR had a problem. Its customers such as department stores and supermarkets did not want to rewire every time after the store's layout was changed.

The Federal Communications Commission made an adjudication at that time, opening up certain frequency bands for free usage. It was this decision that inspired a rule-changing idea. By using wireless connections in unlicensed spectrum (rather than traditional wired connections), electronic cash registers and POS systems could be easily moved around stores without renovations.

Then, NCR assigned the project to an engineering team in Netherlands. This team faced the challenge of creating a wireless communication protocol. They successfully developed the WLAN, which is recognized as the pioneer of Wi-Fi. NCR did not make it as a purely proprietary protocol, but established it as a standard, so this company was able to become the leader after the wireless market rose. By 1990, the IEEE 802.11 working group for wireless communications based on the unlicensed spectrum was formed.

The spread spectrum technology, which was already very innovative at that time, was able to reduce interference and improve signal integrity in noisy environments. The original version of Wi-Fi was finally officially standardized in 1997. The data flow at work at that time was only 2 Mbps, but it laid the foundation for the future development of Wi-Fi.

Although the 802.11 wireless standard was released in 1997, it was not immediately widely adopted. Slow speed and expensive hardware had hindered its attraction to the market for a very long time, but its development process was destined to change. 10 Mbps Ethernet was the network standard at that time, and the IEEE 802.11 research team understood that if 802.11 could also achieve such speed, they could gain a strong advantage. In 1999, they successfully developed the 802.11b wireless standard. This technology used the same 2.4 GHz ISM frequency band as the original 802.11 wireless standard, but greatly increased the data flow to 11 Mbps. Thus, wireless Ethernet finally came into reality.

Soon after the establishment of the 802.11b standard, the IEEE working group released a faster 802.11a standard which no longer works in the increasingly crowded 2.4 GHz band, but runs in the 5 GHz band and supports speeds up to 54 Mbps.

Since 802.11a occupies the 5 GHz frequency band which is far away from the more popular (and therefore more crowded) 2.4 GHz band, it has shown better performance even in noisy communication environments. However, compared with the 2.4 GHz wireless connection, higher frequency also means that its function range is reduced. Since 802.11b has cheaper equipment and a better function range, it proved to be the most popular wireless standard at that time. However, although 802.11b has more cost performance than 802.11a, its price was still too high for average consumers. Besides, wireless routers and network cards still cost hundreds of dollars.

All this was completely changed after a conference call by Jobs. At that time, Apple was launching a new generation of computers and wanted to use radio networks as part of its functions. The terms set at that time were very difficult. Apple's specified purchase price was US$99. Of course, the batch could be very large. At this time, Lucent Technologies had acquired NCR and finally agreed to the terms.

Although all these seemed to be difficult to achieve in the first place, Apple's deal eventually delivered Wi-Fi to consumers and promoted it to the market. Personal computer manufacturers saw the strong market impact of Apple's computers and therefore also hoped to add radio networks to their computers. Soon, major PC hardware vendors, including Dell, Toshiba, HP and IBM, all offered Wi-Fi technology. Microsoft also added support for Wi-Fi through Windows XP, and by cooperating with Lucent engineers, Microsoft built the Wi-Fi connection into their operating system, and users could achieve wireless connections without installing third-party driver software. With the release of Windows XP, Wi-Fi was built into millions of computers around the world, and it was officially announced that it has entered the great era of Wi-Fi.

At the beginning of 2000, as the advantages of wireless connection became more clear, radio networks were rapidly popularized. As businesses and consumers began to realize the application potential of Wi-Fi, which was the prototype of what we now call mobile computing, Wi-Fi hotspots had begun to appear in coffee shops, airports and hotels. Many families had also begun to use broadband Internet to easily achieve wireless connections at home.

Thanks to the efforts of the IEEE 802.11 working team, this special wireless protocol originally designed to connect the cash register has become the basis of radio network standards and is also changing the structure of the entire society.

802.11b came out in 1999, laying the foundation for the widespread application of radio networks. Its cheap price made it easy for consumers to accept, and the speed of 11 Mbps made it fast enough to replace the original Ethernet connection of enterprise users. In the first few years of the 21st century, broadband Internet was quickly accepted by the market, and 802.11b achieved remarkable success. Both ordinary consumers and business users have found that wireless is a better high-speed connection than the previous xDSL, cable or other broadband technologies.

As the high speed provided by broadband Internet becomes normal, consumers' computer usage habits have also changed. Higher bandwidth applications such as music/movie sharing and streaming media have become more and more popular in the consumer domain.

3.1.2 *Enhance Wi-Fi speed*

To meet increasing business requirements (such as VoIP, file sharing, printer sharing, desktop virtualization and mobile working and living), radio networks must provide higher speeds to compete with wired Ethernet networks.

Even in the early 2000s, the speed that 802.11b could support was far from a super high level. But in terms of wired network communication,

10/100 Ethernet was already a common standard. At 100 Mbps, it is nearly 10 times faster than the 11 Mbps speed of the 802.11b standard. Excluding the necessary waste of the 802.11b communication protocol itself means that the theoretical speed that 802.11b can achieve was 5.9 Mbps. But in fact, because 802.11b uses the increasingly popular 2.4 GHz frequency band, the actual speed is seen to be lower than the theoretical value. Interference from microwave ovens, wireless phones and other electronic products means that 802.11b's actual speed usually cannot reach the 5.9 Mbps standard (sometimes it can't even be close to this speed).

To solve this problem, the IEEE 802.11 working group introduced 802.11g in 2003 (see Ref. [7] for details). Although 802.11g uses the same 2.4 GHz frequency band as 802.11b, it can reach speed of up to 54 Mbps. Even after reducing the necessary waste of the communication protocol which causes lower speed, the theoretical maximum speed can reach 31.4 Mbps, which is enough to meet the increasingly high speed requirements of home broadband communication.

In fact, 802.11g is not the first 802.11 wireless standard to reach 54 Mbps. 802.11a had achieved such speed in 1999, but it used another separate 5 GHz frequency band to achieve such speed. Although 5 GHz has the advantage of less electronic products' radio interference, it was not compatible with 802.11b standard. Taking this situation into account and its relatively expensive equipment cost, 802.11a could only get a few orders in the commercial market, which means it was not widely popularized in the home/consumer field.

By using the 2.4 GHz frequency band to realize the speed of 54 Mbps, 802.11g maintains full compatibility with 802.11b while achieving high speed. This is crucial because at this time 802.11b had been established as the main wireless standard for consumer devices and dominated the world. Compared with 802.11a, 802.11g's backward compatibility and cheaper hardware devices were great selling points, so it could quickly enter the consumer field and gradually became a new and faster wireless standard in related commercial applications.

802.11n, launched in 2009, further improved the speed of 802.11g and 802.11a. 802.11n worked in both 2.4 GHz or 5 GHz frequency band (not running at the same time) and improved the transmission efficiency by adopting frame aggregation. It also introduced optional Multiple-Input and Multiple-Output (MIMO) and 40 MHz bandwidth, making the channel width of 802.11n reach that of 802.11g twice.

802.11n could greatly increase the network speed. In low-end applications, if the same type of single antenna operation as 802.11g is used and the same 20 MHz channel width configuration is adopted, the 802.11n network can reach a speed of 72 Mbps. In addition, if a double-width 40 MHz channel is used and multiple antennas are used, the data speed can be even faster, up to 600 Mbps (in a four-antenna configuration), which replaces 802.11g to dominate the Wi-Fi domain.

3.1.3 *GB-level Wi-Fi*

The latest 802.11 wireless technology that has been adopted on large scale is the 802.11ac standard, which extends 802.11n, especially in improving the 5 GHz frequency band, and maintains the 802.11n in the 2.4 GHz frequency band to achieve backward compatibility.

By supporting the 5 GHz frequency band, 802.11ac can achieve a huge bandwidth advantage which is 160 MHz. Such bandwidth is almost impossible to achieve in the crowded 2.4 GHz frequency band. In addition, through beamforming and by supporting up to eight MIMO data streams, the achievable speed is improved. According to different configurations, in the case that both router and user have multiple antennas, the data rate can range from the lowest (433 Mbps) to several Gbps.

If the speed is not fast enough, there is a more advanced 802.11ad standard (which is currently selling on the market). The 60 GHz millimeter wave frequency band used by this new standard can achieve transmission speeds of up to 7 Gbps even without MIMO configuration. As the function range and penetration of 802.11ad are greatly reduced at the frequency of 60 GHz, its main application is limited to indoor small spaces, which is why it has not become the main Wi-Fi standard.

Although 802.11ac's 7 Gbps speed can be realized in controlled environments of laboratory, in fact, users can be frustrated with the data speed when using public Wi-Fi in crowded airport terminals to check emails.

Thus, the Wi-Fi Alliance Standards Organization has revised and introduced 802.11ax standard on the basis of 802.11a/b/g/n/ac. The main purpose of this revised version is to improve the situation mentioned above. Its main functions and features are summarized as follows:

1. Continues to be backward compatible with 802.11a/b/g/n/ac and supports both 2.4 GHz and 5 GHz.
2. In high-density scenarios, such as train stations, airports and stadiums, the average user throughput increases by four times compared to 802.11ac.
3. Similar data rate and channel width as 802.11ac, but uses new modulation and coding scheme based on 1024-QAM (MCS 10 and 11) to make Wi-Fi faster.
4. Specifies downlink and uplink multi-user operations through MU-MIMO and Orthogonal Frequency Division Multiple Access (OFDMA) technology to make Wi-Fi less congested.
5. Larger OFDM FFT (four times larger), narrower subcarrier spacing (nearly four times) and longer symbol time (four times) to improve robustness and performance in multipath fading and outdoor environments.
6. Optimizes data traffic and channel access.
7. *Better power management*: Supports Target Wake Time (TWT). Allows terminal devices to enter the dormancy state when no data is transmitted, so that up to seven times battery power consumption is saved and thus battery life is greatly extended.

These new features will promote excellent mobile experience and longer battery life and will also open the door to various new applications that could not be implemented before Wi-Fi.

Especially after 802.11ax supporting the OFDMA technology, its wireless bandwidth is theoretically increased by 10 times. The theoretical data rate is close to 10 Gbps, which is faster than the current GB level fiber. But in a practical test, it could only increase by about four times.

Many of the above improvements benefited from the evolution of LTE and 4G networks, because they solved many technical problems which Wi-Fi faced a long time ago.

Relevant physical layer parameters and indicators of the 801.11g/n/ac/ax standard are shown in Table 3.1.

The theoretical data rate in the table = single-stream theoretical rate × the number of spatial streams. Single-stream theoretical rate = the number of bits carried by each subcarrier (i.e., symbol bit length) × the number of subcarriers contained in an OFDM × the code rate/OFDM duration (i.e., transmission period). See Section 3.2.4 for specific calculation details.

Table 3.1 A comparison of 801.11g/n/ac/ax physical layer parameters.

Protocol standard	802.11g	802.11n	802.11ac	802.11ax
Frequency band	2.4 GHz	2.4 GHz 5 GHz	5 GHz	2.4 GHz 5 GHz
Channel bandwidth	20 MHz	20 MHz 40 MHz	20 MHz, 40 MHz, 80 MHz, 80 + 80 MHz, 160 MHz	20 MHz, 40 MHz, 80 MHz, 80 + 80 MHz, 160 MHz
FFT	64	64, 128	64, 128, 256, 512	256, 512, 1024, 2048
Subcarrier spacing	312.5 KHz	312.5 KHz	312.5 KHz	78.125 KHz
OFDM symbol duration	3.2us + 0.8us CP	3.2us + 0.8/0.4us CP	3.2us + 0.8/0.4us CP	12.8us + 0.8/1.6/3.2us CP
Highest modulation order	64-QAM	64-QAM	256-QAM	1024-QAM
Highest theoretical rate	54 Mbps	72.2 Mbps (20 MHz) 600 Mbps (40 MHz, 4 SS)	433 Mbps (80 MHz, 1SS) 6933 Mbps (160 MHz, 8 SS)	600.4 Mbps (80 MHz, 1 SS) 9607.8 Mbps (160 MHz, 8 SS)

Note: (1) Spatial Streams (SS) stands for MIMO spatial stream.
(2) When the OFDM symbol duration of the 802.11n protocol is 3.2us + 0.8us = 4us, its theoretical rate is 65 Mbps (20 MHz) and 135 Mbps (40 MHz, 1SS). See Ref. [7] for details.

Compared with previous generations of Wi-Fi standards, 802.11ax incorporates many new technologies. Besides the 1024QAM modulation and coding scheme, 160 MHz bandwidth increase, OFDMA and MU-MIMO, etc., the new generation of Wi-Fi encryption protocol WPA3 (Wi-Fi Protected Access 3) provides a safer and simpler online encryption mechanism. Recently, IEEE has opened a total of 1.2 GHz bandwidth from 5.925 GHz to 7.125 GHz for 802.11ax usage, plus the original 2.4 GHz (2.412 GHz–2.484 GHz) and 5 GHz (5.15 GHz–5.825 GHz). Wi-Fi 6 can use more frequency bands and channels, which makes resource allocation more flexible.

802.11ax also has a great feature, that is, the distributed architecture, which uses the One Pod Per Room design. Such radio networking solution will cause changes in indoor radio networks.

802.11ax has a clear definition and description in the specification for mesh or distributed Wi-Fi network architecture. In the concept of One Pod Per Room, Wi-Fi transmission technology is responsible for the transmission backbone of the entire smart home, using mesh or distributed Wi-Fi network architecture to achieve network coverage without blind spots. So, all information is connected to the home gateway through Wi-Fi transmission and then uploaded to the cloud through the broadband service provided by telecom operators. Thus, a completely seamless connection and indoor roaming full-covered communication environment is achieved through special algorithms and communication protocols.

In order to support the connection and intercommunication with other smart devices, each Wi-Fi router in the distributed Wi-Fi network must have the transmission technology to receive and transmit Bluetooth and Zigbee. Since Bluetooth and Zigbee are both short-distance transmission technologies, it is necessary to ensure that all commands can be transmitted and received completely. Through distributed network architecture, each Wi-Fi router can completely cover every small area and then transmit to the backbone network via Wi-Fi. Each Wi-Fi router is connected to each other to achieve the goal of full coverage.

3.2 Comparison of Wi-Fi (4/5/6) [17]

Wi-Fi is now the most important function in all smartphones. No matter whether a device supports 3G, 4G or commercial 5G, Wi-Fi is a standard configuration, and it is indispensable under the current large-scale popularity of Wi-Fi. However, the "real name" of Wi-Fi is too "professional". Few people can distinguish the difference between 802.11 a/b/g/n and ac, let alone the new ad and ax standards.

Because of this phenomenon, in October 2018, the Wi-Fi Alliance took the opportunity of promoting the new 802.11ax standard to formally rename the Wi-Fi specification generation. The mainstream standard 802.11n was renamed as Wi-Fi 4, the mainstream standard 802.11ac was renamed as Wi-Fi 5, and the new standard 802.11ax was renamed as Wi-Fi 6. Of course, the previous generation 802.11/ 802.11b/802.11 (a/g) can be renamed as Wi-Fi 1/2/3 (see Figure 3.1).

The new naming system that Wi-Fi uses can make the new and old Wi-Fi specifications more intuitive and friendly to a certain extent, but only to a certain extent. The level of Wi-Fi specifications is not entirely determined by its algebra, but it is also affected by multiple technical factors.

The current mainstream Wi-Fi standard is Wi-Fi 5, but it only supports the 5 GHz frequency band, so Wi-Fi 4 is still another mainstream under the 2.4 GHz frequency band.

Agile readers should see that the theoretical bandwidth of Wi-Fi 6 in Figure 3.1 is 9.6 Gbps, while Wi-Fi 5 also has 6.9 Gbps. Is this the only difference between the so-called sixth generation Wi-Fi and the fifth generation? The difference is big! Following is the comparison of protocol, modulation, bandwidth and frequency band, spatial stream and OFDMA (see Table 3.2) between Wi-Fi 4/5/6, respectively, to better understand the differences between these Wi-Fi standards.

3.2.1 *Protocol*

Let's start with the generation of Wi-Fi specifications. The so-called Wi-Fi (4/5/6) is a number of standards or protocols in the IEEE 802.11 specification. Regardless of 11n, 11ac or 11ax, each set of standards consists of a number of technologies, specifications and functions and forms a unified industry specification to ensure the compatibility and consistency of the equipment.

Wi-Fi 4 (802.11n) was born in 2009. It contains some of the most common standards, such as 802.11 a/b/g/n, which we all referred to as the 2.4 GHz Wi-Fi. Their characteristics are universal, cheap, with the highest single antenna rate up to 150 Mbps. With 40 MHz bandwidth and the support for MIMO, it increases the theoretical rate of Wi-Fi from 54 Mbps at

Table 3.2 Wi-Fi (4/5/6) comparison.

Generation	Wi-Fi 4	Wi-Fi 5		Wi-Fi 6
Protocol standard	802.11n	802.11ac		802.11ax
		Wave 1	Wave 2	
Year	2009	2013	2016	2018
Frequency band	2.4 GHz, 5 GHz	5 GHz		2.4 GHz, 5 GHz
Maximum channel bandwidth	40 MHz	80 MHz	160 MHz	160 MHz
MCS range	0–7	0–9		0–11
Subcarrier spacing	312.5 KHz	312.5 KHz		78.125 KHz
OFDM symbol duration	3.2us + 0.8/0.4us CP	3.2us + 0.8/0.4us CP		12.8us + 0.8/1.6/3.2us CP
Highest modulation order	64QAM	256-QAM		1024QAM
Single stream rate	150 Mbps	433 Mbps	867 Mbps	1201 Mbps
Spatial stream	4 × 4	8 × 8	8 × 8	8 × 8
MU-MIMO	Unsupported	Unsupported	Supports downlink	Supports uplink and downlink
OFDMA	Unsupported	Unsupported	Unsupported	Supports uplink and downlink
Theoretical data rate	600 Mbps	433 Mbps	6933 Mbps	9607.8 Mbps

11a/g to 600 Mbps (150 Mbps × 4 spatial streams), and Wi-Fi 4 supports 2.4 GHz or 5 GHz frequency bands at the same time, which eventually replaced the old standard perfectly and dominated the world. The only disadvantage is that it overlaps with the Bluetooth standard channel, and the signal interference is thus serious.

Wi-Fi 5 (802.11ac) was born in 2013. The original version (Wave 1) increased the single-stream Wi-Fi bandwidth to 433 Mbps relying on the 80 MHz bandwidth and 256QAM modulation. In 2016, the second version (Wave 2) referred to some features of 11ax and the bandwidth was doubled to 160 MHz again, plus the 10% increase in modulation efficiency. So, the theoretical rate of Wi-Fi 5 reached 6.9 Gbps (8SS), which produced the so-called MU-MIMO. But this MU-MIMO is not the complete version, it only supports downlink multi-terminal parallel transmission, and its usage is limited (only when all terminals under the same channel are compatible with MU-MIMO, the MU-MIMO mechanism can be effective). Although Wi-Fi 5 theoretically supports eight spatial streams, it can only achieve 4×4 (80 MHz) or 2×2 (160 MHz) in home Wi-Fi market, that is, the theoretical rate is only 1733 Mbps, which is far from the limit rate of Wi-Fi 5 (6.9 Gbps), but it can still increase the Wi-Fi bandwidth to the comparable GB-level home wired network. In addition, 802.11ac has a series of extended standards such as 802.11ah and 802.11ad, but it is not the mainstream.

It is worth noting that Wi-Fi 5 only supports the 5 GHz frequency band and technically cannot completely replace Wi-Fi 4 which supports 2.4 GHz frequency band. Therefore, the so-called Wi-Fi 4 and Wi-Fi 5 are actually parallel standards. While the Wi-Fi Alliance defining them as an iterative relationship is not rigorous enough.

From the perspective of theoretical bandwidth alone, Wi-Fi 6 seems lackluster. The 1024QAM modulation, which is routinely updated, does not improve the single-stream bandwidth by leaps and bounds. The limit rate (1.2 × 8 = 9.6 Gbps) is only about 40% higher than Wi-Fi 5 (6.9 Gbps).

However, the advantage of Wi-Fi 6 is that it supports both 2.4 GHz and 5 GHz frequency bands, which is the sixth-generation Wi-Fi standard in the true sense. Therefore, it is bound to replace Wi-Fi 4 and Wi-Fi 5 and reunite both of them.

Another advantage of Wi-Fi 6 is that it establishes the full version of MU-MIMO, which supports eight terminals uplink or downlink MU-MIMO, and at the same time introduces OFDMA to achieve another parallel transmission capability complementary to MU-MIMO which is more flexible and practical than MU-MIMO.

Now this standard has been seen in some high-end products and commercial occasions, such as a three-band device released by Asus, which supports a theoretical wireless rate of 10.756 Gbps. Of course, this is only a gimmick because this theoretical rate is obtained by simply adding the rates of several frequency bands, which is neither rigorous nor scientific.

3.2.2 *Modulation*

It is not difficult to see from the evolution of the Wi-Fi protocol that modulation and bandwidth have the greatest impact on the single-stream

Table 3.3 MCS strategy.

MCS	Modulation	Bit rate	Wi-Fi 4	Wi-Fi 5	Wi-Fi 6
0	BPSK	1/2			
1	QPSK	1/2			
2	QPSK	3/4			
3	16QAM	1/2	Standard protocol		
4	16QAM	3/4		Standard protocol	
5	64QAM	2/3			Standard protocol
6	64QAM	3/4			
7	64QAM	5/6			
8	256QAM	3/4	Non-standard expansion*		
9	256QAM	5/6			
10	1024QAM	3/4	Non-standard expansion*	Non-standard expansion*	
11	1024QAM	5/6			

Note: (1) Wi-Fi 4's 256QAM is also called TurboQAM. TurboQAM introduces some features of the IEEE 802.11ac standard into IEEE 802.11n. Precisely, it is to change the IEEE 802.11n network under 2.4 GHz into IEEE 802.11ac 256QAM modulation mode, thus the single-stream rate is increased from 150 Mbps in IEEE 802.11n standard to 200 Mbps in IEEE 802.11ac standard.

(2) The 1024QAM of Wi-Fi 4 and Wi-Fi 5 is also called NitroQAM. NitroQAM is equivalent to the second generation of TurboQAM, which increases the modulation and demodulation standard of wireless transmission to 1024QAM. In the 2.4 GHz frequency band, the single-stream rate (40 MHz) is further increased to 250 Mbps, and the 5 GHz frequency band (80 MHz) has reached 541.25 Mbps, which is equivalent to increasing the single stream 200 Mbps of QAM256 in the 2.4 GHz band and the single stream 433 Mbps in the 5 GHz band by 25%.

rate. Modulation is actually the process of converting electrical signals into radio waves, while demodulation refers to the reverse process. Its core technology is the modulation method. The higher the order of the modulation method, the higher the data density during the conversion process. Table 3.3 shows the MCS strategy of the recent three generations of Wi-Fi.

According to the 802.11 standard protocol, Wi-Fi 4 supports up to 64QAM, Wi-Fi 5 up to 256QAM, and Wi-Fi 6 up to 1024QAM. However, some chip/device manufacturers transplant high-level modulation technology to low-level protocols, so that the Wi-Fi 4 protocol can also support 256QAM or even 1024QAM. The single-stream rate is increased from 150 Mbps to 200 Mbps or even 250 Mbps.

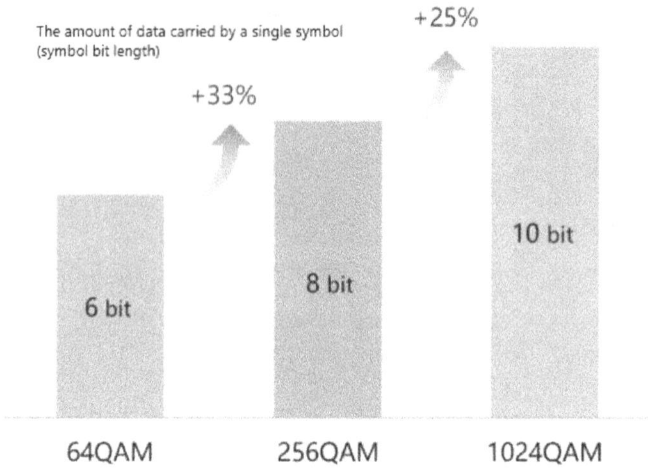

Figure 3.2 Encoding efficiency improvement.

The modulation method determines the data density of a single symbol of the wireless signal subcarrier. The conversion method is very simple. If the QAM value is 2^N, then the symbol bit length is N. Therefore, the 64QAM symbol bit length is six bits, which means that a symbol can transmit six bits of data. The 256QAM symbol bit length is 8 bits, and the 1024QAM symbol bit length is 10 bits (see Figures 3.2 and 3.3). This is the reason that single-stream bandwidth of Wi-Fi 4 can increase from 150 Mbps to 200 Mbps or even 250 Mbps.

In order to ensure the integrity of data transmission, some redundant data needs to be inserted in the modulation process for error correction (which is channel coding). Hence comes the concept of coding rate, which expresses the ratio of valid data to whole data in each transmission in the form of fractions. For example, 1/2 means that only half is valid data and the other half is redundant data; 5/6 means that 5/6 is valid data and 1/6 is redundant data.

Combining the modulation method and the code rate, we can get a Modulation and Coding Scheme (MCS) strategy table. The actual connection rate of the Wi-Fi device is dynamically and adaptively selected in this table. When the wireless signal is strong, MCS will try to choose a high-level combination (high bit and low redundancy). When the wireless signal is weak, MCS will try to choose a low-level combination (low bit and high redundancy). Thus, the Wi-Fi rate of user

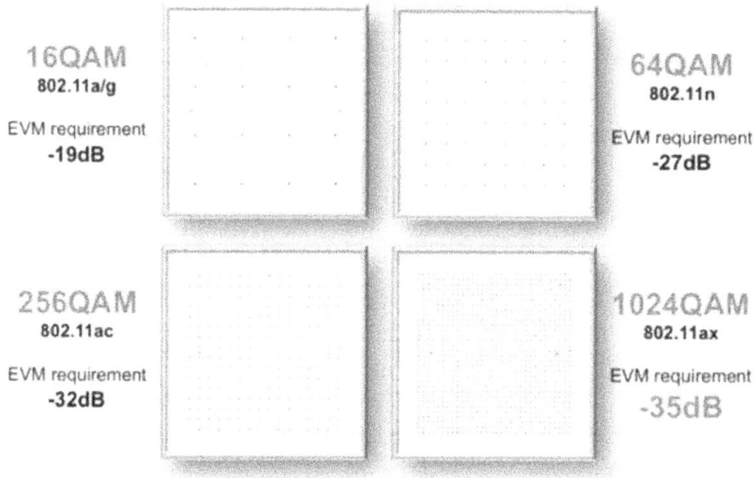

Figure 3.3 Constellation of four kinds of QAM.

terminal will dynamically change between specific values (floating around) because we cannot have high data rate and high data integrity at the same time.

With the increase of data density, the anti-interference ability of digital modulation is declining, which places higher requirements on the quality of wireless signals. Going back to the previous MCS strategy table, the principle of Wi-Fi rate adaptation is as simple as this: when the protocol and bandwidth are fixed, the closer the terminal is and the fewer obstacles the router (or AP) have, the better the signal quality will be, so MCS will automatically select high-level combination. The higher the data density and the bit rate is, the higher the Wi-Fi rate will be.

It is worth noting that the entire MCS dynamic selection mechanism is completely evaluated and selected by the Wi-Fi device based on the current signal quality, and it is cannot be controlled by the user. For example, in the case of poor wireless signal, it is impossible if you want to accept packet loss in exchange for a higher Wi-Fi rate because the 802.11 protocol does not support it.

It should be noted that the single-stream rate and the theoretical rate both refer to the Wi-Fi rate in the highest-order MCS case, that is, the limit air interface rate, which has nothing to do with the wireless signal quality.

3.2.3 *Bandwidth and frequency band*

Unlike MCS, bandwidth is more familiar to consumers because it is one of the core configurations for Wi-Fi devices. Regardless of 2.4 GHz or 5 GHz, the smallest channel frequency band is 20 MHz, referred to as bandwidth. Two adjacent small channels can be aggregated into a large channel (this is the carrier aggregation discussed in Chapter 2, Section 2.4), and the transmission bandwidth is doubled. Thus, single-stream bandwidth of Wi-Fi is multiplied.

Radio waves are transmitted in the form of frames in the channel, and each frame is composed of several subcarriers. The number of subcarriers directly reflects the transmission rate. Taking Wi-Fi 4 and Wi-Fi 5 as examples, the 20 MHz channel supports 64 subcarriers. After deducting the anti-interference subcarrier and pilot subcarrier, the actual subcarriers used for data transmission are 52, while in 40 MHz channel this number is 108, which is 2.08 times that of 20 MHz (not twice as neat). Table 3.4 shows the subcarriers of other bandwidths.

It is interesting that the number of data subcarriers of Wi-Fi 6 at 20 MHz has increased to 234 (see Table 3.4). This has something to do with the frame transmission period.

In Wi-Fi 4 and Wi-Fi 5, each frame is transmitted for 3.2 microseconds and then stopped for 0.4 microseconds (guard interval) and then the next frame is transmitted. So, the transmission period of each frame is 3.6 microseconds.

Wi-Fi 6 standard redesigned the frame structure, and the single frame capacity is quadrupled (256 subcarriers/20 MHz), and the frame transmission time is quadrupled too (12.8 microseconds). But the frame interval is

Table 3.4 Comparison of Wi-Fi (4/5/6) transmission period and subcarriers.

		Number of data subcarriers			
Protocol	Transmission period (us)	20 MHz	40 MHz	80 MHz	160 MHz
Wi-Fi 4	3.2 + 0.4	52	108		
Wi-Fi 5	3.2 + 0.4	52	108	234	468
Wi-Fi 6	12.8 + 0.8	234	468	980	1960
Bandwidth and frequency band		2.4 GHz			
				5 GHz	

only twice that of the former (0.8 microseconds), that is, the transmission period of each frame is 13.6 microseconds.

Wi-Fi 6 uses nearly four times the transmission period, transmitting slightly more than four times the number of data subcarriers. The overall efficiency is improved by about 10%.

It is true that the greater the bandwidth, the more data subcarriers are transmitted in a single frame, and the higher the Wi-Fi rate is. But it still faces the dilemma: the larger the bandwidth is, the worse the quality of Wi-Fi signal is, and the weaker the coverage and compatibility is. Therefore, there are usually bandwidth settings on the wireless router (or AP) (see Figure 3.4), and the users can choose according to their terminal and application.

So, here comes the question, what is the relationship between frequency band and Wi-Fi rate? Is the 5 GHz band always faster than the 2.4 GHz band?

In fact, the frequency band is not directly related to the data rate. The reason why the theoretical rate (or bandwidth) of the 5 GHz band is much

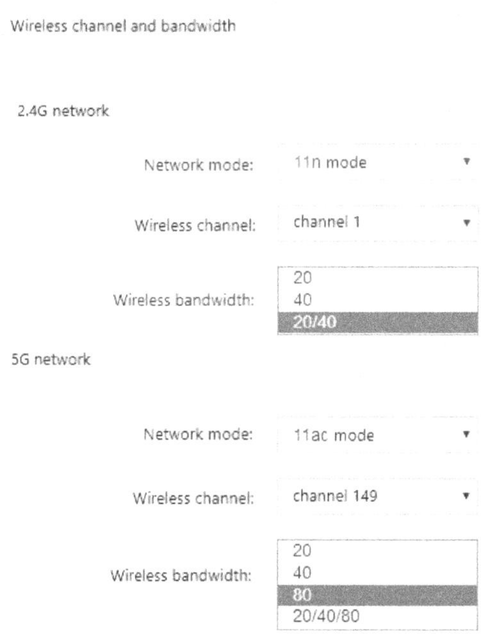

Figure 3.4 Router (or AP) bandwidth setting.

higher than that of the 2.4 GHz band is only due to the inherent advantages in spectrum allocation. The frequency spectrum used for Wi-Fi transmission in the 5 GHz band is much wider than the 2.4 GHz, and the 2.4 GHz band can only aggregate up to 40 MHz bandwidth, while the 5 GHz band can easily reach 80 MHz or even 160 MHz bandwidth.

Assuming under the same protocol, MCS range, bandwidth and spatial stream, the theoretical rate under the 2.4 GHz and 5 GHz frequency bands is actually the same. In terms of function range and penetration ability, the actual 5 GHz rate is not as good as the 2.4 GHz rate.

3.2.4 *Spatial stream*

Spatial stream is derived from MIMO technology. Usually we use I × O to identify the number of receiving and transmitting antennas. The two symbols can be in any proportion. But in Wi-Fi devices, they are basically equal, such as 2 × 2, 4 × 4 or 8 × 8, representing Two Spatial Streams (2SS), Four Spatial Streams (4SS) or Eight Spatial Streams (8SS), respectively. Therefore, when single-stream rate is fixed, the theoretical rate of Wi-Fi device = the theoretical rate of one single-stream × the number of spatial streams.

Note that the spatial stream is low adapted at both ends of the device. Whether a 4 × 4 wireless router is paired with a 2 × 2 terminal, or a 2 × 2 wireless router is paired with a 4 × 4 terminal, the actual running spatial stream is two.

Since spatial stream showed great improvement in the performance of Wi-Fi and Wi-Fi 6 already supports 8SS, why home wireless routers only have 4 × 4 specification at most? Because terminals cannot fit higher spatial streams, most current smartphones or tablets have only 2 × 2 streams at most, and desktops or laptops are basically 2 × 2 too. Only a few high-end computers equip 3 × 3 or even 4 × 4 radio network cards. The reason is simple because the more antennas are, the greater the power consumption is, and the battery capacity of mobile devices is always a drawback.

3.2.5 *Wi-Fi data rate algorithm*

At present, the home Wi-Fi market directly uses the cumulative value of the theoretical rate of each frequency band to calibrate the Wi-Fi

Table 3.5 Wi-Fi (4/5/6) modulation, coding and rate comparison.

	Modulation		Single-stream rate (Mbps)			
Protocol (generation)	QAM	MCS	20 MHz	40 MHz	80 MHz	160 MHz
Wi-Fi 4	64	7	72	150		
	256*	9*	…	200		
	1024*	11*	…	250		
Wi-Fi 5	64	7	72	150	325	650
	256	9	87*	200	433	867
	1024*	11*	…	250	542	1083
Wi-Fi 6	64	7	86	172	360	720
	256	9	115	229	480	961
	1024	11	143	287	600	1201

Note: Light gray stands for non-standard specifications.

specifications. It is simple but rude and extremely imprecise. Since the cumulative value cannot represent the rate composition after decomposition, consumers cannot rely on the cumulative value to determine the level of equipment.

Let's discuss the single-stream rate of the recent three generations of standards under different modulation and bandwidth conditions, which is the familiar single-stream theoretical rate.

In fact, the calculation of Wi-Fi theoretical rate is much simpler than we think. All the key factors have been explained and given the range of values. We just need to select the corresponding values according to the Wi-Fi technical specifications (see Table 3.5) and substitute them into the formula to get the answer.

The theoretical rate of standard Wi-Fi device = the single stream theoretical rate of a standard Wi-Fi device × the spatial stream number of a standard Wi-Fi device.

Single stream theoretical rate = the number of bits carried by each subcarrier (the symbol bit length) × the number of subcarriers contained in an OFDM × the code rate/OFDM duration (the transmission period).

The number of bits carried by each subcarrier is determined by the modulation method in the MCS strategy table. 64QAM is 6 bits, 256QAM is 8 bits and 1024QAM is 10 bits.

The number of subcarriers, specifically the number of data subcarriers, is determined by the protocol and bandwidth. Wi-Fi 4, Wi-Fi 5 and Wi-Fi 6 have different frame structures, and the number of subcarriers is basically proportional to the bandwidth.

The coding rate is determined by the MCS strategy table and is related to the modulation mode. For high-order modulation (64/256/1024QAM), the coding rate is 5/6.

The OFDM duration is determined by the protocol. Wi-Fi 4 and Wi-Fi 5 are set as 3.6 (3.2 + 0.4) ms, and Wi-Fi 6 is set as 13.6 (12.8 + 0.8) ms.

Here are some examples to show how to calculate the standard Wi-Fi theoretical rate:

Example 1: Calculation of the theoretical rate of standard Wi-Fi 4

The theoretical rate of standard Wi-Fi 4 = the single-stream theoretical rate of standard Wi-Fi 4 × the spatial stream number of standard Wi-Fi 4.

The single-stream theoretical rate of standard Wi-Fi 4 = the number of bits carried by each subcarrier of standard Wi-Fi 4 (symbol bit length) × the number of subcarriers contained in one OFDM of standard Wi-Fi 4 × code rate/standard Wi-Fi 4 OFDM duration (transmission period).

Here, the number of bits carried by each subcarrier of standard Wi-Fi 4 (symbol bit length) = 6; the number of subcarriers contained in one OFDM of standard Wi-Fi 4 = 108; code rate = 5/6; standard Wi-Fi 4 OFDM duration (transmission period) = 3.6 (3.2 + 0.4) ms.

Substituting the previous formula, the theoretical single-stream rate of standard Wi-Fi 4 is 150 ((6 × 108 × 5/6) / 3.6)) Mbps.

From the above, the theoretical rate of standard Wi-Fi 4 = 600 (150 × 4) Mbps.

Example 2: Calculation of the theoretical rate of standard Wi-Fi 5

Theoretical rate of standard Wi-Fi 5 = the single-stream theoretical rate of standard Wi-Fi 5 × the spatial stream number of standard Wi-Fi 5.

The single-stream theoretical rate of standard Wi-Fi 5 = the number of bits carried by each subcarrier of standard Wi-Fi 5 (symbol bit length) × the maximum number of subcarriers in one OFDM of standard Wi-Fi 5 × code rate/standard Wi-Fi 5 OFDM duration (transmission period).

Here, the number of bits carried by each subcarrier of standard Wi-Fi 5 (symbol bit length) = 8; the number of subcarriers contained in one OFDM of standard Wi-Fi 5 = 468; code rate = 5/6; standard Wi-Fi 5 OFDM duration (transmission period) = 3.6 (3.2 + 0.4) us.

Substituting the previous formula, the theoretical single-stream rate of standard Wi-Fi 5 is 866.7 ((8 × 468 × 5/6) / 3.6) Mbps.

From the above, the theoretical rate of standard Wi-Fi 5 is 6933.3 (866.7 × 8) Mbps.

Example 3: Calculation of the theoretical rate of standard Wi-Fi 6

The theoretical rate of standard Wi-Fi 6 = the single-stream theoretical rate of standard Wi-Fi 6 × the spatial stream number of standard Wi-Fi 6.

Single-stream theoretical rate of standard Wi-Fi 6 = number of bits carried by each subcarrier of standard Wi-Fi 6 (symbol bit length) × number of subcarriers contained in one OFDM of standard Wi-Fi 6 × code rate/standard Wi-Fi 6 OFDM duration (transmission cycle).

Here, the number of bits carried by each subcarrier of standard Wi-Fi 6 (symbol bit length) = 10; the number of subcarriers contained in one OFDM of standard Wi-Fi 6 = 1960; code rate = 5/6; standard Wi-Fi 6 OFDM duration (transmission period) = 13.6(12.8 + 0.8) us.

Substituting the previous formula, the theoretical single-stream rate of standard Wi-Fi 6 is 1200.98 ((10 × 1960 × 5 / 6) / 13.6) Mbps.

From the above, the theoretical rate of standard Wi-Fi 6 is 9607.8 (1200.98 × 8) Mbps.

3.2.6 *OFDMA*

According to the calculation of Wi-Fi data rate above, the theoretical rate of Wi-Fi 6 is 9607.8 Mbps, while the theoretical rate of Wi-Fi 5 is 6933.3 Mbps. But the so-called new generation of Wi-Fi 6 is supposed to have more than just faster data rate.

From Table 3.6, in terms of specifications, Wi-Fi 6 has no obvious advantage over the popular non-standard Wi-Fi 5 and Wi-Fi 4.

In the 2.4 GHz frequency band, the single-stream bandwidth of non-standard Wi-Fi 4 can reach 250 Mbps, which is not much different from the 287 Mbps of Wi-Fi 6. Within a short time, the Wi-Fi terminal will have 4SS at most, which cannot reflect the difference of Wi-Fi 6 and Wi-Fi 4.

In the 5 GHz frequency band, non-standard Wi-Fi 5 has a single-stream bandwidth of 1083 Mbps, which is also close to the 1201 Mbps of

Table 3.6 Comparison of Wi-Fi (3/4/5) on frequency band, bandwidth and other parameters.

Frequency band	2.4 GHz		5 GHz	
Protocol	802.11n (Non-standard Wi-Fi 4)	802.11ax (Wi-Fi 6)	802.11ac (Non-standard Wi-Fi 5)	802.11ax (Wi-Fi 6)
Maximum bandwidth	40 MHz	40 MHz	160 MHz	160 MHz
MCS range	0–11	0–11	0–11	0–11
Highest modulation order	1024QAM	1024QAM	1024QAM	1024QAM
Single-stream rate	250 Mbps	287 Mbps	1083 Mbps	1201 Mbps
Maximum spatial stream	4 × 4	8 × 8	8 × 8	8 × 8
MU-MIMO	Not supported		Downlink only	Uplink/ Downlink
OFDMA	Not supported			Uplink/ Downlink

Wi-Fi 6, and both of them support 8SS. So, Wi-Fi 5 has a similar performance to Wi-Fi 6.

However, in terms of parallel transmission, Wi-Fi 5 is an incomplete version of MU-MIMO, while Wi-Fi 6 is a standard version of MU-MIMO.

Why do we need to introduce parallel transmission? This has something to do with the MIMO technology.

Strictly, the original MIMO is also called SU-MIMO. Although it supports simultaneous transmission of multiple antennas, the wireless router can only communicate with one terminal at the same time, that is, serial transmission.

Assuming that the router supports 4SS and three terminals are connected to channel 149 (5 GHz band), which are a 2 × 2 laptop, a 1 × 1 mobile phone A and a 1 × 1 mobile phone B. Then at a certain moment, the router can only choose one of these three to communicate. If a notebook is selected, other terminals will be queued to wait for the selected device to finish its communication. Even if a 2 × 2 notebook only

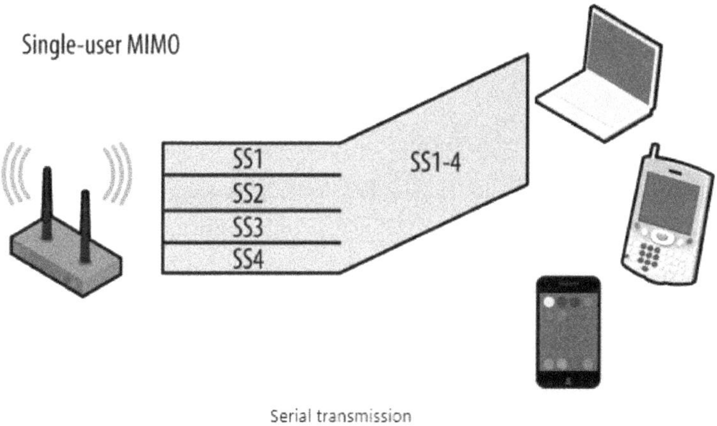

Figure 3.5 Single-user MIMO.

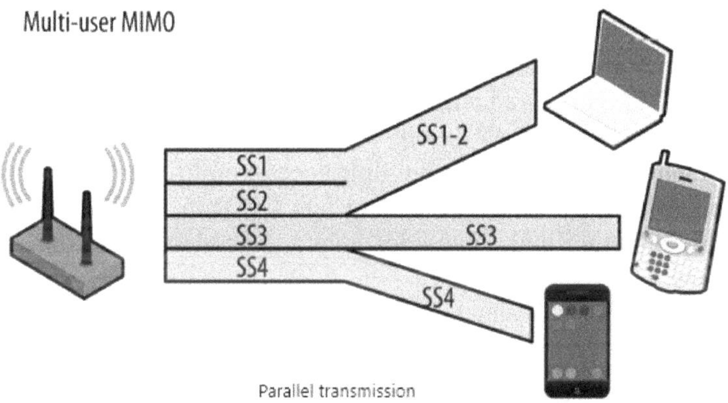

Figure 3.6 MU-MIMO.

occupies two of the 4SS, the remaining two cannot be allocated to connected to the two mobile phones (see Figure 3.5).

As a result, we need MU-MIMO. On the basis of SU-MIMO, it adds a multi-terminal synchronous transmission mechanism to improve the utilization of space streams. When MU-MIMO is supported, 4SS are just enough for the three terminals mentioned above to transmit at the same time without speed reduction, so that the high-end router finally reaches full efficiency (see Figure 3.6).

4 spatial streams	Terminal 1	Terminal 2	Terminal 3	Terminal 4	Terminal 5
Combination 1	1SS	1SS	1SS	1SS	
Combination 2	2SS	1SS	1SS		
Combination 3	2SS	2SS		MU-MIMO restricted	
Combination 4	3SS	1SS			
Combination 5	4SS				

Figure 3.7 Terminal space stream combination.

However, MU-MIMO is not perfect, its operating state is not stable enough, and it is easily affected by the terminal. For example, four spatial streams can only satisfy 4SS terminals for a perfectly running MU-MIMO, basically the following five combinations (see Figure 3.7), supporting at most four terminals. Once the number of terminals exceeds four, the following terminals will be queued. Once the total load exceeds 4SS, the transmit speed will be reduced.

For a long time, Wi-Fi has always adopted OFDM as the core transmission scheme. Wi-Fi 6 adds multiple access (multi-user) technology on the basis of OFDM, thereby evolving into OFDMA. OFDMA redesigns the frame structure and subdivides it into a number of Resource Units (RU) to serve multiple users.

Take a 20 MHz channel as an example. In the OFDM scheme (i.e., Wi-Fi 4/5), each frame consists of 52 data subcarriers. This group of subcarriers can only serve one terminal. If the data packet transmitted by the terminal is small (i.e., message), it can't be loaded with 52 subcarriers at all, and the other empty subcarriers cannot be allocated to other terminals.

In the OFDMA scheme (Wi-Fi 6), each frame is composed of 234 data subcarriers, but the second grouping is performed within the frame. All the 26 subcarriers are defined as an RU (Resource Unit), and each RU can serve one terminal. Then, each frame is divided into nine parts, which can serve nine users at the same time.

It's more intuitive to explain with trucks pulling goods. The OFDM scheme is to dispatch the vehicles according to the order, no matter how much the goods are. One truck is used to send the goods of one order, even if the truck is not full (see the upper part of Figure 3.8). While the

OFDMA scheme aggregates multiple orders and tries to fully utilize the truck (see the lower part of Figure 3.8), which greatly improves the transmit efficiency.

Readers might believe that OFDMA is similar to MU-MIMO, but they are quite different. Although both of them are parallel transmission solutions, they are neither an iterative relationship nor a competitive relationship, but a complementary relationship. Their technical principles are different, and the applicable scenarios are also different. They should be applied depending on the application type of the service.

OFDMA is suitable for parallel transmission of small data packets (see the upper part of Figure 3.9), which improves the channel utilization and transmission efficiency of a single spatial stream and reduces application delay and user queuing. The running state is stable and not easily affected by the terminal.

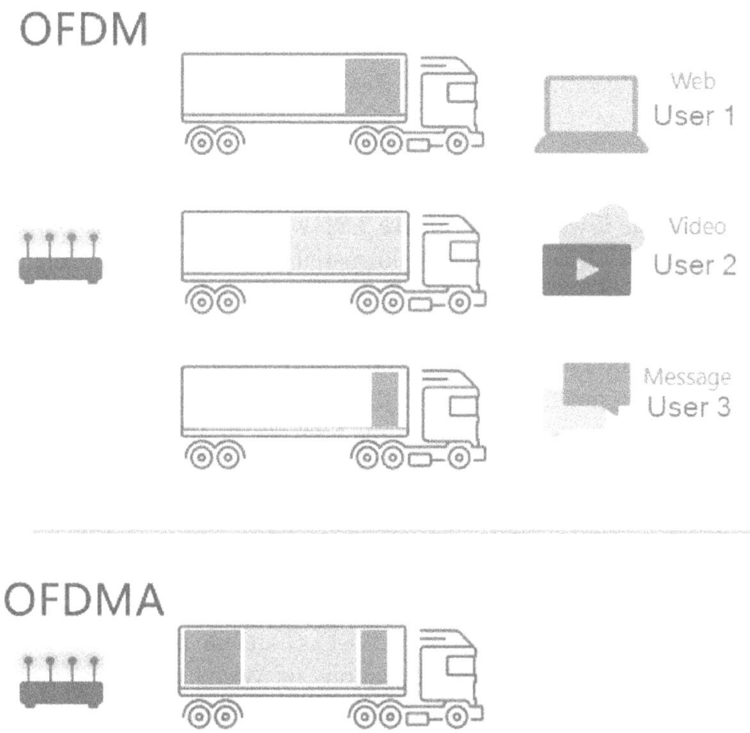

Figure 3.8 Comparison of OFDM and OFDMA.

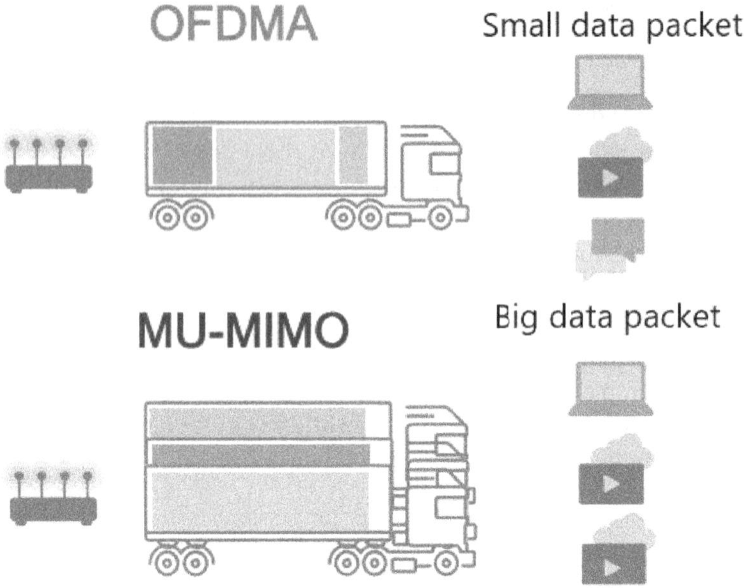

Figure 3.9 Comparison of OFDMA and MU-MIMO.

MU-MIMO is suitable for the parallel transmission of large data packets (see the lower part of Figure 3.9), which improves the utilization rate and system capacity of multiple spatial streams, increases the effective bandwidth of a single user and also reduces delay. The running state is not stable enough and is easily affected by the terminal.

However, the two solutions do not conflict and can even be superimposed. Users do not need to know the mechanism behind parallel transmission. They only need to see that the network is smooth no matter how many terminals are connected.

Unfortunately, both solutions require the support of Wi-Fi devices, and only when all terminals under the same channel support Wi-Fi 6, the running state of parallel transmission is perfect, otherwise the effect will be seriously compromised.

Of course, the new features of Wi-Fi 6 are far more than this. The development of the first five generations of Wi-Fi standards is mainly dedicated to the improvement of wireless bandwidth. When Wi-Fi bandwidth is equal to the wired network and reaches the GB level, the Wi-Fi standard begins to develop horizontally, and Wi-Fi 6 focuses on

Figure 3.10 Wi-Fi terminal application scenario.

improving multiple terminals for user experience. Especially in crowded areas, the end user experience of Wi-Fi 6 is even better.

In the following section, we talk about how to choose home Wi-Fi equipment. In a word, the terminal decides everything (Figure 3.10).

It is not difficult to see from the evolution of Wi-Fi that the real bottleneck is not the wireless router (or AP), but the user terminal. No matter how high the specifications of the router are, the performance depends on the terminals. In theory, high-end routers can indeed carry more terminals, but all terminals must support MU-MIMO or OFDMA first. Only when the router (or AP) and the terminal are all qualified, can the performance reach the best.

3.3 Home Wi-Fi Router Penetrates Wall and Networking Solution [18]

This section mainly discusses how the Wi-Fi signal penetrates the wall and how to design the family's radio network.

3.3.1 *The problem of Wi-Fi penetrating walls* [18]

Wi-Fi has become the third largest basic requirement after water and electricity. There is a classic question, "why my router, even with so many antennas and high wall penetration ability, cannot get signals a few steps away?"

3.3.2 *Wi-Fi protocol and spectrum*

Let's first talk about problems about Wi-Fi protocols and spectrum.

As mentioned in the previous sections, since the Wi-Fi standard was proposed by IEEE, after 20 years of development, one step at a time, one version after another, the download rate has been improved greatly since Wi-Fi 4 (802.11n). The latest version of Wi-Fi has reached the GB level.

It can also be seen from the discussions in the previous sections that currently Wi-Fi has two operating frequency bands: 2.4 GHz and 5 GHz. These two frequency bands are unlicensed frequency bands, but as long as they comply with national regulations, they can be used without authorization. In other words, installing a wireless router at home does not require anyone's approval.

3.3.2.1 *2.4 GHz frequency band*

The 2.4 GHz spectrum ranges from 2.4 GHz to 2.4835 GHz, with a total bandwidth of 83.5 Mpbs, divided into 13 channels, and each channel is 20 Mpbs wide.

However, 20 Mpbs multiplied by 13 is 260 Mpbs. These 13 channels require a total bandwidth of 260 Mpbs, while the 2.4 GHz spectrum is only 83.5 Mpbs. Why is that?

The answer is like getting on a crowded bus: "Squeeze, move back, and get on again!" There are many people who want to get on the bus but the bus is small. There is no personal space and being stepped on is inevitable.

It can be seen from Figure 3.11 that the 13 channels of the 2.4 GHz spectrum overlap closely, and interference is inevitable. Note that the 14th channel is not allowed in China.

To what extent does the channel overlap? It can be seen intuitively from Figure 3.12 that in these channels, only 1, 6, 11 or 2, 7, 12 or 3, 8, 13 don't overlap at all. Yes, we can see the congestion level of the 2.4 GHz band. It is like a very narrow road, but there are a lot of cars on it, and frequent traffic jams will inevitably cause a decrease in traffic speed.

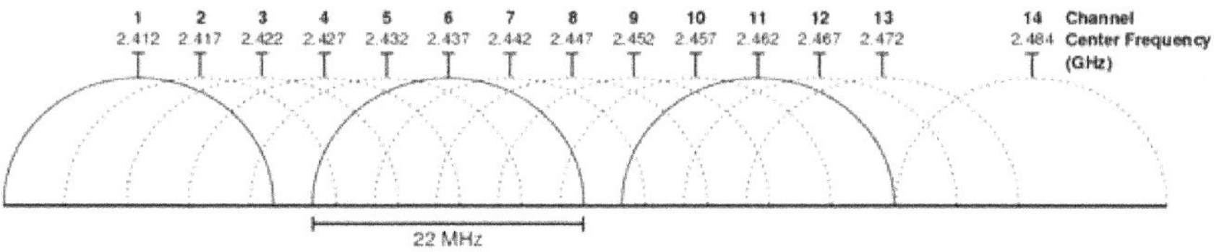

Figure 3.11 2.4 GHz spectrum and channel.

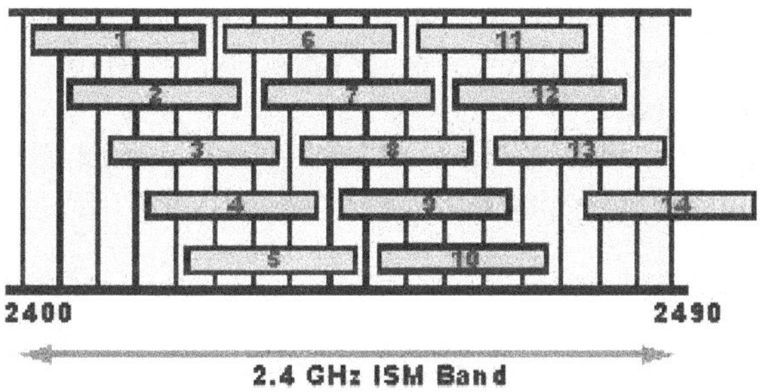

Figure 3.12 2.4 GHz non-overlapping channel distribution.

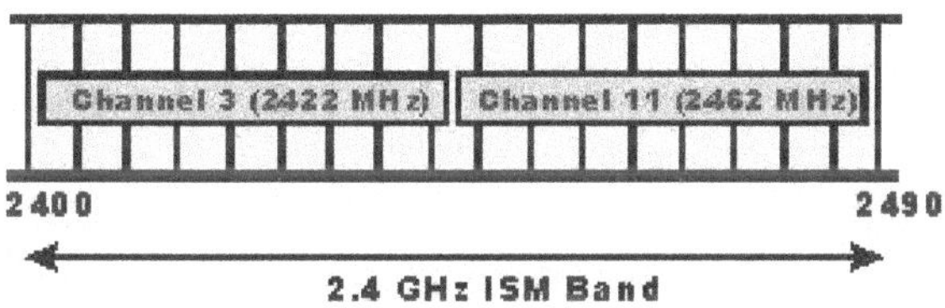

Figure 3.13 2.4 GHz 40 Mpbs bandwidth channel.

In Wi-Fi 4, users can use 40 Mpbs channels. But the 2.4 GHz band still has a total bandwidth of 83.5 Mpbs, which can only accommodate two channels (see Figure 3.13). Therefore, only in the dead of night when the network is idle, a single user can use the 40 Mpbs channel. Coupled with the interference from the next door home, the high-speed rate of Wi-Fi 4 is difficult to achieve to a large extent.

3.3.2.2 *5 GHz frequency band*

Since the 2.4 GHz frequency band is so crowded, IEEE experts have to look for another unlicensed frequency band which is 5 GHz.

Although Wi-Fi 4 can work on 5 GHz, this version of the protocol is old after all, and there are still a large number of mobiles, TVs and computers that do not support the 5 GHz frequency band. So, the actual products are still using 2.4 GHz, and the advantage of 5 GHz is reflected by Wi-Fi 5 (802.11ac).

The 5 GHz spectrum ranges from 4.910 GHz to 5.875 GHz and has a bandwidth of more than 900 Mpbs, which is more than 10 times of 2.4 GHz. Since this spectrum is very wide, the internal frequency is divided into several sub-bands, and some have not yet been allocated. Other few bands are reserved for other purposes. Even 4G LTE has taken some parts of this band.

The usage of this section of spectrum varies greatly among different countries. The parts that can be used in China are shown in Table 3.7.

Table 3.7 5 GHz spectrum available in China.

Channel number	Frequency (MHz)	Available in China
36	5180	Available
38	5190	Available
40	5200	Available
42	5210	Available
44	5220	Available
46	5230	Available
48	5240	Available
52	5260	Available, need DFS/TPC to avoid interference
54	5280	Available, need DFS/TPC to avoid interference
56	5300	Available, need DFS/TPC to avoid interference
58	5320	Available, need DFS/TPC to avoid interference
149	5745	Available
153	5765	Available
157	5785	Available
161	5805	Available
165	5825	Available

Note: (1) DFS stands for Dynamic Frequency Selection.
(2) TPC stands for Transmit Power Control.

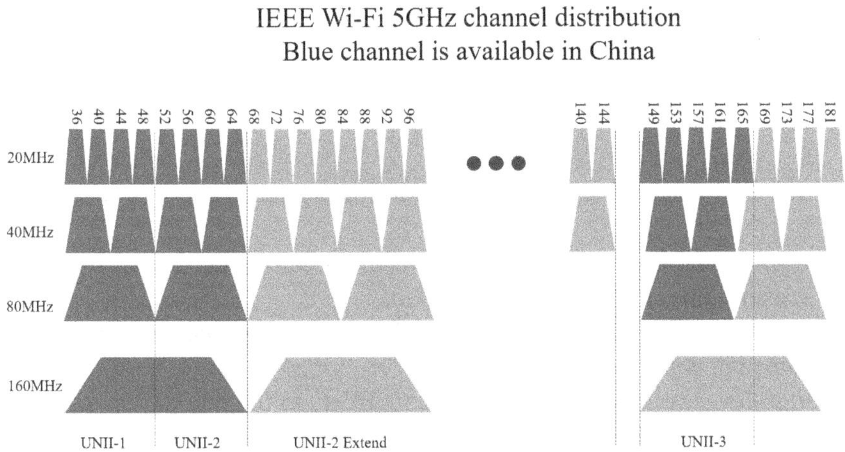

Figure 3.14 China's 5 GHz channel distribution.

Figure 3.14 is the distribution of China's 5 GHz available channels based on this table. Routers and mobile phones that support Wi-Fi 5 can freely combine these available channels and use 20 Mpbs, 40 Mpbs, 80 Mpbs or even 160 Mpbs bandwidth channels. Numerous channels, large bandwidth and low interference result in the download speed fast.

3.3.3 *Router's data rate estimation*

What download rates can Wi-Fi achieve in 2.4 GHz and 5 GHz?

For the 2.4 GHz band, we use Wi-Fi 4 to estimate. This version is also the latest protocol version that supports 2.4 GHz. For the 5 GHz band, we use the newer Wi-Fi 5 to estimate, and this protocol version only supports 5 GHz.

Wi-Fi 4 introduces the following features:

- It supports MIMO, that is, you can use up to four antennas to receive and transmit data, and the download rate is doubled. Therefore, there is no doubt that routers with multiple antennas can support at least Wi-Fi 4.
- On the basis of supporting standard 20 Mpbs bandwidth channel, a double bandwidth has been added, which is 40 Mbps. The theoretical peak download rate of a single user has been doubled again.
- Therefore, it is very simple to estimate the maximum download rate of Wi-Fi 4 by counting the number of antennas. One antenna with

40 Mpbs channels can reach a maximum rate of 150 Mbps, and two antennas reach 300 Mbps, and so on. Four antennas can reach 600 Mbps. While the maximum rate of one single antenna has been discussed in the previous section.

Wi-Fi 5 has been enhanced in the following features:

- *MIMO enhancement*: Up to eight antennas can be used to receive and transmit data, and the rate is doubled.
- *Channel expansion*: Supports 80 MHz–160 MHz channel bandwidth. Thus, the data rate is doubled too.
- *High-level modulation*: Supports 256QAM, so the rate is increased by 30%.

With the addition of these features, Wi-Fi 5's single antenna can reach 867 Mbps, two antennas can reach 1734 Mbps, three antennas can reach 2601 Mbps, four antennas can reach 3468 Mbps and eight antennas can reach 6933 Mbps. However, this is only a theoretical calculation. In fact, the number of antennas of the routers supporting Wi-Fi 5 is no more than Wi-Fi 4.

3.3.4 *The emergence of dual-band routers*

Why did the "dual-band router" turn out?

Since 5 GHz and Wi-Fi 5 is good enough, do future routers and mobile phones no longer need to support 2.4 GHz? Can we only use 5 GHz and Wi-Fi 5?

One obvious reason is compatibility. There is still a large number of old mobile phones, computers and TVs which only supported the 2.4 GHz. To prevent these old devices from being unable to access the Internet, 2.4 GHz support is necessary.

Another reason is the coverage. The 2.4 GHz frequency is low and the wavelength is large, so the signal loss is small and the penetration is strong. Compared with the 5 GHz frequency band, it has certain advantages such as coverage (see Figure 3.15). We will discuss this issue later.

Since both 2.4 GHz and 5 GHz have their advantages and disadvantages, we can combine 2.4 GHz and 5 GHz into a dual-band Wi-Fi. Thus, the dual-band router comes into being.

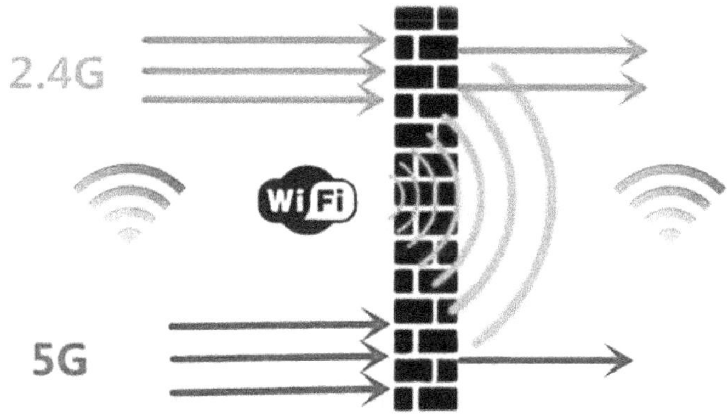

Figure 3.15 2.4 GHz penetration capability is stronger than 5 GHz.

Figure 3.16 Dual-band router.

Since 2.4 GHz and 5 GHz signal indicators are quite different, their antennas are generally independent. As is shown in Figure 3.16, they are separated to make sure that the antennas of the same frequency band are as far as possible to ensure the best performance.

To estimate the maximum download rate supported by the dual-band router, we still need to count the antennas. The difference is that the 2.4 GHz and 5 GHz rates must be calculated separately and then added together. In fact, router manufacturers have already calculated the rate for us and printed it on the package. But their calculation is not precise.

After discussing the frequency bands and data rate, the following section discusses the penetration ability, which is the signal loss of Wi-Fi.

3.3.5 *Why Wi-Fi signal is often not strong enough?*

This is because Wi-Fi signal has the following conditions after encountering obstacles during propagation (see Figure 3.17):

Refraction: When the signal passes through glass or water, the signal path is deflected. The antenna and the router are all in the air, so the effect of refraction can be ignored.
Reflection: When the signal passes through a smooth surface, the signal will reflect, similar to the reflection of light.
Diffraction: The signal will spread when passing through concrete walls. When it encounters an area that can be passed through, it will continue to propagate straight in the original direction.
Diffusion: When blocked, part of the signal will spread out and continue to extend up, down, left and right along the wall.
Penetration: Part of the energy of the signal is absorbed by obstacles, and the remaining energy continues to spread through the obstacles.

Under normal circumstances, in an indoor environment, the Wi-Fi signal received by our mobile phone is mainly a superposition of the four

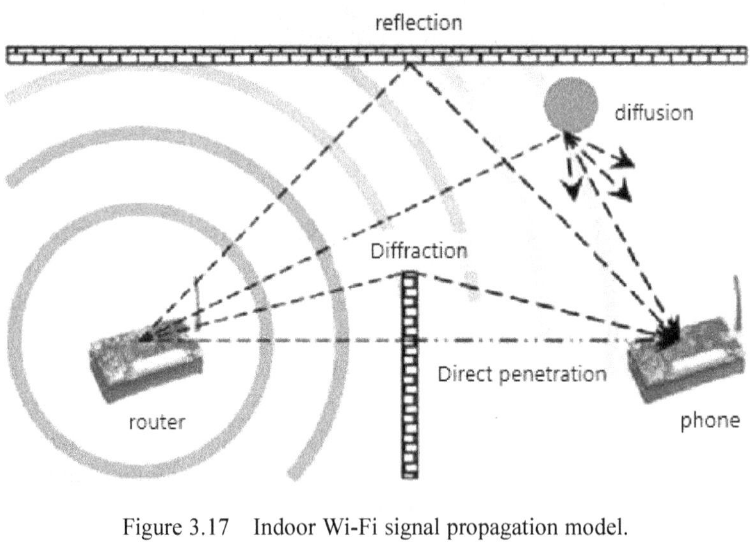

Figure 3.17 Indoor Wi-Fi signal propagation model.

effects of reflection, diffraction, diffusion and penetration. The latter three effects are closely related to the frequency of the signal. The higher the frequency is, the worse the diffraction and penetration ability will be.

Regardless of whether it is 2.4 GHz or 5 GHz frequency band, it belongs to the higher frequency band in wireless communication. Although the diffraction and penetration ability of 2.4 GHz is better than that of 5 GHz, it is still weak. See Table 3.8 for the empirical value of loss after 2.4 GHz signal penetrates the wall.

We usually use the unit of dB to express the signal's difference. We only need to understand the following relationships to have an intuitive understanding of these data:

- A reduction of 3 dB means that the signal strength is reduced to 1/2 (half) of the original.
- A 10 dB reduction means that the signal strength is reduced to 1/10 (one–tenth) of the original.
- A 30 dB reduction mcans that thc signal strength is reduced to 1/1000 (one–thousandth) of the original.

Combining the above table, we can draw the following conclusions:

- Obstacles such as wood and glass do not strongly block Wi-Fi signals. They only bring about half signal loss.
- Brick and concrete walls will reduce the signal to one–thousandth of the original.

Table 3.8 2.4 GHz band penetration loss.

Obstacle	Penetration loss
8 mm board	1–1.8 dB
38 mm board	1.5–3 dB
40 mm board	2–3 dB
12 mm glass	2–3 dB
25 cm concrete wall	20–30 dB
Brick wall	–15 dB
Floor block	–30 dB
Elevator block	20–40 dB

- Metal obstacles such as elevators have the greatest influence on Wi-Fi signals. They will reduce the signal to one–ten–thousandth of the original.
- Is there anything to do to help the weak Wi-Fi signal?

(i) *Is there a so-called "high-power router?"* It is easy to come up with the idea that if we want the signal to pass through the wall and have a strong penetration ability, we can simply increase the power. But in the case of Wi-Fi signals, it is impossible to simply increase the transmit power. Since the transmit power of wireless router is regulated by the state, the maximum power is 100 mW, which is 0.1 W. Products that do not meet this requirement cannot be put on the market. This regulation is made by China's Ministry of Industry and Information Technology for people's health. Therefore, it is impossible to improve the Wi-Fi signal by increasing the transmit power. The maximum power of routers is defined by each National Radio Regulatory Agency.

(ii) *Is "high-gain antenna" feasible?* Since the transmission power is limited, we can increase the power gain of antennas to make the energy transmission more concentrated and farther. This is possible in theory. However, the price of high gain and energy concentration is the narrowing of the beam, so that the coverage of Wi-Fi signals will be uneven, which means that in some directions the signal is strong but in other directions the signal is weak. If multiple people use it at different locations at the same time, they cannot all get the best performance (see Figure 3.18).

Moreover, the antenna gain of a general router is only 9 dBi, which cannot satisfy the penetration loss of tens of dB when penetrating.

3.3.6 *How can I improve the Wi-Fi signal?*

So how can we improve the Wi-Fi signal?

3.3.6.1 *Scene 1*

"I have only one router. I want to make the Wi-Fi signal better, but I don't want to spend any money. What should I do?"

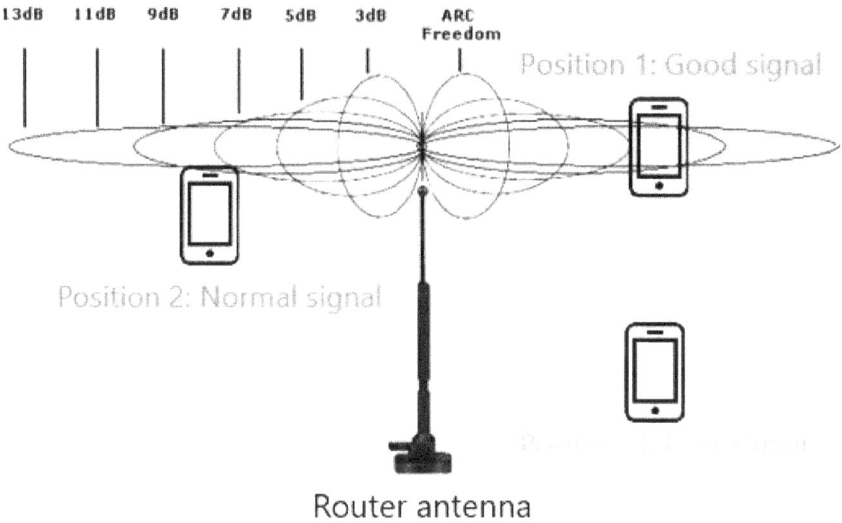

Figure 3.18 Antenna gain and radiation pattern.

Method 1: Optimize the placement of routers

Since the so-called "Wall Through King" and "Single Through Three Walls" routers do not exist at all, it is inevitable that a single router cannot completely cover every corner.

If you want to solve the problem at the lowest cost, please remember the following rules:

- The shorter the propagation distance is, the better the signal is. So, try to put the router in the center of the house.
- Wooden doors, glass doors, etc., have little influence on the signal. Try to make the signal go through doors rather than go through walls.
- The location of the weak current box is generally at the door, and the surrounding panels will also shield the wireless signal, so try not to put the wireless router in the weak current box.
- Place the router in a higher, open place and avoid obstacles such as refrigerators, washing machines, and various furniture.

Can you tell which position is better in Figure 3.19, position one or position two?

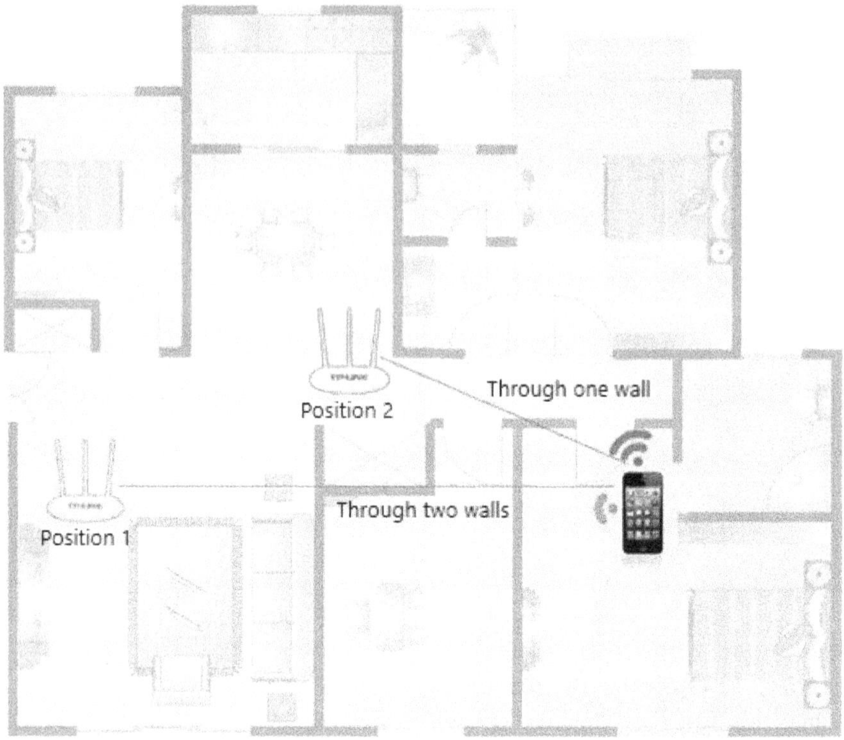

Figure 3.19 The influence of router placement difference on signal.

Please see Figure 3.20 and choose which position is better, position one or position two?

I believe the answers to these two questions are quite easy.

Method 2: Adjust to a reasonable antenna angle

General home routers use omnidirectional antennas. The wireless signal distribution of omnidirectional antennas is similar to that of a horizontally flattened tire, so the wireless signal coverage is better in the direction perpendicular to the antenna. As shown in Figure 3.21.

So, adjusting the antenna of the wireless router to be perpendicular to the ground and pointing it straight toward the roof can get the best performance. There is no need to adjust to other angles (see Figure 3.22).

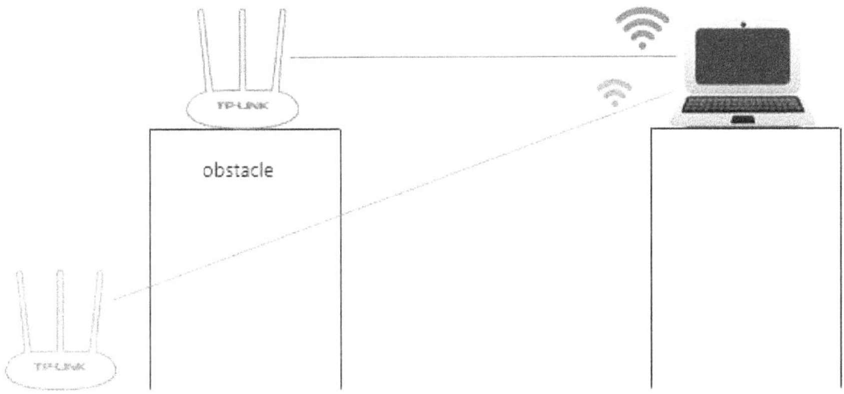

Figure 3.20 The influence of the height of the router on the signal.

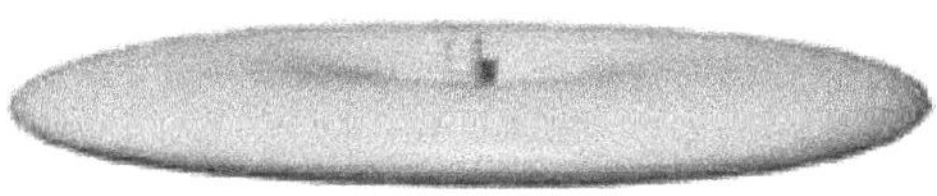

Figure 3.21 Radiation pattern of omnidirectional antenna.

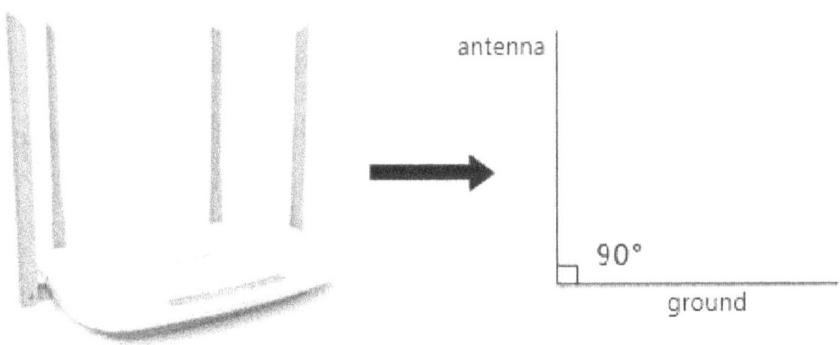

Figure 3.22 Suggested antenna angle adjustment for router.

Method 3: Avoid environmental interference

1. Use a dual-band router and use the 5 GHz frequency band. As shown above, the 2.4 GHz frequency band is very congested and naturally suffers more interference. The 5 GHz band has relatively less interference. Although sometimes it seems that the 5 GHz signal is a little weaker than 2.4 GHz, it may be faster.

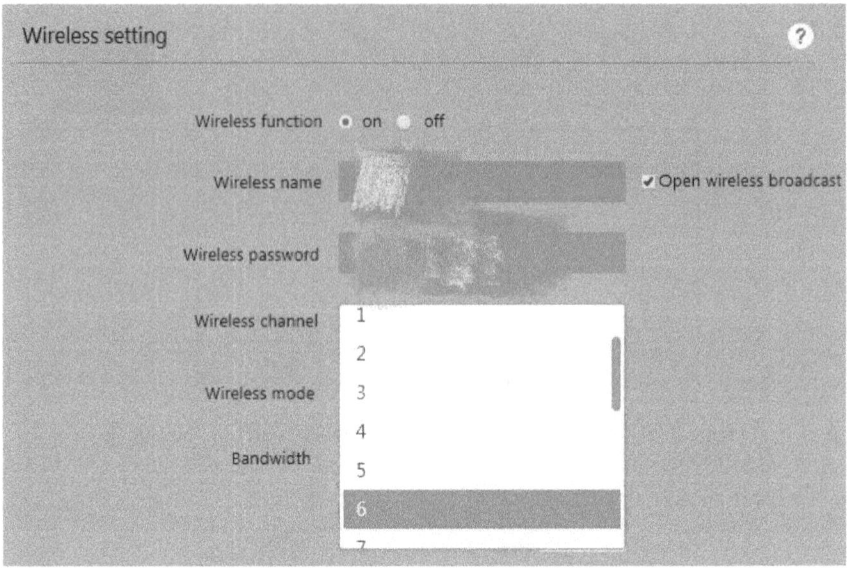

Figure 3.23 Modify 2.4 GHz channel settings.

2. Put the router as far away as possible from microwave ovens, Bluetooth devices, wireless mouse and keyboard and other devices. These things all work on 2.4 GHz, which causes interference to routers.
3. Change the channel configuration of the router. As mentioned before, there are only three independent non-overlapping channels on the 2.4 GHz frequency band, generally 1, 6 and 11. Try to change it and use channels which are better (see Figure 3.23).

3.3.6.2 *Scene 2*

"I'd like to spend more money for faster Internet connection, what should I do?"

We should discuss the home networking solution.

Solution 1: Use a wireless extender to expand and amplify wireless signals (see Figure 3.24).
This is the current mainstream solution for Wi-Fi signal extension for apartments and houses. Plugging an extender in the place which has a

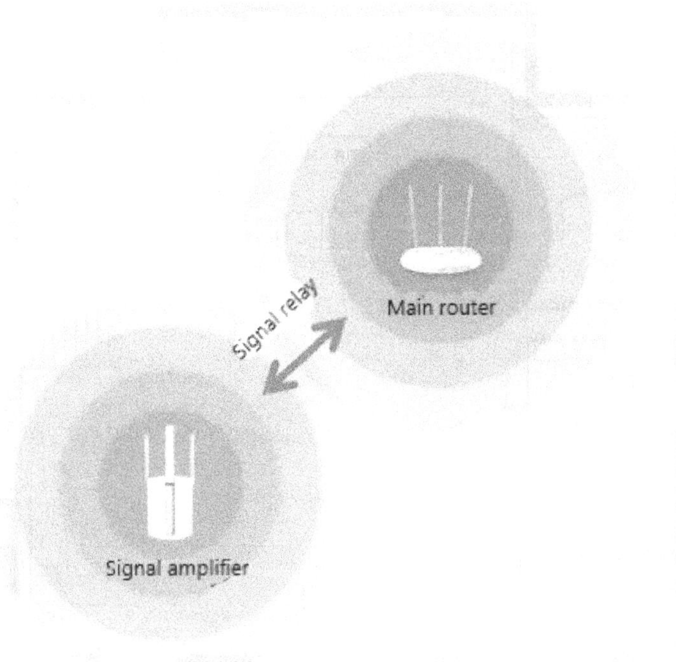

Figure 3.24 Wireless extender.

weak signal without using the network cable can increase the wireless signal. This solution is called signal relay. As the name suggests, it means that the signal amplifier receives the signal from the main router, amplifies it and then transmits it. Therefore, the signal at home is still the same, and our mobile devices can connect to the signal which has better quality.

Application: Applicable to any environment.

Solution 2: Use a wireless router to bridge and expand the signal (Figure 3.25).

This solution uses the wireless bridging function of the router to connect multiple wireless routers wirelessly, and there is no need for network cable connection between the routers.

After wireless bridging, the Wi-Fi signal realizes a relay and plays a role in amplification. Of course, the secondary router generally has four

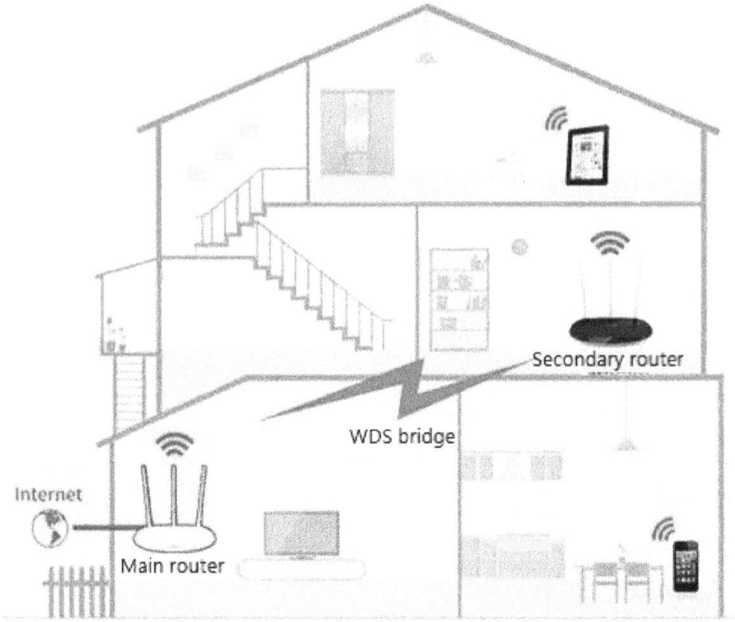

Figure 3.25 Bridging two routers.

network ports, which can be used to connect to wired computers, which is also one of the advantages of this solution.

So, what's the difference between Option 2 and Option 1? The difference is the two routers that are bridged actually work independently, and different Wi-Fi names and passwords can be set.

Application: Have already got multiple wireless routers and want to connect devices to the secondary router.

Solution 3: Use network cables to connect two (or more) routers (see Figure 3.26).
Use a network cable to connect multiple routers, among which the broadband connected router is the main router, and the others are the secondary routers. The primary and secondary routers are connected through their respective LAN ports. At this time, the secondary router is equivalent to a wireless switch. Through configuration, the entire network having only one signal and automatic roaming can be realized.

This solution requires that each router be connected by a network cable. The biggest advantage is good stability, there is no wireless loss

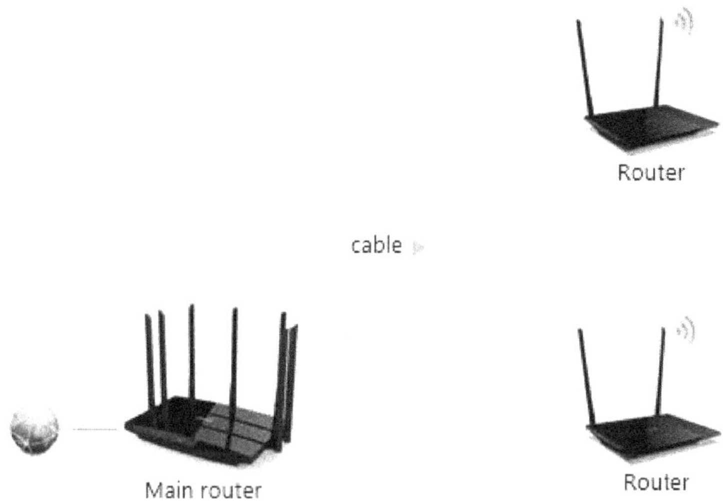

Figure 3.26 Wired connection to multiple routers.

between routers, and the speed of each router is very high. But there is also an obvious limitation, that is, all routers must pass through a cable connection.

Application: Network cables have already been laid.

Solution 4: Use HyFi package (child–mother router) to build a full coverage radio network (Figure 3.27).

HyFi (Hybrid Wi-Fi), also known as electric cat, means that the same product adopts both "wired" and "wireless" technologies, using the ubiquitous "wire" in the room as the backbone and "wireless" as the access. This product has line stability, reliability and wireless mobile convenience.

TP-Link's HyFi package consists of two parts: HyFi router and HyFi extender, both of which transmit data through the wire at home. After the router is configured, the wireless settings will be automatically pushed to the extender, thus forming a radio network with only one signal.

The HyFi wireless solution is convenient and fast. It requires no wiring for transmission through wires and has fast transmission speed and good stability. The wireless configuration is simple, only by configuring the router, the signal of the extender will change accordingly. Therefore, this solution is very extensible, and multiple extenders can be configured

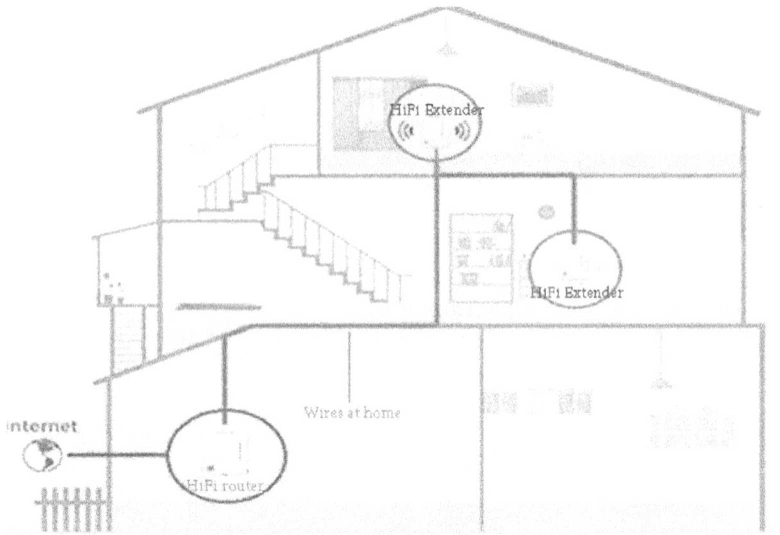

Figure 3.27 HyFi package transmitting signals through wires.

according to actual needs. The appearance is also very atmospheric, beautiful and simple.

Huawei's solution is also called the child-mother router. It is more straightforward compared to the name HyFi. It is a combination of router and electric cat, which can transmit signals through the wires at home.

Compared with the previous ones, the cost of this solution is relatively large, and the previous router needs to be replaced.

Application: Any home environment

Solution 5: Use AP/AC to extend wireless signal (see Figure 3.28).
If you prefer a clean and tidy environment and do not want to see the cables on the sockets and obtrusive router antennas, while all rooms must be covered by signals, then the AC plus wireless AP method should be used to build an Internet café. This can perfectly satisfy your wishes.

AP: The radio access point is simply a Wi-Fi signal transceiver, which is equivalent to a wireless router without routing and management functions.

AC: The access controller is used to manage multiple wireless APs. It is responsible for issuing configurations, modifying related

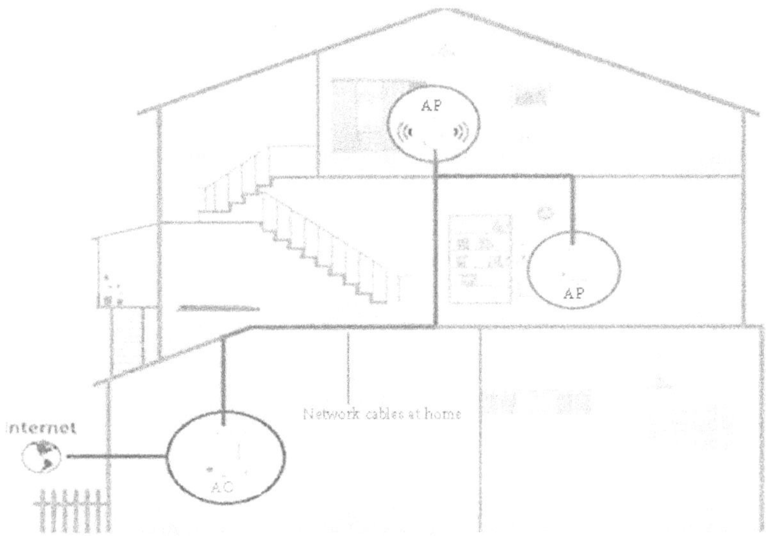

Figure 3.28 AC/AP management network.

Table 3.9 Comparison of various networking schemes.

Solutions	Extender	Wireless bridging	Cable	HyFi	AC/AP
Main router requirements	Any router	Any router	Any router	HyFi	Any router
Connect method	Wireless	Wireless	Cable	Electricity	Cable
Coverage	Good	Good	Best	Best	Best
Connect quantity	Normal	Normal	High	High	High
Scalability	Few	Few	Unlimited	Less than seven	According to AC
Stability and speed	Normal	Normal	Good	Good	Good
Reusability	High	High	High	Low	Low
Cost	Low	Low	Low	Medium	High
Aesthetics	Medium	Medium	Low	Medium	High

Note: Routers all refer to wireless routers, and reusability refers to the utilization of existing routers.

configuration parameters, radio frequency intelligent management and accessing security control.

The solution requires all APs and ACs to be connected in a wired manner, and all wireless APs are uniformly configured by the AC. The entire radio network has a unified signal and can realize wireless roaming.

The AP selected for this solution is a network cable panel that can be embedded in the wall, powered by AC through the network cable. It does not require an external power supply, so it does not take up space. But it's relatively more expensive.

Application: Large family type, high coverage requirements, clean and tidy, all network equipment are invisible.

Table 3.9 is a comparison of the various schemes.

3.4 Enterprise Wi-Fi Networking Scheme and Applications [19]

In the previous section, we introduced routers and home Wi-Fi networking solutions and finally proposed several solutions for different families to choose from, one of which is AC plus AP. In fact, this solution can also be applied to all types of enterprises. However, the configuration of AC and AP should be determined according to the specific situation of enterprises. Besides, small or micro enterprises with low security requirements can use other home Wi-Fi networking solutions mentioned above. Of course, enterprises should choose solutions according to its conditions.

3.4.1 *Overview of enterprise Wi-Fi networking and applications*

With the rapid development of enterprise informatization, wireless local area network technology is becoming more and more mature. Compared with wired networks, radio networks are simpler to install, more flexible to use, lower cost for wiring and has stronger ability to expand. The application scenario is becoming more and more diverse, and the service is becoming more and more complicated.

Next, we will introduce a Wi-Fi networking solution for enterprises with certain scale and requirements. Since enterprise conditions can vary from one to another, it is impossible to cover every type of enterprises.

Therefore, our solution is aimed at a common type of enterprise. For example, an enterprise with offices, storages, dormitories and production lines. We referred to this type of enterprise as enterprise A.

Enterprise A conducts intelligent construction for the entire facilities. While improving the original wired network, it needs to use the radio network to carry out the business requirements of offices, storages, production lines, dormitories and other scenarios. At the same time, the entire radio network is required to be safe, easy to manage, fast and reliable.

If enterprise A is a company that is neither an ICT company nor a Wi-Fi development company, the best way is to find a Wi-Fi product supplier (or Wi-Fi product agent) to complete the Wi-Fi networking and application project. Otherwise the company's IT engineers must do the system integration work by themselves. Although it can save some money for the first time, it is difficult to achieve perfection in various aspects and later maintenance and upgrades can lead to bigger costs. Therefore, we suggest introducing a supplier B to build enterprise A's Wi-Fi networks and related applications.

The way to find suppliers is complicated but also simple, which is bidding. Firstly, we need to determine the requirements and principles of network construction and its applications and pass the requirements and principles to relevant suppliers where they can provide solutions and related test products. Only through the feasibility evaluation of the scheme and comprehensive testing of its products, the supplier can enter the bidding list. After meeting the needs of enterprise A, the supplier can enter the bidding and finally determine the supplier with the highest cost performance. Of course, you can also negotiate with suppliers in accordance with pre-determined bidding rules to reduce costs. The following networking scheme introduction is based on the Wi-Fi products of supplier B to complete the establishment of the Wi-Fi network and its applications of enterprise A.

3.4.2 *Wi-Fi construction requirements*

The Wi-Fi construction of enterprise A requires that the radio network can be used in specific scenarios such as offices, storages, production lines and dormitories, etc., and can carry out applications such as wireless office, mobile code scanning, production line system and employee Internet access. The radio network is required to provide radio access and being able to be secure, easy to manage, fast access, stable signal, and reliable wireless, etc.

(i) *Security requirements*: It is required that the construction of a radio network can provide secure access authentication methods for different scenarios and provide different levels of security protection for employees, users and devices.
(ii) *Stability requirements*: The entire radio network is required to have good stability and to meet the high concurrency requirements of office and dormitory scenarios so as to prevent wireless jams, dropped connections, data interruption, etc. In storage and production line scenarios, the equipment is required to provide good terminal roaming and prevent the loss of data package.
(iii) *Reliability requirements*: Reliability is the overall requirement for the entire radio network. The radio network is required to have a redundant and recovery function to prevent the overall radio network from being unavailable due to equipment failures and therefore affecting the entire business.
(iv) *Easy management requirements*: Radio network management is required to be convenient and easy to operate, so that network managers can easily and promptly get information such as network topology, network performance statistics, network failures, etc., and can easily configure and adjust the network to ensure that the network is working in good condition.
(v) *Wiring and installation requirements*: In factory area, the network cables are laid in place by hidden wires or pipes. APs can be ceiling-mounted or wall-mounted and can be installed using the installation attachments that come with the equipment. The APs are powered by POE (Power over Ethernet) switches.
(vi) *Others*: In addition to the abstract requirements mentioned before, specific requirements such as the number of personnel and terminals in each scenario, the size of the venue and the layout are also required, so that suppliers can make detailed plans. In addition, the after-sales service needs to be clarified.

Combining the above requirements of enterprise A radio networks and the characteristics of specific products of suppliers, they can develop a set of overall enterprise smart factory Wi-Fi network solutions for enterprise A.

3.4.3 *Principles of network construction*

According to the specific situation of the network, the following principles in network design should be followed:

(1) *Advanced principle*: Adopt advanced design ideas and advanced network equipment to keep the network technologically advanced for a certain period in the future.

(2) *Openness principle*: Network design and network equipment selection should comply with international standards and industrial standards, so that the network has a high level of openness and the technical compatibility of provided equipment.

(3) *Scalability principle*: While fully considering the current situations, the network design should consider the needs of development in a relatively long period of time in the future, leaving sufficient possibilities for upgrading and expansion. Make full use of existing communication resources to provide sufficient spaces for future expansion to higher data rate. On the other hand, there must be sufficient spaces for network expansion.

(4) *Security principles*: The design of the network system must implement the security principle to prevent various damages from inside or outside the network. The implementation of security principle is reflected in the following aspects:

   (i) The equipment adopts spread spectrum technology.
   (ii) Provides encryption for wireless data transmission.
   (iii) Users can achieve higher security by setting their own APs or adding independent encryption devices.
   (iv) Security measures such as authorization, authentication, control, audit, etc., should be adopted within the network in order to prevent illegal access and damage to the network resources by users inside the network.

(5) *Reliability principle*: The design of the network system must implement the principle of reliability to make the network system highly available. The principle of reliability is reflected in the following aspects:

   (i) Selects network equipment with advanced technology, maturity and high reliability.

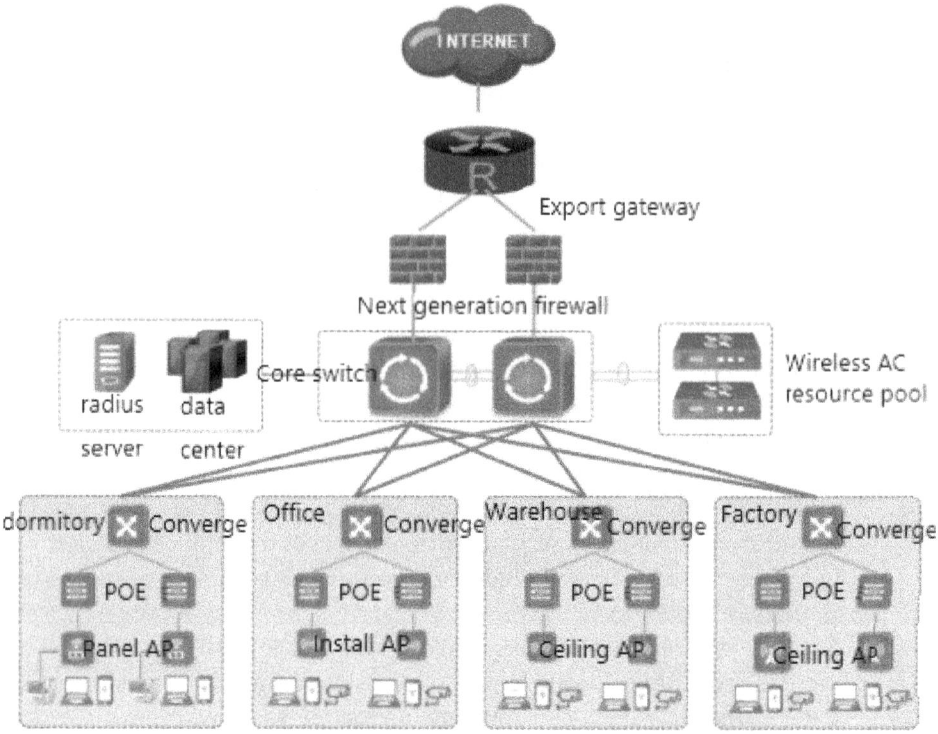

Figure 3.29 Enterprise Wi-Fi networking architecture.

(ii) The system has strong redundancy and fault tolerance.
(iii) The maintainability of the network should be good.

(6) *The principle of manageability*: The network system should have good manageability, so that network managers can easily and promptly grasp information such as network topology, network performance statistics, network failures, etc. It should also be possible to easily configure and adjust the network to ensure that the network is working in good condition.

3.4.4 *Network architecture*

According to user's radio network requirements and combining technical characteristics of specific Wi-Fi products, the design should be based on the AP plus AC structured Wi-Fi network architecture (see Figure 3.29) in order to meet the user's need of building a high-speed, stable, safe, reliable and easy-to-manage radio access network.

The entire network architecture adopts thin AP network architecture. On the basis of not changing the original network links, the wireless AC is side-mounted on the core switch of the computer room, and through the LAN link (connect via optical fiber lines when the distance between the computer room and other areas is too far) the wireless APs in each area are managed uniformly. Besides, firewalls are deployed at the network exit to ensure the network security of the entire enterprise network exit links.

3.4.5 *Scenario-based solutions*

3.4.5.1 *Wireless office scenario*

3.4.5.1.1 Secure access authentication

For companies which use radio networks, the authentication for visitors and the authentication for employees should be introduced, so that visitors are isolated and prohibited from accessing the internal network through isolation methods.

(1) *Employee access certification*: Supports internal user identity authentication via authentication servers. You only need to configure the information on the configuration page to communicate with Lightweight Directory Access Protocol (LDAP), Active Directory (AD) domain, Remote Authentication Dial in User Service (RADIUS) and other enterprise internal user identity databases to perform fast identity verification, which is both safe and reliable. Enterprise authentication supports local internal database, and the local database supports authentication on AC, which can satisfy small- and medium-sized enterprises which don't have internal identity authentication servers. When connecting to the radio network for the first time, the user name and terminal can be automatically bound to help complete identity binding quickly. If the user has multiple Internet terminals, the administrator can also flexibly approve the binding of subsequent newly added terminals manually. Therefore, enterprises can divide different access permissions according to user's organizational structure to avoid unauthorized access.

(i) *802.1X authentication*:
Products support 802.1X WEP authentication and provide enterprise-level security authentication for enterprises. While ensuring authentication security, in some special scenarios, such as the case of no RADIUS server, 802.1x WEP authentication can directly connect to the AD domain to use user names directly for authentication, saving the cost of deploying RADIUS server. At the same time, when deploying radio networks which are based on certificate authentication (EAP-TLS, Extensible Authentication Protocol-Transport Layer Security) or username and password authentication (PEAP-MSCHAPv2, Protected Extensible Authentication Protocol-Microsoft Challenge Handshake Authentication Protocol version 2), the radio network configuration methods are inconsistent as the types of connected terminals are numerous. For convenient deployment, 802.1X authentication can conduct one-key configuration to ensure both security and convenience.

(ii) *YYY certification*:
YYY certification is supplier B's first proposal in industry, which combines mobile OA software platforms such as Pocket Assistant, DingTalk, WeChat Enterprise Account and employee Internet access management for certification to achieve enterprise authentication for Pocket Assistant, DingTalk and WeChat. This authentication here can be simply understood as account and password authentication. This can help enterprises save manpower and resources to build other systems to manage employees' online information and deploy and maintain authentication servers and can use these mobile platforms to uniformly manage employees' online information and achieve access authentication.

(a) *Visitor authentication*:
- *QR code audit and certification*:

When suppliers, customers, business partners and related leaders come to visit the enterprise, they all need convenient access to the radio network. Enterprises should provide a more concise authentication method. Visitors only need to connect to the company's radio network and then open the browser to automatically display a QR code. The internal reception staff can scan with their own terminals to pass the authentication (see Figure 3.30). The QR code authentication technology is certified by the Ministry of Public Security and can be traced to the visitors' behavior.

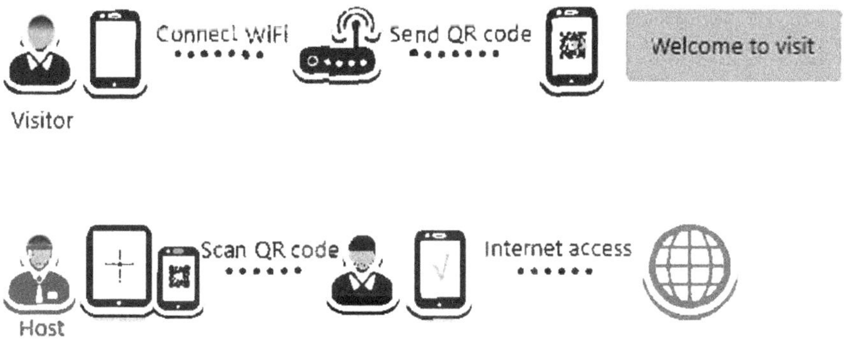

Figure 3.30 Visitor authentication.

• *WeChat smart authentication*:

Through WeChat's authentication function, visitors can be directed to the designated prompt page and be prompted to follow the WeChat account and gain access to the Internet when entering the enterprise and accessing the radio network. The visitor can access the Internet after following the WeChat account, which greatly draws the attention of visitors and is also convenient for enterprise advertising, business push and publicity.

3.4.5.1.2 Wi-Fi intelligent attendance

Compared with traditional attendance systems such as fingerprint attendance, clock-in attendance and RFID attendance, the use of office OA software is more accurate and user-friendly, and there is no need to queue for attendance. Pocket Assistant will remind you to go to work on time every day for attendance, and employees only need to open their software and Pocket Assistant can automatically identify and conduct attendance (see Figure 3.31).

(1) Click the attendance button of the APP, and APP will transmit the terminal Wi-Fi connection information (connected SSID, connected AP location name, AP MAC address, terminal MAC address, etc.) to the APP server.
(2) Pocket Assistant will check the terminal Wi-Fi connection information in conjunction with the wireless AC. If the check is successful, it will return the message "Sign in successfully" to the terminal. Otherwise, it will return "Please connect to the company Wi-Fi to sign in".

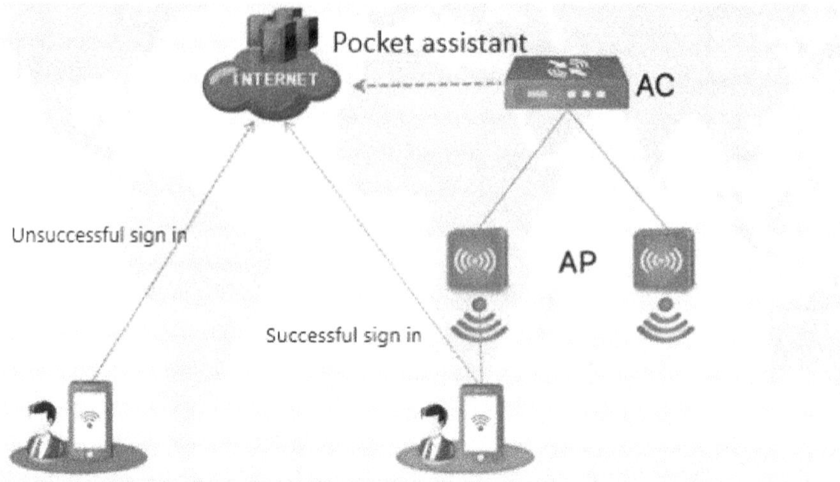

Figure 3.31 Smart attendance.

3.4.5.1.3 Internal information security assurance

Due to the openness of radio network and the diversity of Bring Your Own Device (BYOD) accessed terminals, the legality and security of terminals connected to the radio network cannot be fully guaranteed. For example, due to the employees' own devices diversity, especially the openness of Android terminals, these devices are easier to be attacked and be embedded with Trojan horses and hacker tools. Such terminals may cause internal business information to leak when connected to the internal network. Although the access control policy can be used to make permission partition for accessed terminals, a more robust solution should be used to filter out most attacks on the access level.

(a) *Secure access for binding terminals*:

- *Terminal MAC address binding*:

The user can directly bind the MAC address of the terminal by entering user name and password when accessing the radio network for the first time. Administrators do not need to intervene, which not only guarantees the credibility of the access device but also reduces the burden on administrators. If the user name and password are stolen, and the MAC address of the illegal terminal is bound, the user can find such an error during normal access and the administrator can promptly make the illegal terminal offline and add it to the blacklist and manually bind the MAC address of the legal terminal address. Thus, hackers can no longer log in with the stolen user name and password, since the MAC address is incorrect.

- *Security certificate authentication*:

Security certificate authentication provides a safer certificate authentication method, and a built-in CA certificate server does not require the support of a third-party server. Enterprises distribute certificates to trusted mobile terminals (mobile phones, PADs, laptops) via email or mobile storage devices. After users install the certificate, they need to perform encrypted two-way certificate authentication with AC to pass the verification. In this way, even if the user name and password are leaked, hackers are still unable to access the radio network because they do not have the certificate, resulting in higher security.

- *MAC address whitelist*:

For companies which distribute smart mobile terminals, the MAC addresses of terminals can be collected in batches to form a whitelist, and only terminals in this whitelist can access the radio network to prevent illegal terminal access.

- o *VLAN pool*: In the environment of radio network access, mobile terminals can communicate with each other, which poses great security risks. Transferring a large number of files between terminals consumes AP's limited bandwidth resources and reduces radio network performance. Arbitrary mutual visits between terminals could lead to data being stolen and even the risk of a virus being spread.

Therefore, when deploying a radio network, enabling the user isolation option in VLAN can prohibit communication between users in the same VLAN to reduce the broadcast messages between wireless terminals in the same VLAN, which improves radio network performance and security. At the same time, it avoids the risk of virus spreading and ensures safety and efficiency of the user's wireless experience to the greatest extent.

- o *RF control strategy*: In normal conditions, the warehouses of enterprises and factories are not working in the evening. Then the radio network can automatically turn off the radio frequency signal to save electricity and prevent illegal users from invading the network at night, which ensures the security of wireless data. In addition, to be more user-friendly, an additional SSID-based radio network is added and can be turned off selectively.

3.4.5.1.4 Internet behavior control

Network applications are extremely abundant, especially with the emergence of social network applications. But bringing personal network behaviors into the office can cause various management and security issues.

- *Employee Internet access management*: Comprehensive identification helps administrators to thoroughly understand the status of network conditions and user behaviors to ensure management effects. A third-party application recognition library can be used. AC can accurately identify more than thousands of mainstream Internet applications. The third-party application recognition library is regularly maintained and updated by a professional R&D team to ensure that the library is in the latest state, which improves the accuracy of application recognition. In addition to the accurate application identification library, a rich and flexible control strategy should also be applied. Enterprises can control employees' Internet access permission for different users, locations and time periods. Through application-based access control, flexible employees' Internet access control can be realized. Each employee can only access authorized systems to prevent illegal and unauthorized access. Compared with traditional ACL access control strategy based on MAC, IP address or port, this strategy is more accurate and powerful (see Figure 3.32).
- *Allow access to specified websites*: Network restrictions are often imposed on the R&D department and employees are not allowed to access the Internet. However, in many cases searching for information on the Internet is required. At this time, you can configure the global exclusion address in the access control policy, such as setting the domain "baidu.com", which allows the R&D group to search data via Baidu. It can be used in a variety of wireless deployments. Users are not allowed to access the Internet before authentication when accessing radio networks. However, the official website can be configured to be

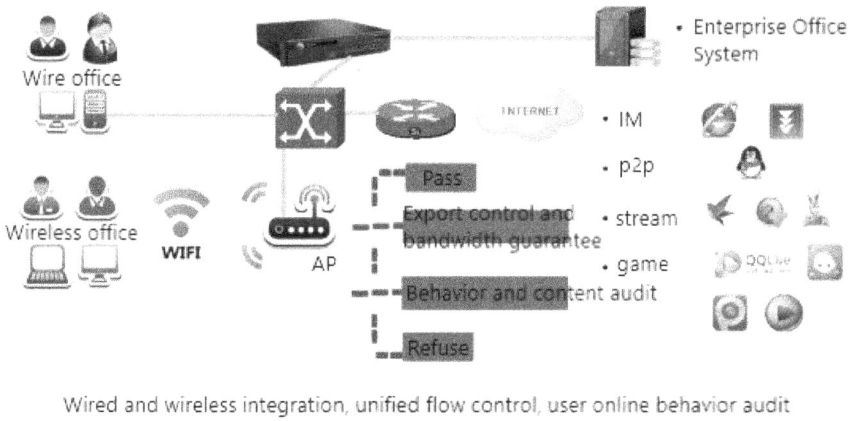

Figure 3.32 Internet behavior control.

an exclusion website to allow users to browse the official website or a certain website.

- *Employee Internet behavior audit*: The Internet access audit function supports the audit of the network behavior and content of wired and wireless users, including but not limited to HTTP outgoing content, visited websites and downloads, email, FTP, TELNET, identified or unidentified network applications, web page content, ACL denial behavior, and data traffic and duration of users. By configuring the audit policy, referencing it to a certain role and assigning corresponding roles to users, user audit is realized.
- *Data center report function*: Complete user access and authentication for enterprise wired and radio networks, audit user network behavior and content and store these results in the data center. The data center report function supports two retention methods: built-in data center and external data center.

3.4.5.1.5 Flow control/network optimization

- *Borrow free bandwidth*: When channels with different priorities compete with each other, the channels with higher priority can get all the free bandwidth. Same priority channels are guaranteed based on the rate of their bandwidth.

- *Intelligent bandwidth allocation*: Bandwidth is allocated between users who have data traffic passing through the device at a certain moment. If the user has no traffic at this moment, it does not participate in the allocation. Bandwidth is gradually allocated to users who have traffic passing through at this moment. If a user's bandwidth demand has been met, it will be ignored and continue to allocate to other users in need, until the bandwidth is allocated.
- *Flow control*: There are three ways of flow control strategy:

  (1) User-based traffic speed limit strategy
  (2) WLAN-based bandwidth guarantee
  (3) Application-based bandwidth guarantee

As the name implies, by prohibiting the use of applications that reduce work efficiency and consume high bandwidth, the working efficiency and network access experience can be improved.

In order to ensure that Internet traffic related to important business can take priority under limited bandwidth resources, flow control based on user, terminal type and access AP location is conducted to restrict the uplink and downlink bandwidth.

3.4.5.1.6 Wired and wireless integration

In addition to wired networks, enterprise A could still have many wired devices that need to connect to the Internet. The wired and wireless integration system can be configured through AC to manage wireless and wired users together, complete flow control, management and audit, and the ports of AC can be used to complete authentication of wired users. By configuring authentication policies for physical interfaces, aggregate interfaces and VLAN interfaces, filter conditions including IP groups and MAC addresses can be set to complete the authentication control of wired users.

3.4.5.1.7 Smart security network card

The smart security network card is suitable for office networks in enterprises, governments and financial industries that have strict network security requirements. It is suitable for laptops and desktops. The smart security network card has a built-in certificate issuing center. After inserting the card into laptops or desktops, the driver is installed, that is, the card writes the digital certificate into the PC, and users can enter their Personal Identification Number (PIN) code to directly connect to the company's radio network.

3.4.5.2 *Wireless warehouse and production workshop scenario*

In traditional factory warehouse and production lines, some mature technologies such as Bluetooth, Zigbee, RFID, etc., all have the drawback of data transmission lagging. Therefore, in modern smart factories and under the scenario of real-time information transmission and business diversity, Wi-Fi is bound to rise and will become an irreplaceable part of the construction of smart factories. The Wi-Fi of smart factories can not only provide radio network for workers in the factory to realize personnel positioning, attendance and other functions but also can be applied to the factory's mobile code scanning devices, wireless cameras, mobile forklifts, tag readers, Automated Guided Vehicle (AGVs) or Manufacturing Execution System (MES) to improve factory production efficiency.

3.4.5.2.1 Device access security

- *Terminal access recognition technology*: Terminal access recognition technology is based on powerful terminal type recognition. Through Organizationally Unique Identifier (OUI), DNS fingerprint characteristics and other technologies, the type of device can be identified once it is connected without installing third-party applications. In order to ensure the stability and data security of the radio network used by mobile scanners, forklifts, monitoring and other devices in the storage, the terminal access authority can be set to prohibit mobile phones, PC, Pad and other terminals to connect to the network after the device type is identified to ensure the effect of storage radio network usage.
- *Radio air interface encryption technology*: Wireless data is transmitted in the air and carries various data of the enterprise. An efficient and reliable encryption mechanism is required to prevent data from being cracked or tampered. The radio air interface encryption technology supports multiple data encryption methods of international standards, ensuring that the business data of the enterprise is transmitted safely and reliably.

- *IoT security certification*: IoT is regarded as the third revolution of the information industry after computer and Internet. Behind smart homes, Internet of Vehicles, artificial intelligence and others is the acceleration and maturity of IoT. The arrival of IoT era is beyond no doubt. But the U.S. network disconnection incident that shocked the world was only caused by hackers using web cameras and related Digital Video Recorder (DVR) to launch the IoT DDoS attack. For a time, well-known American websites such as Twitter, PayPal, Spotify, Netflix, Airbnb, GitHub, Reddit and the New York Times were not available, and the screens were full of "404 errors".

For many connected mobile scanners, monitoring, tag readers and other devices, some high-end mobile terminal devices support PSK (Pre-Shared Key) authentication. But because the PSK password is relatively simple and all devices use the same password, once a password is cracked, the entire IoT network can be easily controlled by hackers, causing serious problems for data security.

(1) *IoT secure access authentication technology*:
Since IoT devices such as monitoring, code scanners, etc., use one password to access the network, it means that hackers can invade any terminal and use the same method to crack other devices, which brings danger to the entire network.

IoT security access authentication configures different Internet passwords according to the MAC address of each terminal to ensure the security of each IoT terminal and prevent hackers from cracking a single terminal device password to control the entire network terminal devices.

(2) *IoT automatic classification management*:
A large number of IoT devices are connected to the network, such as cameras, sensors, code scanners, collectors, etc. The connected network is not grouped, and different security devices adopt a unified management policy, which makes the management of IoT devices complicated showing a lack of grouping and more accurate management.

IoT automatic classification management can identify the type of terminals through MAC and DHCP. By grouping and issuing different strategies to different devices, it can carry out accurate management.

(3) *Security of IoT cloud platform*:
Using cloud platform for account isolation, it can be ensured that hacking a single account will not cause the leakage of other account information. At the same time, adopting database backup and HTTPS encrypted access will prevent hackers from exhaustive cracking.

- *Illegal hotspot countermeasure*: In addition to powerful behavior management and application control, wireless invasion detection system (WIDS) and wireless invasion prevention system (WIPS) should also be included for monitoring the operating status of the network and system through software and hardware and discovering as much attacks as possible to ensure the confidentiality, integrity and availability of the network. Using WIPS technology, illegal access points can be detected and countered. It can also launch a regular scan and detect wireless signals with certain characteristics, such as specific SSID (same or similar to the original network), MAC (physical address) of illegal APs or signals emitted by non-wireless AC managed APs. The detected malicious Wi-Fi can be countered after it is detected, and the wireless terminal is guided not to access the malicious Wi-Fi.

3.4.5.2.2 Mobile zero-roaming networking with the same frequency

In traditional warehouse scenarios, the biggest drawback of using radio network data backhaul is that there is a situation of packet loss and data loss caused by terminal devices roaming between APs. A better solution is to use the zero-roaming networking with the same frequency.

In zero-roaming network, scanning will not be interrupted while moving. Mobile code scanning is a typical mobile terminal application. In traditional wireless solutions, the moving between multiple APs will cause packet loss due to roaming problems, causing network freezing and interruptions. Through optimization, virtualized AP is used to merge all APs into one AP. Different APs use the same channel and use a special algorithm to solve the problem of same frequency interference, so that no matter how the connected terminal is moving, the signal is never lost and no repeat authentication is needed, achieving roaming without disconnection and a smoother factory business.

3.4.5.2.3 Forklift wireless monitoring

(1) *Networking*: Compared with wired monitoring, wireless monitoring has the characteristics of easy installation, no wiring and flexible deployment. In a radio network scenario, an integrated solution of wireless cameras and radio network coverage can be used. The scenario of warehouse wireless monitoring is mainly applied on mobile forklifts. Compared with traditional wired surveillance, wireless surveillance is simpler, cost saving and more convenient for data transmit. At the same time, the back-end is perfectly compatible with the Network Video Recorder (NVR) of mainstream manufacturers and can connect to the cloud platform according to customer needs (see Figure 3.33).

(2) *Forklift wireless monitoring deployment design*:

- *Front-end*:

The front-end encoder can use LAN, SFP (Small Form-Factor Pluggable/Mini-GBIC, Gigabit Interface Converter), EPON (Ethernet Passive Optical Network), RRPP (Rapid Ring Protection Protocol), WLAN and other methods flexibly to access the bearer network, replacing the traditional optical transceiver access method, saving a lot of fiber laying costs. If the WLAN ring network access method is

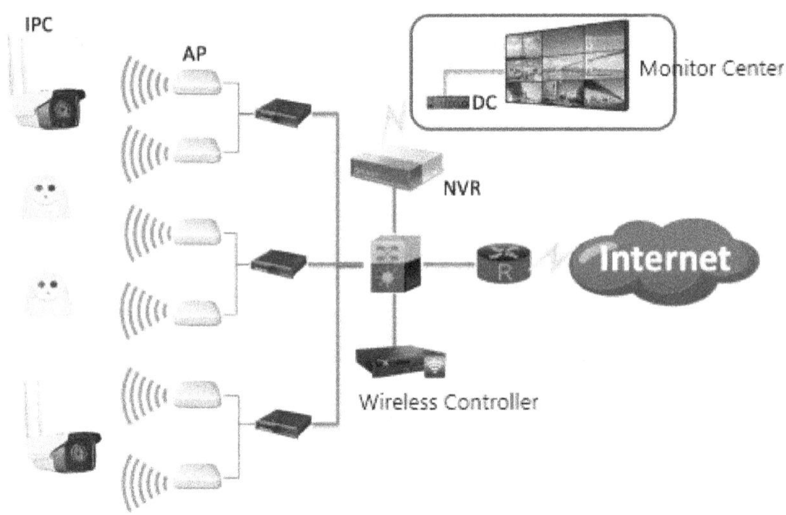

Figure 3.33 Forklift wireless monitoring.

adopted, not only does the wired network no longer need to be deployed separately, but also investment is saved.

- *Bearer network*:
  Because the flow model of surveillance video stream is quite different from the traditional network data flow model, the traditional method adopts the construction of private network to divide the line. This requires us to form a private network for the surveillance network. While wireless surveillance adopts a 5 GHz link to transmit data, which combines the radio network and the video network into one network, and transmits data to the back-end storage device through the radio network, saving the deployment cost of the surveillance network. At the same time, for non-video bearer networks, we can use application layer optimization to ensure stable transmission of video data without affecting the video data transmission.

(3) *Real-time monitor*: The wireless monitoring system has the characteristics of easy deployment, no need for cables and flexible deployment, which saves the cost of deploying network cables and solves various deployment problems for customers. After testing, one AP can guarantee the normal data transmission of seven wireless monitors at the same time, and the data transmission is very stable. Due to the complex environment of storages and the relatively difficulty of network deployment, wireless monitoring can be deployed on a forklift to realize flexible deployment of monitor equipment in a radio network environment. The 1080P high-definition image quality makes the monitoring clearer, and with H.264 image compression technology, the transmit bandwidth consumption is greatly reduced.

3.4.5.2.4 Wi-Fi smoke detector system

To address the fire safety problems in many cases, traditional fire smoke detectors have problems such as low information transmission efficiency and transmission lagging. For a special scenario in a factory, we consider the dual problems of Wi-Fi and fire safety and combine Wi-Fi and fire smoke detectors into an AP of IoT smoke sensor.

- *Transmit speed*: Smoke sensor AP has a built-in matrix smart antenna array that supports 5 GHz (11ac/a/n) and 2.4 GHz (11b/g/n) dual-band concurrency. The maximum rate of the whole device can reach 1167 Mbps, which can provide faster radio Internet access and larger coverage area and can easily meet the use of various wireless services, such as multimedia (video and voice) and provide intelligent radio frequency, service quality assurance, seamless roaming, etc.

- *Fire monitor and alarm*: The smoke sensor AP has a built-in smoke sensor that can monitor the concentration of smoke to prevent fires. The smoke alarm uses an advanced photoelectric smoke sensor inside, which is stable and reliable. The AP also has a built-in buzzer alarm, which can give an alarm as soon as it detects a fire hazard and can connect to the IoT cloud platform to upload data and transmit notifications.
- *Linkage with other equipment*: The smoke sense AP has a USB port and supports USB flash drives, Bluetooth iBeacon, RFID and other applications. The smoke sense AP can be connected to the relevant IoT cloud platform to achieve links with IoT systems such as broadcasting, monitoring and power sockets.

3.4.5.2.5 Wi-Fi voice system

In office and production workshop scenarios, in order to strengthen the management of personnel and the convenient delivery of notifications, companies usually deploy a set of voice systems by themselves. For this scenario, the radio network and the voice system are integrated into a speaker AP, which combines the original two systems into one and reduces the overall construction cost while ensuring fast Internet access. At the same time, it reduces the operation and maintenance management work and facilitates the issuance of notifications.

- *Transmit speed*: The speaker AP has a built-in matrix antenna array that supports 5 GHz (11ac/a/n) and 2.4 GHz (11b/g/n) dual-band concurrency, and the maximum rate of the whole machine can reach 1167 Mbps, which can provide faster radio Internet access and larger coverage to meet the various use of wireless services, such as video, voice and other multimedia services and provide intelligent radio frequency, service quality assurance, seamless roaming, etc.
- *Voice play function*: Speaker AP supports playing a pre-recorded audio in APP and also supports uploading recorded audio to AC for play,

redirecting audio to AP by PC software or APP. Through rich and flexible recording methods, enterprises can simplify and diversify the use of speaker AP.

3.4.5.2.6 Integration with MES system

- *Problem of production workshop scenario*: In the production workshop scenario, due to the frequent changes of production lines, if covered by wired network, future changes to the network will be very difficult to conduct. Using radio network to combine the radio network with production line MES system and transmitting the data of the production line equipment through radio network can solve unnecessary network wiring problems caused by production line changes.
- *Wi-Fi link data transmission*: Through wireless scanning gun, RFID and standard-based WLAN network technology, each process of production is recorded in real time to the back-end database. Through the analysis and mining of database, the production process can be optimized to improve efficiency, prevent error and reduce or even avoid quality problems, and problems can be traced back to improve user satisfaction.

3.4.5.2.7 RFID asset tracking management

The communication industry assets have the characteristics of large quantity, wide distribution and frequent changes. Asset management currently uses traditional manual methods, which is inefficient and has high error probability. At the same time, the information of asset and the information of management system cannot be synchronized in real time. Using RFID technology and combining wireless transmission technology to integrate the daily asset management activities with the asset management system can achieve real-time synchronization of asset information and system information.

(1) *System flow*: Asset management includes operations such as addition, allocation, idleness, scrapping, maintenance and inventory of assets. It includes the entire process of equipment from putting into use to scrapping (see Figure 3.34).

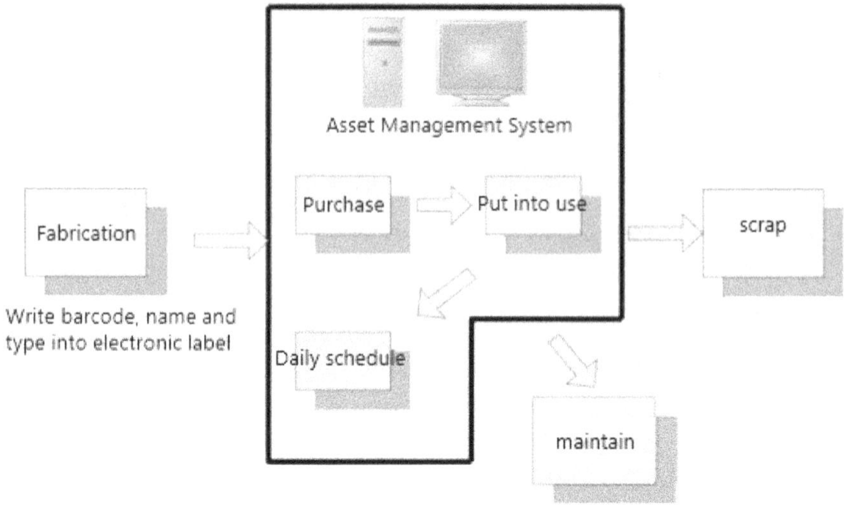

Figure 3.34 Asset management flow.

(2) *Principle*: When the equipment leaves the factory, an electronic tag will be installed, and the barcode number, name, equipment type and other information of the asset will be written in the tag, and the information will be recorded in the central server. The working status of the reader is controlled by the assembly line system sensor. When entering and exiting from the base station or warehouse, the reader starts to work and reads the electronic tags on the assets and then transmits the tag information to the central server through the wireless link.

(3) *Asset management system* (see Figure 3.35):
- It has the characteristics of long-distance rapid identification, high reliability, high confidentiality, easy operation and easy expansion.
- It is possible to establish safe and reliable registered asset files, strengthen asset supervision through high-tech, rationally allocate resources, reduce resource waste and prevent asset loss.
- It can make full use of RFID automatic collection and GPRS wireless remote transmission functions to achieve real-time consistency of asset change information and system information.
- All asset data are input at one time, and the system automatically judges the asset status (add, allocation, idle, scrap, etc.) according to the data collected by different base stations and regional

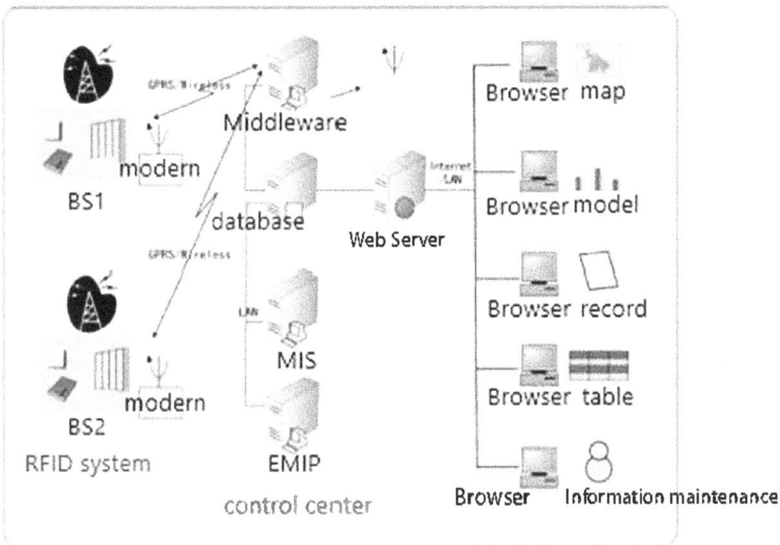

Figure 3.35 Asset management.

RFID readers. Users can use the visual GIS operation interface at any time and view statistics or conduct query through browsers anywhere.

3.4.5.3 *Staff dormitory scenario*

3.4.5.3.1 Coverage

A typical feature of scenarios such as dormitory buildings and apartment buildings is that the room structure is uniform, the walls are densely distributed, and the wireless signal attenuation is serious. If the traditional corridor installation method is adopted, the AP is deployed in the corridor. The signal has to penetrate the wall and enter the room, so the coverage effect is greatly reduced.

Solution 1: Deploy the AP in the corridor, and the signal penetrates the wall and enters the room (Figure 3.36).

Solution 2 (Recommended solution): Use new type 86 panel AP for wireless deployment (see Figure 3.37) and directly replace the panel cassette

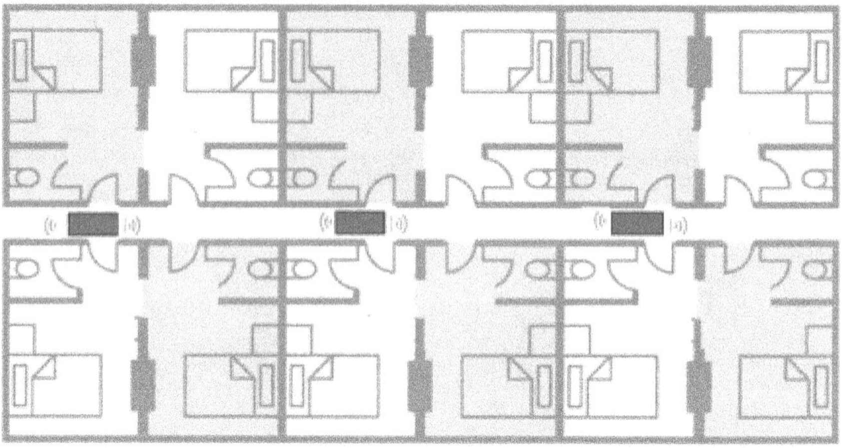

Figure 3.36 Corridor AP deployment.

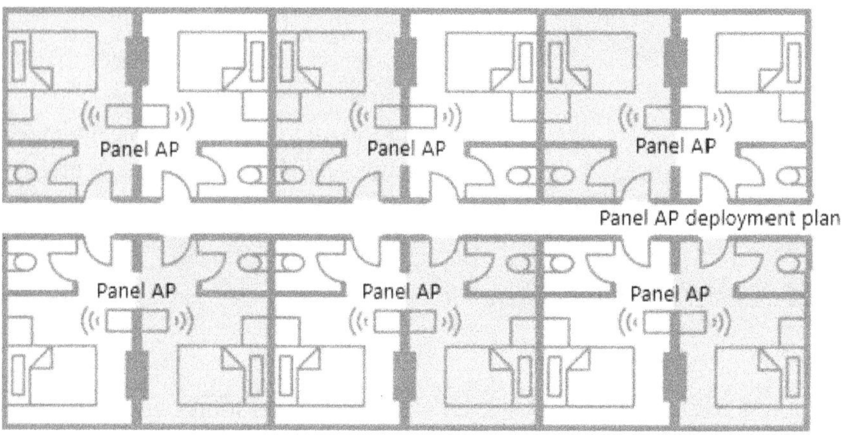

Figure 3.37 Panel AP deployment.

in the room to deploy radio network quickly. This solution ensures that the signal of every dormitory is full and reduces the construction cost. At the same time, it does not destroy the room environment design.

Solution 3 (Alternative solution): Use the shared feeder solution (Figure 3.38) to introduce the antenna into the room through the feeder to improve the signal coverage effect. An AP can cover four to eight rooms and ensure that the signal of the room is full.

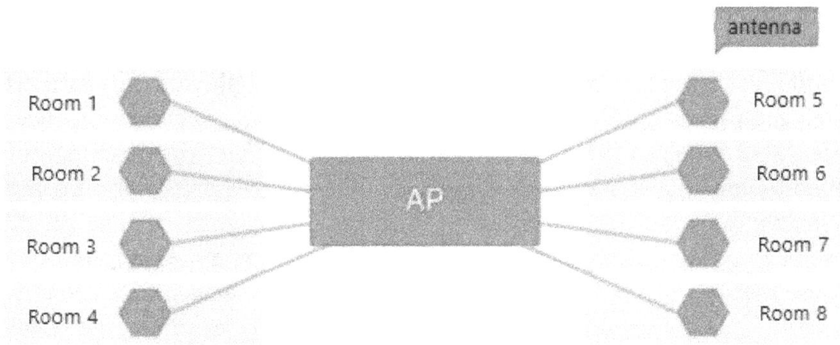

Figure 3.38 AP shared feeder deployment.

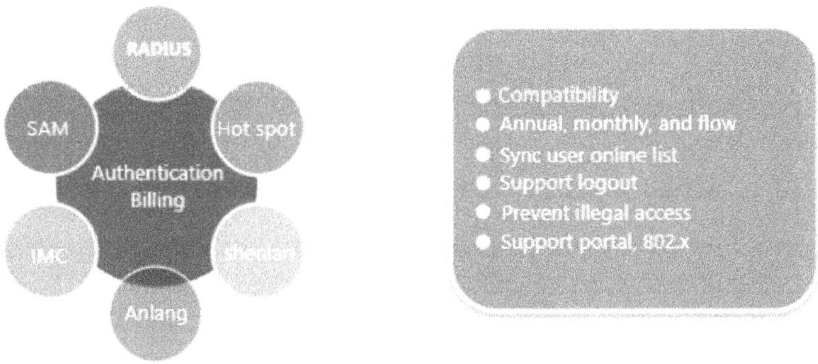

Figure 3.39 Certification and charging.

3.4.5.3.2 Certification and charging scheme

YYY AC can be combined with existed certification and charging systems, such as Urban Hotspot, Shenlan, Anlang, Ruijie SAM, etc., to synchronize organizational structure, online user list, certification success and cancellation information, etc., to achieve certification charging (see Figure 3.39).

3.4.5.4 *Smart energy management solution*

For big enterprises, electricity expenditure and safety have always been the most difficult issues that they face. On the one hand, enterprises have to pay huge electricity bills monthly and are unable to reduce them even if they want to advocate energy and electricity saving. Although rules ask employees to turn off computers after work, some employees only turn off their monitor and do not turn off their host, and arranging patrol inspections will increase labor costs. Similar conditions are, for example, air conditioners, fans, water dispensers, etc. The above problems increase consumption of electricity. On the other hand, the aging of lines, illegal use of electricity and sudden restoration of power after a power outage can cause the server of enterprise computer room to be damaged by an electrical fire. It is not uncommon in major enterprises. For example, the warehouse of SF Express and the office building of Unicom suffered severe fires due to electricity consumption, and the property losses were too big to calculate. In order to be able to effectively detect and manage electricity consumption, enhance electricity safety, prevent fires and provide safe usage of electricity for enterprises at the same time, enterprise-level IoT smart socket is introduced in energy consumption management system to save electricity expenses and prevent electrical fires.

3.4.5.4.1 Network architecture (see Figure 3.40)

Smart sockets support wireless link and IoT LoRa (Long Range Radio) protocol data transmission and unified configuration and management of sockets through third-party cloud IoT platforms. In the initial configuration step, public cloud or private cloud and enterprise account need to be chosen. If public cloud is chosen, the socket will use the third-party cloud platform domain name. If private cloud is chosen, private cloud IP is used to address (private cloud server address) the enterprise IoT platform. Then, a data communication connection is established between the socket and the enterprise according to the account.

3.4.5.4.2 Electricity consumption statistics and data analysis

Smart sockets can transmit electricity usage and data back to the relevant cloud IoT platform through Wi-Fi, LoRa and other links for unified electricity usage statistics and data analysis. Administrators can view real-time power and electricity usage on the cloud platform, as well as electricity bills, socket power, power off and offline status. They can analyze the changes of electricity consumption within a week or a month through the various horizontal and vertical comparison

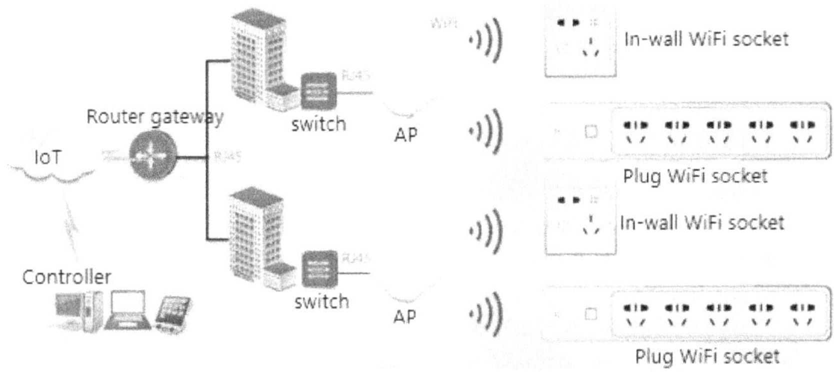

Figure 3.40 Intelligent energy consumption management network architecture.

of graphs, real-time power, power consumption statistics and data reports.

3.4.5.4.3 Power consumption strategy

(i) *Electrical Safety*:

• *Automatic/remote control switch*:

For corporate office, warehouses and dormitory scenarios, after using equipment, users can sometimes forget to turn off the equipment, resulting in electricity waste, and manual inspections increase labor costs. To solve this problem, smart sockets can be pre-configured with automatic switch strategy. For example, the water dispenser is turned off at the end of work every day and automatically turned on at work, and computers can be remotely turned off if forgotten.

• *Automatic alarm/power failure for high-power consumption*:

In enterprises, factories, warehouses and dormitories, irregular electricity use and high-power electricity consumption are usually the causes of fire. In these scenarios, in order to ensure the safe use of electricity, we can set outlet rating of sockets by policy configuration, buzzer alarms when the threshold is exceeded, APP and SMS notification and automatic power-off strategies through configuration.

• *Delayed power-on protection*:

Equipment and lines burnt out are often due to excessive electric power load after power outage. Configuring the background delayed power-on protection and setting a delay for power on can prevent the instantaneous load from being too big to burn out lines.

3.4.5.4.4 Data security guarantee

- *Transmission security guarantee*: At the transport layer, Message Queuing Telemetry Transport (MQTT) uses TLS not only to encrypt communications but also to use X509 certificates to authenticate devices, therefore, ensuring data security. At the application layer, it supports customer identification, password and X509 certificate to verify the device at the application layer.
- *Data storage guarantee*:
  Public cloud: Data are encrypted and stored using independent encryption algorithms, allowing users to control their own data.
  Private cloud: Supports the deployment of private cloud platform, and all data are stored locally on the private cloud platform.
- *Operation and maintenance management guarantee*: Supports anti-hacking and anti-cracking for accounts. At the same time, data transmitted on APP or web platform is encrypted through https to ensure the security of data link transmission.

3.4.6 *Wireless stability solution*

3.4.6.1 *Smart radio frequency*

In the 2.4 GHz frequency band, only three non-overlapping channels can be accommodated at most. Generally, 1, 6 and 11 are selected. The same channel will cause co-channel interference, and different overlapping channels will cause adjacent channel interference. With the popularity of WLAN, more and more radio networks are built and the air is full of wireless signals, which causes more channel interference, and only smart radio frequency can solve this problem.

3.4.6.1.1 Smart RF power adjustment

After turning on smart radio frequency, YAC (a diversified, high-performance AC product, which integrates wireless controller, user authentication, marketing push, customer flow analysis, user profile, Internet behavior management, traffic control, Internet behavior audit, VPN, Firewall and other functions in one) conducts real time monitoring of the surrounding radio frequency environment. If the environment changes, the system will adjust the AP's transmit power. When the AP detects that the signal strength is weak in a certain area, the AP will automatically adjust the signal power to increase the signal transmission power in that direction, so as to meet the Internet and roaming needs of the area. At the same time, when the enterprise does not use the radio network at night or on weekends, the AP can automatically adjust and reduce the transmit power to meet the requirements of energy saving and environmental protection.

3.4.6.1.2 Smart channel optimization

Smart channel optimization is a smart channel adjustment. YAC supports dynamic and intelligent channel optimization functions, which can adjust AP working channels regularly and automatically, so that adjacent AP channels can be separated to avoid channel conflicts.

3.4.6.2 *Smart load balancing*

3.4.6.2.1 Frequency band load balancing

Access to the 5 GHz frequency band should be prioritized, so that mobile terminals that support 2.4 GHz/5 GHz dual-frequency can access the 5 GHz frequency band first, which can balance the utilization of the 2.4 GHz/5 GHz network, improve terminal users' Internet experience and maximize the performance.

3.4.6.2.2 Load balancing between access points

The terminal access according to the load situation of the AP and other conditions (such as the number of access people, channel utilization, etc.) should be controlled to achieve load balance between APs and improve network throughput and service quality. In addition, it also supports flexible settings of different load balancing parameters according to different scenarios and different areas to achieve the best roaming and wireless experience. When the terminal requests access to an AP, YAC decides whether to allow new clients to access according to the load of the AP and neighboring APs. The terminal can automatically connect to other idle APs to achieve the effect of load sharing.

3.4.6.2.3 Dynamic load guidance (anti-terminal sticking)

When the terminal moves between different areas, in order to allow the terminal to receive the radio access point with the best signal and obtain a good Internet experience, the anti-sticky function can be set according to the actual situation to achieve better roaming effect. When the access point finds out that the signal strength of the terminal is less than the threshold, and the wireless traffic of the terminal is less than the traffic threshold, the access point will actively let the terminal roam.

3.4.6.3 *Radio air interface optimization*

YAC supports the allocation of air interface resources based on application and SSID and supports 802.11e/WMM (Wi-Fi Multimedia) priority setting to fully guarantee the rational use of radio air interface resources.

- *SSID-based air interface resource management*: Through SSID-based air interface resource allocation, multiple SSIDs on the same AP are allocated according to percentages. For example, the leader SSID is allocated 50%, the employee SSID is allocated 40% and the visitor SSID is allocated 10%. The bandwidth is allocated dynamically to ensure the priority traffic of important SSID.
- *Air interface resource management based on business type*: Based on the business type of radio air interface resources, traffic channels can be divided according to different business types, and the radio air interface resources can be piped for fine management, such as 30% for related business systems, 30% for web browsing, 20% for video downloads and 20% for others. Reasonable allocation of radio air interface resources can ensure priority transmission of important applications.
- *Equal distribution of bandwidth among users*: The bandwidth is evenly allocated between users, and the terminal is fairly scheduled (time fair algorithm) to avoid the problem of low overall network performance caused by low-speed terminals occupying radio air interface resources for a long time and improve the overall Internet experience.
- *WMM/802.11e*: IEEE 802.11e adds QoS features to the WLAN system based on the 802.11 protocol. Through IEEE 802.11e standard, the QoS priority can be set to guarantee the priority processing

capabilities of important business applications, such as voice and video.

3.4.6.4 *High-density optimization*

For some areas such as meeting rooms and exhibition halls with a large number of people, the terminals are distributed densely, so that the user experience is reduced. However, deploying more APs in this limited space couldn't solve the multi-user problem, but only brought more interference roaming problems. The high-density optimization can optimize radio network performance for scenarios where terminals are distributed densely, reducing the consumption of wireless performance space due to low-rate terminal transmitting probe response frames, thereby improving the wireless Internet experience of users in high-density scenarios.

3.4.6.5 *Broadcast optimization*

According to the characteristics of wireless communication, the broadcasting optimization technology is introduced by optimizing the TCP/IP protocol stack:

1. Address Resolution Protocol (ARP) broadcast to unicast.
2. Prohibits the broadcast of DHCP requests to wireless terminals.
3. Improves the broadcasting speed.

After the broadcast is optimized, unnecessary "junk" messages can be reduced to spread the radio network, the overall capacity of the radio network is increased, the user's Internet experience is improved, and the number of concurrent wireless users is also increased.

3.5 Wireless Customer Premise Equipment

At present, mobile phone users generally use mobile cellular communication networks outdoors and Wi-Fi Internet access indoors. This has actually formed a situation where mobile cellular networks and Wi-Fi networks are combined with the main and auxiliary networks. If the indoor access bandwidth can be fully guaranteed, the quality of indoor

Internet access may be better than that of outdoor. But when the 5G network is put into use, this difference does not exist, and even mobile cellular networks will be better than indoor Wi-Fi networks.

But when most user equipment is still 4G terminals or users are even in places where 4G signal coverage is weak, how can they enjoy the performance of 4G or the newly commercialized 5G network? At this time, the wireless CPE came into being.

Wireless CPE is a mobile signal access device that receives 4G or 5G mobile signals and forwards them with wireless Wi-Fi signals. It is also a device that converts 4G or 5G signals into Wi-Fi signals. It can support a large number of simultaneous Internet access, but external power supply is required. CPE can be widely used in rural areas, towns, hospitals, units, factories, communities and other radio network access, which can save the cost of laying wired networks.

Of course, there are indoor and outdoor types of CPE. The indoor type has a transmission power of 500–1000 mW, and the outdoor type has a transmission power of up to 2000 mW. It can also adapt to more severe environments.

People always hope that the larger the wireless coverage of the base station, the better, but the transmission distance of the wireless system is also related to the technical indicators of the client terminal equipment. The uplink power of the standard Wi-Fi client is limited (usually 15 dBm). No matter how much the downlink power of the base station is, the uplink distance of the standard Wi-Fi client is limited by the uplink power, and the distance is very limited. The wireless CPE is a dedicated customer front-end equipment or a secondary router. As its uplink power is higher than that of a standard Wi-Fi client, the wireless outdoor CPE can be used to receive signals from a base station (such as a 5G base station) farther than a standard Wi-Fi client. It can reach four times more than that of the standard Wi-Fi client.

This not only solves the cost and construction difficulties caused by fiber at home, but at the same time, the wireless CPE can also accommodate a larger number of terminals and the use of longer-distance communication equipment (normally the distance between CPE user and base station is 1–5 km).

In fact, wireless CPE has already existed in the 4G era, and wireless CPE is used to solve areas with poor 4G information coverage. So, before the real commercial arrival of 5G, wireless CPE can also be used to build a transitional 5G network. At present, domestic and foreign operators are trying to use wireless CPE to let users experience the extremely fast experience of 5G networks.

Compared with the difficult construction and high investment of mobile 5G network services, there may be more 5G CPEs for home or office use in the future. At present, there are 5G CPE products on the domestic market, such as Huawei's mobile router 5G CPE Pro. The theoretical peak download rate can reach 2.3 Gbps, and it supports 5G, 4G and wired broadband access.

3.6 Looking to the Future of Wi-Fi

The current 4G network is already well known by the majority of users and has been used on a large scale. With the rapid development of 4G networks, the 5G era has come, and some users have already tried the high-speed experience brought about by 5G networks in advance, which also indicates that in the future there will be faster networks and will continue to accelerate the progress of the entire mobile terminal network. We can see that in the 4G era, the primary and secondary integration with Wi-Fi will continue, and both network forms and network terminal equipment will be further introduced.

Entering 2019, the next-generation mobile communication system 5G will also gradually enter commercial use. Major countries including China have carried out pilots and promotions to meet the needs of mobile business traffic growth. Some countries and international organizations have also carried out 6G research.

Network forms are evolving toward diversification, broadbandization, integration and intelligence. In particular, future data services will be mainly distributed indoors and hotspots, which makes ultra-dense networks one of the main means to realize 5G. As the construction density of base stations in urban areas is already high, it is difficult to continue to deploy macro or micro stations, and 5G may only deploy femto stations and hotspots. In this situation, the layout of 5G base stations and Wi-Fi hotspots and the function of absorbing data traffic have become consistent.

Now that 5G is on the starting line, it is more difficult for 5G high-band signals to enter indoors. At the same time, operators' traditional indoor distribution systems will also face large-scale upgrades or replacements in the 5G era. Therefore, facing the continuous development of video and IoT business, the world will need stronger Wi-Fi to offload data traffic and fill the gap in 5G coverage.

In addition, after Wi-Fi has achieved GB data rates and gradually merged with 4G/5G, the IEEE 802.11 working group has seen that low-speed, low-energy communication is actually an area with great development potential. Although Wi-Fi is traditionally a relatively power consumption standard, future protocols will be able to give radio network connections more new features, so as to provide more energy-efficient communications to IoT (discussed in Chapter 4) and others.

Chapter 4

Evolution of the Internet of Things

5G called IMT-2020 by ITU is a highly integrated network of the Internet of People and the Internet of Things. The Internet of People uses the architecture of the cellular mobile network, and the connected objects are people. It has two distinct characteristics. First, the network structure is a honeycomb, which is a measure to achieve large-area network coverage. Second, the user (terminal) is mobile. The network is for consumer Internet services. According to the requirements of consumer Internet services (such as high-definition video, online games, VR and AR signals), the main emphasis is on data transmission rate (broadband). Therefore, the evolution process of the Internet of People from 2G→3G→4G to eMBB in 5G is actually a process of gradually increasing the data transmission rate of the network.

The Internet of Things is a network where everything is connected, and the connected objects are mainly things (smart objects). It can be used in vertical industries and scenarios such as cities, agriculture, industry, energy (electricity), scientific research, education, logistics and medical care. Borrowing the notion that the Internet of People serves the consumer Internet, we can see that the Internet of Things serves the industrial Internet. Therefore, the Internet of Things will gradually develop with the demands of the Industrial Internet.

Let's first look at the origin of the Internet of Things.

4.1 The Past and Future of Internet of Things

4.1.1 *The origin of the Internet of Things*

First, let's take a look at Xerox's Internet Coke vending machine in 1990, which was on the campus of Carnegie Mellon University (see Figure 4.1). There was a group of programmers who used to buy cans of ice Cola from the Coke vending machine downstairs every time they typed out the codes. But many times, they only encountered the empty Coke machine, which made them very distressed. So, they connected the vending machine to the Internet and wrote a piece of code to monitor how much Coke was left and, of course, to see if it was iced.

Here is another example, the Troy coffee pot in 1991, which is also recognized as the origin of the Internet of Things (IoT). Scientists at the Troy Computer Laboratory at the University of Cambridge often went downstairs to see if the coffee was ready, but they were afraid that it would affect their work. To solve this problem, they wrote a set of programs. A portable camera was installed next to the coffee maker. Using image capture technology of the terminal computer, the images were transmitted to the laboratory computer at a rate of three frames per second, so that the staff could check whether the coffee was done at any time. This is the earliest prototype of the IoT. In 1993, as the first X-Windows system

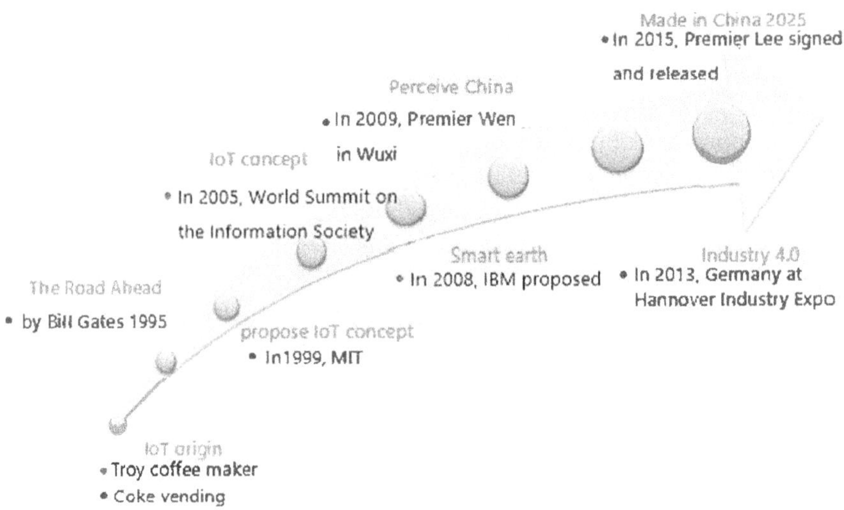

Figure 4.1 Origin of the Internet of Things.

case, the "Troy Coffee Pot Server" was also uploaded to the Internet and received nearly 2.4 million hits.

But the real IoT appeared in 1994. In 1994, Kevin Ashton, one of the founders of MIT's Auto-ID Center, was the first to use the IoT. Ashton's thoughts on the IoT focused on the fact that Radio Frequency Identification (RFID) technology is used to connect devices together. This is similar to today's IoT. But it is different from the extensive information that depends on IP networks for devices to exchange. The functions provided by RFID tags are relatively limited because radio networks were still in the initial stage in 1994. Just like cellular networks, it had not yet switched to a completely IP-based configuration. In this case, it was difficult to imagine that all devices in the Internet have a unique IP address. In addition, in the case of Internet Communication Protocol Version 4 (IPv4), if all devices join the network, there will not be enough IP addresses for allocation. However, because RFID did not require every device to have an IP address or be actually connected to the Internet, it was a cheaper and feasible solution.

In 1995, Bill Gates mentioned the concept of the IoT in The Road Ahead. For example, he wrote the following words in his book: "If a user loses a camera, it can automatically transmit a message to tell the user where it is. Even when it is already in a different location, it can also tell the user where it is now".

In 1999, the Massachusetts Institute of Technology first proposed the concept of the IoT.

On November 17, 2005, at the Information World Summit held in Tunisia, the International Telecommunication Union released the "ITU Internet Report 2005: Internet of Things", which officially put forward the concept of "Internet of Things". The report pointed out that the ubiquitous IoT communication era is coming. All objects in the world, from tires to toothbrushes, from houses to paper towels, can be actively exchanged through the Internet. RFID technology, sensor technology, nanotechnology and intelligent embedded technology will be more widely used. It can be seen that the definition and scope of the IoT have changed and greatly expanded, and it is no longer just the IoT based on RFID.

4.1.2 *America: Smart Earth*

In 2008, IBM first proposed the concept of "Smart Earth". The construction of smart cities came into being, and a new wave has been set off in global city construction in recent years. IBM believes that building a smart planet requires three steps.

First, various innovative sensing technologies have begun to be embedded in various objects and facilities, thus making the material world extremely digitized and providing massive data sources.

Second, with the highly developed network, people, data and various things will be connected to the network in different ways.

Third, advanced technology and supercomputers can sort, process and analyze these piles of data, transform the huge data into real insights and help people make correct decisions.

At the same time, IBM proposed to establish smart action plans in six major areas, which are smart electricity, smart medical care, smart city, smart transportation, smart supply chain and smart bank.

The IoT is a basic concept among these so-called smart infrastructures. Now the "smart earth" has risen to the national strategy of the United States.

4.1.3 *Germany: Industry 4.0*

The German government proposed the "Industry 4.0" strategy and officially launched it at the Hannover Industrial Expo in April 2013. Its purpose is to improve the competitiveness of the German industry and take the lead in the new round of industrial revolution. This strategy has been widely recognized by German scientific research institutions and industry. The Fraunhofer Institute introduced the concept of Industry 4.0 to its research institutes in six to seven production areas. Siemens had already begun to introduce this concept into its industrial software development and production control systems.

Since its official launch at the Hannover Industrial Expo in April 2013, Industry 4.0 has quickly become another label of Germany and has triggered a new round of industrial transformation competition in the world.

4.1.4 *China: Perceive China and made in China (2025)*

On August 7, 2009, Premier Wen Jiabao visited the Wuxi High-tech Micro-Nano Sensor Network Engineering Technology R&D Center of the

Chinese Academy of Sciences. After learning that the core technology of domestic sensor networks has not reached the most advanced level in the world, Premier Wen said, "When the computer and Internet industries are developing on a large scale, we have gone through some detours because we did not master the core technology. In the development of sensor networks In China, we must plan the future earlier and break the core technology earlier." Premier Wen pointed out that at least three things can be done as soon as possible.

First, the sensor system needs to be combined with TD (refers to TD-SCDMA) technology in 3G.

Second, the development of sensor networks in major national science and technology projects should be accelerated.

Third, China's sensor information center should be established as soon as possible, which is also referred to as the "Perceive China" center.

On November 3, 2009, Premier Wen delivered an important speech entitled "Let Science and Technology Lead China's Sustainable Development". In this speech, the IoT was listed as one of the country's five emerging strategic industries (new energy, new materials, biological sciences, information networks and space ocean development). It is required to "Strive to break through the key technologies of sensor networks and the Internet of Things. Realize early deployment of related technology research and development in the post-IP era. Make the information network industry an engine that promotes industrial upgrading and moves toward an information society".

On March 5, 2010, Premier Wen pointed out at the Third Session of the Eleventh National People's Congress that "We must develop new energy, new materials, energy conservation and environmental protection, biomedicine, information networks, and high-end manufacturing industries. Actively promote new energy vehicles and the integration of the "three networks" to make substantial progress. And accelerate the development and application of the Internet of Things. Enhance the investment and policy support for emerging industries."

This is the first time the "Internet of Things" has been included in a government report. It also means that the development of the IoT has entered the national perspective and been promoted to the national strategy. People usually referred to 2010 as the first year of the IoT. It is reported that China has planned to invest 3.86 trillion yuan in the research and development of the IoT before 2020.

In August 2009, after Perceived China's speech of Wen Jiabao, the research and development of the IoT was pushed to a climax. Wuxi City has taken the lead in establishing the Perception China Research Center. The Chinese Academy of Sciences, operators and many universities have established the IoT Research Institute in Wuxi, and Jiangnan University in Wuxi has also established the country's first physical IoT Business School.

On May 8, 2015, Li Keqiang signed and announced Made in China 2025.

In 2016, the IoT once again received the attention of the State Council. On March 5, 2016, Premier Li Keqiang emphasized "promoting the widespread application of big data, cloud computing, and the IoT" in his government report. On October 31, Premier Li Keqiang sent a congratulatory letter to the World IoT Summit in Wuxi to send congratulations to the convening of the World IoT Expo and hoped to use the Expo platform to exchange innovative ideas, deepen mutual cooperation, drive innovation and benefit human society.

In April 2017, China's Ministry of Industry and Information Technology held an Narrow-band IoT (NB-IoT) work promotion meeting to jointly cultivate the NB-IoT industry chain. It is also required to build a large-scale outfield based on the standard NB-IoT at the end of the year. Looking forward to 2020, NB-IoT technology will be mature enough to become ubiquitous cellular IoT coverage. The good prospects of NB-IoT have infinitely expanded the commercial field of information and communication.

4.2 Overview of the IoT

4.2.1 *Definition of the IoT*

The IoT is the "Internet of Every Things Connected". It is an extension and expanded network based on the Internet. It combines various information sensing equipment with the Internet to form a huge network, which realizes the interconnection of people, machines and things at any time and any place. From this, the following two points can be derived: (1) the core and foundation of the IoT is still the Internet; (2) its user end is extended to the information exchange and communication between any item and the user. Therefore, the IoT uses information sensing equipment such as radio frequency identification, infrared sensors, global positioning systems, and laser scanners, to connect any item to the Internet for information exchange and communication according to an agreed protocol, so as to realize a network which can realize intelligent identification, positioning, tracking and management of items.

4.2.2 *Architecture of the IoT*

As seen in the above definition, the basic architecture of the IoT is shown in Figure 4.2. It can be seen from this figure that the basic architecture consists of three levels, which are perception layer, network layer and application layer. The perception layer is the bottom layer of the IoT. It is used to obtain data of various items through a variety of different sensing devices. Therefore, the perception layer is actually a variety of different information sources in the physical world. The network layer is the middle layer of the IoT. It is used to solve the problem of long-distance safe and reliable data transmission which is acquired by the perception layer. In the case of wireless connection, it is a wireless communication network composed of a radio access network and a core network. The application layer is the interface between the IoT and users (people or things). The interface varies with the application scenarios of the IoT. The intelligent application of the IoT in different application scenarios is realized by analyzing big data through the interface, and smart applications are the ultimate goal of the IoT.

4.2.3 *Classification of the IoT*

The IoT can be applied in many scenarios, but in terms of coverage, it can be divided into two categories.

(1) *WLAN IoT* (*short range*): It is represented by Wi-Fi, Bluetooth (BT) and Zigbee.
RFID, wireless local area network Wi-Fi and BT that have been used in daily life are non-cellular short-range IoT. The main characteristics of its network are short communication distance and small coverage.

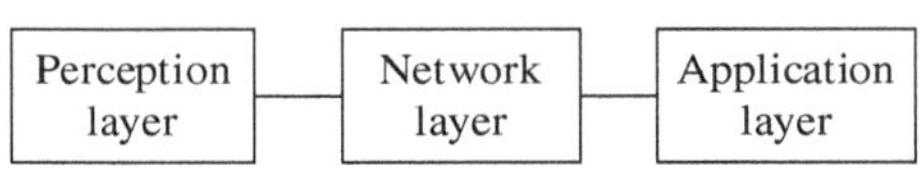

Figure 4.2 The architecture of IoT.

(2) *Cellular IoT* (*long distance*): It is represented by NB-IoT, enhanced Machine Type Communication (eMTC), LoRa and Sigfox.
It is known from the Internet of People that the transmission power of terminal equipment is less than the transmission power of the base station, and the power of both (especially the terminal) is limited. Therefore, the coverage of the cell is determined by the terminal performance, that is, the uplink performance. For this reason, one way to expand network coverage is to adopt a cellular network structure. There is a type of IoT that needs to be able to communicate under long-distance conditions that use this network structure. The main characteristics of this type of IoT are long communication distance and large coverage.
The IoT can also be divided into three categories from the perspective of data transmission rate, that is, low, medium and high data rate. If the coverage area is taken as the horizontal axis and the data rate is taken as the vertical axis, the location of these IoT mentioned above is shown in Figure 4.3.
This method of classifying according to the size of the coverage area is actually similar to the two methods of accessing the Internet via mobile phones. While using Wi-Fi to access the Internet, the coverage area is small. In contrast, while using mobile data, the coverage area is large.

4.2.4 *Typical applications of the IoT*

Typical applications of the IoT are shown in Figure 4.4.

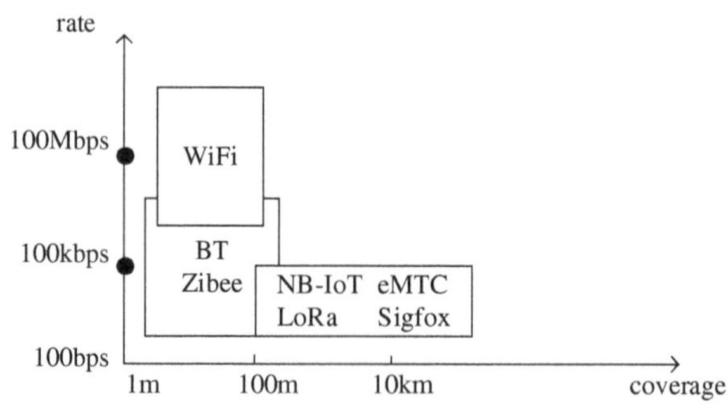

Figure 4.3 Classification of several IoT.

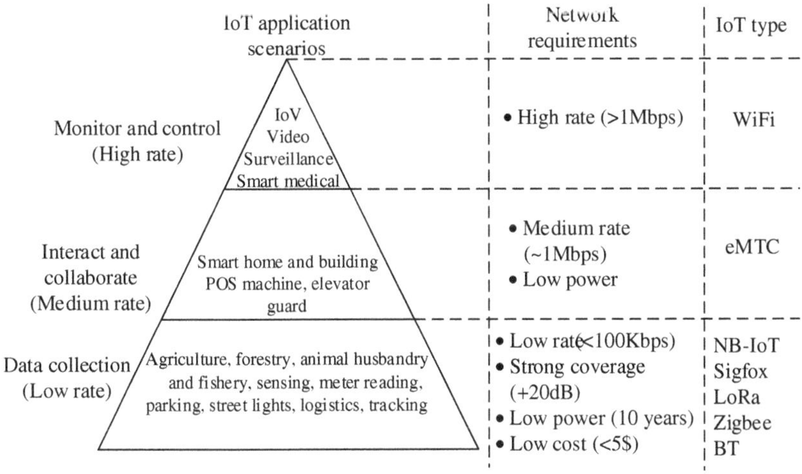

Figure 4.4 Typical applications of the IoT.

In these IoT, LoRa and Sigfox are private standards formulated by companies. The rest are standards promulgated by ITU. Among them, Wi-Fi has been discussed in Chapter 3. BT has been widely used in mobile phones. The IoT in 5G is developed on the basis of NB-IoT (NB) and eMTC. Therefore, this chapter focuses on the NB and eMTC, especially the NB. The following section provides a brief introduction to Zigbee, LoRa and Sigfox.

4.3 Brief Introduction to Zigbee, LoRa and Sigfox

4.3.1 *Zigbee*

Zigbee is a short-distance (10–75 m), low-complex, low power-consuming, low-speed and low-cost two-way wireless communication technology. It uses 2.4 GHz frequency band (unlicensed frequency band) worldwide. It has the following characteristics:

- *Low power consumption*: Two AA batteries can last for six months to two years.
- *Low cost*: The Zigbee module is estimated to be US$1.5–2.5.
- *Short time delay*: The communication delay and the delay from sleep to activation are both very short. The typical device searching delay is

30 ms, the sleep activation delay is 15 ms, and the active device channel access delay is 15 ms.

- *Large network capacity*: A star-shaped Zigbee network can accommodate up to 254 slave devices and a master device. A maximum of 100 Zigbee networks can exist simultaneously in an area, and the networking is flexible.
- *Reliability*: It can adopt a collision avoidance strategy and a data packet retransmission technology.
- *Safety*: Data packets with Cyclic Redundancy Check (CRC) are used for error detection. It supports authentication and authentication and uses the AES-128 encryption algorithm (Advanced Encryption Standard (AES)).

Zigbee adopts a mesh topology, which is easy to manage and control in the industry, and is suitable for the field of a smart home.

4.3.2 *Long Range Radio* (*LoRa*)

Its biggest feature is that under the same power consumption conditions, the propagation distance is 3–5 times longer than other wireless methods, achieving the unity of low power consumption and long distance. LoRa characteristics are as follows:

- *Transmission distance*: It can reach 2–5 km in towns and 15 km in suburbs.
- *Working frequency*: ISM frequency band, including 433, 868, and 915 MHz.
- *Standard*: IEEE 802.15.4g
- *Modulation*: Chirp Spread Spectrum (CSS)
- *Capacity*: One LoRa gateway can connect thousands of LoRa nodes.
- *Battery life*: Up to 10 years
- *Safety*: AEC-128 encryption
- *Transmission rate*: Several hundred to several tens of kbps, the lower the rate, the longer the transmission distance.

LoRa is suitable for fields such as smart street lights and smart homes.

4.3.3 *Sigfox*

It is a radio network launched by the French company Sigfox (founded in 2009) to connect low-power devices. It uses ultra-narrowband (UNB) technology, that is, the advantages of communication distance (long distance) and power consumption (low power consumption) are achieved at the cost of reducing data rate (the principle will be discussed in NB). Its characteristics are low speed, low power consumption, low cost and long distance. Sigfox is used in applications such as smart meter reading and smart manufacturing.

4.3.4 *Wi-Fi*

Wi-Fi has a large number of applications in the IoT scenarios with medium or high speed and short distance. Wi-Fi can be used as a wireless transmission technology in scenarios such as household appliances and factories' production line storage. Wi-Fi has already been introduced in Chapter 3.

Wi-Fi has achieved great success indoors, but with the introduction of the IoT, Wi-Fi hopes to achieve success outdoors too. In September 2016, the Draft 9.0 version of the 802.11ah standard led by IEEE was completed. In December, it completed the approval procedure of the Standards Committee and was named HaLow (but so far, there is no commercial Wi-Fi HaLow access point or router on the market). It uses the unlicensed 900 MHz frequency band with a transmission distance of 1 km and a data rate of 150 kbps to 347 Mbps.

IEEE also plans to use 802.11af technology in the white space frequency band 54–790 MHz of TV, to provide low power consumption and longer transmission distance. However, according to the specifications of HaLow, there is still a gap in transmission distance compared to other LPWAN technologies that can reach a distance of dozens of kilometers. Although it can be extended to several kilometers through a multi-point relay, the industry chain is still weak due to its late initial time.

The advantage is that the network construction of HaLow is not difficult. It can be done by simply upgrading the equipment. Currently, it is only regarded as a supplement to NB, and in order to truly achieve wide-area network coverage, we still need the aid of NB.

4.3.5 *Ultra-Wide Band*

Ultra-Wide Band (UWB) (for simplicity, it is not marked in Figures 4.3 and 4.4) is a wireless technology that transmits data at a high speed (hundreds of Mbps to several Gbps) within a short distance (about 10 meters) with very low power. It is a non-carrier communication technology, that is, instead of using a sinusoidal carrier, it uses nanosecond (<1 ns) non-sine wave narrow pulses to transmit data. Therefore, the occupied frequency spectrum is very wide. According to the Federal Communications Commission (FCC), the frequency band of UWB ranges from 3.1 GHz to 10.6 GHz.

UWB technology has been one of the combat technologies used by the US military for many years. For example, the triangular positioning principle can be used to achieve precise positioning. It is also suitable for short-distance digital audio and video wireless links, short-distance broadband, high-speed radio access and other related civilian fields, such as home digital entertainment.

4.3.6 *Cat 1*

Cat 1, also known as LTE Cat 1, is also not marked in Figures 4.3 and 4.4 for simplicity. In fact, in the very early 3GPP LTE standard E-UTRAN specification (3GPP Release 8), a device called Category 1 (Cat 1, see Table 1.8 in Chapter 1) is included. Category 1 is the user terminal level for the lowest version of the parameter configuration, allowing the industry to design "low version" 4G terminals at low cost. The design of this kind of equipment is very simple and power-saving, but its speed can only reach 10 Mbps. In order to reduce complexity, Cat 1 is the only device type that can use a single antenna in 2008, which means it does not support MIMO.

In 2016, China Telecom released its Cat 1 commercial plan and launched a subsidy program for Cat 1 modules. However, in the past few years, Cat 1 has not reached the expected large-scale connection due to reasons such as cost. However, in November 2019, China Mobile started to promote Cat 1 IoT applications. This is mainly because, in the past few years, Chinese telecom operators have built the world's largest 4G network, which is far ahead of overseas telecom operators both in breadth and depth of network coverage. With the help of 4G network infrastructure, Cat 1 can provide IoT users the opportunity to access it anytime and anywhere. It can be said that its network infrastructure is not weaker than 2G/3G and NB-IoT. Laying a good foundation for the migration of IoT connections to Cat 1 is mainly on the 4G base station side. In addition, eMTC domestic progress has not met expectations due to funding and other reasons. However, there are not many alternatives for the connection scheme of the medium- and low-speed cellular IoT, and Cat 1 may be the final choice in the end.

4.3.7 *C-V2X and LTE-V2X*

Cellular Vehicle-to-Everything (C-V2X) and Vehicle-to-Everything (V2X) both refer to the exchange of information between cars and the outside world. C-V2X is a car networking technology based on cellular networks. For simplicity, they are also not marked in Figures 4.3 and 4.4.

C-V2X refers to the smooth evolution from LTE-V2X to 5G V2X. It not only supports existing LTE-V2X applications but also supports new applications for future 5G V2X. It is based on a strong 3GPP ecosystem and complete cellular network coverage, which can significantly reduce the deployment costs of autonomous driving and the Internet of Vehicles in the future.

Unlike sensors such as radar and lidar, V2X can be regarded as a solution for wireless sensor systems. It allows vehicles to share information with each other through communication channels. It can detect hidden threats and expand the perception scope of autonomous driving to foresee what happens next, thereby further improving the safety, efficiency and comfort of autonomous driving. C-V2X is considered to be one of the key technologies to realize autonomous driving.

4.3.8 *(2/3/4/5) G IoT*

2G is currently the most important connection method for the global cellular IoT, accounting for 42% of all connections (see Table 4.1). 3G comes next, accounting for 22% of all connections (see Table 4.1).

As a large number of telecom operators around the world have completed 2G withdrawals, and the telecom operators which are currently serving have also introduced 2G withdraw plans, 2G and 3G connections will migrate to NB in the next five to eight years. It is estimated that by 2025, the number of global 2G and 3G cellular IoT connections will be less than 1% (see Table 4.1), and NB will account for 45% (see Table 4.1).

Table 4.1 Migration prediction of cellular IoT connection from (2/3/4) G to NB+4G.

Connection method	2018	2025
2G	42%	0%
3G	22%	0%
NB	4%	45%
eMTC	1%	6%
4G	31%	39%
5G	0%	10%

Source: Counterpoint.

4G is still one of the main forces of cellular IoT connections. In 2025, the number of connections will be close to 40% (see Table 4.1).

For high-bandwidth and low-latency applications, 4G can meet the requirements of most scenarios. 5G will play a role in some key scenarios, and the number of connections will account for about 10% in 2025 (see Table 4.1).

In 5G, the two major application scenarios of mMTC and URLLC are the IoT scenarios, which are produced by the aforementioned evolution of the IoT.

4.4 NB-IoT

4.4.1 *Evolution of NB*

From the previous discussion, we know that the evolution process from 1G to 4G is actually a process of continuously increasing network transmission data rates, which is also the guideline for the development of the Internet of People. But in the 4G era, the guideline for 5G evolving has changed, as shown in Figure 4.5. It can be seen from the figure that there are two branches in the development of mobile communication networks. The evolutionary guideline of the above branches is the same as the evolution from 1G to 4G. The other branch is the development process of the IoT, and its evolutionary guideline is different from the Internet of People.

The technical standards formulated by the International Standardization Organization 3GPP are carried out in the post-4G era according to the development branch shown in Figure 4.5. In order to meet the needs of broadband multimedia applications, 3GPP has

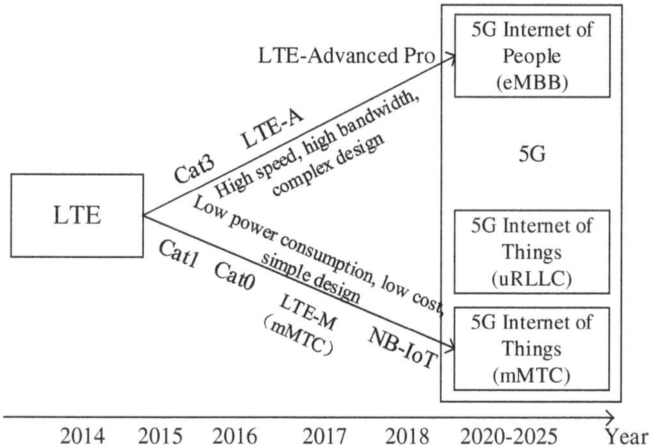

Figure 4.5 Evolution of 4G→5G.

formulated Cat 3, Cat 4 and Cat 5 terminal levels in the R8 (Release 8) version (see Table 1.8 in Chapter 1). LTE-A was formulated in the R9 version. In the R13 version, LTE-Advanced Pro was developed (Pro is Professional, so LTE-A Pro can be referred to as the advanced version of LTE-A) and is called 4.5G.

In order to meet the requirements of the IoT, 3GPP has formulated Cat 1 terminal levels in R8. The Cat 0 terminal is established in R12 (see Table 1.8 in Chapter 1). The eMTC and NB technical standards were formulated in R13. The eMTC standard and the LTE standard with a bandwidth of 1.4 MHz are exactly the same (see Table 1.8 in Chapter 1). Although NB is also derived from LTE, it is redesigned. Therefore, we can only say that NB and LTE are compatible, but they are not equal.

NB is actually developed based on the continuous improvement of the research and development of various countries (especially in China and the United States). The evolution process is shown in Figure 4.6.

4.4.2 *Typical application of NB*

From the perspective of application scenarios, the IoT can be divided into three categories:

- *Type 1*: Strong mobility, frequent switching and small data volume, such as fleet tracking management.

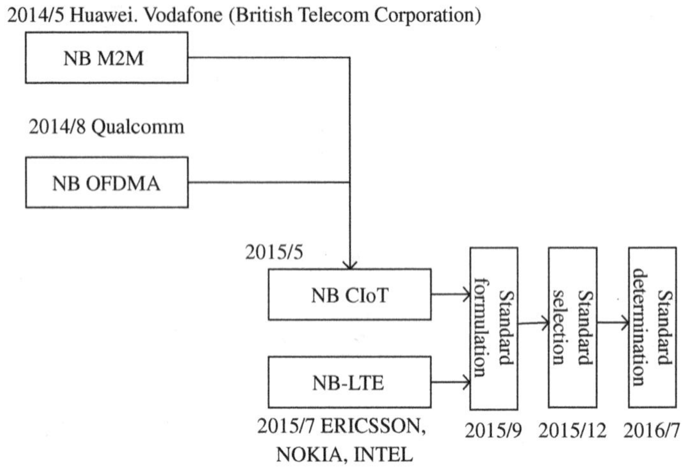

Figure 4.6 NB evolution process.

- *Type 2*: No mobility, large data volume (uplink) and higher bandwidth, such as urban surveillance cameras.
- *Type 3*: No mobility, small data volume and insensitive to delay, such as smart meter reading.

NB belongs to the third category, and it has the following 10 typical applications:

1. Intelligent manhole cover monitoring system
2. Intelligent smoke and air management system
3. Intelligent street light management system
4. Two-wheeler (such as shared bicycle) management system
5. Intelligent parking management system
6. Intelligent power management system
7. Intelligent water supply pipe network monitoring system
8. Intelligent fire hydrant monitoring system
9. Intelligent water supply pipe network detection system
10. Smart door lock system

It should also be pointed out that both NB and eMTC are cellular IoT derived from 4G LTE. This type of IoT is also known as Low-Power

Wide-Area Network (LPWAN) technology. LPWAN has the characteristics of longer coverage distance, lower power consumption, higher safety and reliability. The emergence of this type of new IoT has greatly enriched the application scenarios of IoT and has also met the requirements of more and more industry users, providing opportunities for the explosive development of IoT. As we all know, human beings have made great progress in the consumer Internet and meeting the needs of communication between people. Therefore, the potential for the demographic dividend is much smaller than that of the IoT. For this reason, we should focus on the industrial Internet and the government enterprise Internet. We hope to help traditional industries carry out digital, network and intelligent transformations through advanced IoT technology, communication technology and IT technology, in order to realize the transformation of the entire industrial chain. This is the "Internet +" model advocated by the Chinese government, which is Internet plus various traditional industries. But such a model is not a simple addition of these two, it is actually the use of information and communication technology and the Internet platforms to deeply integrate the two and create a new development ecology.

4.4.3 *NB design ideas*

Based on the knowledge of the above applications, NB should have the following four key technical requirements: enhanced coverage, low power consumption, low cost and large connections. The first three requirements are called the "3C" characteristics of the cellular IoT, and eMTC also needs to have these "3C" characteristics. However, the existing LTE network cannot fully meet the above requirements. Taking the water meter as an example, its wireless transmission environment is poor. Compared to smartphones, the water meters are placed on the ground while mobile phones are held in our hands. The height difference between this two makes the signal strength of the water meter 4 dB worse than that of a smartphone. Besides, the water meter has to be covered, which causes another 10 dB loss. Therefore, the smart water meter coverage needs to be increased by around 20 dB. In terms of large-scale connections, there is a massive amount of devices in the IoT. If the existing LTE network is used to connect them, it will cause the network to overload and paralyze. In addition, if smart devices are composed of mobile phone chip components, the power consumed by mobile phones is also unacceptable. For the above reasons, 3GPP can only formulate NB standards in R13 according to the characteristics of NB (such as the above four technical requirements), and NB does not need to transmit delay-sensitive data packets, so there is no concept of QoS to meet the needs of the IoT at this stage. At the same time, a terminal category Cat NB1 is added. Although NB and LTE are closely related and can be integrated into the existing LTE system, many places are optimized and designed specifically for the IoT based on LTE. Therefore, from a technical point of view, NB is still an independent new air interface technology. This air interface design requirement should be able to meet the technical requirements of NB through improvement and optimization on the premise of being compatible with the LTE air interface. This guideline must be implemented in the design of NB network side and terminal side.

4.4.4 *NB network*

The content of this section mainly refers to the R13 version, and the service data transmitted is IP data or non-IP data (NIP data).

4.4.4.1 *NB network architecture*

The 4G LTE/EPC process has two serious problems for NB terminals.

(1) In order to transmit or receive a few bytes of data (small data packets), the terminal enters the connected state from the idle state, and the network signaling overhead consumed is much greater than the size of the data itself, and its transmission efficiency is very low.
(2) Based on the complex signaling process of LTE/EPC, the power consumption of the terminal is also very large.

In order to solve the above two problems, 3GPP has improved (enhanced) the network architecture and process. The schemes of the control plane and the user plane are optimized, respectively. The former is referred to as CP scheme, and the latter is referred to as UP scheme. NB-IoT must support the CP scheme.

(i) *Principle of CP scheme*: It uses the signaling in the control plane between the terminal and the core network to transmit IP data or NIP data between the terminal and the network. According to this scheme, the terminal carries NAS data packets in the Signaling Radio Bearer (SRB) and IP data or NIP data in the NAS data packets when requesting the RRC connection. This means that the service data is transmitted along with the NAS signaling, instead of first transmitting the NAS signaling to establish a connection to the core network like 4G and then transmitting the service data through the user plane. Therefore, it has the following advantages compared to 4G:

- The transmission efficiency of the system is improved.
- The signaling overhead is reduced, and the power consumption of the terminal is reduced.

(ii) *UP scheme*: The UP scheme introduces the RRC suspend and resume process, which is shown in Figure 4.7. It can be seen from the figure that NB and 4G still support two RRC states, namely the idle state and the connected state. But the idle state of NB is different from 4G. When the RRC connection suspension procedure is performed between the NB terminal and the eNB, the NB terminal transitions from the connected state to the idle state. But at this time, the UE/eNB/MME still saves the terminal RRC context information (see Section 1.11.3.4 in Chapter 1 for details), or the RRC connection still exists. For this reason, when the terminal performs the recovery process, the RRC connection request process can be omitted. This saves AS signaling overhead, thereby reducing terminal power consumption. But the core network establishment process is still needed. Only after the core network is established, the NB terminal

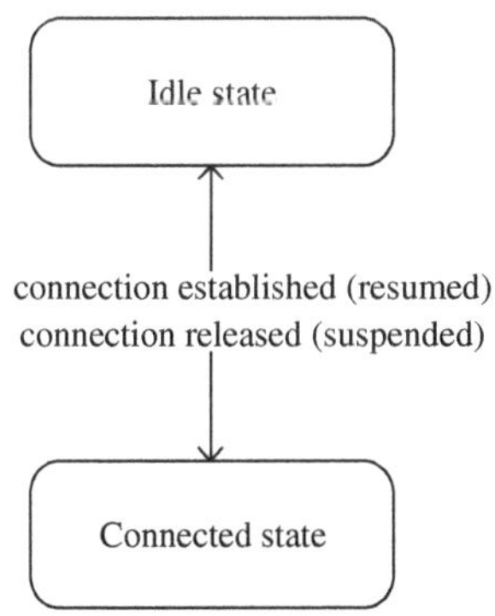

Figure 4.7 RRC suspend/resume process.

can migrate from the idle state to the connected state. It can be seen that the RRC suspend/resume process is different from the RRC inactive state in 5G (see Figure 5.22 in Chapter 5). The difference between the two is as follows:

NB's RRC suspend/resume process involves

- RRC connection establishment,
- core network that is not connected.

The inactive RRC of 5G involves

- RRC connection establishment,
- core network connection establishment.

According to the above two optimization schemes, the network architecture of NB-IoT can be obtained as shown in Figure 4.8. Compared to the 4G network architecture (see Figure 1.56 in Chapter 1), it adds three network elements: Service Capability Exposure Function (SCEF), Service Capability Server (SCS) and Application Server (AS). Correspondingly, three interfaces of S6t, T6 and API are introduced. Next, we use Figure 4.8 to illustrate the data transmission path under the conditions of the CP scheme and the UP scheme.

4.4.4.2 *NB service data transmission path*

(i) *CP scheme*: It can be seen from Figure 4.8 that the CP scheme has the following two transmission paths:

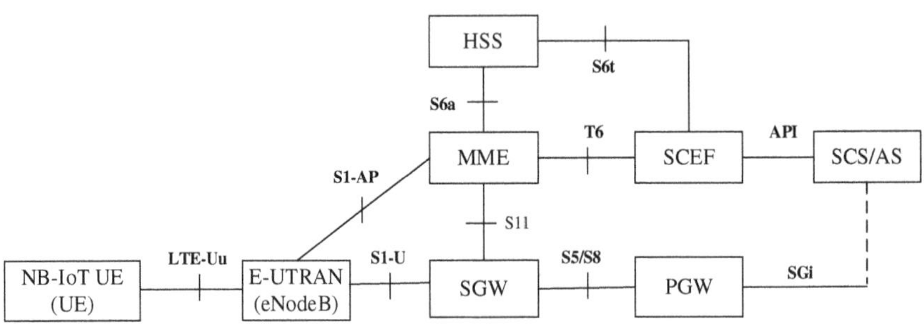

Figure 4.8 NB-IoT network architecture.

path.1 UE↔eNB↔MME↔SCEF↔SCS/AS
path.2 UE↔eNB↔MME↔SGW/PGW↔SCS/AS

(ii) *UP scheme*: It is similar to the 4G transmission path:
path.3 UE↔eNB↔SGW/PGW↔SCS/AS

But unlike 4G, the RRC suspend/resume process is applied in the UP scheme, which reduces signaling overhead and terminal power consumption.

(iii) *NIP data transmission scheme*: From Figure 4.8, we can get the following two schemes for NIP data transmission:

Scheme 1: Transmission via SCEF (CP scheme uses path.1, and UP scheme uses path.3)

The MME extracts the user's NIP data from the NAS signaling, transmits it to SCEF through the T6 interface and then transmits it to SCS/AS.

There are two data exit points in this scheme, that is, all IP data goes out from PGW and NIP data goes out from SCEF, so there will be two charging points in the network.

Scheme 2: Transmission via PGW (CP scheme uses path.2, and UP scheme uses path.3)

The NIP data extracted from the MME is transmitted to the PGW through the connection shown in Figure 4.8 and then transmitted from the PGW to the SCS/AS through a dedicated tunnel.

The advantage of this scheme is that there is only one data exit point, that is, the IP/NIP data are all transmitted out from the PGW. This allows unified billing and unified data processing.

4.4.4.3 *CP scheme protocol stack*

The protocol stack of the CP solution is shown in Figure 4.9. Compared to 4G (see Figure 1.58 in Chapter 1), there is no direct interaction between the PDCP layer, the RLC layer and the RRC in the CP solution protocol stack. This is because there is no need to establish a DRB bearer (data radio bearer) between the UE and the eNB, and there is no user processing, and the NAS layer can be responsible for data encryption, so the PDCP layer can be removed. In this solution, the uplink data is carried in the NAS message included in the uplink RRC message. The

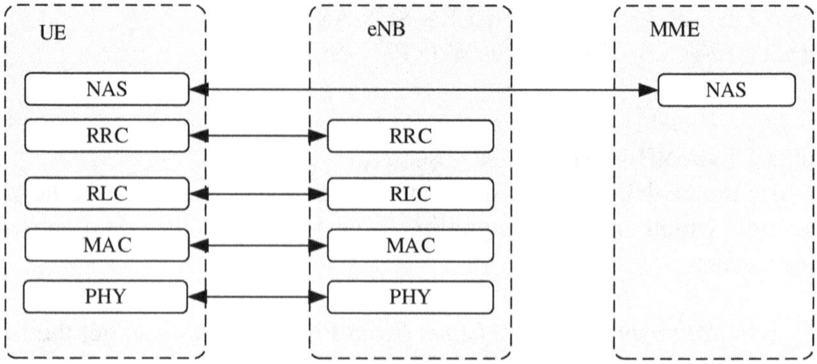

Figure 4.9 Protocol stack of CP solution.

downlink data is carried in the NAS message included in the downlink message.

4.4.4.4 *UP scheme protocol stack*

The protocol stack of the UP scheme is shown in Figure 4.10, which is the same as 4G (see Figure 1.58 in Chapter 1). Both uplink and downlink data need to be carried by the DRB bearer. The PDCP layer in the air interface protocol stack needs to be enabled to provide the AS layer security mode.

However, it should be pointed out that compared to the MAC in 4G, the MAC layer in the R13 version of NB has made a lot of simplifications. It only supports functions such as random access, scheduling request, buffer status report, power headroom report, discontinuous reception (DRX), DPR reporting and HARQ.

4.4.4.5 *Short message transmission scheme*

The R13 version of NB does not support voice services, so short message service (SMS) is a very important service for NB. SMS transmission is shown in Figure 4.11. It has two transmission schemes.

(i) *SGs short message*: The terminal is not attached to the joint, and the MME acts for the joint attachment of the terminal to the CS domain. The short message is transmitted from the terminal to the MME through NAS signaling, then transmitted to the Mobile Switching Center (MSC)

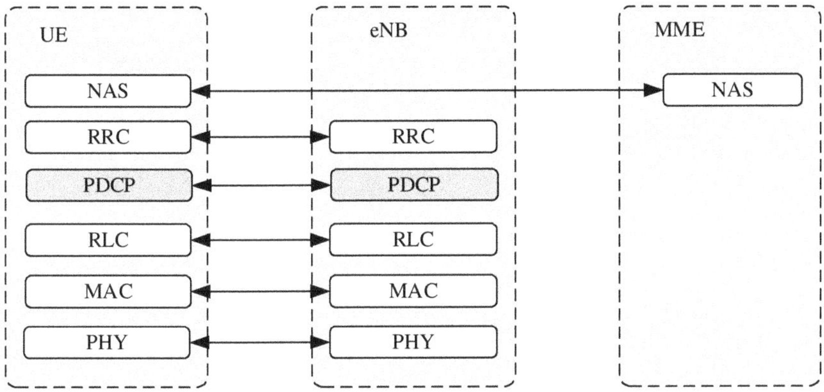

Figure 4.10 UP scheme protocol stack.

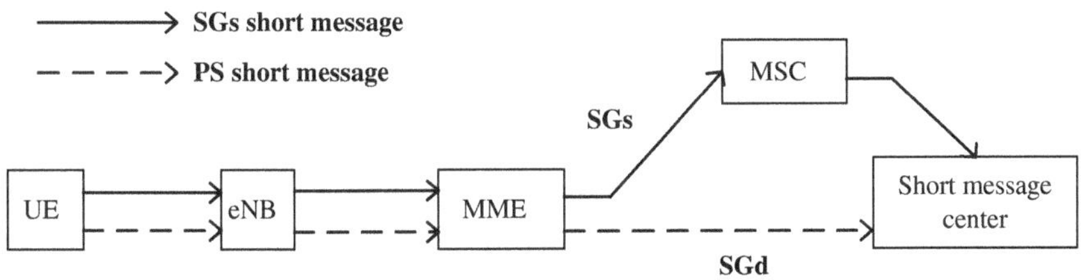

Figure 4.11 SMS transmission.

through the SGs interface and finally transmitted to the short message center by the MSC. The ingenuity of this scheme is that although the terminal is non-joint attachment, it will not fall back to 2G (CSFB). However, because the MME in the 4G core network and the MSC in the 2G core network are connected through the SGs interface, the existing 2G short message technology can still be used to directly realize the transmitting and receiving of short messages under the 4G network. So, this scheme is called SMS Over SGs.

(ii) *PS short message*: The terminal is not attached together, and the short message is transmitted to the MME through the NAS and then transmitted to the short message center through the SGd interface. It can be seen from Figure 4.11 that the program actually uses 4G network to transmit short messages.

Table 4.2 NB working frequency band.

Band number	Uplink frequency range (MHz)	Downlink frequency range (MHz)
1	1920–1980	2110–2170
2	1850–1910	1930–1990
3	1710–1785	1805–1880
5	824–849	869–894
8	880–915	925–960
12	699–716	729–746
13	777–787	746–756
17	704–716	734–746
18	815–830	860–875
19	830–845	875–890
20	832–862	791–821
26	814–849	859–894
28	703–748	758–803
66	1710–1780	2100–2180

4.4.4.6 *Working frequency band (radio frequency)*

In order to be able to use the original LTE antenna feed system and radio frequency circuit, the NB operating frequency band must be selected from the LTE operating frequency band (see Table 1.9 in Chapter 1). 3GPP selects 14 frequency bands as shown in Table 4.2 for NB (R13 only supports half-duplex FDD and does not need to support TDD). The carrier in each frequency band in the table occupies a bandwidth of 180 KHz in the frequency domain. It is equal to the bandwidth of LTE Physical Resource Block (PRB). There are 12 subcarriers in a PRB and the subcarrier spacing is 15 KHz. So, bandwidth = 12 × 15 KHz = 180 KHz. This creates a condition where the NB signal is directly arranged in the LTE network.

Finally, it should be pointed out that the NB network should also include the core network. In order to highlight the key points, the core network is not discussed here.

4.4.5 ***Physical layer***

The technical parameters of the NB physical layer are shown in Table 4.3. The multiple access mode, sub-frame length and LTE in the table are the

Table 4.3 NB physical layer parameters.

Physical layer parameters	Downlink	Uplink
Multiple access	OFDMA	SC-FDMA
Subcarrier spacing	15 KHz	15 KHz/3.75 KHz
Transmit power	43 dBm	23 dBm
Sub-frame length	1 ms	1 ms
TTI length	1 ms	1/8(0.125) ms
SCH modulation (low order)	QPSK	BPSK
SCH modulation (high order)	QPSK	QPSK
Symbol repetition	32	32

Note: TTI, Transmission Time Interval; SCH, Shared Channel.

same. Some of the subcarrier spacing and TTI length are the same as LTE (15 KHz, 1 ms) and some are fractional times of LTE (15 KHz/4 = 3.75 KHz, 1/8 = 0.125 ms). Therefore, they are compatible with LTE. The rest of the parameters in the table are determined according to the needs of NB and are not restricted by LTE (but when NB downlink signal is deployed in the LTE band, its downlink transmission power will be limited by LTE power). The "number of symbol repetitions" in the table refers to the maximum number of retransmissions allowed for the symbol when the receiving end finds a symbol error through error detection.

SC-FDMA in the table is single-carrier frequency division multiple access (Section 1.7.1.2 of Chapter 1).

4.4.5.1 *Working mode*

(i) *NB deployment method*: As mentioned above, the 180 KHz bandwidth occupied by NB is the same as the bandwidth of one PRB in the LTE frame structure. Therefore, NB has three deployment methods as shown in Figure 4.12.

- *Standalone mode*:

Independent deployment is suitable for recultivating the GSM frequency band. The channel bandwidth of GSM is 200 KHz (see Table 2.1 in Chapter 2), which leaves space for the 180 KHz bandwidth of NB and a 10 KHz protection interval on both sides.

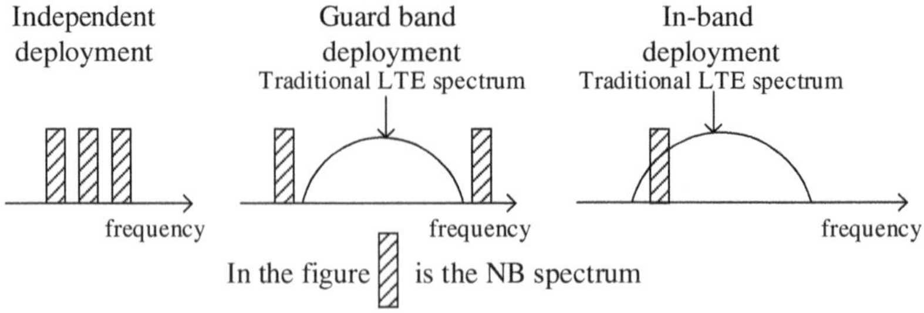

Figure 4.12 NB deployment method.

- *Guard band deployment (guard band) mode*:

It arranges the 200 KHz bandwidth of NB in the LTE edge guard band.

- *In-band deployment mode*:

It uses a certain PRB in the LTE band to place the signal frequency band of the NB. However, some PRBs in the LTE band cannot be occupied.

It should be pointed out that if the NB is deployed independently in the downlink direction, its transmit power can be independently configured. The transmit power of the other two deployment modes is limited by the power of LTE (because of mutual interference between NB and LTE). There is basically no difference between the three deployment modes in the upstream direction.

It should be pointed out that if NB is deployed independently in the downlink direction, its transmit power can be independently configured. The transmit power of the other two modes is limited by the power of LTE (because of mutual interference between NB and LTE). However, in the uplink transmission, there is basically no difference between these three modes.

(ii) *NB uplink Coverage Enhanced Level (CE Level)*: There are three levels of CE Level, namely 144 dB, 154 dB and 164 dB coupling loss. The MCL is 164 dB. The NB terminal selects the number of retransmissions of the uplink information according to the required CE Level. The concept of coupling loss will be discussed later.

(iii) *Duplex mode*: NB only supports half-duplex FDD. Half-duplex (HD) means that it cannot receive when transmitting, and cannot transmit when receiving. While LTE has two modes which are Frequency Division Duplexing (FDD) and Time Division Duplexing (TDD).

4.4.5.2 *Downlink*

(1) According to the characteristics of NB, the downlink physical channel required is less than that of LTE, and only the following three types are defined:

(i) Narrowband Physical Broadcast Channel (NPBCH)
(ii) Narrowband Physical Downlink Control Channel (NPDCCH)
(iii) Narrowband Physical Downlink Shared Channel (NPDSCH)

In addition, two physical signals are defined.

(i) Narrowband Reference Signal (NRS)
(ii) Narrowband Primary Synchronization Signal (NPSS) and Narrowband Secondary Synchronization Signal (NSSS)

(2) Structure of frame, time slot and physical resource block
NB downlink frame and time slot structure are exactly the same as LTE.

- Slot length is 0.5 ms.
- Two time slots form a sub-frame (SF), and the length of the sub-frame $= 2 \times 0.5$ ms $= 1$ ms.
- A radio frame (RF) is composed of 10 sub-frames, and the frame length $= 10 \times 1$ ms $= 10$ ms.

The radio frame structure is shown in Figure 4.13.

(3) Physical Resource Block (PRB)
PRB is composed of time domain and frequency domain, and its structure is exactly the same as LTE. The length of the time domain is 0.5 ms, which is equal to the total length of seven OFDM symbols (each OFDM symbol's length is calculated after adding the cyclic prefix). The bandwidth of the frequency domain is 180 KHz, which is the total bandwidth ($12 \times 15 = 180$) of 12 consecutive subcarriers (the subcarrier spacing is 15 KHz). Therefore, the two-dimensional structure of NB's PRB is shown

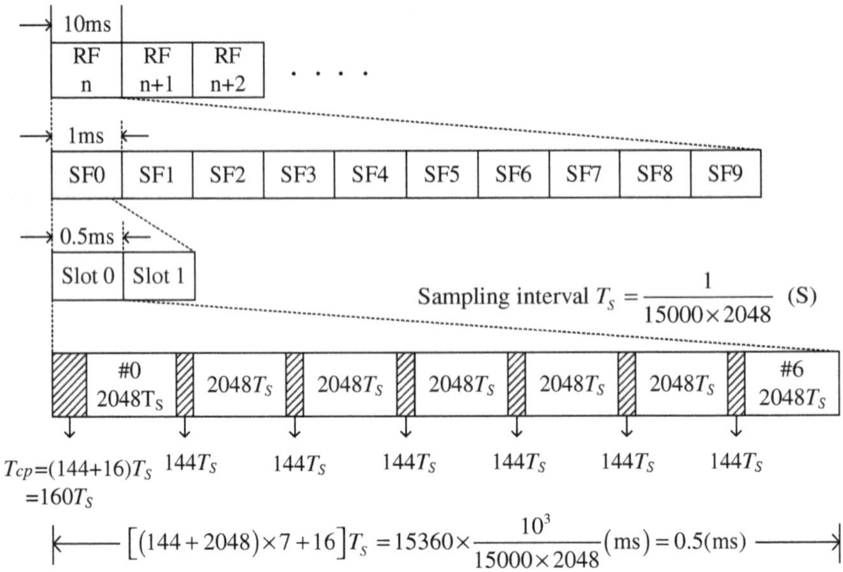

Figure 4.13 NB downlink frame structure.

in Figure 4.14. The smallest physical resource in the figure is called a resource element (RE). The RE occupies one subcarrier spacing and one OFDM symbol in the frequency domain and time domain, respectively. A PRE is composed of 84 (12 × 7) RE.

The above-defined NB parameters are the same as LTE, but according to the needs of NB (discussed in Section 4.4.6.2), it adds the superframe concept, which is defined as a frame composed of 1024 system frames and a system frame composed of 1024 radio frames. In this way, the length of superframe = 1024 × 1024 × 10 ms = 2.9 h, which is almost three hours. Although the superframe is not used in LTE, its length is still an integer multiple of the LTE radio frame. Therefore, both of them are compatible.

4.4.5.3 *Uplink*

(i) According to the requirements of uplink NB, two physical channels and one physical signal are defined:

- Narrowband Physical Uplink Shared Channel (NPUSCH)
- Narrowband Physical Random Access Channel (NPRACH)
- Demodulation Reference Signal (DMRS)

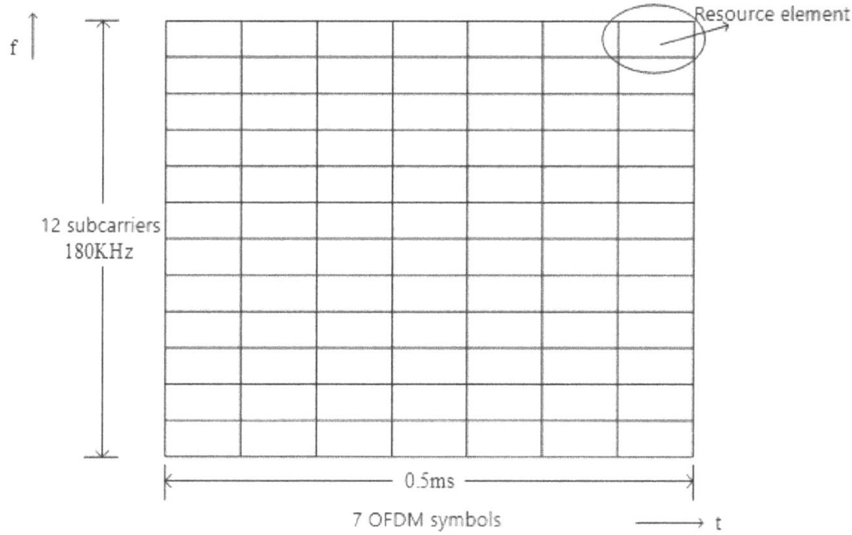

Figure 4.14 NB's PRB structure.

(ii) Time slot and frame structure:
It can be seen from Table 4.2 that the NB downlink only uses the 15 KHz subcarrier spacing. But the uplink can use two kinds of subcarrier spacing, which are 15 KHz and 3.75 KHz (15/4 = 3.75), and the former is four times that of the latter. When the subcarrier spacing is 15 KHz, its time slot and frame structure are the same as the downlink, as shown in Figure 4.13. But when the subcarrier spacing is 3.75 KHz, its time slot length and frame structure are different from the downlink.

• *Time slot*:
Since the physical resources of NB are two-dimensional, that is, frequency domain and time domain, we call the product of these two resources the resource volume. When the subcarrier spacing is 15 KHz and the time slot is 0.5 ms, the resource volume is $15 \times 0.5 = 7.5$ (KHz·ms). When the subcarrier spacing is 3.75 (15/4) KHz, in order to get the same value of resource volume, it is required to increase the time slot by four times, which is 2 ms and the resource volume is still $15 \times 0.5 = 3.75 \times 2 = 7.5$ (KHz·ms). In this way, we can find out that when the subcarrier spacing of the uplink is 3.75 KHz, the time slot length of the uplink is 2 ms.

• *Frame structure*:

In order to make sure the uplink radio frame length is still 10 ms, it is composed of five time slots, and one time slot is still composed of seven OFDM symbols, as shown in Figure 4.15. It can be seen from the figure that the sampling frequency is 1.92 MHz at this time, which is different from the downlink. It can be seen from Figure 4.13 that the downstream sampling frequency $f_s = 1/T_s = 15000 \times 2048$ Hz = 30.72 MHz, which is 16 times the upstream sampling frequency (30.72/1.92 = 16).

(iii) Resource Unit (RU):

The downlink resource scheduling unit is the same as LTE, which is called a resource block pair, and its resource volume is 180 × 0.5 = 90 (KHz·ms). The uplink resource scheduling unit is called resource unit (RU). Since RU has several different values in the frequency domain and the time domain, RU has multiple values as shown in Table 4.4.

- The single-frequency working mode in the table means that an uplink user only occupies one subcarrier. In this case, the minimum uplink bandwidth is 15 KHz or 3.75 KHz. Multi-frequency operation means that an uplink user can occupy several consecutive subcarriers, and

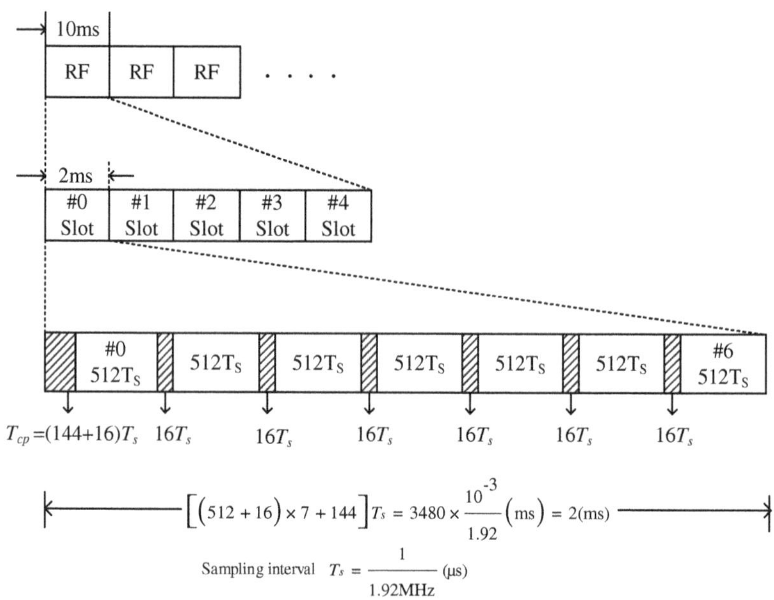

Figure 4.15 Frame structure when subcarrier spacing is 3.75 KHz.

Table 4.4 Uplink RU of NB.

NPUSCH format	Subcarrier spacing	Number of subcarriers	Number of time slots per RU	Duration of each time slot (ms)	Duration of each RU (ms)	Each resource volume (KHz·ms)	Way of working
Format 1	3.75 KHz	1	16	2	32	120	Single-tone
(Transmit uplink data)	15 KHz	1	16	0.5	8	120	
		3	8		4	180	Multi-tone
		6	4		2	180	
		12	2		1	180	
Format 2	3.75 KHz	1	4	2	8	30	Single-tone
(Other use)	15 KHz	1	4	0.5	2	30	

the user bandwidth is wider. It can be seen from the table that only single-frequency operation is supported when the subcarrier spacing is 3.75 KHz. When the subcarrier spacing is 15 KHz, both single-frequency and multi-frequency operations are supported.

- All RU durations are powers of 2.
- All Rus have only three resource capacities of 30, 120 and 180.
- Regardless of single-frequency operation or multi-frequency operation, the multiple access modes for different uplink users are Single-Carrier Frequency Division Multiple Access (SC-FDMA). The downlink is Orthogonal Frequency Division Multiple Access (OFDMA). Both SC-FDMA and OFDMA are orthogonal multiple access methods.

4.4.6 *How to realize the four key technical requirements of NB*

4.4.6.1 *Enhanced uplink coverage*

Before discussing the implementation method, we want to first introduce a parameter that determines the coverage strength, which is called MCL. For uplink, it refers to the path loss from the transmitting antenna of the terminal to the receiving antenna of the base station. When the antenna gain, feeder loss and various other gains (or losses) in the transmission path are not considered, the MCL expression can be written as

$$\mathrm{MCL(dB)} = 10\log_{10}\frac{P_o(\mathrm{mw})}{P_i(\mathrm{mw})} = P_o(\mathrm{dBm}) - P_i(\mathrm{dBm}) \tag{4.1}$$

where P_o is the output power of the terminal transmitter and P_i is the base station receiver sensitivity.

It should be noted that there are also MCL parameters for the downlink. However, P_o and P_i have different meanings in equation (4.1). P_o refers to the base station transmitter power and P_i refers to the terminal receiver sensitivity. Since the cell coverage is determined by the uplink, our main concern is the uplink MCL.

Receiver sensitivity means that under the condition that the system bit error rate P_b is guaranteed (for example, 5G requires transmission reliability of 5 nines, that is, 99.999%, and the corresponding system bit error rate $P_b = 1 \times 10^{-5}$), the minimum signal power is required at the receiver. It can be seen from the physical meaning of P_o and P_i that the input signal power at the input of the uplink transmission path (that is, the output of the terminal transmitter) is P_o. After transmission, the minimum signal power provided at the output of the path (the input of the base station receiver) must be P_i. Therefore, the difference between these two powers P_o(dBm)–P_i(dBm) is the maximum allowable path loss (MAPL) of the transmission path. It can be seen that MCL = MAPL (dB). The larger the MAPL, the stronger the coverage and vice versa. Therefore, the MCL parameter can be used to describe the coverage strength of the cell.

The minimum signal power required at the input of the receiver is related to the equivalent interference power of the receiver. Under the condition of Additive White Gaussian Noise (AWGN), the receiver equivalent interference power is

$$Ne(\text{dBm}) = 10\log_{10} kT + 10\log_{10} B + NF(\text{dB}) + M(\text{dB}) \qquad (4.3)$$

where k is Boltzmann's constant.

$k = 1.38 \times 10^{-23}$ J/K $= 1.38 \times 10^{-23}$ w/Hz · K $= 1.38 \times 10^{-20}$ mw/Hz · K

$T = 290$ K (Room temperature)

$$kT = 1.38 \times 10^{-20} \times 290 \text{ mw/Hz}$$

$$10\log_{10} kT = 10\log_{10} 1.38 \times 290 \times 10^{-20} \cong -174(\text{dBm/Hz})$$

B (Hz) is the signal bandwidth. NF (dB) is the noise figure, which is used to express the influence of the receiver internal noise on the total receiver noise. M (dB) is the uplink interference margin.

Therefore, (4.3) can be written as

$$Ne(\text{dBm}) = -174 + 10\log_{10} B(\text{Hz}) + NF(\text{dB}) + M(\text{dB}) \tag{4.4}$$

Thus, the minimum signal-to-noise ratio (also called the demodulation threshold) required by the receiver is

$$SNR(\text{dB}) = 10\log_{10}\frac{P_i(\text{mw})}{Ne(\text{mw})} = P_i(\text{dBm}) - Ne(\text{dBm}) \tag{4.5}$$

The receiving sensitivity can be obtained by using (4.5) as follows:

$$P_i(\text{dBm}) = SNR(\text{dB}) + Ne(\text{dBm}) \tag{4.6}$$

Substituting (4.4) into (4.6), and then substituting (4.6) into (4.1), we obtain

$$\begin{aligned} MCL(\text{dB}) &= P_0(\text{dBm}) - [SNR(\text{dB})] - 174 + 10\log_{10}B(\text{Hz}) + NF(\text{dB}) + M(\text{dB})] \\ &= P_0(\text{dBm}) + 174 - 10\log_{10}B(\text{Hz}) - NF(\text{dB}) - M(\text{dB}) - SNR(\text{dB}) \end{aligned} \tag{4.7}$$

It can be seen from the above equation that the only two parameters that can be changed in the engineering design are the bandwidth B occupied by the signal and the demodulation threshold SNR of the receiver. For this reason, there are two ways to increase the MCL value.

- Improve uplink Power Spectrum Density (PSD)

The uplink power spectral density can be expressed as

$$PSD_{UL} = \frac{\text{Terminal transmit power}}{\text{Signal bandwidth}} = \frac{P_o}{B}(\text{mw/KHz}) \tag{4.8}$$

According to the definition of equation (4.8), Figure 4.16 shows the power spectrum density of the uplink (UL) signal for GSM (traditional GPRS), eMTC/LTE and NB protocols. It can be seen from the figure that the power spectrum density of the uplink signal of the NB-IoT is higher than that of the original Internet of People GSM and LTE. The fundamental reason for this result is that the data rate of NB's uplink transmission has dropped, so the bandwidth occupied by the uplink signal has become narrower, which is why NB-IoT is called the narrowband IoT. So, how can

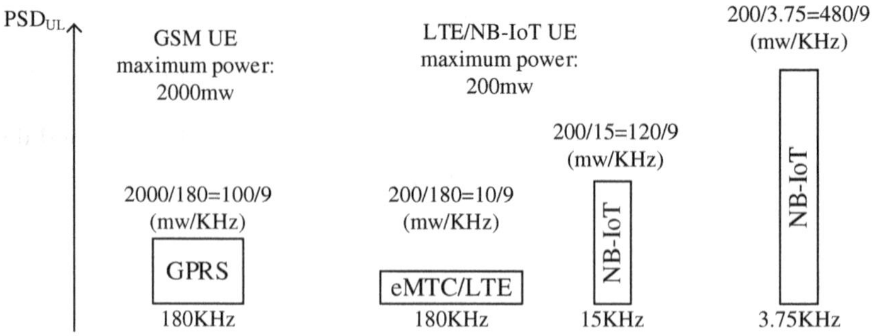

NB-IoT in the figure has two signal bandwidths (15/3.75KHZ)

Figure 4.16 Diagram of several uplink signal PSD.

we increase the MCL and the coverage by increasing the power spectral density of the uplink signal?

According to (4.8) $PSD_{UL} = \frac{P_o}{B}$, where B is the bandwidth occupied by the signal, that is, the frequency band occupied by the symbols transmitted in the channel. If the symbol time width is T_s, the bandwidth occupied by the symbol sequence can be considered as B = $1/T_s$, thus

$$PSD_{UL} = \frac{P_o}{1/T_s} = P_oT_s = E_{s,i} = nE_{b,i} = (\log_2 \mathrm{M})E_{b,i} \tag{4.9}$$

where $E_{s,i}$ is the energy contained in one symbol at the input of the channel. N is the number of binary bits contained in a symbol in the case of a multi-system. M is the order of high-order modulation. $E_{b,i}$ is the energy contained in a bit at the input of the channel. The subscript i represents the input.

It can be seen from the above equation that the power spectral density PSD_{UL} of the uplink signal is the energy contained in one symbol at the input of the channel. In this way, when PSD_{UL} increases, $E_{b,i}$ also increases and vice versa. Figure 4.17 shows the diagram of the receiver demodulating the symbol sequence. It can be seen from the figure that the bit error rate P_b of the system is actually determined by the energy signal-to-noise ratio of one bit $E_{b,o}/N_o$, where N_o is the power spectral density of white noise (that is, $N_o = KT$) and is a constant. Therefore, when PSD_{UL} increases, $E_{s,p}$ and $E_{b,p}$ increase, which makes the signal-to-noise ratio of

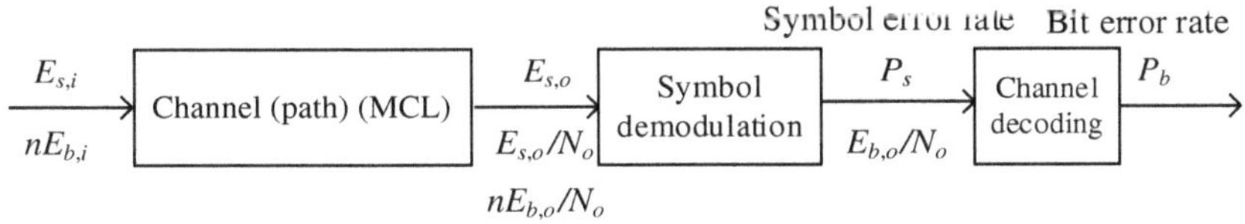

Figure 4.17 Demodulation diagram.

Note: The subscripts *i* and *o* (except N_o) in the figure indicate input and output.

one bit energy during demodulation $E_{b,o}/N_o$ increase, so the system bit error rate P_b decreases. In this way, under the condition of ensuring the system's bit error rate, the maximum allowable path loss MCL is increased when PSD_{UL} increases. The gain produced by increasing MCL is directly invisible in equation (4.7). Therefore, such gain can be integrated into the demodulation threshold SNR (dB), and it is considered that the decibels of the decrease in SNR (dB) are equal to the decibels of the increase in the power spectral density (dB) of the uplink signal. After such a process, such gain can be directly reflected.

- *Data retransmission technology for uplink and downlink*: Retransmission means that the same group of data is transmitted repeatedly. Therefore, the bit error rate P_b of the system is decreased, so that the MCL is increased while ensuring the bit error rate P_b of the system. The gain obtained by retransmission is also merged into the demodulation threshold SNR (dB) of equation (4.7), even if the decibel of the SNR (dB) decreasing is equal to the decibel obtained by retransmission. The maximum number of NB uplink and downlink retransmissions is 128 and 2048, respectively.
- Theoretically speaking, a retransmission makes the SNR drop by 3 dB (because it is equivalent to only ½2 of the original data rate), and four retransmissions make the SNR (dB) drop by 6 dB (because the equivalent data rate is only ¼4 of the original). By analogy, the gain obtained by retransmission is $10\log_{10}$ (number of retransmissions) (dB). In fact, as the number of retransmissions increases, the gain obtained by retransmission is increasingly small and is smaller than the theoretical value.

In addition, as the number of retransmissions increases, the data transmission rate decreases and the spectrum efficiency decreases. If the

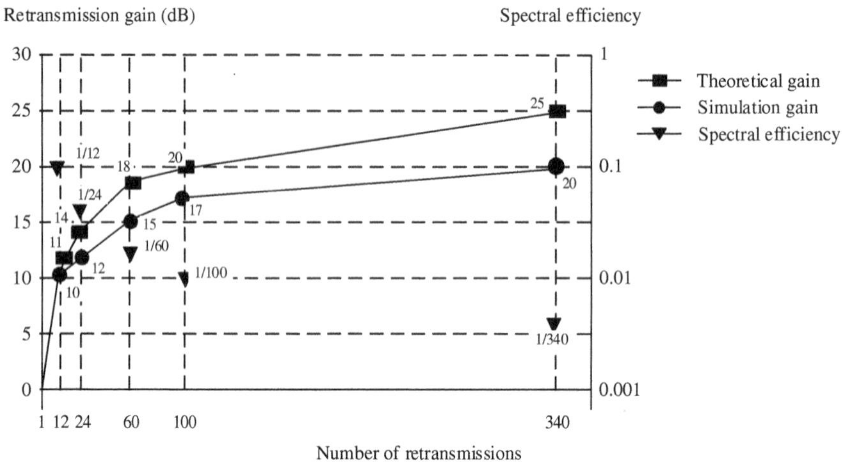

Figure 4.18 Retransmission gain.

minimum data transmission rate allowed at the coverage edge is 100 bps, and the data rate without retransmission is 3.2 kbps, then the maximum number of retransmissions allowed by the system is 32. It can be seen that the maximum allowable number of retransmissions is limited.

Figure 4.18 [8] shows the relationship curves between theoretical gain, simulation gain, spectral efficiency and retransmission times.

The protocol requires the uplink coverage of NB which is 20 dB to be greater than that of GSM. The following discusses how this 20 dB is obtained. NB uplink transmission has two bandwidths of 3.75 KHz and 15 KHz. First, when B is 3.75 KHz, the related parameters of GSM and NB are shown in Table 4.5. It can be seen from this table that the uplink power spectral density of NB is larger than that of GSM by $10\log_{10} 5.33 \approx 7$(dB). If NB allows 24 retransmissions, it can be seen from Figure 4.18 that the retransmission gain obtained is 12 dB. The sum of the above two gains is $7 + 12 = 19$ (dB), which is basically close to 20 dB. When B is 15 KHz, the uplink and downlink parameters of both GSM and NB are shown in Table 4.6. The SNR value in the table is obtained by simulation. It can be seen from the table that the demodulation threshold SNR of NB is much lower than that of GPRS, which is mainly caused by three gains, that is, the gain of increasing the terminal transmission power spectral density, the gain of 32 retransmissions and the low-order modulation of NB (BPSK) gain.

Table 4.5 Parameters of NB (bandwidth 3.75 KHz) and GSM.

Standard	Terminal transmit power (mw)	Occupied broadband (KHz)	PSD_{UL} (mw/KHz)	Ratio
GSM	2000	200	10	/
NG-IoT	200	3.75	53.33	5.33

Table 4.6 Parameters of NB (bandwidth 15 KHz) and GSM.

	Traditional GPRS		NB	
Parameter	DL	UL	NPDSCH In-Band Power (2 × 1.6 w)	NPUSCH 15 KHz
Data rate (bps)	20 k	20 k	670	160
(1) The transmitting power P_o (dBm) of terminal in the occupied bandwidth	43 (20 w)	33 (2000 mw)	35 (3000 mw)	23 (200 mw)
(2) Thermal noise density KT (dBm/Hz)	–174	–174	–174	–174
(3) Occupied bandwidth B (KHz)	180	180	180	15
(4) Receiver noise figure NF (dB)	5	3	5	3
(5) Interference margin M (dB)	0	0	0	0
(6) Equivalent interference power Ne(dBm) = (2)+$10\log_{10}$ ((3)) + (4) + (5)	–116.4*	–118.4	–116.4	–129.3
(7) Required SNR (dB)	10.4	7.4 (consider processing gain)	–12.6	–12
(8) Receiver sensitivity P_i(dBm) = (6) + (7)	–106	–111	–129	–141.3
(9) MCL (dB) = (1) – (8)	149	144	164	164.3

Note: *(6) = –174 + $10\log_{10} 180 \times 10^3$ + 5 + 0 = –116.4.

It can be seen from Table 4.6 that the uplink MCL of NB is 20.3 Db (164.3 – 144 = 20.3) higher than GPRS, and it also meets the requirement of 20 dB greater coverage than GSM. Here we also explain the reason for NB to adopt two kinds of bandwidth (3.75 KHz and 15 KHz) for the uplink. The main reason is that when the 15 KHz bandwidth is selected, in the case of multiple retransmissions, the terminal covering the edge can still have a higher uplink data rate.

4.4.6.2 *Terminal low power consumption*

The terminal of NB-IoT and the terminal of Internet of People (such as a mobile phone) have different working characteristics. The latter needs to be ready to receive/transmit signals at any time (randomly) and because it is in a mobile state, the geographic location of the terminal is also uncertain. However, the former terminal usually receives/transmits signals regularly, and the geographic location is usually fixed. In this way, when the terminal side and the network side of the NB have strict time synchronization (synchronization timer is set on both sides), these differences between the terminals can be fully utilized. On the basis of the original LTE power-saving mode, some new measures are added to make the power consumption of the NB terminal lower. NB has added the following three technical measures:

- *Power-Saving Mode* (*PSM*): Mobile phones need to be on standby at all times, which means that they need to constantly monitor the network, which is very power-consuming. But unlike mobile phones, IoT terminals are in a sleep state most of the time. After staying in the idle state for a period of time to report one or two messages every day or even a week, it enters the sleep state without monitoring air interface messages. For this reason, the NB terminal adds a new sub-state in the LTE idle state, that is the PSM state. In this sub-state, the terminal radio frequency is turned off, which is equivalent to the sleep state. Therefore, the power consumption of the terminal can be greatly reduced. However, the core network side also retains the user context, and when the user enters the idle state and then goes to the connected state, it can be activated without reattachment. The transition between the three working states of the terminal is shown in Figure 4.19.

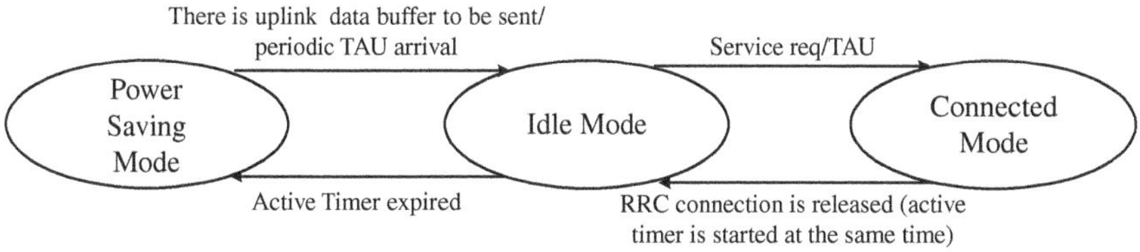

Figure 4.19 NB terminal working status.

It can be seen from the figure that after the terminal enters the idle state from the connected state, the active timer is started. When the activation timer expires, the terminal enters the PSM state from the idle state. After entering the PSM state, the terminal no longer receives paging messages. It seems that the terminal is disconnected from the network, but in fact, the terminal is still registered in the network (because the core network side retains the user's context). In the PSM state, the terminal will exit the PSM state when the following two situations occur:

- *Main call*: When there is data/signaling to be transmitted in the uplink (that is, there is a calling service), the terminal immediately exits the PSM state and enters the connected state by triggering the random-access process.
- *Called*: For the called service, because the terminal can only perform the service during the Active Timer period, the terminal will not exit the PSM state until the periodic Tracking Area Update (TAU) timer expires and needs to execute the periodic TAU. At this time, the called service can be carried out. In the PSM state, after the downlink data arrives at the MME, the MME notifies the SGW to buffer the user data and delay the triggering of paging.
- *Extended discontinuous reception (eDRX)*: LTE mobile phones can receive signals intermittently (non-continuously) to save power. NB has extended this intermittent interval to make it more power-efficient. As shown in Figure 4.20, the maximum 2.9 hours in the figure is the superframe duration discussed earlier.
- *Extended period timer*: Flexible configuration of long-period location update timer RAU/TAU (Route Area Update (RAU); Tracking Area Update (TAU)) should be performed to reduce the number of wake-ups.

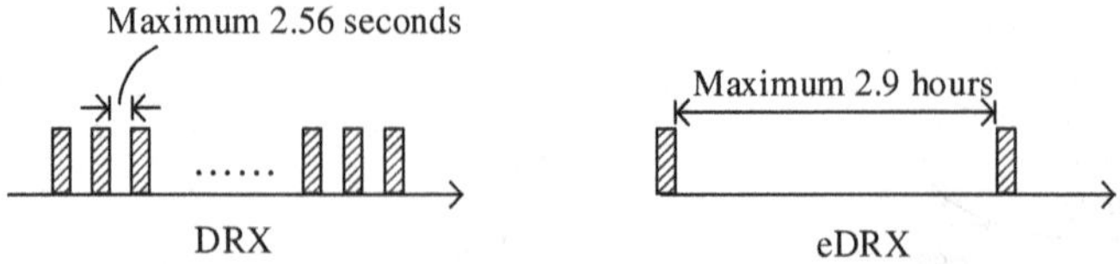

Figure 4.20 Comparison of eDRX and DRX.

4.4.6.3 *Terminal low cost*

In order to reduce the cost of the NB terminal, the following two improvements have been made to the terminal in the design of NB:

- *Simplified software*: Because NB has features such as narrowband and low speed, the redesigned protocol stack and various channels are simplified compared to LTE (such as simplifying the MAC layer, removing the PDCP layer from the control plane and reducing some channels), thereby greatly reducing chip cost. As shown in Figure 4.21, the details of software simplification are not discussed here.
- *Simplified hardware*: Figure 4.22 shows a diagram of hardware comparison between the mainstream LTE Cat 4 terminal (see Table 1.8 in Chapter 1) and the NB terminal. The slashed part in the figure is the simplified hardware part, which has the following parts:
    - The NB terminal supports one antenna, and one antenna can be removed.
    - The NB terminal supports 1TX/1RX, and the 1RX part can be removed.
    - LTE is full-duplex FDD or full-duplex TDD, and NB only supports FDD and is half-duplex FDD. Half-duplex means that during the same time period, it cannot receive when transmitting and cannot transmit when receiving, so the duplexer can be omitted.
    - It has a low sampling rate (1.92 MHz) and low rate (160 bps), which makes the cache Flash/RAM capacity requirement small (28KB).
    - It directly simplifies the IMS protocol stack, which means that NB will not support voice services, but still support SMS services.

By the way, there are two ways to support voice services in LTE: (1) the IMS voice, such as the VoLTE voice mentioned earlier; (2) the

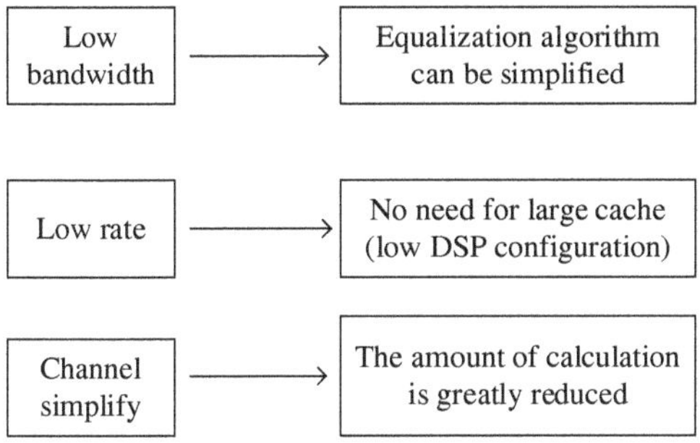

Figure 4.21 The benefits of software simplification.

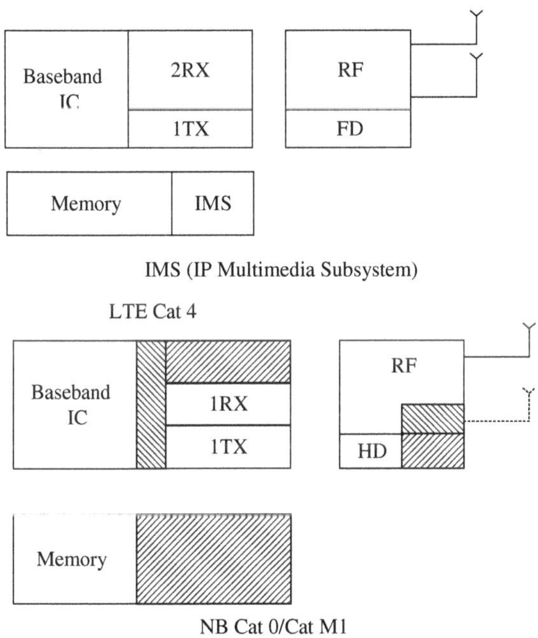

Figure 4.22 Comparison of terminal hardware.

voice of Circuit Switched (CS), such as the CS Fallback (CSFB) mentioned earlier, which indicates that the LTE network falls back to the 2G network when making a call.

4.4.6.4 *Large connection*

The coverage strength mentioned above refers to the coverage strength of uplink radio waves under the premise of ensuring certain technical requirements. The connection discussed here refers to the number of users (terminals) that can be accommodated in a cell. These users are connected to the base station wirelessly. When there are more users, there are more wireless connections and vice versa. The so-called large connection refers to the number of users allowed to be accommodated in the cell, which increases the number of wireless connections in the cell. It is required that the capacity of NB in a cell be larger than 50K, which means that each cell must be able to accommodate at least 50,000 users.

The following first point discusses why 50K capacity is needed in a cell. Obviously, it was proposed based on actual needs. The actual capacity of a cell can be estimated as follows.

- *The area covered by a cell*: In order to achieve truly seamless coverage in the cellular network, each cell must be a regular hexagon, as shown in Figure 4.23. Given that the distance between two adjacent base stations (ISD) is 1732 m, the radius of the circumscribed circle of the regular hexagon is R = ISD/3 = 577.31(732/3) m = 0.5773 km, thus the area of the regular hexagon is obtained as

$$S = \frac{1}{2} nR^2 \sin\alpha = \frac{1}{2} \times 6 \times 0.5773^2 \times \sin(60)$$

$$= \frac{1}{2} \times 6 \times 0.5773^2 \times \frac{\sqrt{3}}{2} \cong 0.87\text{km}^2$$

- *Household density (number of households per square kilometer)*: 3GPP estimates that the household density is 1517 according to the situation in London, England.

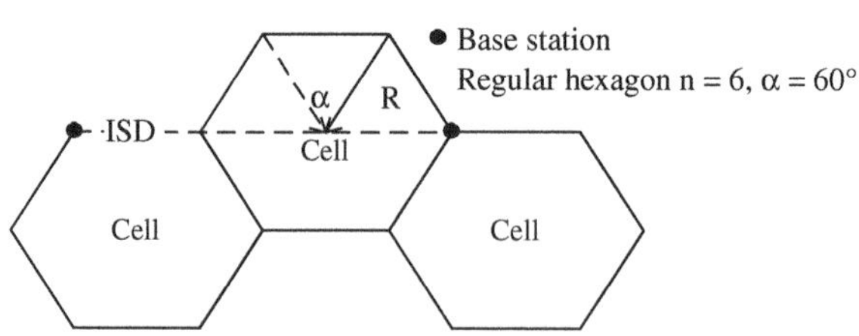

Figure 4.23 A regular hexagonal cell in a cellular network.

- *Number of users in each household*: 3GPP assumes that the number of users (terminals) in each household is 40, and the number of users in a cell can be obtained from the above values:

$$= 1517 \times 40 \times 0.87 \cong 52792 \cong 52\text{ K}$$

Therefore, it is reasonable to determine the cell capacity is larger than 50K.

Next, we discuss how the 50K connections in a single cell are achieved. NB's base station is designed based on the IoT model. IoT users work differently from those with mobile phones. The number of users of the IoT is large, but the data packets transmitted by each user are small, and the transmitted packets are not sensitive to the delay requirements. The design idea of 2G, 3G and 4G base stations is to ensure that users can communicate simultaneously and meet the delay requirements. Based on this approach, the number of user connections can only be controlled at about 1K (the typical number of users in a single cell is 400). But for NB, since it is not sensitive to delay, it can accept more user access and save more user context. In this way, about 50K users can be in one cell at the same time. A large number of users are in a sleep state, but the context information is maintained by the base station and the core network. Once data is transmitted, it can quickly enter the active state. It can be seen that although the number of users on the IoT is 50–100 times the number of users on the Internet of People, it can still work normally.

4.5 eMTC

The design idea of the IoT is determined according to its application scenarios. The application scenarios of the eMTC IoT to be discussed here are different from the NB-IoT. Therefore, the design ideas of the two are also different.

4.5.1 *Typical applications of eMTC*

There are three typical applications for eMTC.

1. Smart wearables, such as smartwatches. Such applications need to provide services such as positioning, voice and low-speed video interaction.

2. Vehicle management, including car management, bus management and logistics tracking. Such applications also need to provide services such as positioning, voice or low-speed video.
3. Electronic advertising, such as elevator guards. Such applications require positioning and low-speed video.

It can be seen from the above three types of typical applications that eMTC must support the following three points:

1. The eMTC terminal is mobile, so the eMTC network must support mobility.
2. The eMTC network needs to support VoLTE voice.
3. The eMTC network needs to support low-speed video.

The NB network cannot support these three points. But although eMTC and NB have the above three differences, it is still an IoT. From the perspective of IoT, it still needs to meet the four key technology requirements of strong coverage, low power consumption, low cost and large connectivity just like NB.

4.5.2 *The design idea of eMTC*

It is known from the above discussion that the requirements for eMTC are the same as for LTE. Therefore, the network architecture of eMTC is the same as that of LTE, so that the LTE network can be directly used as the eMTC network. As mentioned in Chapter 1, LTE supports a total of six channel bandwidths of 1.4 MHz, 3 MHz, 5 MHz, 10 MHz, 15 MHz and 20 MHz. So which bandwidth should be selected for eMTC and why? As mentioned earlier, the IoT must meet the requirements of four key technologies, and these requirements are the basis for selecting LTE bandwidth. Accordingly, a narrow bandwidth should be selected, that is, the 1.4 MHz should be selected.

Choosing a narrowband LTE network will greatly reduce the complexity of the entire system, thereby bringing the following benefits. First, the cost is reduced. Second, power consumption is reduced and battery life is longer. Third, the power spectrum density of the transmitted signal is increased and thus the coverage strength of the network is increased. Fourth, the data rate decreases, so the number of users that the network allows to accommodate increases. These characteristics of narrowband LTE networks are exactly what IoT needs.

According to this, we should choose an LTE network with a channel bandwidth of 1.4 MHz as the eMTC network. In other words, an LTE network with a channel bandwidth of 1.4 MHz is a suitable network for eMTC.

4.5.3 *Evolution of eMTC*

In the first version R8 of LTE formulated by 3GPP in 2008, in addition to Cat 3, Cat 4, and Cat 5 terminal levels that meet broadband multimedia applications, there is also the terminal level Cat 1 with an uplink peak rate of only 5 Mbps (see Chapter 1, Table 1.8), which is suitable for low-rate applications of the IoT. In the early stages of LTE development, Cat 1 was not paid attention to by the industry. However, with the gradual popularity of wearable devices, it was gradually been paid attention to by the industry. However, at this time, the Cat 1 terminal still needs to use two antennas, which is obviously unacceptable for wearable devices such as smart-watches. For this reason, 3GPP later formulated a new terminal class Cat 0 in R12. It only uses one antenna and uses half-duplex FDD. The peak data rate is reduced to 1 Mbps (see Table 1.8 in Chapter 1), and the terminal transmission bandwidth is reduced to 1.4 MHz. In this way, the terminal complexity is reduced to 40% of that of ordinary LTE terminals, which initially meets the cost requirements of the IoT. At this time, although the Cat 0 terminal transmit bandwidth is 1.4 MHz, the bandwidth received by the terminal is still 20 MHz (∵The maximum bandwidth in 4G is 20 MHz). As it is obviously too big, it is unnecessary. Therefore, after 3GPP, a Cat M1 terminal was added to R13, and the uplink and downlink channel bandwidths are both 1.4 MHz. The terminal complexity is further reduced, and Cat M1 is also eMTC. There is also a Cat NB1 grade terminal added to R13 at the same time, which is the NB discussed above.

Through the above review, we can see that the evolution of eMTC is as shown in Figure 4.24.

4.5.4 *Comparison of eMTC and NB system performance*

From the previous discussion, both eMTC and NB are cellular IoT derived from 4G LTE. But the design ideas of the two are different. NB is

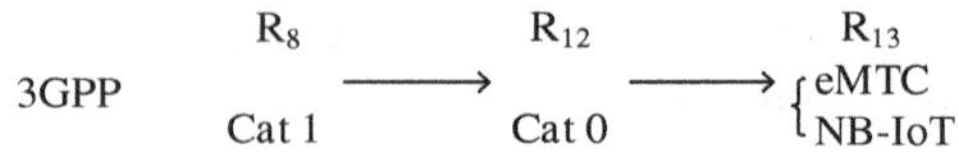

Figure 4.24 eMTC evolution process.

redesigned and optimized on the basis of LTE. Therefore, NB is a new air interface technology. It can only be said that NB and LTE are compatible (because of the compatibility, the NB design can be much simpler.). But it cannot be said that they are the same, and eMTC itself is narrowband LTE. So, it can be said that these two are the same. This shows that the performance of the two systems is different. A comparison of their performance is shown in Table 4.7.

It can be seen from Table 4.7 that the performance requirements related to the four key technologies of the IoT(LPWAN) are both the same. But the other performance requirements are not the same.

Compared to NB, eMTC has five advantages:

1. *High data rate*: Since eMTC uplink and downlink peak rate is 1 Mbps, low-speed video and voice transmission can be guaranteed.
2. *Supports mobility*: NB has poor mobility. It only supports reselection and handover of the terminal in the idle state rather than the connected state. Therefore, it is generally used in scenarios that do not require movement, such as water meters, street lights and manhole covers. But eMTC supports mobility in the connected state and ensures that terminals can be switched seamlessly, so it is more suitable for wearable devices such as smartwatches.
3. *Supports positioning*: TDD-based eMTC is measured by PRS (Positioning Reference Signal) on the base station side. Positioning can be performed without adding a GPS (Global Positioning System) chip. This is beneficial to the application of eMTC in logistics tracking, item tracking and other scenarios.
4. *Supports voice*: eMTC supports VoLTE voice, so it can be widely used in IoT devices related to emergency calls.
5. *Supports LTE network multiplexing*: eMTC can be directly upgraded and deployed based on the existing LTE network by co-siting with the existed LTE base station and antenna system, thereby reducing network construction costs.

Table 4.7 Comparison of eMTC and NB system performance.

Comparison item	NB	eMTC
Frequency band	LTE frequency band	LTE frequency band
Deployment	LTE in-band, guardband, independent	LTE in-band and independent
Duplex mode	HD	HD/FD
Number of antennas	1/2(RxD)	1/2(RxD)
Carrier bandwidth	180 + 20 = 200 KHz	1.4 MHz
Uplink coverage gain	20 dB greater than GSM	15 dB greater than GSM
Uplink MCL	164 dB	156 dB
Peak rate	UL: 250 kbps(MT)/200 kbps(ST) DL: 250 kbps	UL: 1Mbps(FD)/375 kbps(HD) DL: 1Mbps(FD)/300 kbps(HD)
Subcarrier spacing	UL: Single-frequency 15/3.75 KHz Multi-frequency 15 KHz DL: 15 KHz	UL: 15 KHz DL: 15 KHz
TTI	1 ms	1 ms/8 ms
Modulation	BPSK,QPSK	QPSK,16QAM
Multiple access	UL/DL: SC-FDMA/OFDMA	UL/DL: SC-FDMA/OFDMA
Mobility	Not supported, cell reselection	Support, cell handover
Time delay	Second level	100 ms level
Voice	Not supported	Supported (VoLTE)
SMS	Supported	Supported
Cell capacity	Objective: 50 k/cell	Objective: 50 k/cell
Positioning	Supported	Supported
CP optimization	Required	Not required
Power consumption	PSM, eDRX (Period 20.48 s to about 2.9 h)	PSM, eDRX (Period 5.12 s to about 44 min)
Chip cost	Objective <$1 (below eMTC)	Objective $1–2
Module cost	Objective $2–5	Higher than NB
3GPP version	R13, 2016	R13, 2016

Notes: For better understanding, the following details are given for Table 4.7:

HD: Half-Duplex; FD: Full-Duplex; RxD: Receive Data; MT: Multi-Frequency; ST: Single Frequency; TTI: Transmission Time Interval; CP: Cyclic Prefix; PSM: Power-Saving Mode; eDRX: extended Discontinuous Reception

Table 4.8 Comparison of eMTC and NB terminal performance.

	Release 8	Release 8	Release 13	Release 13
Terminal type	Cat 4	Cat 1	Cat M1 (eMTC)	Cat NB1 (NB-IoT)
Downlink peak rate	150 Mbps	10 Mbps	1 Mbps	170 kbps
Uplink peak rate	50 Mbps	5 Mbps	1 Mbps	250 kbps
Number of antennas	2	2	1	1
Duplex mode	FD	FD	FD/HD	HD
Terminal receiving bandwidth	1.08–18 MHz	1.08–18 MHz	1.08 MHz	180 KHz
Terminal transmit power	23 dBm	23 dBm	20/23 dBm	20/23 dBm
Multiplexing with LTE	Yes	Yes	Yes	Yes/No

However, eMTC is worse than NB in terms of coverage capacity and module cost, since the bandwidth of eMTC is greater than NB.

Finally, it should be pointed out that China is currently mainly developing NB IoT. Countries such as Europe, the United States and Japan are mainly developing eMTC IoT.

4.5.5 *Comparison of eMTC and NB terminal performance*

A comparison of the performance of user terminals corresponding to these two networks is shown in Table 4.8.

4.5.6 *Comparison of four types of cellular IoT performance*

The performance comparison of the four types of cellular IoT is shown in Table 4.9.

4.6 NB Application Example: NB Smart Water Meter

At present, traditional water meters and original smart water meters have many problems in solving the disadvantages of water supply enterprises. The NB-IoT smart water meter has the characteristics of high security, wide coverage, large connection and low power consumption, so it can solve the problems of traditional water meters and original

Table 4.9 Comparison of four kinds of IoT performance.

IoT	Spectrum	Channel bandwidth	Throughput	Capacity	Coverage (Downlink)	Time delay	Module cost	Battery life	Network construction
NB	Licensed spectrum	180 KHz	<250 kbps	>50 K	MCL = 164 dB	10 s	<$5	5–10 years	LTE software upgrade
eMTC	Licensed spectrum	1.4 MHz	<1 Mbps	>50K	MCL = 156 dB	<10 ms	$5–10	5–10 years	LTE software upgrade
LoRa	Unlicensed spectrum	125/500 KHz	<50 kbps	—	MCL = 156 dB	—	<$5	5–10 years	New network
Sigfox	Unlicensed spectrum	100 Hz	<100 bps	—	—	—	<$5	5–10 years	New network
				MCL					

smart water meters and meet the development requirements of the water industry.

Thus, Shenzhen Water Affairs Group, China Telecom and Huawei Technologies Co. Ltd. Have jointly launched the world's first smart water business project based on NB technology and released the "NB-IoT Smart Water Meter White Paper" to comprehensively analyze the common disadvantages of traditional smart water meters and the advantages of NB smart water meters. This paper summarized the situation of the NB smart water meter project from the collaborative development of end plane, management plane, and cloud plane and looked forward to the development trend of smart water affairs in order to promote future cooperation and a win-win situation.

With the development of smart cities and the widespread application of new technologies such as the IoT, cloud computing and next-generation Internet, smart water services based on NB technology will become one of the signs of the informatization level of smart city management, which promotes industry innovation, enhances people's livelihood and social value and protects environmental resources.

4.6.1 *NB smart water meter*

4.6.1.1 *Common disadvantages in the water industry*

At present, the business operation model of the water supply industry is water enterprise to water meter to customer. As an important link between water enterprise and customer, the safety, reliability, accuracy and scientific management of water meters are very important.

For a long time, the water supply industry had many disadvantages in operation and management, such as leakage caused by failure of water supply facilities, trade disputes caused by water meter failure, increased operating costs caused by rising labor costs, and so on.

Firstly, poor production and sales constitute the biggest disadvantage of water supply companies. The high gap between production and sales will directly affect the economic benefits of water supply enterprises. According to the 2016 "Urban Water Supply Statistical Yearbook" of the China Urban Water Supply and Drainage Association, in 2015, the average production–sales gap ratio of 603 domestic water supply enterprises was 20.72%, which is calculated on the basis of the water supply volume of 41.5 billion cubic meters in 603 cities. In 2015, the difference between production and sales of water in 603 organized cities (i.e., pipe network leakage, metering loss, stolen water, and useless water) was approximately 8.6 billion cubic meters. This is equivalent to losing one-fifth of the capacity of the Three Gorges Reservoir in one year. The difference in production and sales mainly comes from several aspects. First is physical leakage, which is mainly the leakage of the water supply pipe network. This type of leakage is difficult to detect in time because the pipe network is mostly buried underground. Second is the apparent leakage, which is mainly the measurement error of water meter. Third is free and illegal use of water.

Although in response to the problems of poor production and sales, many domestic water supply companies have adopted many prevention and control measures. However, due to the large water supply area, high cost of manual monitoring and lagging management, the situation of poor control of production and sales is still severe.

Secondly, there is the issue of trade settlement. The water meter serves as an important basis for trade settlement between water companies and customers. In the event of failure or manual data transcription errors, it is inevitable to cause damage to the interests of customers or water supply companies.

At present, most water supply companies manually read the water meter at a cycle of one month, two months or even longer. Thus, it takes a long time to discover if there are any problems, and the historical water use situation cannot be restored, which requires huge manpower and resources to resolve disputes. At the same time, due to the long manual management cycle of water meters, illegal water use becomes easier, which increases the gap in production and sales of water supply enterprises.

Thirdly, rising operating costs is also a problem.

In order to solve the above problems, many water supply companies have begun to use smart water meters.

4.6.1.2 *Problems of traditional smart water meters*

In the process of using traditional smart water meters for water supply companies, problems such as data transmission and data security, high power consumption of smart water meter equipment and weak radio network coverage have increased the difficulty and cost of management. The details are covered in the following sections.

4.6.1.2.1 Data transmission security issues

At present, many water supply companies use small wireless technologies to transmit data from smart water meters. Since it is an ad-hoc network established using an unlicensed spectrum, its anti-interference and data management technology levels are weak, and the security of data transmission and data management is worrying.

4.6.1.2.2 Power consumption issues

The high-power consumption of traditional smart water meters is an urgent problem for water supply companies. If external power is required, communication costs will increase. If the battery is built-in, it will need to be replaced frequently due to high power consumption, which increases the maintenance costs and management difficulty. At present, the national mandatory periodic replacement time for domestic water meters is six years. However, the batteries of many smart water meters can only be used for two or three years. If the water supply company has a high requirement for the frequency of data transmission, the battery may need to be replaced every year or even after a shorter period.

4.6.1.2.3 Network coverage issues

At present, many domestic water meters are installed in corridors, indoors or underground, so the installation environment is complicated. Therefore, the network requirements are higher. To ensure the transmit effect, it is often necessary to install a signal amplifier. But its result is not satisfactory.

4.6.1.2.4 Access problems with a large number of water meters

In the future, water meters in water supply companies will gradually become intelligent. The number of water meters in Shenzhen alone will exceed three million in the future. If such a large number of water meters use small wireless methods for data transmission, it will be difficult to construct this intelligent water meter data communication network. If we

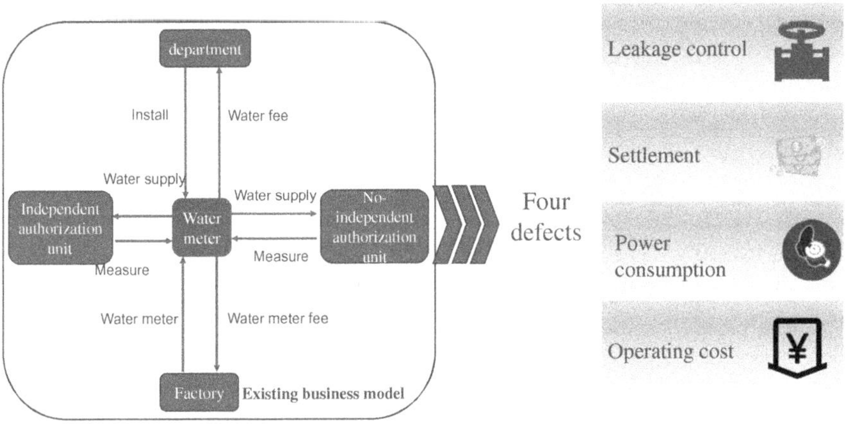

Figure 4.25 Current situation of traditional smart water.

use small areas to manage them separately, it will increase the management difficulty and costs. On the other hand, the communication methods of smart water meters produced by different suppliers are not unified. When water supply companies use smart water meters on a large scale, they need to spend a lot of manpower, material and financial resources to solve the interconnection problems between smart water meters (Figure 4.25).

4.6.1.3 *NB smart water meter practice*

In order to solve the disadvantage of water meter reading and traditional smart water meters, on May 31, 2016, Shenzhen Water Affairs Group, China Telecom and Huawei Technologies Co. Ltd. Signed the "Smart Water Strategic Cooperation Framework Agreement" in Shenzhen, announcing the official launch of NB's smart water strategic cooperation. NB smart water meter aims to provide a reproducible solution, which provides high security, wide coverage, low power consumption and massive connection of wide-area communication technology of NB to realize remote operations such as real-time smart water meters information collection, equipment status monitoring and control command issuance. It analyzes and processes the collected water meter data and status information in time, so that it can achieve more targeted and scientific dynamic management and improve the management efficiency and service level of smart water meters.

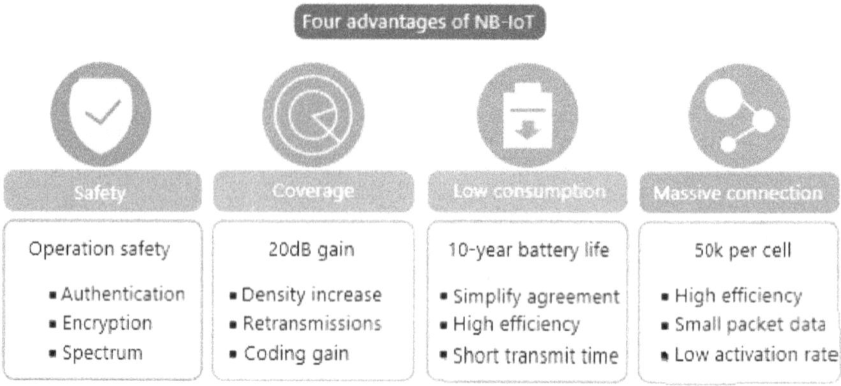

Figure 4.26 The four advantages of NB-IoT.

NB smart water meter has the following characteristics (Figure 4.26):

- It adopts the advanced NB wide-area communication technology and radio spectrum resources to realize real-time collection of urban water supply terminal information.
- End-to-end security technology guarantees data security and access security.
- Information sharing platform can build an intelligent table management platform based on the IoT and cloud computing.

4.6.1.4 *Technical advantages of NB smart water meter*

The technical characteristics of the NB smart water meter help to solve the above-mentioned problems faced by traditional smart water meters and bring high scalability in subsequent application scenarios. They are as follows (Figure 4.27):

- It improves the efficiency of operation, maintenance and management.
- In response to the common disadvantages of water meter reading, NB smart water meter helps reduce meter reading costs, real-time data analysis, scientific meter management, timely troubleshooting, reduce operating costs and improve operating efficiency.
- In response to the data security issues of traditional smart water meters, the end-to-end security management solution of NB ensures data

reliability. The telecommunications network is based on an authorized frequency spectrum, and its anti-interference ability, data security, technical service and other aspects have high-security guarantee. At the same time, it is easy to promote and provides a reliable guarantee for the overall security of the water meter data.

- In response to the power consumption of traditional smart water meters, coupled with low-power signal acquisition units, NB technology significantly reduces energy consumption. Tests have shown that the power consumption of NB water meters can meet the specifications of continuous use for six years at the frequency of data collection twice a day.
- In response to the network coverage problems of traditional smart water meters, wide-area low-power NB technology has the advantages of wide coverage, large capacity and high reliability. It can help solve network coverage problems such as dense or scattered installation of water meters and complex installation location conditions.
- In response to the large-scale connection problem of traditional smart water meters, a unified IoT platform can be established to unify different IoT applications, solve the compatibility of multiple supplier agreements for water company applications and simplify the integration of

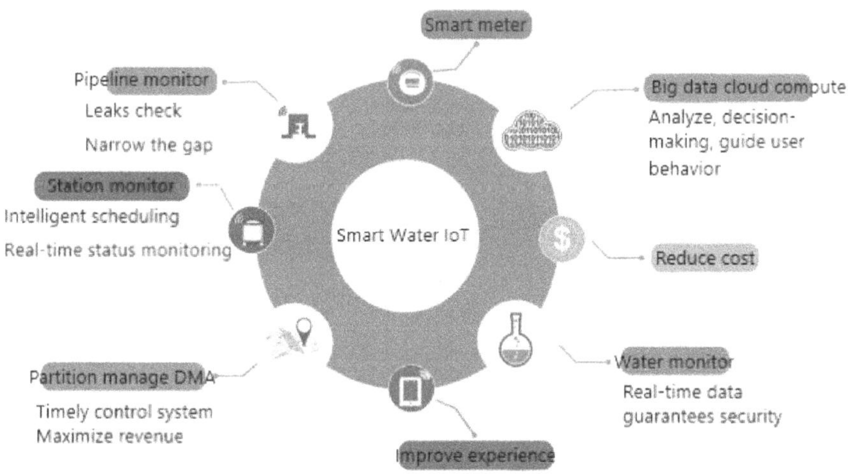

Figure 4.27 Technical advantages of NB smart water meter.

different suppliers, which meets the requirements for the convenience of water meter terminal access and solves the large-scale connection problem of interconnection and interchange.

It can be expanded to other areas of water affairs in the future.

While satisfying the application of smart water meter business, it can be expanded to smart water-related services such as smart pipe network, leakage monitoring, water quality monitoring and other relative business.

In the future, it can be expanded or reproduced to improve the smart water meter business application of the national water company.

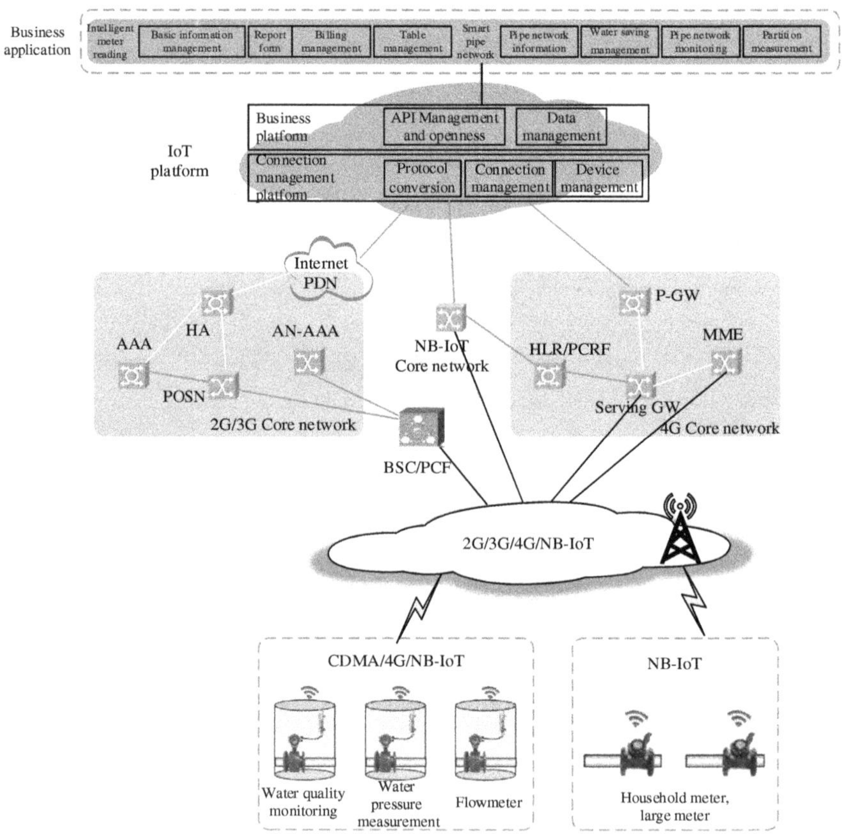

Figure 4.28 NB smart water meter overall solution.

4.6.2 *NB smart water meter solution*

4.6.2.1 *NB smart water meter overall solution* (*Figure 4.28*)

The NB smart water meter solution is an end-to-end industry solution built around the needs of water companies for refined operation and management. With the help of technologies such as the IoT, cloud computing and big data, massive information is collected, analyzed and processed in time to help water companies improve their operational management efficiency, manage the entire production, and service process of the water system in a more refined and dynamic way and realize the internal management innovation and external service innovation of water enterprises so as to realize intelligent management and decision-making.

The NB smart water meter solution is built according to the system architecture of cloud plane, pipe plane and end plane to meet the needs of the future evolution of ICT. The solution includes several levels such as the terminal layer, network layer, cloud platform, IoT platform and water application layer. Through the IoT, cloud computing, big data and other technologies, all levels are integrated into an organic whole to support the construction and rapid launch of smart water applications.

4.6.2.1.1 Terminal layer

Terminal equipment is the basic carrier of the IoT. With the development of the IoT, terminals have gradually evolved from the original dumb terminals to smart terminals. By adding various sensors and communication modules, the terminals are controllable, manageable and interoperable, including smart water meters, water pressure meters, flow meters, water quality monitoring and pumping station monitoring equipment. The terminal equipment realizes the communication capability by integrating the NB standard module and connecting with the NB base station. The smart terminal uploads information to the IoT platform through the NB base station.

4.6.2.1.2 Network layer

The network is the communication basis of the entire IoT. China Telecom is based on the 800 Mbps frequency band NB network to carry water services such as meter reading. It has the characteristics of low speed, low power consumption, low cost and wide coverage, which meets the needs of IoT communications. The ability to use a narrow frequency band to achieve wide coverage and large connections will greatly reduce the cost compared to traditional LTE networks. Compared to other LPWA technologies, the NB network builds a network based on authorized spectrum, and its anti-interference ability, data security, technical services and other aspects have high-security guarantees.

4.6.2.1.3 Cloud computing platform

The cloud computing platform has the characteristics of elastic scalability, high reliability and economy. Through the cloud computing data center, it supports the data aggregation and storage of the IoT and supports the construction and reliable operation of smart water applications. The cloud computing platform is hosted by China Telecom Tianyi Cloud.

- Taking advantage of the resource pool layout of Tianyi Cloud 2 + 31 + X, applications can be deployed nearby to ensure data transmission quality.
- Utilizing the advantages of China Telecom's DCI (Data Center Interconnect) network, data collection points can be deployed nearby. Through the cloud-to-cloud high-speed service, it provides stable bandwidth services that can be activated immediately on demand and realizes stable high-speed data transmission between data collection points and platforms.
- Using China Telecom's cloud security services can resist DDOS attacks worldwide and ensure the stability of data collection.

4.6.2.1.4 IoT platform

The IoT platform provides basic functions such as connection, device, data management and capability opening. It supports the rapid access of multiple IoT terminal devices and the rapid integration of water applications and supports customers in focusing on applications and business innovation. The platform provides connection status queries and functions, such as connection awareness, connection diagnosis and connection control. The access of different terminals is realized through unified protocols and interfaces. The upper-level industry applications do not need to care about the specific physical connection and data transmission of the terminal equipment and realize the terminal object management. The platform provides flexible and efficient data management, including data collection, classification, structured storage, data calling and usage analysis. It also provides analytical business customized reports and has a business modular design to ensure flexible business logic arrangement, which meets the rapid development needs of industry applications.

4.6.2.1.5 Application layer

IoT applications are the upper control core of the IoT business. Based on the NB smart pipeline and IoT platform, the water industry can focus on its own application development, so that IoT can be better embodied. The smart water application system obtains data from the device layer through the IoT platform and manages various water equipment such as water meters and flow meters. The IoT platform can shield the differences of device types and connection protocols and use unified standards and protocols to connect to the application layer.

4.6.2.2 *Advantages of NB smart water meter solution*

4.6.2.2.1 Better signal coverage in 800 Mbps frequency band

China Telecom's 800 Mbps frequency band has great advantages in signal penetration and coverage and can fully guarantee the stability and reliability of data signal transmission for services such as smart water meters in complex application environments. China Telecom has started 800 Mbps NB pilot work. It is expected that the NB end-to-end business operation test and verification would be completed by the end of June 2017, and it would be the first to conduct commercial business trials nationwide.

4.6.2.2.2 Private network services ensure safe and reliable data transmission

China Telecom's IoT business is based on dedicated network elements for the IoT mobile core network, which is separated from traditional networks to prevent malicious network attacks. It can help enterprises achieve terminal access control and flow control, as well as the isolation of internal and external networks and provide customers with reliable and stable networks. At the same time, China Telecom provides IoT wireless Virtual Private Dial Network (VPDN) value-added services. Layer 2 Tunneling Protocol (L2TP) or Generic Routing Encapsulation (GRE) tunnel technology is used to build a virtual private network isolated from the public Internet for IoT customers to meet their needs to access their internal IoT networks.

4.6.2.2.3 Intelligent network is perceivable and manageable

By integrating communication network capabilities and IT operation capabilities, it provides water companies and water meter manufacturers with perceivable, diagnosable and controllable intelligent networks, which can meet the needs of customers for real-time independent query and management of the working status and communication status of the terminal. At the same time, in order to meet the different use requirements of the IoT customers in the production test phase, inventory phase and formal use phase of the terminal manufacturing and sales process, it provides the functions of one activation period, silent period and secondary activation period of the number.

4.6.2.2.4 IoT platform enables rapid construction and deployment of water applications

The rich protocol adaptation capabilities of the IoT platform can support the access of a large number of diverse terminal devices, based on unified protocols and unified interfaces. The IoT platform can realize the integration of different brands, different models of water meters and different application platforms for access, management and interconnection. At the same time, the open API can realize the rapid integration of water affairs applications and realize rapid construction and deployment.

4.7 NB Becomes a 5G Technical Standard

On July 9, 2020, the ITU formally incorporated NB technology into the 5G standard system in the ITU-R WP5D #35 meeting.

As a streamlined version of LTE, how did NB suddenly become a 5G technical standard?

This is still because of the application scenarios of 5G.

As we all know, the ITU has defined three major application scenarios of 5G (see Chapter 5 for details) as reference targets for 5G indicators. In other words, the real 5G must have the ability to cope with these scenarios.

Among the three scenarios, the one that has received the most public attention and is most familiar to us is eMBB.

In the past few years, 4G has achieved great success, and actually, 4G is MBB (Mobile Broadband). It cooperates with smartphones to thoroughly stimulate users' mobile Internet demand, promotes the explosive growth of traffic and also promotes the prosperity of the entire information and communication industry chain.

The 5G eMBB is a further upgrade of bandwidth and speed on the basis of 4G MBB.

Because of this, the industry has high hopes for eMBB and regards eMBB as the main part of 5G. Many people believe that the high speed and large bandwidth of eMBB can provide a better user experience and can stimulate further prosperity of the digital age, and commercial construction should be promoted first.

But the real significance of 5G lies in the IoT scenario. Its main service targets are vertical industries including industrial Internet and smart cities.

This leads to a very embarrassing situation. Those who design 5G hope that people can pay more attention to URLLC and mMTC. But the majority of people are interested in eMBB.

In reality, the situation of URLLC and mMTC is indeed a little bit subtle. Everyone knows that they are in demand, but no one knows where the explosive demand is and when it will appear.

For URLLC, it is a bit unique since its key requirement is a delay, which is impossible to achieve with the existing technology, and it must be met by 5GNR and slicing technology.

The most troublesome is mMTC, as its situation is completely different from the previous two.

The capacity of NB is 50,000 terminals per cell. Test data shows that NB has a coverage radius of about 2 kilometers. After calculation, we can see that it can support around 4000 terminals per square kilometer.

In 5G mMTC scenario, the ITU's indicator requirement is one million terminals per square kilometer. There seems to be a big gap between these two, and in reality, there aren't many users who need to access one million terminals in one square kilometer.

According to the current commercial network, there are only 1000 terminals in one square kilometer at most. In other words, NB is basically enough to support those terminals. Since the demand for mMTC is not clear, it is unnecessary to develop a brand new technology since the cost and investment are too high. Therefore, 3GPP hopes to use the existing NB technology and it is the most reasonable choice for the entire communications industry behind 3GPP.

It has been more than three years since the NB standard was introduced in 2016. The expected explosive growth of IoT applications did not come. Under this circumstance, insisting on promoting the upgrade of NB is unnecessary.

On March 19–22, 2018, at the 79th plenary meeting of the 3GPP radio access network held in Chennai, India, the delegates made it clear that the R16 version of the 5G new air interface would not be researched and standardized for low-power wide-area (LPWAN) IoT use cases. LPWAN IoT use cases would continue to rely on the evolution of NB.

In other words, 3GPP postponed the standardization of mMTC in R16 and incorporated the existing NB standard into the 5G IoT.

At the same time, 3GPP has been working hard to prove to the ITU that the NB standard can meet the ITU's vision and demand for low power consumption and wide coverage of the 5G IoT.

At the 3GPP meeting held in Athens, Greece, relevant companies submitted an assessment report on the demand for NB and eMTC to meet the 5G mMTC connection density. The conclusion of the related report shows that NB and eMTC meet the 5G mMTC connection density requirements proposed by the ITU, which lays the foundation for the inclusion of NB into the 5G IoT family.

In order to allow NB to better integrate into 5G, 3GPP has also done a lot of work, such as implement NB support in 5GNR deployment (that is, allow NB to use 5G frequency bands) and allow NB to use 5G network system-side equipment (such as base stations and core networks).

It can be said that NB is fully equipped with the basic conditions to integrate into the 5G network after a smooth upgrade.

This is the reason why the ITU officially incorporated NB technology into the 5G standard system.

4.8 Future of the IoT

At the "2020 Zhiwu·AioT Industry Leaders Summit" held in Shanghai on August 26, 2020, Huawei's Radio Network Strategy and Business Development Department and Chief Strategy Officer Xu Weizhong said that NB-IoT has developed well in the Chinese market and has accumulated over 100 million.

Xu Weizhong said that 5G provides the possibility of the interconnection of everything and will bring five major industry opportunities. The addition of cloud, computing power and AI through 5G will provide unlimited possibilities for the entire fourth industrial revolution. The so-called five major opportunities refer to the five major opportunities that 5G will drive, which include connectivity, AI, cloud, computing and industry applications.

He pointed out that 5G technically supports multiple types of applications on one network. It also serves four types of connections with different scenarios, namely massive IoT, broadband IoT, critical IoT and industry IoT. It is worth mentioning that the NB-IoT technology in massive IoT has officially become a 5G standard at the ITU meeting on July 9, 2020 (Figure 4.29). At present, NB-IoT has been applied in hundreds of application

Figure 4.29 Conference site.

fields and has achieved tens of millions of connections in the four fields of water meter, gas meter, smoke detector and electric vehicle.

After the efforts of the past few years, NB-IoT is developing well in the Chinese market. As of February 2020, the cumulative number of NB-IoT connections in China has exceeded 100 million. It is expected that the total number of new NB-IoT this year will exceed the number of 2G IoT connections. In 2021, it will fully replace the original 2G IoT connection, and it will exceed one billion connections by 2025.

At the same time, Xu Weizhong also said that although 5G has a bright future, 5G industry-wide scale development still faces many challenges. To this end, he made three suggestions. First, the price of 5G modules is high, and it is recommended that policies and industries promote industrial-grade low-cost modules. Second, there is a lack of 5G industry integration standards, and it is recommended that industry management departments and alliances lead unified industry standards. Third, the current 5G ecological chain is still immature and the number of partners is small. It is recommended that various industries should establish a partner circle.

At the same time, he said that Huawei has established 5G Open Lab certification laboratories in Xi'an, Beijing, Chengdu, Shanghai and Shenzhen, opening the 5G to B partnership circle, joining hands with the industry, arousing the potential of 5G and driving the digital transformation of the entire industry.

4.9 NB-IoT Standard Progress

From 2015 to 2020, 3GPP Rel-13 formally established the NB-IoT project. NB-IoT has been in the 3GPP Rel-14, Rel-15 and Rel-16 versions in terms of positioning, broadcasting, power-saving and improving the efficiency of the uplink and downlink transmission. These versions have been highly enhanced and the progress of the standard is shown in Figure 4.30.

Rel-13 NB-IoT was established at the RAN#69 meeting in September 2015. In June 2016, the standardization of the core part was basically completed. Rel-13 NB-IoT downlink adopts the Orthogonal Frequency Division Multiplexing (OFDMA) scheme based on 15 KHz subcarrier spacing. The uplink adopts SC-FDMA technology and supports

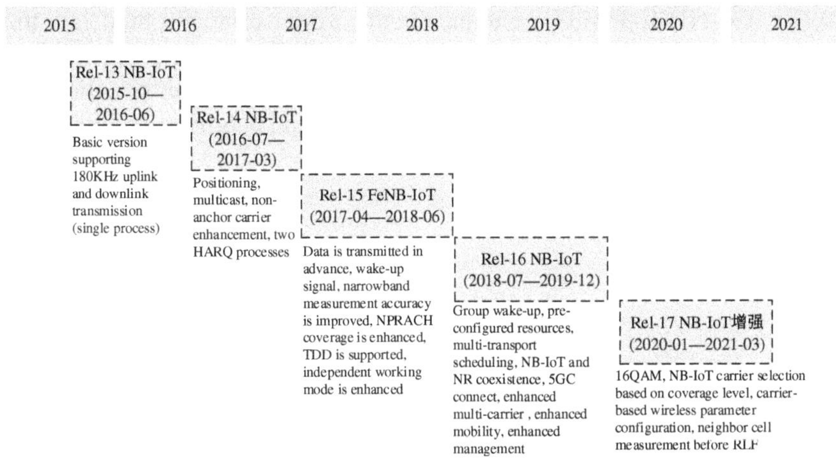

Figure 4.30 NB-IoT standard progress.

single-tone transmission and multi-tone transmission. The terminal needs to indicate the ability to support single-tone transmission and multi-tone transmission.

Rel-14 NB-IoT enhancement work was approved at the RAN#72 meeting in June 2016. The standardization of core parts was completed in March 2017. Rel-14 NB-IoT enhancement mainly introduces positioning function, multicast function, non-anchor carrier enhancement and two HARQ processes support.

Re-15 NB-IOT enhancement work was approved at the RAN#75 meeting in March 2017. The standardization of the core parts was completed in June 2018. Re-15 NB-loT enhancement mainly introduces data early transmission function, wake-up signal, narrowband measurement accuracy improvement, NPRACH coverage enhancement, TDD support and independent working mode enhancement.

Re-16 NB-IoT enhancement work was established in the RAN#80 meeting in June 2018. The standardization of the core parts will be completed in December 2019. Rel-16 NB-IoT enhancement mainly introduces group wake-up signals, pre-configured uplink resources, multi-transport block (TB) scheduling, NB-IoT and NR coexistence, connection to 5GC, enhanced multi-carrier operation, enhanced mobility and enhanced network management tools.

Re-17 NB-IoT/MTC enhancement work was initiated at the RAN#86 meeting in December 2019. It mainly introduces unicast uplink and downlink 16QAM, coverage level-based NB-IoT carrier selection and carrier-based wireless parameter configuration and neighbor cell measurement before Radio Link Failure (RLF).

Chapter 5

5G Overview

As mentioned before, starting from 2G, the ITU has been formulating standards for mobile communications, and the official name, technical indicators and application scenarios of 5G have been confirmed.

However, ITU is mainly responsible for standard project establishment and achievement appraisal, not for specific technical research. The real standard research and formulation work is completed by 3GPP.

3GPP was established in 1998 to study and formulate WCDMA, which is the most successful 3G standard and later introduced the 4G LTE standard.

The formulation of 5G standards is divided into three stages:

(1) Phase 1 standard is Release 15 (R15 for short) released by 3GPP, focusing on eMBB. In other words, this standard is to meet the requirements of Internet connection. The R15 version was introduced in September 2018, and the relevant standards have been officially released.
(2) Phase 2 is Release 16 (R16 for short), mainly for URLLC application scenarios. The relevant standards were also completed in June 2020.
(3) Phase 3 is the R17 version. It will be developed for mMTC application scenarios.

5.1 Three Application Scenarios of 5G Network

With the development of society, the mobile Internet and IoT need to be able to transmit new services, and they cannot be transmitted by the current 4G network due to performance limitations. In order to solve this problem, the 4G network must be further developed. It is expected that by 2020 the fifth generation (5G) of cellular mobile communication networks will be put into commercial use. The 5G network integrates the Internet of People and IoT, and the integration of these two networks is called the 5G network. From the perspective of application, 3GPP divides the services into three categories: enhanced Mobile BroadBand (eMBB), massive Machine Type Communication (mMTC) and Ultra-Reliable and Low Latency Communication (URLLC). Therefore, 5G must be able to work under these three application scenarios. Due to the emergence of these brand new services, the mobile Internet will enter the era of intelligent Internet. Among them, eMBB is a business on the Internet of People, and mMTC and URLLC are IoT businesses. From the perspective of application, the fastest commercial scenario is eMBB, which is expected to be officially commercialized in 2020. South Korea and the United States have announced their commercialization in April 2019. Affected by the timing of communication standards and the maturity of the industry chain in the mMTC and URLLC scenarios, the full version of 5G will not be commercially available until 2021 at the earliest, and it will take at least three to five years to be relatively mature. This is because the purpose of the comprehensive coverage of the 5G network lies in uninterrupted network services such as the autopilot vehicle mentioned in the following. If the deployment of 5G is only completed in the city, then when cars drive out of the city into suburbs or rural areas, there will be no network coverage. In addition, technology and law need to be adapted to the development of 5G. Taking autopilot vehicle as an example, if a traffic accident occurs due to technical loopholes, how to determine the responsibility? From this point of view, even if 5G technology is mature, without comprehensive coverage and legal cooperation, it will still take a long time to realize the full commercial use of 5G.

5.1.1 *eMBB*

From the above discussion, it is known that 4G networks can be classified as the MBB application scenario. Therefore, the 5G network working in the eMBB scenario is obviously an evolution of 4G networks, that is, an upgraded version of 4G LTE. In the eMBB scenario, 5G needs to transmit new services, mainly three-dimensional (3D) video and

Ultra-High-Definition (UHD) video and other high-traffic mobile broadband services.

5.1.1.1 *3D video* (*VR and AR*)

The sounds in the real (physical) world are all stereo, and the scenes that we see in the real world are all 3D. In order to reproduce (imitate) the stereo effect, people invented two-channel stereo and surround sound technology, so that the information that we receive has changed from monophonic to stereo. The reproduced sound is closer to the sound of the real world. In the same way, in order to reproduce (imitate) 3D vision, we invented VR and AR technology. Thus, the information that we receive has changed from 2D (for example, the currently used 4G phones can only show 2D images) to 3D, which makes the reproduced image closer to the real world. Borrowing the concept of "stereo" in hearing, we can call the image reproduced by VR and AR a "stereo image".

VR: Virtual reality is the use of computer simulation to generate a 3D virtual world, providing users with visual, auditory, tactile and other sensory simulations, so that users can observe things in 3D space in a timely and unrestricted way as if they were on the scene (so-called immersion and telepresence). It can be seen that VR simulates people's experiences in the real world. The sense of immersion it provides will only make our consciousness believe that we are actually in there. But other parts of our body are not in that environment. So, VR is the virtual world, and the world seen through VR is all virtual. The typical device to experience VR is the Oculus Rift helmet (head-mounted display) launched by Facebook. When putting on this helmet, we can enter the virtual world, where we can interact with others, play games and even give a high five.

AR: Augmented reality is to use computer technology to apply virtual information to the real world and superimpose the real environment and virtual objects on the same screen in real time or both can exist in the same space. AR uses digital information that generates virtual objects to enhance the real world. So, AR is the real world plus digital information. In other words, part of the world seen through AR is real and part of it is virtual. The typical device to experience AR is Google Glass. When you wear this glass and look at a person in the real world, relevant information about this person is displayed on this glass, such as name, age and other information, which enhances the understanding of this person.

For VR and AR, the complete information chain should include five parts: acquisition, processing, storage, transmission and display. The mobile Internet consisting of 5G networks must have these five parts:

1. *Information acquisition*: This involves the extraction of useful information from a given environment background (such as noise). For people, obtaining information mainly relies on the eyes, ears, nose, tongue and other organs. In particular, vision accounts for 90% of the information obtained. Due to the development of computer science and digital technology, it is not a problem to obtain VR and AR signals. These two signals can be generated by Internet operators, and AR can also be generated by dual-camera phones.
2. *Information processing*: This involves the processing of the acquired information (such as reducing redundancy or digital transformation) in order to obtain the signal that we need. There is no difficulty in processing VR and AR signals with the microprocessor or cloud computing.
3. *Information storage*: This involves the usage of various media to transfer information from one point to another in the time domain. Books are an ancient method of information storage. People have invented a variety of storage media, such as massive cloud storage and large-capacity memory chips, so there is no problem with the storage of VR and AR information.
4. *Information display*: This involves the reproduction of information in a certain way for human hearing or visual perception. It requires accuracy and completeness when reproducing the information. For VR and AR information display, the "Ding stereo glasses" made by Professor Ding Shouqian of China using the principle of non-paraxial optics can be used. This product was exhibited at the 2012 Beijing International Communication Exhibition. It can be used on mobile phones, laptops, TVs or other screens to enjoy 3D shows with high definition, brightness and full color.
5. *Information transmission*: This involves the transfer of information from one point to another in the space domain through certain methods. The flames of war are an ancient means of transmission. VR and AR signals need to be transmitted by mobile Internet, so that VR and AR data can reach the user terminal (such as a mobile phone) from the

source (server or cloud) through the Internet and 5G network. Since both VR and AR are 3D video signals, the corresponding video resolution needs to reach the resolution of the human eyes. Therefore, the data rate of VR and AR is much higher than the 2D video signal transmitted in the 4G network. In addition, the immersive experience is also required, which is not needed in 4G networks. Due to the above two reasons, the network speed of the 5G network must be above 300 Mbps, and the end-to-end delay should be less than 7–12 ms. These two requirements cannot be achieved by the 4G network. For this reason, the following two requirements are proposed for the eMBB scenario of 5G. First, in order to ensure more than 300 Mbps speed, the 5G network downlink data peak rate is required to be 20 Gbps. Second, in order to ensure less than 7–12 ms end-to-end delay, the air interface one-way delay (wireless delay) of 5G is required to be less than 4 ms. The end-to-end delay refers to the total time cost for an IP data package (user plane) from the source node to its destination and being successfully received. The above-mentioned 7–12 ms end-to-end delay includes the sum time cost for the data package (such as a frame of data) transmitting to the destination through 5G core network and radio access network, which is related to the length of data, throughput, wired and wireless time delay and other related factors. The above-mentioned 4 ms air interface delay is only the wireless delay from the source to the destination between two MAC layers of the base station and the terminal.

5.1.1.2 *UHD video*

4G network transmits 2K video with a resolution of 1920×1080, where 1920 represents the number of pixels contained in a line of the display screen. In the binary domain, $2K = 2^{11} = 2048$, and 1920 is close to 2048, so it is called 2K video. Among them, 1080 refers to the number of lines scanned in one screen of the display screen, so the number of pixels in one screen of 2K video is $1920 \times 1080 = 207.36 \times 10^4 = 2$ (million). Each pixel is composed of three signals of brightness Y, two color differences CB and CR. Each signal is coded with eight bits. The PAL TV outputs 25 frames per second. In this way, the data rate of the 2K video signal can be obtained as $(200 \times 10^4) \times (8 \times 3) \times 25 = 12 \times 10^8 = 1.2 \times 10^9 = 1.2$ Gbps.

5G network needs to transmit 4K and 8K video, and 4K is the standard configuration. The 4K video resolution is 4096×2160, so the number of pixels in a screen is $4096 \times 2160 = 884.736 \times 10^4 = 8$ (million).

According to the same method above, the data rate of 4K video can be obtained as $(800 \times 10^4) \times (8 \times 3) \times 25 = 4.8$ Gbps.

The resolution of 8K video is 7680×4320, so the number of pixels in one screen is $7680 \times 4320 = 3317.76 \times 10^4 = 32$ (million). According to the same method as above, the data rate of 8K video can be obtained as $(3200 \times 10^4) \times (8 \times 3) \times 25 = 19.2$ Gbps.

According to the above results, it can be seen that in order to transmit 4K video signals, the 5G network downlink data peak rate should be higher than 5 Gbps. In order to transmit 8K video signals, the 5G network downlink data peak rate should be 20 Gbps.

In summary, in order to transmit 3D video and ultra-high-definition video, 3GPP defines the following two technical requirements for the 5G air interface (5G NR):

1. The peak downlink data rate under normal conditions should be 10 Gbps and 20 Gbps under certain scenarios (VR and AR), both of which are much higher than the 4G peak downlink rate. Such rate brings an immersive, fiber rate-like experience and is capable of supporting ultra-high-definition video, VR and AR applications and mobile cloud services.
2. The one-way delay (wireless delay) of 5G NR is 1 ms, which is much smaller than that of 4G. It should be pointed out that in fact the one-way delay in the eMBB scenario only needs to be 4 ms. Here, the 1 ms requirement is proposed for the URLLC scenario.

In order to meet the requirements of the 5G network transmission data rate in the eMBB scenario, in addition to adopting high-order QAM and carrier aggregation, the following two measures can also be adopted:

(i) *Increase the channel bandwidth of the system*: In September 2018, 3GPP introduced the first version of the 5G standard in R15. It is the air interface standard in the eMBB scenario, which is called 5G NR (5G New Radio). Its frequency band is divided into two major Frequency Ranges (FR):

FR1 (450–6000 MHz): The frequency band number is from 1 to 255 (5G uses some parts of this frequency band). It is usually called 5G sub-6 GHz, which means that the frequency band used by 5G is below 6 GHz, and the maximum single carrier channel bandwidth is 100 MHz.

FR2 (24250–52600 MHz): The frequency band number is from 257 to 511 (5G also uses some parts of this frequency band). It usually means that 5G uses the millimeter wave frequency band (mmWave), where the maximum channel bandwidth of a single carrier is 400 MHz. Strictly, the mmWave frequency band should be higher than 30 GHz. But here the frequency band below 30 GHz in FR2 is said to be a millimeter wave frequency band.

It can be seen that the maximum channel bandwidth of FR1 and FR2 is much wider than 4G.

It should also be pointed out that the first version of the 5G NR standard mainly supports eMBB (some of which can also be used for URLLC), and the air interface standards of mMTC and URLLC will be announced in the following versions.

The detailed frequency band allocation scheme of FR1 and FR2 in FDD and TDD will be discussed later.

(ii) *Adopt large-scale antenna technology to improve system spectrum efficiency and increase network capacity*: The massive antenna technology is called massive MIMO (mMIMO). MIMO has been used in 4G to install a small number of antennas (LTE: 2×2 and 4×4; LTE-A: 8×8) in both the base station and user terminal to form several independent parallel data channels in the space domain, which can increase the transmission data rate. This method is called spatial multiplexing. However, it is conditional to truly form multiple independent parallel data channels, that is, the cross-correlation between the signals transmitted by each antenna of base station or terminal on different paths in the space domain is required to be small. For this reason, the distance between adjacent antennas of the base station usually requires about ten times the wavelength (10λ), and the distance between adjacent antennas of the terminal is required to be about half a wavelength (0.5λ).

When the working frequency band used by 5G increases (especially mmWave), its wavelength becomes shorter, so that more antennas can be installed in base stations and terminals. According to the 5G NR standard, 256 antennas can be used in the base station. For this reason, compared to 4G MIMO, MIMO in 5G is called mMIMO. It can be seen that 5G can form more independent parallel data channels in the space domain, thereby greatly increasing the data rate of transmission. However, it should be pointed out that when the operating frequency increases, the path loss of wave propagation increases and the penetration ability becomes weak. Moreover, the higher the frequency, the greater the path loss and the weaker the penetration capability, which will cause problems in cell coverage. In order to solve this problem, 5G needs to use beamforming technology.

The principle of beamforming is shown in Figure 5.1. The figure consists of several non-directional (omnidirectional) antennas to form an antenna array (multi-antenna array only requires that the distance between the antennas is more than half a wavelength). Among them, the data information transmitted by each antenna is the same, but the phase of the radio frequency signal added to each antenna is adjustable. In this way, although the pattern of each antenna is omnidirectional (antenna gain $G = 1$, as shown in Figure 5.2(a)), as long as the phases of n different signals are adjusted correctly, it is possible to superimpose n omnidirectional patterns with different radio frequency phases, and the total equivalent pattern becomes a directional pattern, that is, a directional antenna, as shown in Figure 5.2(b). Directional antennas are not non-directional, but directed, and the transmission power is mainly concentrated in the specified direction. Therefore, the antenna gain is $G > 1$. The gain obtained by the antenna can be used to compensate for the path loss, thereby creating conditions for solving the problem of cell coverage.

It can be seen from Figure 5.2(b) that the directional antenna can only cover the part of the cell to which the antenna points. In order to cover the entire cell, multiple directional antennas with different directions must be used, that is, multiple antenna arrays with different directions (as shown

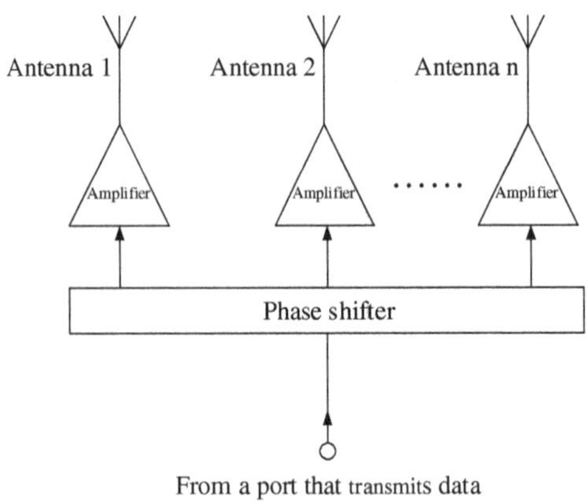

Figure 5.1 Beamforming principle.

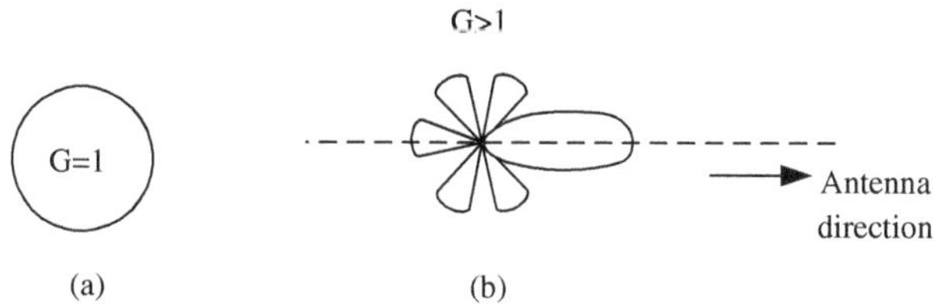

Figure 5.2 Antenna pattern: (a) omnidirectional antenna; (b) directional antenna.

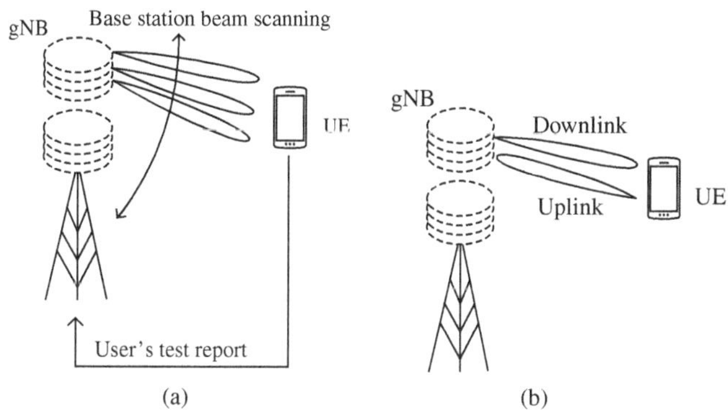

Figure 5.3 Beam scanning and communication: (a) base station beam scanning; (b) communication between user and base station.

in Figure 5.3(a)). In the downlink process, the base station sequentially uses beams of different directions to transmit wireless signals. This process is called beam sweeping. At the same time, users in the cell perform beam measurement on the received signal strength of the transmitted beams. The measurement result is transmitted to the base station through the uplink channel (called beam reporting). The base station determines the best transmit beam based on the beam reporting (called beam determination). After the base station finds the user through beam scanning, the user and the base station use service beam signals for communication (as shown in Figure 5.3(b)).

When beamforming technology is used, there are two types of MIMO, as shown in Figure 5.4. Figure 5.4(a) is called single-user MIMO (SU-MIMO). Since there is only one beam in the cell, the beam can only serve one user, that is, all time-frequency resources can only be allocated

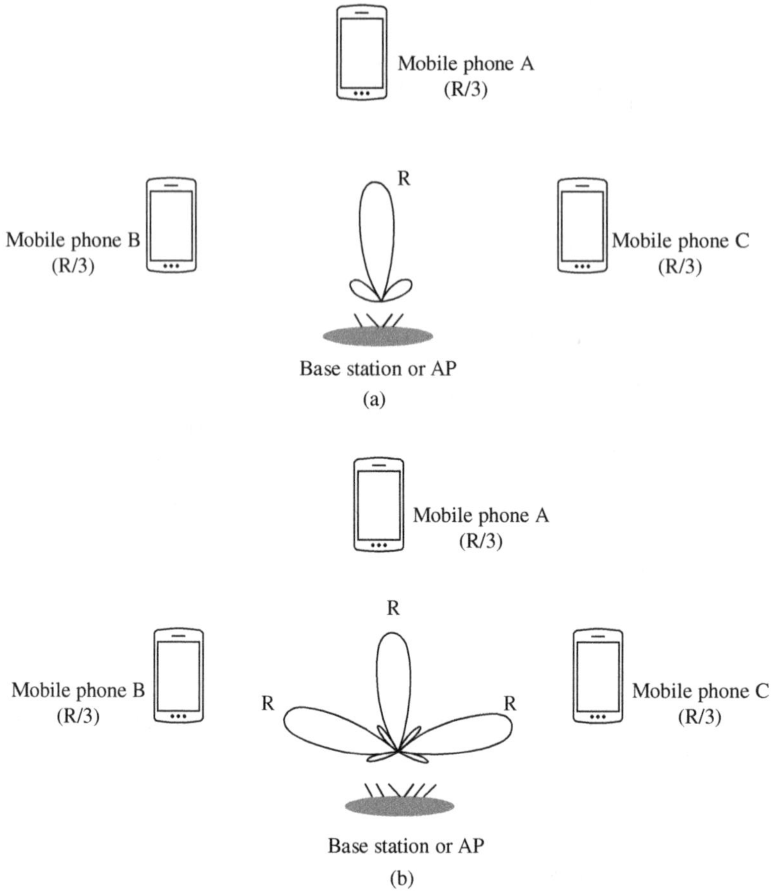

Figure 5.4 Schematic diagram of two MIMOs: (a) SU-MIMO; (b) MU-MIMO.

to the user who is aimed by the beam. Like the mobile phone A shown in Figure 5.4(a), phones B and C can only communicate when the beam is aimed at them. If the downlink data rate of the base station or Access Point (AP) is R, the actual average data rate of all three users in the cell is only R/3.

Figure 5.4(b) is called multi-user MIMO (MU-MIMO). Since there are multiple beams in the cell, such as three beams, the base station or AP can communicate with three mobile phones at the same time, that is, the same time-frequency resources can be used at the same time to serve multiple users. If the downlink data rate is R, then the data rate of each user is R rather than R/3. It can be seen that MU-MIMO can improve spectrum efficiency and expand network capacity.

The working principles and technical requirements of the 5G network used in the Internet of People, namely in the eMBB scenario, are discussed above. Next, we will discuss the working principles and technical requirements of 5G networks used in IoT, namely in the two scenarios of mMTC and URLLC. Compared to 4G, the problems involved at this time are unprecedented, not just the increase in network speed. More importantly, it has changed people's lifestyles. It is no exaggeration to say that whoever masters 5G will master the next era.

In addition, the issue of how to reduce network latency will be discussed specifically in the following section.

5.1.2 *URLLC*

Typical applications in the URLLC scenario are unmanned driving (such as auto-driving cars and airplanes), telemedicine (such as remote surgery) and industrial automation (such as robotic-arm operation). Obviously, the controlled objects in these applications need to be connected wirelessly. The role of the wireless channel is to transmit various control commands, so that the controlled object can perform correct operations according to the commands. Obviously, in order to implement these operations safely and correctly, the following two requirements should be put forward in the URLLC scenario. First of all, the data transmitted by the wireless channel must be very reliable (with very few errors). Moreover, the network system formed by these wireless connections must also be safe and reliable, and a powerful firewall must be set up to avoid hacking. Secondly, because the controlled objects in these applications are in high-speed condition, the delay must be short enough. If these two requirements are not satisfied, the result will be catastrophic. Here is an example for illustration.

Auto-driving cars require a wireless connection between the control center and cars, as well as cars and cars. If the car needs to brake when it is driving at a high speed, then the braking command must be transmitted to the car instantly and respond immediately, otherwise the car will crash. If the car travels at a speed of 50 km/h, the typical end-to-end delay of a 4G network is usually 50–100 ms. When the 4G network is used to transmit this braking command, the distance that the car rushes out in the 100 ms delay period is

$$50 \times 10^3/3600 \times (100 \times 10^{-3}) = 1.4 \text{ m}$$

A serious car accident may happen in such a case with a distance of nearly 1.4 m. But if the end-to-end delay of the network is shortened to 5 ms, the distance that the car rushes out at the same speed is only 1.4/20 = 1.4 × 100/20 = 7 cm. In this case, a car accident can hardly happen. This is especially true for auto-pilot aircraft. For example, hundreds of auto-pilot aircraft are flying in formation, and a small deviation will cause the aircraft to collide with each other.

China's IMT-2020 promotion group put forward two requirements for URLLC scenarios:

(1) The ideal end-to-end delay of the user plane is 5 ms, and the typical delay is about 5–10 ms.
(2) Within 1 ms delay of user plane, the reliability of transmitting 32-byte packets at one time is 99.999% (corresponding bit error rate P_b is 10^{-5}).

In order to meet the above requirements, what technical measures should be adopted in the 5G network?

5.1.2.1 *Technical measures adopted to achieve ultra-low latency*

Judging from the above definition, latency is the sum of the delays generated when data passes through multiple paths. It can be seen that in order to meet the requirement of 5 ms ultra-low delay, a variety of measures are needed. But in general, there are three main ideas:

(1) Greatly reduce the air interface transmission delay, which is wireless delay.
(2) Minimize the number of forwarding nodes in the data transmission process and shorten the distance between the source node and the destination node.
(3) Change some design parameters in 4G LTE.

The three measures mentioned above are as follows:

(i) *Methods to reduce air interface delay*: The mobile communication network is composed of a radio access network (air interface) and a core network. The process of data transmission in the network is shown in Figure 5.5. It can be seen from the figure that data communication in the core network is mainly wired communication. The transmission medium is a wired medium such as copper wire or optical fiber. The transmission quality and the data rate on wired media are very quick. For example, the maximum speed of a single fiber can reach 26 Tbps. Therefore, the delay of the core network is very small. The communication of data in radio access network is wireless communication. Wireless communication is different from wired communication. Data is not transmitted on physical materials but in the air. The transmission quality of this wireless channel will be affected by many factors, resulting in a lower data transmission rate. Therefore, the time delay (radio time delay) of the radio access network is relatively large. Compared to the core network delay, wireless delay can be said to be a bottleneck for reducing end-to-end delay.

We know that the wireless channel in mobile communication is a typical multipath fading channel. In order to meet the requirements of the bit

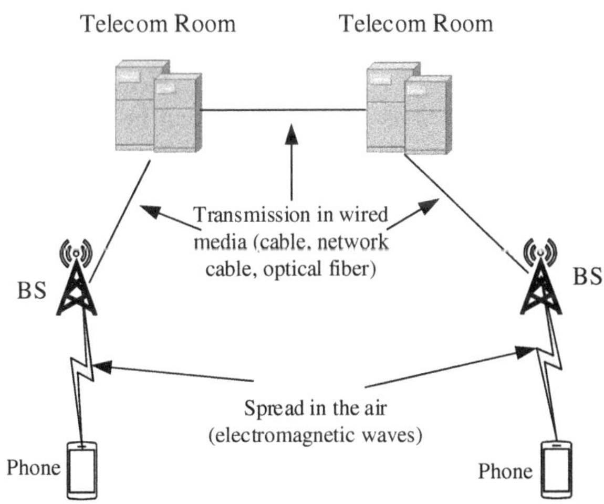

Figure 5.5 Data transmission.

error rate after decoding at the receiver, the corresponding modulation and coding scheme MCS must be correctly selected in the physical layer, that is, according to the determined modulation method (determined according to the required transmission data rate and spectral efficiency) to select the matching channel coding scheme (see Ref. [7]). It should be pointed out that no matter which MCS scheme is selected, the receiver must perform channel decoding after demodulation. No matter what kind of channel decoding algorithm is used, it takes a certain amount of time. Moreover, the delay caused by the entire baseband processing mainly comes from the delay of the decoder.

It can be seen that in order to reduce the wireless delay, the channel decoding time must be shortened. The length of this time is related to the decoding algorithm, and the decoding algorithm is determined by the type of channel coding. In 3G and 4G, channel coding adopts turbo code, which was proposed by the French in 1993. The current Wi-Fi network uses Low-Density Parity-Check (LDPC) code, which was proposed by the Americans in 1963. Therefore, the available channel codes in 5G include Turbo2.0, LDPC and polar. Polar code is called polarization code, and it was proposed by Turks in 2007. A comparison of these three codes is as follows:

(a) When the bit error rate $P_b \leq 10^{-5}$ after decoding, the 1-bit energy signal-to-noise ratio required by both LDPC and polar is very close to the theoretical value of C. E. Shannon. However, the difference between turbo codes is about 0.7 dB. It can be seen that comparing the performance of bit error rate or power efficiency, LDPC and polar are both superior to turbo.
(b) Due to the different structures of these codes, the decoding methods are different. The decoding of turbo code is based on serial decoding of mother code, which leads to a long decoding delay; LDPC codes can use high decoding parallelism. Some studies show that the throughput of LDPC code is 8.4 times that of turbo code under the assumption of the same decoding complexity, so the decoding delay of LDPC code is only 1/8 of that of turbo code. It can be seen that LDPC code is suitable for 5G working in high data rate and low delay scenarios.

It can be seen from the above comparison, when 5G is in eMBB and URLLC scenarios, LDPC codes or polar codes should be used for channel coding, while turbo applied in 3G and 4G can only be eliminated.

Therefore, 3GPP determines that the data channel in the 5G eMBB scenario uses LDPC codes and the control channel uses polar codes. LDPC codes are mainly promoted by Qualcomm, and polar codes are mainly promoted by Huawei. After adopting this scheme, the wireless delay can be 1 ms, but it also means that in the 5G eMBB scenario, the channel encoder/decoder must use two different sets of software and hardware. It should be said that this is a disadvantage. If both the data channel and the control channel use LDPC codes or polar codes, the channel encoder/decoder only needs a set of software and hardware, thereby reducing the complexity and cost of the equipment.

(ii) *The following two methods can be used to reduce the forwarding node in the data transmission process and shorten the distance between the source node and the destination node*:

(a) *Adopt D2D (Device to Device) method*:
The working principle of D2D is shown in Figure 5.6(b). It can be seen that when two users under D2D (such as two cars) communicate with each other, their working mode is different from that of non-D2D (see Figure 5.6(a)). Except that the signaling is still connected through the base station, the data between them is no longer transferred through the base station. But there is direct communication between the users.

(b) *Using Mobile Edge Computing (MEC) technology*:
The basic idea of MEC is to migrate the cloud computing platform from within the mobile core network to the edge of the mobile access network to realize the elastic use of computing and storage resources. This concept deeply integrates traditional cellular mobile networks and Internet services. This kind of integration is to move the business down to the base station side and complete the data processing in the nearby base station, thereby reducing the end-to-end delay of mobile service delivery.

(iii) *Change some design parameters in 4G LTE*:

(a) By using a shorter time slot structure and a shorter Transmission Time Interval (TTI), the frame structure in 4G LTE is fixed. The subcarrier spacing is 15 KHz, corresponding to a time slot of 0.5 ms. Two slots form a basic transmission unit sub-frame, that is, the length of the

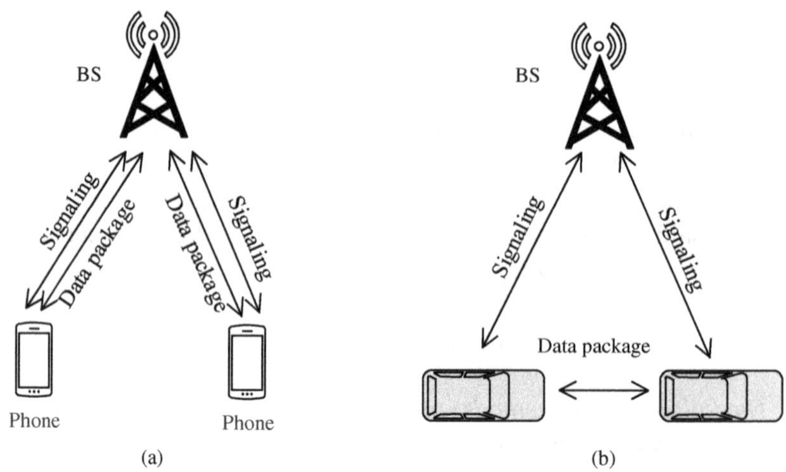

Figure 5.6 D2D communication: (a) non-D2D, (b) D2D.

sub-frame is 1 ms. However, the 5G NR frame structure is variable, and its subcarrier spacing Δf is variable. When the subcarrier spacing is larger, the time required to transmit the same number of OFDM waveforms is also shorter (see Figure 5.31).

(b) *Use shorter scheduling units*:
The scheduling units of 4G LTE and 5G NR are both time slot pairs composed of two adjacent time slots. Therefore, the scheduling duration is related to the time slot duration. The length of the time slot in 4G LTE is fixed at 0.5 ms. However, the length of the time slot in 5G NR is variable, and the smallest time slot is only 0.125 ms (see Table 5.7).

(c) *More flexible scheduling*:
For downlink services, 4G LTE is based on scheduling and different business resources cannot overlap. It takes at least 4 ms from the beginning of the data transmission to the confirmation. 5G NR is based on scheduling. But URLLC resources can overlap with other services (eMBB services). Through preemption instructions to improve the terminal decoding success rate and introduce a self-contained slot structure, data transmission and reception confirmation can be completed in one slot, that is, the fastest data transmission can be completed within 0.125 ms. For uplink services, 4G LTE is based on scheduling. Different business resources cannot overlap. It takes at least 10 ms from the initiation of data scheduling to the

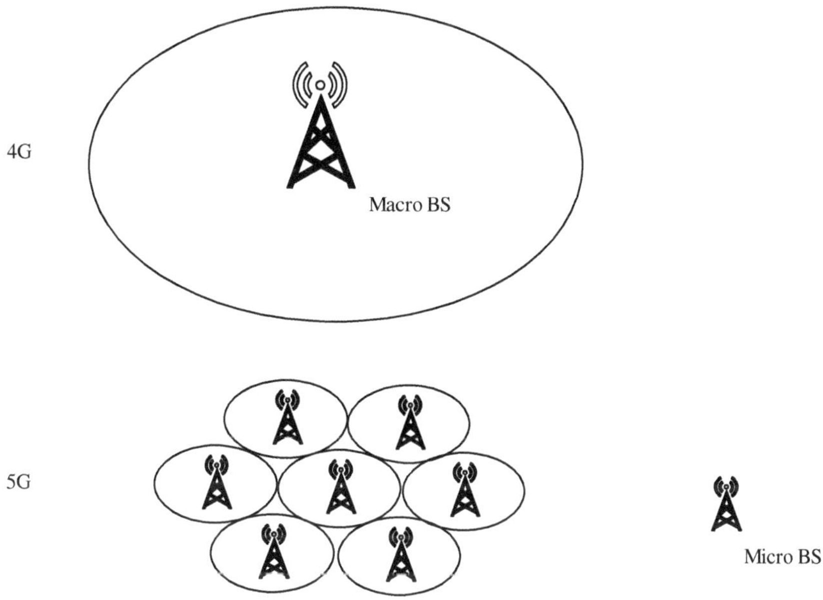

Figure 5.7 5G and 4G coverage.

confirmation of data (it can be shortened to 4 ms if semi-static scheduling is used), and 5G NR can be based on non-scheduling. It can reserve resources for URLLC services, and data transmission can be completed within at least 0.25 ms.

(d) *Reduce the time delay of 5G URLLC terminal during mobile network cell handover*:
When the terminal switches between different 4G LTE cells, signaling interaction between the terminal and the base station is required, which results in a handover delay. For 5G URLLC terminals, for example, in the case of auto-driving, if 5G works in a high-frequency band, its coverage capability is weakened because the path loss of the high-frequency band is larger than the mid- or low-frequency band. Therefore, to cover the same area size, the number of 5G base stations is much bigger than that of 4G, as shown in Figure 5.7. This kind of network consisting of large numbers of micro base stations is called Ultra-Dense Network (UDN). Due to the small coverage of micro base stations and the increase in the number of cells, cell switching of terminals is very frequent. Therefore, some measures must be taken for cell handover algorithms to shorten the handover delay.

5.1.2.2 *Measures to achieve ultra-high reliability*

The reliability of 5G should include data transmission reliability (bit error rate) and network security, which are discussed separately in the following.

(i) *Data transmission reliability*: The user plane reliability of the five 9s defined above refers to the reliability of data transmission. In order to achieve this requirement, the following measures can be adopted:

(a) Adopt more robust multi-antenna diversity technology.
(b) Adopt a robust modulation and coding scheme (MCS).
(c) Adopt a robust channel estimation algorithm.
(d) If the time delay permits, retransmission (HARQ) technology can also be used.

(ii) *5G network security*: The mobile Internet composed of 4G and Internet is mainly used to achieve information transmission. However, the intelligent Internet composed of 5G and Internet must not only realize information transmission but also establish a new mechanism and new system for society and life. Therefore, in addition to convenience and efficiency, the most important requirement for the intelligent Internet is that it must be safe and reliable. If an intelligent Internet security system is not guaranteed, it can lead to a disastrous situation.

If the auto-driving car system is compromised, the cars on the road will be controlled by hackers. If the smart health system is compromised, the health information of a large number of users will be leaked. If the smart factory system is compromised, the factory will lose secrets or stop production. If the smart home is breached, the safety of the home will not be guaranteed.

In order to prevent such problems, in the construction of 5G networks, security issues should be addressed at the bottom layer. From the start of network construction, security mechanisms such as information encryption and authentication should be added. The network should not be open, and special security mechanisms need to be established for special services. The network should not be completely neutral and fair. For example, ordinary users surfing the Internet may only have one system to ensure the smooth flow of their network, and users may face congestion. However, the intelligent transportation system requires multiple systems to ensure its safe operation and ensure its network quality. When the network is congested, the network of the intelligent transportation system must be smooth.

5.1.3 *massive Machine Type Communication (mMTC)*

5.1.3.1 *mMTC application examples*

Both mMTC and URLLC are IoT, but there is a difference between the two. Among them, two very demanding requirements are put forward for the data transmitted in the URLLC scenario, that is, ultra-reliable and ultra-low latency. However, the data transmitted in the mMTC scenario does not have these two requirements, or various applications in this scenario are not so sensitive to delay and reliability. But in the mMTC scenario, the number of wireless connections (used to connect to terminal devices) required to realize the Internet of Everything will be massive. It is expected that by 2020, it will reach a magnitude of 100 billion (of course, the number of terminal devices is also massive). Therefore, mMTC is called massive machine type communication.

mMTC can be widely used in various industries such as the following:

- infrastructure (roads, industrial environments, ports, etc.) monitoring;
- smart parking;
- fleet management (taxi or shared bicycle);
- logistics tracking;
- patient monitoring and care;
- smart home (control lighting, heating, ventilation, air conditioning, rice cooker, etc.);
- smart grid;
- smart city.

Smart city refers to the connection of public infrastructure in the city into a network, such as street lights, water meters, trash cans, and traffic lights, connected by a wireless network and these public devices will become terminals in the network. In this way, city managers can quickly and accurately know the status of each infrastructure to ensure immediate processing. For example, if the light on the road is broken or if the water pipe is leaking, or the trash can is full, the city managers will know about it immediately, thereby improving the operation efficiency of the city and people's quality of life.

5.1.3.2 *Features of mMTC IoT*

From the mMTC application examples, it can be seen that this type of IoT has the following characteristics:

- The data transmitted in this IoT is mainly various monitoring states (such as normal or abnormal status) and various sensor parameters (such as temperature and humidity).
- The upstream data flow is large, and the downstream data flow is small.
- No matter whether it's upstream or downstream, it is still small packet data.
- A low data rate is noted.
- There are a large number of terminal devices in this IoT, so the number of wireless connections is also large.

5.1.3.3 *Technical requirements for mMTC IoT*

Since the number of wireless connections and terminal devices in the mMTC IoT are massive, they have the following technical requirements:

- The price of terminal equipment must be very low.
- The terminal equipment must have low power consumption. It is required that the battery does not need to be replaced during the life of the device, that is, the terminal device must have high energy efficiency or high power efficiency. For example, based on AA (5000 mAh) batteries, the service life can exceed 10 years. Energy efficiency can be expressed in bit/J (bits/joule), which means that each joule of energy consumed corresponds to the number of data bits that can be transmitted. Energy efficiency can also be converted to power efficiency expressed in b/s/w (bits/second/watt), which represents the data rate that can be transmitted for every 1 watt of power consumed. To reduce the power consumption of terminal equipment, the measures adopted by NB-IoT in Chapter 4 can be adopted.
- New multiple access methods need to be adopted.

When there are multiple terminals working simultaneously in the same cell, it is called user multiple access. In the case of multiple access, the network must assign different methods of access to different users. FDMA, TDMA, CDMA and OFDMA used in 1G to 4G are all orthogonal multiple access (OMA), where CDMA is quasi-orthogonal. In this way, one-dimensional frequency domain, time domain, code domain and two-dimensional time-frequency resources are divided orthogonally, and the divided address resources are allocated to different users by signaling. From the perspective of network capacity, delay, power consumption and other technical performance, the OMA method is not optimal. In order to achieve optimal performance, Non-Orthogonal Multiple Access (NOMA) is adopted in 5G.

5.2 Eight 5G Performance Indicators

The three major application scenarios of 5G and their corresponding technical requirements are discussed above. Putting them together is the technical requirement for the entire 5G. For this reason, ITU-RWP5D (5D Working Group of the Wireless Communications Department of the International Telecommunication Union) has formulated eight performance indicators that 5G must have.

5.2.1 *Peak rate (Gbps)*

The basic requirement is 10 Gbps, and the specific scenario (VR and AR) is 20 Gbps.

5.2.2 *User experience rate (Mbps or Gbps)*

The basic requirement is 100 Mbps, and some scenarios are 1 Gbps.

It should be noted that the peak rate refers to the maximum data rate transmitted by the source. The user experience rate refers to the rate actually achieved by the user or device in the coverage area. It is not only related to the source rate but also related to the network speed (related to the number of users working at the same time and wireless channel transmission characteristics) and other factors. For example, when the mobile phone is on Wi-Fi network, we can see the uplink/downlink speed

(MBPS), signal strength (dBm, which is the received signal power), connection rate (Mbps, which is the source rate) and network delay (ms, it is the end-to-end delay). Among them, the connection rate is adaptively changed with different signal strengths, and the connection rate that may change is shown in Tables 3.1 and 3.2 in Chapter 3. The user experience rate determines the smoothness of the user when surfing the Internet. The higher the rate, the smoother it will be, otherwise unpleasant phenomena such as the connection being unsmooth or even stuck, or with pauses.

5.2.3 *Spectrum efficiency* (*bps/Hz*)

Compared to IMT-A (which is also formulated by ITU-RWP5D, 4G LTE-A is made according to the IMT-A standard), the basic requirements should be increased by three times, and some scenarios can be higher than three times (such as five times).

5.2.4 *Mobility* (*km/h*)

The maximum moving speed that can be supported under the condition of meeting the nominal QoS requirement is 500 km/h.

5.2.5 *Air interface delay* (*ms*)

One-way wireless delay is 1 ms (ideal end-to-end delay is 5 ms).

5.2.6 *Connection density* (*number of devices/km²*)

The number of devices that can be connected in a unit area is required to reach $10^6/km^2$ (1 million per square kilometer).

5.2.7 *Energy efficiency* (*b/J is the number of bits per joule*)

On the air interface network side, it refers to the number of data bits that can be transmitted or received per unit of energy consumed by the network (such as base stations). On the terminal side, it refers to the number of data bits that the communication module can transmit or receive per unit of energy consumed. The energy efficiency is required to be 100 times higher than that of IMT-A.

5.2.8 *Traffic density (Mbps/m²)*

It refers to the sum of all traffic within the geographic area served by the network. It is required that the data rate should reach 10 Mbps in an area of 1 square meter, that is, the traffic density should reach 10 Mbps/m².

Finally, it must be pointed out that the above eight performance indicators are based on 5G network security (multiple firewalls are used to prevent hackers from intruding and attacking) and reliability (reliability is 99.999%, that is, the system's bit error rate P_b is 1×10^{-5}).

The correspondence between the three major application scenarios of 5G and the eight major performance indicators is shown in Table 5.1. A check (√) in the table indicates a close relationship between the two.

The comparison of the eight major performance indicators of 5G with 4G, 3G and 2G is shown in Figure 5.8.

5.3 5G Working Frequency Band

According to the 5G air interface standard 5G NR, the 5G radio frequency band is divided into two major spectrum ranges, as shown in Tables 5.2 and 5.3, where *nx* represents the 5G frequency band number. In 4G, *Bx* represents the frequency band number. In the table, *n* is number, *B* is Band and *x* is the band number. The two tables are explained in the following sections.

Table 5.1 Relationship between 5G application scenarios and performance indicators.

Scenarios / Indicators	eMBB	URLLC	mMTC
Peak rate	√		
User experience rate	√		
Spectrum efficiency	√		
Mobility	√	√	
Air interface delay	√	√	
Connection density			√
Energy efficiency			√
Traffic density	√	√	√

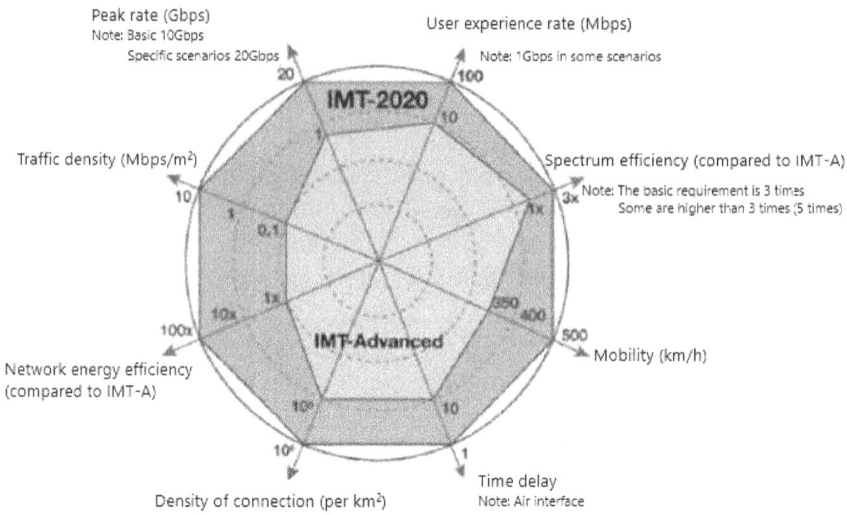

Figure 5.8 The eight major performance indicators of 5G.

5.3.1 *Relationship between 5G frequency band number and 4G frequency band number*

The relationship between the frequency band number of 5G and the frequency band number of 4G is not a one-on-one correspondence. For example, 4G LTE B42 (3.4–3.6 GHz) and B43 (3.6–3.8 GHz), in 5G NR, the two are combined in n78 (3.3–3.8 GHz).

5.3.2 *The reason for n77 including n78*

In 5G NR, n77 also includes n78. There are two reasons for this:

(i) Makes the n77 channel bandwidth wider.
(ii) Supports global roaming.

Since telecom operators in different countries have different available frequency ranges in the C-band (4–8 GHz), n77 simply included all these ranges. The global available C-band is shown in Table 5.4. The solid square in the figure indicates that it is confirmed to be available, while the open square indicates that there is a plan but not confirmed. Using this broadband method to define frequency bands has formed a few globally unified frequency bands, which can greatly reduce the complexity of the terminal (such as mobile phones) supporting global roaming.

5.3.3 *Supplementary frequency band*

Supplementary Downlink (SDL) and Supplementary Uplink (SUL) in the table are supplementary bands. Among them, SUL is uplink supplementary bands, and SDL is downlink supplementary bands. The use of the supplementary band can be illustrated in Figure 5.9. The transmission power of the mobile phone is usually less than the transmission power of the base station, so the size of the coverage area of the base station is actually determined by the uplink transmission power. For this reason, the mobile phone 1 in Figure 5.9 is close to the base station, so the same 3.5 GHz frequency (TDD) can be used for both uplink and downlink. The mobile phone 2 in this figure is far away from the base station. If the uplink frequency is still 3.5 GHz, it cannot communicate due to the large uplink path loss. In order to solve this problem, we can simply lower the uplink frequency. As after the frequency is reduced, the path loss of the radio wave propagation is reduced, and the penetration ability is enhanced. The reduced uplink frequency band is the uplink supplementary band, like the 1.8 GHz band shown in Figure 5.9. In this way, for the mobile phone 2, the downlink frequency is still 3.5 GHz, and the uplink

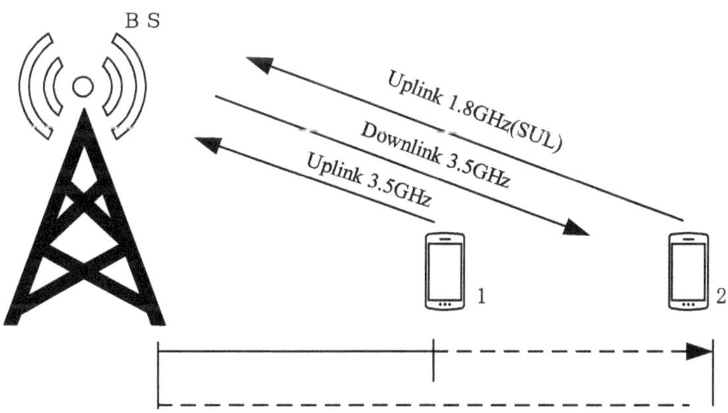

Figure 5.9 SUL.

frequency is 1.8 GHz. The use of the downlink supplementary frequency band is similar to that of the uplink supplementary frequency band. However, it should be pointed out that in order to achieve the function of the uplink/downlink supplementary band, carrier aggregation or dual connection mode (what dual connection is will be discussed later) is required to cooperate.

5.3.4 *Frequency bands that 5G NR will preferentially use*

5G NR will preferentially use the frequency bands n77, n78 and n79 (their channel bandwidths are wider), as shown in Table 5.2. Among them, n77 and n78 are C-bands, which are currently the globally unified 5G NR frequency bands in Table 5.4. n79 may also be used for 5G NR. It can be seen from Table 5.4 that the main driving countries are China, Russia and Japan.

n28 is commonly referred to as 700 MHz. Due to its good coverage performance, this frequency band has been identified as a priority candidate frequency band for global mobile communications at WRC-15 (World Radiocommunication Conference 2015). n71 is 600 MHz. At present, the US operator T-Mobile has announced the use of 600 MHz to build a 5G network.

5.3.5 *Description of mmWave frequency band*

Regarding the millimeter wave frequency band in Table 5.3, the following should be explained:

(i) At present, the United States, Japan and South Korea are testing the 28 GHz frequency band of 5G. Its purpose is to replace the last few hundred meters of communication with fiber at homes with 5G fixed wireless access. Therefore, it may become the first commercial frequency band. But 28 GHz is not within the scope of ITU WRC consideration. Although 3GPP has included this frequency band (n257), it will eventually require ITU approval.
(ii) The n258 band may also be adjusted. This is because it may affect the satellite communication system or need sufficient guard band.

According to the 5G NR standard, the test frequency bands allocated by China's Ministry of Industry and Information Technology to the three

Table 5.2 FR1.

Band number	Uplink (MHz)	Downlink (MHz)	Bandwidth (MHz)	Duplex mode
n1	1920–1980	2110–2170	60	FDD
n2	1850–1910	1930–1990	60	FDD
n3	1710–1785	1805–1880	75	FDD
n5	824–849	869–894	25	FDD
n7	2500–2570	2620–2690	70	FDD
n8	880–915	925–960	35	FDD
n20	832–862	791–821	30	FDD
n28	703–748	758–803	45	FDD
n38	2570–2620	2570–2620	50	TDD
n41	2496–2690	2496–2690	194	TDD
n50	1432–1517	1432–1517	85	TDD
n51	1427–1432	1427–1432	5	TDD
n66	1710–1780	2110–2200	70/90	FDD
n70	1695–1710	1995–2020	15/25	FDD
n71	663–698	617–652	35	FDD
n74	1427–1470	1475–1518	43	FDD
n75	N/A	1432–1517	85	SDL
n76	N/A	1427–1432	5	SDL
n77	3300–4200	3300–4200	900	TDD
n78	3300–3800	3300–3800	500	TDD
n79	4400–5000	4400–5000	600	TDD
n80	1710–1785	N/A	75	SUL
n81	880–915	N/A	35	SUL
n82	832–862	N/A	30	SUL
n83	703–748	N/A	45	SUL
n84	1920–1980	N/A	60	SUL

Note: N/A (not available). FR1 (5G medium- and low-frequency band): 450–6000 MHz, frequency band number from 1 to 255.

major telecom operators in December 2018 are all 5G mid- and low-frequency bands. As shown in Table 5.5, the frequencies are all below 6 GHz, so they are collectively referred to as sub-6 GHz. China Telecom and China Unicom configure the 3.6 GHz frequency band, and China

Table 5.3 FR2.

Band number	Uplink/Downlink (MHz)	Bandwidth (MHz)	Duplex mode
n257	26500–29500	3000	TDD
n258	24250–27500	3250	TDD
n260	37000–40000	3000	TDD
n261	27500–28350	850	TDD

Note: FR2 (5G high-frequency band): 24250–52600 MHz, frequency band number from 257 to 511.

Table 5.4 C-bands available in the world.

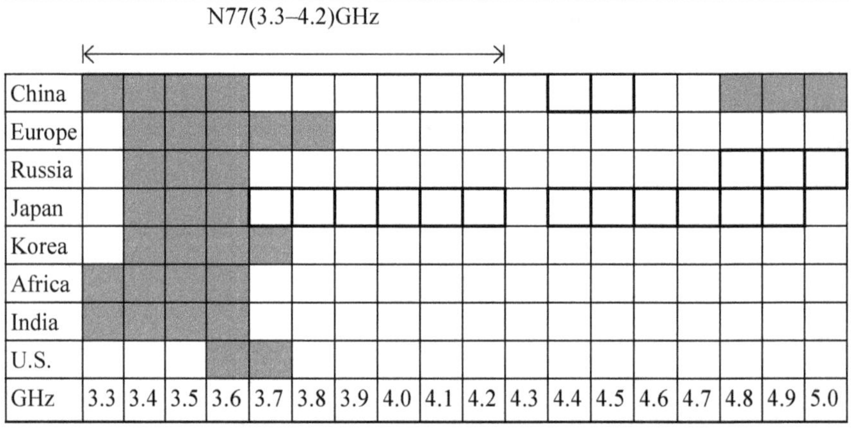

Table 5.5 Working frequency bands of 4G-LTE in China.

	TDD		FDD		
Telecom	Frequency band (MHz)	Bandwidth (MHz)	Frequency band (MHz)	Bandwidth (MHz)	Total (MHz)
China Mobile	1880–1900	20			130
	2320–2370	50			
	2575–2635	60			
China Unicom	2300–2320	20	1955–1980	25	90
	2555–2575	20	2145–2170	25	
China Telecom	2370–2390	20	1755–1785	30	100
	2635–2655	20	1850–1880	30	

Mobile configures the 2.6 GHz and 4.9 GHz frequency bands. The specific frequency range is as follows:

China Telecom: 3400–3500 MHz, bandwidth 100 MHz (n78).
China Unicom: 3500–3600 MHz, bandwidth 100 MHz (n78).
China Mobile: 2515–2675 MHz, bandwidth 160 MHz (n41) and 4800–4900 MHz, bandwidth 100 MHz (n79).

Among them, 2515–2675 MHz is divided into the following three frequency bands: 2515–2575 MHz, 2575–2635 MHz and 2635–2675 MHz. The above 2515–2575 MHz, 2635–2675 MHz and 4800–4900 MHz are newly added frequency bands, and 2575–2635 MHz is the frequency band already used by China Mobile in 4G, so this frequency band is called the reused frequency band. Due to the use of the reused frequency band, China Mobile can make full use of 4G base stations and supporting resources as well as upgrade the antenna feeder system and main equipment to achieve mMIMO. Therefore, China Mobile will have more advantages in the speed of 5G construction.

Table 5.5 shows the working frequency bands used by China's three major operators for 4G. It can be seen from the table that the 2575–2635 MHz frequency band is the reused frequency band in 5G.

5.4 5G Networking Solution

The 4G system architecture mainly includes the radio side (LTE) and the network side (SAE, System Architecture Evolution). 3GPP refers to the 4G system architecture as the EPS (Evolved Packet System). EPS refers to a complete end-to-end 4G system, which includes user equipment (UE), radio access network (called air interface in LTE) and core network (known as EPC, Evolved Packet Core). 3GPP calls 5G system as 5GS. The radio access network (air interface) is called 5G NR, and the core network is called 5GC (5G Core). From 3G to 4G is an overall evolution, that is, the access network and the core network together evolved from 3G to 4G as a whole. But from 4G to 5G, 3GPP did not adopt an overall evolution method. Instead, the access network and the core network were separated and evolved independently. This is because 5G networks are not only used for mobile broadband (MBB in 4G and eMBB in 5G) but also in scenarios such as URLLC and mMTC. The access network and core network required for these different scenarios are different. This means that the 5G access network and core network are not one form but multiple forms. For this reason, it is very necessary to separate the two networks in 5G for independent evolution. In this way, 4G core network (EPC), 4G access network (LTE), 5G core network (5GC) and 5G access network (5G NR) can be mixed and matched to form a variety of 5G networking solutions. It can be divided into two categories: independent networking (SA, Standalone) and non-independent networking (NSA, Non-Standalone). SA refers to the use of all 5G core networks and all 5G base stations to build a 5G network, as shown in Figure 5.10. Among them, option 2 is completely separate from the existing 4G network when the network is built. Option 5 is slightly different from option 2. It connects the ng-eNB (next generation-eNB) base station to the 5GC, and the ng-eNB is a 4G base station after eNB upgrade. It can be seen that the air interface can use millimeter waves because option 2 uses a 5G base station. Option 5 only uses an upgraded 4G base station, so the air interface cannot use millimeter waves. In this way, the advantage for option 5 of using the millimeter wave frequency band is not available.

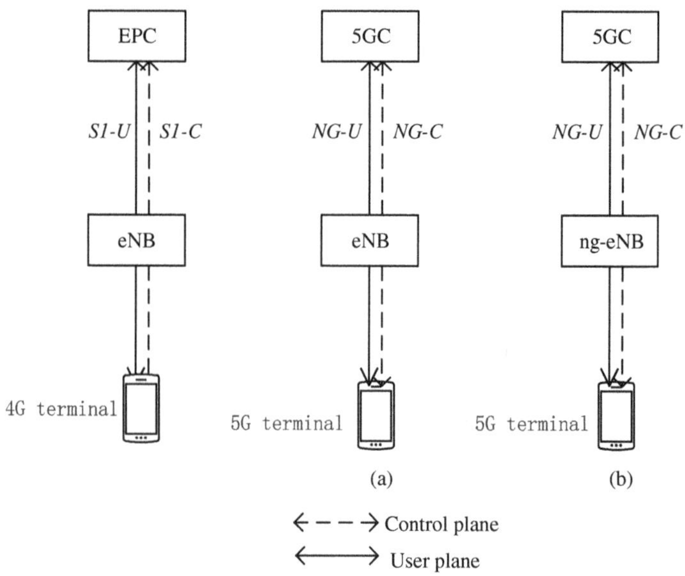

Figure 5.10 Two schemes of the SA networking: (a) option 2; (b) option 5.

The SA networking solution is simple, refreshing, easy to maintain and easy to use. But the capital investment of telecom operator is too high, especially the 5G core network. The increase in the number of 5G users after the completion of the 5G network requires a process. In this long process, there will be a gap between the telecom's revenue and cost. For operators, this gap is unbearable. For this reason, 3GPP proposed the NSA solution. The guiding ideology is to use the existing 4G base station and core network as much as possible or make appropriate transformations when forming a 5G network, so that the investment will be greatly reduced. With the gradual increase in the number of 5G users, telecom operators will gradually move closer to the independent networking and finally achieve independent networking solutions. It can be seen that from 4G to 5G is a process of smooth evolution, long-term coexistence and slow replacement. 4G will continue to play its role for a relatively long period of time.

The three concepts of dual connectivity, user plane and control plane will be involved in the NSA solution, which will be discussed separately in the following.

5.4.1 *LTE dual connectivity architecture*

In a heterogeneous wireless system, when different types of base stations cooperate in networking, because the radio resources and coverage of a single base station are limited, the radio resources of multiple cells or base stations are concentrated to provide services for users, which makes it easier to meet the capacity and coverage requirements. This method is called multiple connections. In the LTE system, commonly used multi-connection methods include Carrier Aggregation (CA), Coordinated Multi-point Transmission (CoMP) and dual connectivity. Among them, LTE dual connectivity means that the user terminal UE can use the radio resources of the LTE base station and the 5G base station at the same time in the connected state. Therefore, the radio resource utilization rate is improved, the system switching delay is reduced, and the user and system performance are improved. One of these two base stations is the Master Node (MN). In addition to transmitting user plane data, it is also responsible for transmitting control plane data (various signaling). The other is the Secondary Node (SN). It is only responsible for transmitting user plane data. From the above discussion, we know that to realize the non-independent networking, it must be supported by LTE dual connectivity. In the case of dual connectivity, there are different methods for using these two base stations (nodes) to transmit data on the two planes. These different methods are collectively referred to as LTE dual connectivity architecture. When the architecture is different, there will be different NSA solutions. These different programs are shown in Figures 5.11 and 5.12 and are classified as series 3, series 7 and series 4. There are a total of eight NSA schemes in these three categories, which will be discussed separately later.

5.4.2 *User plane*

It refers to the data of various services transmitted by users. 5G users transmit large data packets with a high data rate and a large number of bytes, so 5G user data cannot all be transmitted by 4G base stations.

5.4.3 *Control plane*

It refers to the various calls and management signaling required in the transmission network of user data, such as signaling for initial terminal access, paging and mobility management. Usually, all kinds of signaling are small data packets, so 4G base stations can directly transmit 5G signaling.

The eight NSA schemes are discussed in the following sections.

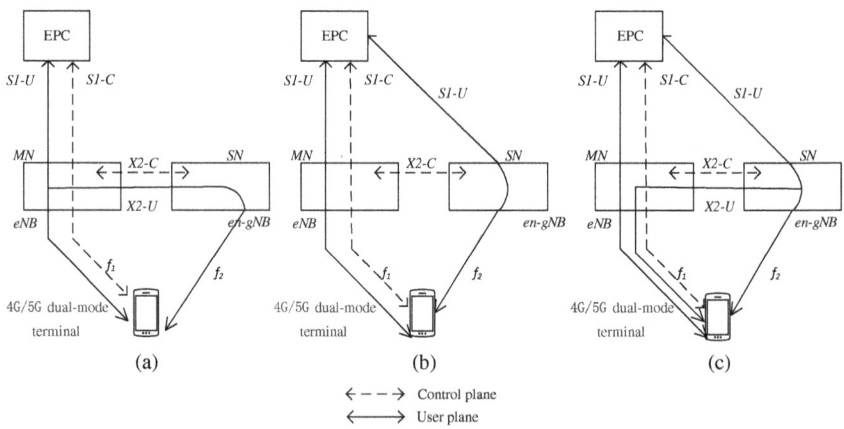

Figure 5.11 Series 3 NSA networking solution: (a) option 3, (b) option 3*a*, (c) option 3*x*.

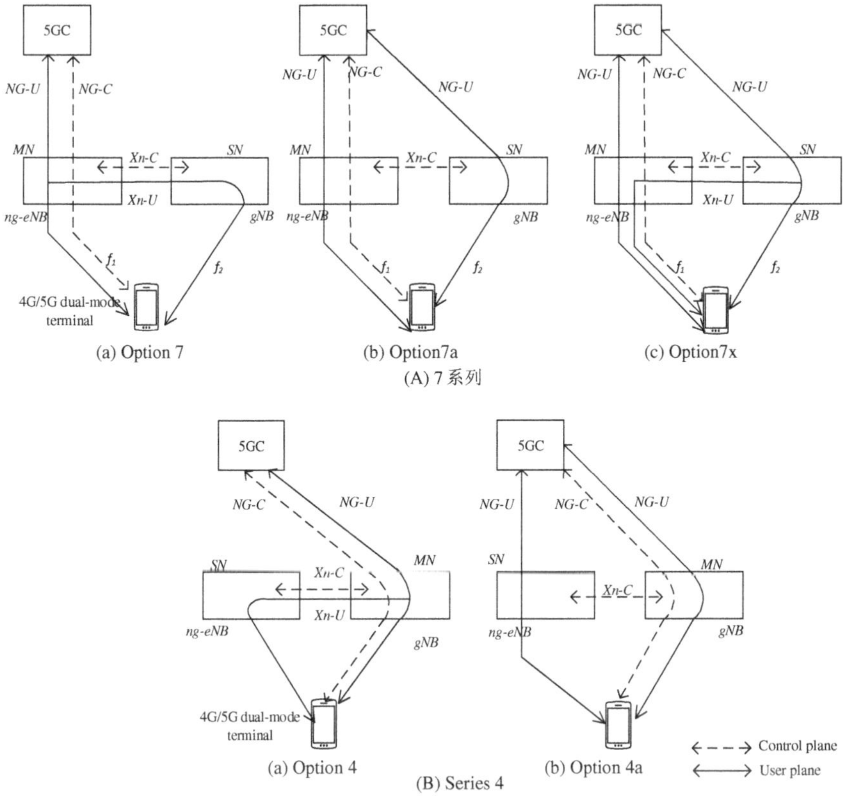

Figure 5.12 Series 7 and Series 4 NSA networking solutions.

5.4.4 *Series 3*

As shown in Figure 5.11, in addition to building a new 5G base station (which is much less invested than a 5G core network), its biggest feature is that there is no need to build a 5G core network, and a 4G core network is still used. Thus, a lot of money can be saved. It can be seen from this figure that the series 3 NSA networking solution has the following characteristics:

- The primary node is eNB, and the secondary node is en-gNB. en-gNB is a 5G base station under the non-independent networking architecture of the series 3. It can connect with EPC and eNB.
- This architecture is due to eNB (E-UTRA) and en-gNB (NR) dual connectivity, and the eNB is the master node. Therefore, this dual connection is called EN-DC (E-UTRA NR Dual Connection).

- In option 3, the user plane data is offloaded at the eNB. In option 3*a*, the user plane data is offloaded at the EPC. Option 3*x* is a combination of option 3 and option 3*a*. Its user plane data is offloaded at EPC and en-gNB.

The series 3 NSA solution is currently the favorite solution of operators because it saves money and builds a 5G network the fastest. The preference for the three options should be $3x > 3a > 3$.

5.4.5 *Series 7*

As shown in Figure 5.12(a), its architecture is similar to the series 3 in Figure 5.11. But the biggest difference between the two is that the core network is different. The 5G core network is used in the series 7, so the program has a large capital investment. The series 7 has the following characteristics:

- The master node is ng-eNB. As mentioned above, it is a 4G base station upgraded by an eNB. Under the series 7 non-independent networking architecture, it can be connected to 5GC and gNB; the auxiliary node is the 5G base station gNB.
- In this architecture, ng-eNB (NG-E-UTRA) and gNB (NR) are dual connected, and ng-eNB is the master node. Therefore, this kind of dual connection is called NGEN-DC (NG-E-UTRA NR Dual Connection).
- In option 7, user plane data is offloaded at the ng-eNB. In option 7*a*, user plane data is offloaded at 5GC. Option 7*x* is a combination of option 7 and option 7*a*, and its user plane data is offloaded at 5GC and gNB.

5.4.6 *Series 4*

As shown in Figure 5.12(b), it has the following characteristics:

- The primary node is gNB, and the secondary node is ng-eNB.
- This architecture is due to dual connections between gNB and ng-eNB, and gNB is the master node. Therefore, this kind of dual connection is called NE-DC (NR E-UTRA Dual Connection).
- In option 4, the user plane data is offloaded at gNB. In option 4a, user plane data is offloaded at 5GC.

It is known from the above analysis that in order to quickly put the 5G network into use and save the initial investment, most telecom operators will choose the 3x solution in the NSA. It can be seen from Figure 5.11(c) that this scheme consists of two parts. The left part is the 4GLTE network, which is an already established 4G network. The right part is 5G NR. There are two options for 5G NR. Therefore, there are currently two ways to deploy 5G networks.

(1) *NSA: 5G NR in millimeter wave band plus LTE network*: This approach is represented by Verizon and AT&T in the United States. It deploys 5G NR millimeter waves on existing LTE networks to supplement coverage of edge areas (see Chapter 7).
(2) *NSA: 5G NR in the sub-6 GHz band plus LTE network*: This method can quickly achieve better 5G NR coverage (due to the use of frequency band being less than 6 GHz). However, there are technical complications such as the interface between 4G LTE and 5G NR and carrier aggregation. At present, Chinese operators adopt this method.

It should be pointed out that there are 4G and 5G base stations in both solutions A and B. There are two ways to select the site of these two base stations (see Section 5.6.7), namely co-site (two addresses are the same) and non-co-site (two addresses are different). Among them, plan A is on-co-site (see Figure 5.45), and plan B is co-site (see Figure 5.46).

It has been pointed out at the beginning of this section that the current NSA solution to deploy 5G networks is only a transitional solution. As the number of 5G users grows, it must eventually evolve from NSA to SA. This is because NSA's non-independent networking mode cannot support future application scenarios such as IoT or auto-driving. So, the best way to evolve is:

$$\text{Option } 3x \text{ of NSA} \rightarrow \text{Option 2 of SA.}$$

First, the 5G radio access network should be deployed (5G access network must be built in the series 3), then the 5G core network should be deployed (5G core network must be built in option 2), and finally the 5G radio access network should migrate to the 5G core network.

5.4.7 *Voice processing methods in 5G*

Voice service (call) is a communication service that must be present in mobile phones. All packages launched by telecom operators include voice services. For this reason, the mobile communication network must have the function of transmitting voice services and ensure the continuity of voice calls.

In order to meet the above requirements, different voice transmission schemes have been adopted in various generations of mobile communication networks, as shown in Figure 5.13. These schemes are discussed separately in the following sections.

5.4.7.1 *2/3G*

2G and 3G network services are mainly voice-based, so that Circuit Switched (CS) voice scheme is adopted. Since IMS (IP Multimedia Subsystem) is Packet Switching (PS), the CS core network in Figure 5.13 contains the conversion of CS domain and PS domain (it is not shown in this figure).

5.4.7.2 *4G*

In the 4G all-IP era, there is only PS in the packet domain and no longer supports the traditional CS circuit domain voice solution. In this case, in order to meet the above voice continuity requirements, the 4G network adopts two voice transmission schemes:

(i) *VoLTE*: Through the introduction of IMS, the LTE network directly provides IP-based voice services. When using VoLTE and considering that the 4G network has not yet reached full coverage, the quality of the voice signal will decrease or even end when the 4G signal coverage is not good. In order to ensure the quality of voice calls, Single Radio Voice Call Continuity (SRVCC) technology was used to seamlessly switch the VoLTE call mode to the 2/3G CS domain call mode. Of course, the mobile phone must be a 2G/3G/4G multi-mode phone. It can be seen that the use of SRVCC technology is a measure to improve the voice quality of VoLTE.

(ii) *CSFB*: CS Fallback (CSFB) is called circuit switched fallback. It means that when the mobile phone initiates or receives a voice call in the 4G network, it falls back to the 2/3G network, uses the CS domain of the 2/3G network to complete the voice call and then returns to the 4GLTE

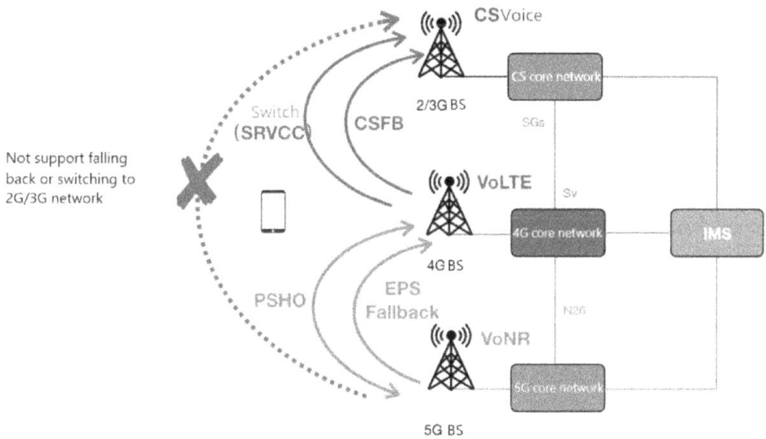

Figure 5.13 Voice transmission schemes of mobile communication network.

network after the call. This circuit domain fallback will occur in the following two situations:

The 2/3/4/G multi-mode mobile phone resides in the 4G LTE network, but the LTE network does not have the VoLTE function. In order to ensure the voice call service can still function normally, only 2/3G network can be used.

The multi-mode mobile phone is moved to a coverage area where there is no 4G network but only a 2/3G network. At this time, the mobile phone resides in the 2/3G network, so voice calls can only be made through the 2/3G network.

It can be seen that the CSFB technology is used to ensure that 4G voice calls can be continuous and uninterrupted. Therefore, CSFB is a mandatory solution in 4G, while VoLTE is only an optional solution.

5.4.7.3 *5G*

The 5G voice solution is related to the 5G networking mode and will be discussed separately in the following.

(i) *Non-independent networking (NSA)*: It can be seen from Figure 5.11 that in the case of NSA, the 4/5G dual-mode mobile phone is dual connected with LTE and NR. 4G base stations and 5G base stations share the 4G core network (EPC), so there are the following two voice solutions:

(1) PS voice is directly carried by EPC and 4G base stations without going through 5G base stations. This is the VoLTE voice solution in 4G.
(2) PS voice is directly carried by EPC and 5G base stations without passing through 4G base stations. This is the VoNR voice solution for 5G in NSA. Compared to VoLTE, VoNR will give users a better experience. For example, voice calls can be used simultaneously with high-speed data services, with a shorter voice connection time.

In the current situation, 4G network coverage performance is better than 5G network. In order to ensure the continuity of voice calls, VoLTE should be selected as a mandatory solution, while VoNR is an optional solution.

(ii) *Independent networking (SA)*:
In the case of SA, there are the following two voice solutions:

(1) *VoNR*:
It can be seen from Figure 5.10 that PS voice is directly carried by 5GC and 5G base stations. This is the VoNR voice solution for 5G in SA. Taking into account that before the 5G network is fully covered, when the 5G signal coverage is not good, the voice quality will be reduced or even dropped. In order to ensure the quality of the voice call, Packet Service Handover (PSHO) technology is used to smoothly switch from VoNR to VoLTE.

(2) *EPS Fallback*:
If the 5G network does not have the above VoNR function, when the 4/5G dual-mode mobile phone needs to initiate or receive voice calls, the 5G network can only be dropped back to the 4G network at this time. The voice call is made by VoLTE which has been set up in 4G. The technology to achieve the above-mentioned network fallback is called Evolved Packet System Fallback (EPS Fallback).

From the above, no matter whether 5G is independent or non-independent, in order to ensure the continuity of voice calls, the 4G network must be equipped with a complete VoLTE function.

It should also be pointed out that the R15 version does not define a mechanism to directly fall back from 5G to the 2/3G network, so only 4/5G dual-mode mobile phones are required. In this case, the above-mentioned voice call continuity only refers to the voice continuity between 5G and 4G and does not support the voice continuity between 5G and 2/3G.

5.5 5G Network Architecture [11,22]

From the previous discussion, the main application scenario of 4G network is mobile broadband (MBB). However, 5G networks have a variety of application scenarios. In addition to eMBB, they also need to be applied in mMTC and URLLC scenarios. As 5G needs to be applied in a variety of scenarios, the design guidelines for 5G networks are different from 4G. In order to meet the requirements of applications in multiple scenarios, it is obvious that the 5G network should be designed as a flexible and changeable network.

Like 4G, the entire network is also composed of a core network (5GC) and an access network (5G NR, NR — New Radio). Among them, the core network and EPC in 4G are quite different, while the access network and LTE in 4G are basically the same.

5.5.1 *5G core network architecture*

5G can be applied to a variety of scenarios. The requirements in each scenario are different, and the corresponding core network structure is also different. Therefore, the 5G core network should be designed as a flexible and variable network. According to this requirement, the 5G core network does not adopt a 4G-level topology network architecture, but uses a Service-Based Architecture (SBA), which means that the core network emphasizes service requirements. The core network architecture based on the service architecture is shown in Figure 5.14. In addition to the three function blocks of user equipment, RAN and Internet, the remaining functional blocks constitute the core network.

AMF, Access and Mobility Management Function
SMF, Session Management Function
UPF, User Plane Function
NEF, Network Exposure Function
NRF, NR Repository Function
PCF, Policy Control Function
UDM, Unified Data Management

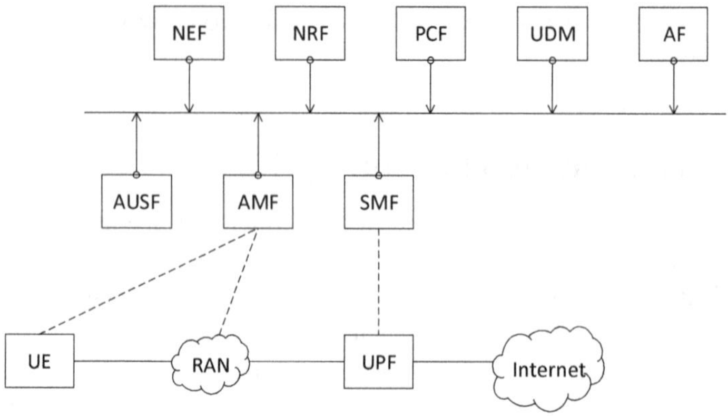

Figure 5.14 5G core network architecture.

AF, Application Function
AUSF, Authentication Server Function

It can be seen from the figure that each independent function block is connected to the system through a standard interface, which is very similar to the bus structure in the computer. The SBA architecture enables the 5G core network to have the following characteristics:

(i) *Facilitates network expansion and upgrade*: This is because each node of the SBA architecture is independent. Therefore, it is easy to access or exit the system regardless of hardware or software.
(ii) *Supports network slicing technology*: In the following discussion, we can see that the UPF functional block in the core network can perceive service requirements. Based on this, functional blocks can be selected, configured and combined accordingly to form a logical core network that can serve various requirements. When the requirements are different, different core networks can be formed. For example, it can be a core network serving in an enhanced mobile broadband application scenario, or a core network serving in a low-latency application scenario. This creates conditions for network slicing.
(iii) *Supports MEC technology*: Mobile Edge Computing (MEC) is defined as "providing IT service environment and cloud computing capabilities at the edge of the mobile network." It is emphasized to be close to users, so that network operation and service delays are reduced, thereby improving user experience. Each functional block in

the SBA architecture can adopt a distributed architecture, and their geographic locations may not be together. Obviously, this distributed architecture also creates conditions for MEC.

(iv) *The user plane function and the control plane are separated*: The advantage of this is to make the two functions independent of each other, allowing independent evolution, and the two can be centralized or separated geographically (distributed architecture).

(v) *User plane function (transfer data)*: User plane function is referred to as UPF. It is a gateway, which replaces S-GW and P-GW in 4G (see Figure 1.56 in Chapter 1). It can connect RAN with external networks such as the Internet. This is the end-to-end connection mentioned earlier.

An important function of UPF can be illustrated in Figure 5.15. In the figure, UPF maps IP data packets to Quality of Service (QoS) flows according to the QoS request of the transmitted service. These requests can be requirements for delay or data rates, and so on. RAN maps QoS flows to radio bearers. It can be seen that in 5G, the core network is used to perceive the needs of the transmitted service, which creates conditions for the core network to adopt network slicing.

(2) *Control plane*: The control plane is composed of two functional blocks.

(a) SMF is responsible for terminal IP address allocation, policy implementation management (billing rule management) and general session management functions.

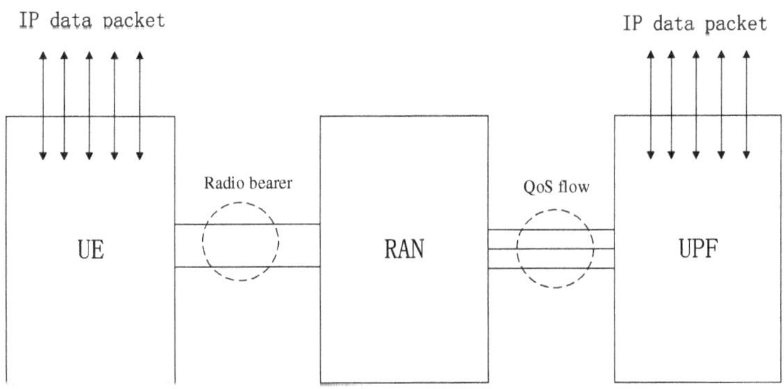

Figure 5.15 QoS flow and radio bearer.

(b) AMF is responsible for the control signaling between the core network and the terminal, security key management, authentication and the transition of terminals from idle state to active state. The function between the terminal and the core network is usually called Non-Access Stratum (NAS) function. The function between the terminal and the radio access network is called Access Layer (AS) function.

The functions of SMF and AMF mentioned above correspond to the functions of MME in 4G (see Figure 1.56 in Chapter 1).

5.5.2 *5G radio access network architecture*

The 3GPP NR Release 15 formulated in September 2018 is the first version of the 5G radio access network. Any new version formulated after 2018 must be backward compatible with the R15 version, so the content discussed in the following refers to the R15 version.

The 5G radio access network architecture is the same as LTE, and it also uses a hierarchical flat architecture, in which there is only a single type of node gNB, while there are various nodes in the 5G core network. gNB (or ng-eNB, it is an enhanced eNB in 4G, and its physical implementation is an enhanced 4G base station or 5G base station) is responsible for all wireless-related functions in one or more cells, and it is a logical node. Its physical realization usually has two methods, one is to use a three-sector base station, which is responsible for processing the signal transmission of the three cells. The other is a distributed structure, which consists of a baseband processing unit connected to several radio frequency units, and the radio frequency units are geographically separate from the baseband units. These remote radio frequency units are, for example, a large number of indoor cells or several cells deployed along a highway. This distributed structure can reduce network transmission delay.

The following section first discusses the interface between gNB and 5G core network and user terminals. Then, the protocol stack of gNB is discussed next.

5.5.2.1 *Access network interface*

(i) *Access network and core network interface*: The two interfaces of the 5G independent networking are shown in Figure 5.16. It can be seen from

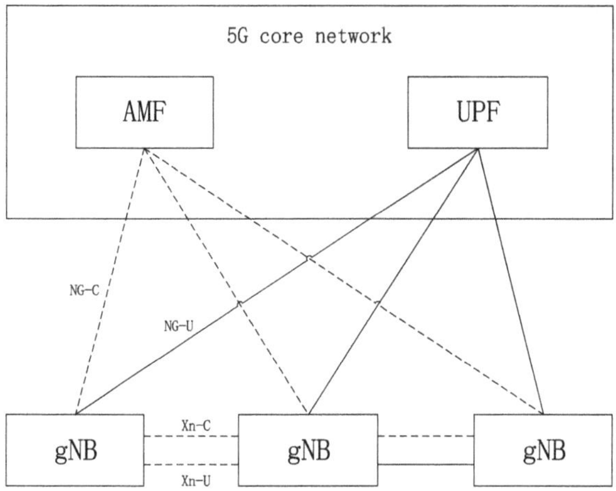

Figure 5.16 Access network and core network interfaces.

the figure that the gNB is connected to the core network through the NG interface, that is, it connects to the UPF of the core network through the NG user plane NG-U and connects to the AMF of the core network through the NG control plane NG-C. The NG interface corresponds to the S1 interface in 4G (see Figure 1.57 in Chapter 1). gNB and UPF/AMF can be connected to each other in the many-to-many manner through the NG interface. The interface between gNBs is Xn. Among them, Xn-U is the user plane interface, and Xn-C is the control plane interface. The Xn interface corresponds to the X2 interface in 4G (see Figure 1.57 in Chapter 1).

(ii) *Access network and terminal connection mode*: When communicating, the terminal and access networks have the following three connection methods:

(1) Connect with a cell in the access network.
(2) When the carrier is aggregated, it will be connected to multiple cells.
(3) The terminal connection of the 5G non-independent networking (NSA) is shown in Figure 5.17. This kind of connection is called dual connectivity. In dual connectivity, the terminal is connected to two cells (or nodes) at the same time: one is called the primary cell, and

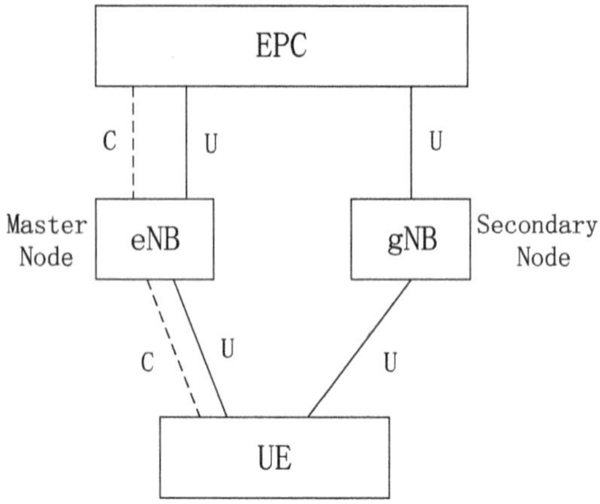

Figure 5.17 Access network and terminal interface.

the other is called the secondary cell. The example in Figure 5.17 is the series 3 scheme in non-independent networking.

5.5.2.2 *Radio access network protocol architecture*

(i) *Radio access network protocol stack*: The 5G NR protocol stack is shown in Figure 5.18. In the figure, gNB includes two protocol stacks, user plane and control plane. Among them, the four protocol entities of PDCP, RLC, MAC and PHY are common to these two individual protocol stacks. The difference is that the user plane above PDCP is the SDAP protocol entity. The control plane is the RRC protocol entity. The two protocol stacks are briefly introduced as follows:

(ii) *User plane protocol stack*: The user plane protocol stack is shown in Figure 5.19. Compared to the user plane in 4G (see Figure 1.59 in Chapter 1), the following differences can be seen:

(1) *Service Data Adaptation Protocol (SDAP) protocol entity is added on top of PDCP*:
It can be seen from Figure 5.15 that UPF maps IP data packets to QoS flows. PDCP interfaces with the Data Radio Bearer (DRB) (see Figure 1.59 in Chapter 1). Therefore, SADP needs to be used to map the QoS flow to DRB so that it can interface with PDCP. The mapping

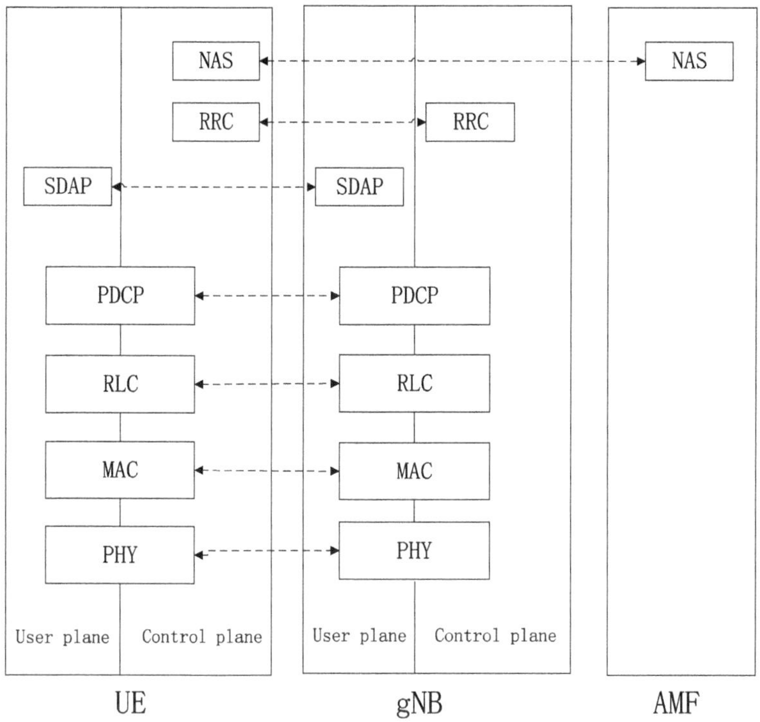

Figure 5.18 5G NR protocol stack.

process is shown in SDAP in Figure 5.20 is an example of user plane data flow.

(2) The data diversion point in the case of dual connectivity (series 4 in Figure 5.12(a)) is at the PDCP layer.

(3) In order to reduce the network delay, the following measures are adopted in the general four-layer protocol:

- *RLC*:

(a) Only the RLC SDU is divided and not cascaded.

(b) It is not ensured that SDUs are delivered to the upper PDCP in order.

- *MAC*:

The MAC layer multiplexes multiple RLC PDUs (i.e., MAC SDUs) and adds a MAC header. It can be seen from Figure 5.20 that a header is added before each RLC PDU is participating in multiplexing. But

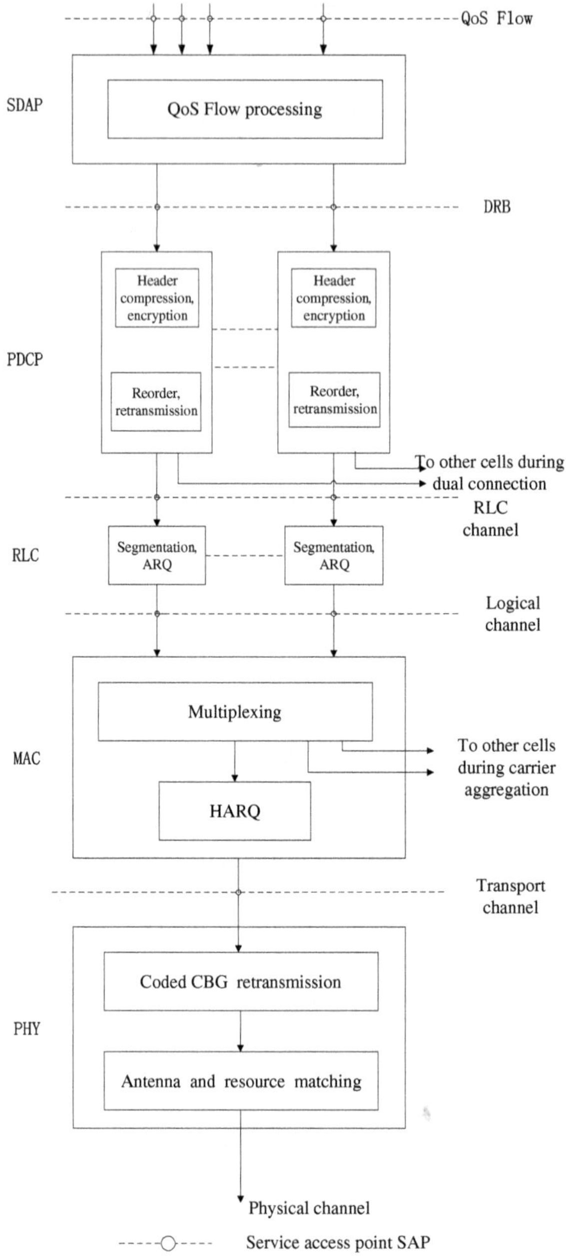

Figure 5.19 NR downlink user plane protocol stack.

Note: Code Block Group (CBG).

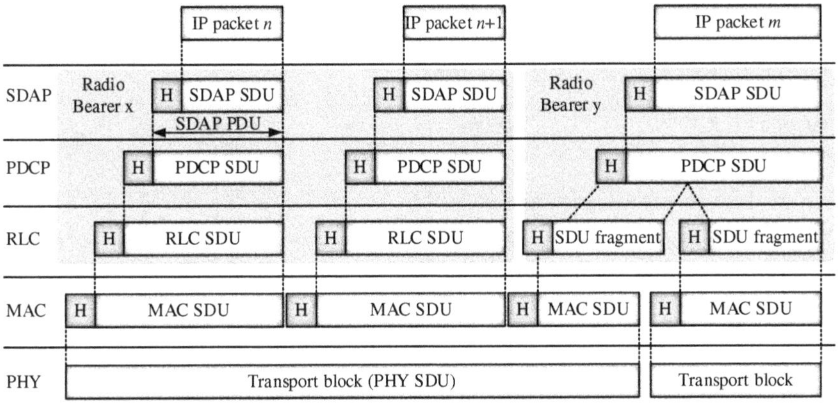

Figure 5.20 Example of user plane data flow.

in 4G (see Figure 1.60 in Chapter 1), just a header is added before all RLC PDUs are participating in multiplexing. The advantage of this is that this structure allows MAC PDUs to be assembled "online", thereby reducing network delay.

(4) *Channel mapping*:
The mapping between logical channels, transport channels and physical channels is shown in Figure 5.21. It can be seen from the figure that it is only a little different from the channel mapping in 4G (see Figure 1.62 in Chapter 1), that is, there is no MBMS service (multimedia broadcast multicast service) in 5G. Therefore, the channels related to it (MTCH, MCCH, MCH and PMCH) no longer exist.

(5) *Scheduler:*
In order to reduce the network delay, the following measures are adopted in the scheduler:

- The time slot in 4G is a fixed length of 0.5 ms, while the length of the time slot in 5G is variable, and the minimum time slot is 0.125 ms.
- Use mini-slots (see Figure 5.33).

(iii) *Control plane protocol stack:* The main function of the control plane protocol stack is basically the same as that of 4G. In order to reduce the network delay, a new state is added to the RRC state machine, namely the

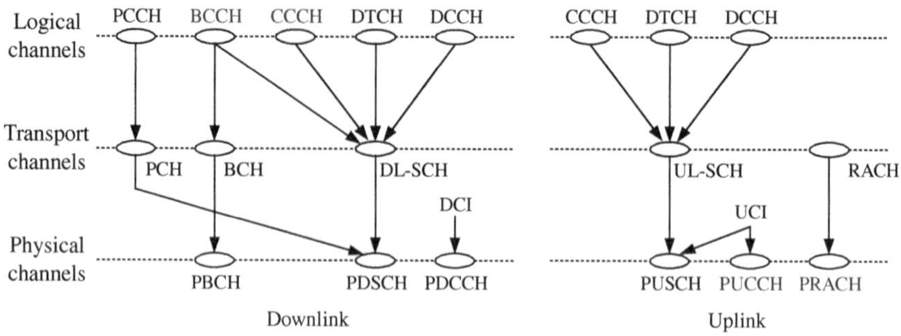

Figure 5.21 Channel mapping.

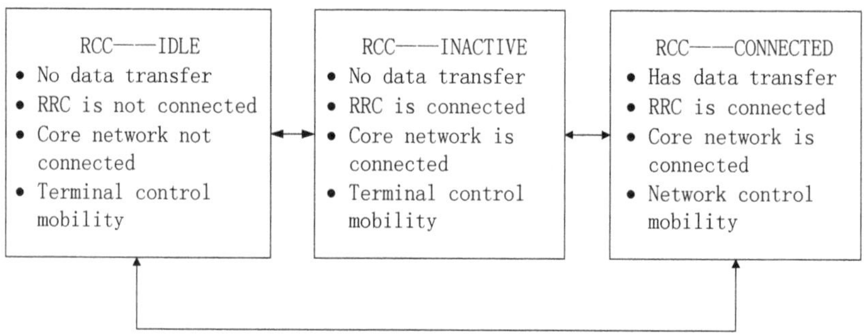

Figure 5.22 5G RRC status.

inactive state (RRC-INACTIVE). The sleep mode is similar to the sleep mode in the idle state. The three states of RRC can be converted to each other, as shown in Figure 5.22.

5.5.3 *5G mobility management*

It can be seen from Figure 5.22 that when the terminal is in different working states, its mobility management methods are different. When the terminal is in an idle or inactive state, the terminal controls the mobility (that is, the mobility is controlled by the terminal in the figure), which is called cell reselection. When the terminal is in the connected state, the mobility is controlled by the network (that is, the mobility is controlled by the network in the figure), which is called cell handover. The purpose and methods of these two types of mobility management are different and need to be discussed separately.

5.5.3.1 *Mobility in idle state and inactive state*

Since the terminal is in a dormant state in these two states, while the terminal will move, the purpose of mobility management in this case is to ensure that the network can contact the terminal. For this reason, a paging message needs to be transmitted to the terminal. Then which cells in the network should be transmitting paging messages to find the so-called terminal? Obviously, there are three methods to choose from.

(1) Send paging messages to all cells in the network. This will definitely find the terminal. But this method has too much overhead for transmitting paging messages because only one paging message is useful. Obviously, this method is not advisable.
(2) Only transmit paging messages to the cell where the terminal is located, which of course will surely find the terminal. However, this method requires the terminal to notify the network of the cell where it is located by signaling every time it moves to a cell. Obviously, this method has too much signaling overhead, so it is also not advisable.
(3) Only transmitting paging messages to some cells is a compromise between the above two methods. 5G adopts this method, and for this, it is necessary to divide the area covered by the network to a certain extent.

(i) *Tracking area and radio access area*: The network coverage area is divided into several tracking areas and radio access areas. The cells contained in one of the tracking areas and radio access areas are shown in Figure 5.23. It can be seen from the figure that several cells

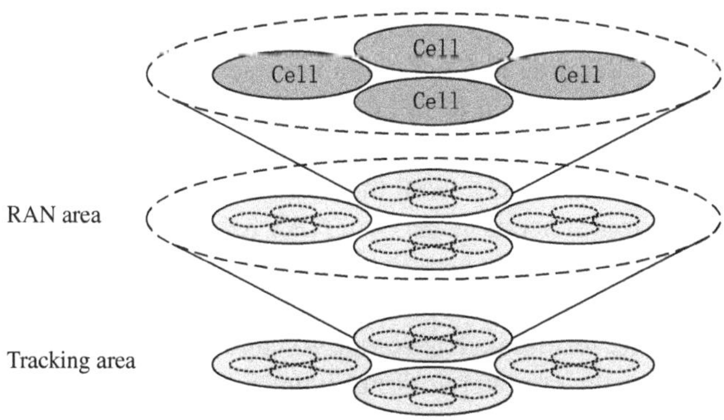

Figure 5.23 Tracking area and radio access area.

form a radio access area (RAN area). Each RAN area is identified by a RAN area identifier. The RAN area forms a larger Tracking Area (TA). Each tracking area is identified by a Tracking Area Identifier (TAI). Therefore, each cell belongs to both a RAN area and a tracking area.

Some people may ask, why should we divide the two areas? The reason for dividing the two areas is that the 5G network is composed of a core network and an access network. The tracking area is used for the core network to track the terminal, and the RAN area is used for the access network to track the terminal.

(ii) *The tracking process of the terminal*: We will not discuss the tracking process of the network to the terminal, but briefly explain the tracking process from the basic principle.

(1) When the terminal is turned on, the terminal searches and measures the received power of the Synchronization Signal Block (SSB) from the PBCH channel of the neighboring cell (see Figure 5.21) and selects the system information 1 of the cell with the larger received power (SIB1, System Information Block 1). In this way, information about which tracking area and RAN area the cell belongs to can be obtained.
(2) The terminal uses random access (see Figure 1.66 in Chapter 1) to establish a connection with the network and notify the core network (AMF) and access network (gNB) of the obtained tracking area and RAN area, respectively.
(3) The core network allocates a UE Registration Area to the terminal according to the tracking area where the terminal is currently located. The registration area is a list containing several tracking area identifiers. When the terminal is still moving within the cell range included in the list, there is no need to update the UE registration area. But when the terminal enters a cell that does not belong to the registration area, the terminal needs to perform NAS registration update. The core network registers the new cell of the terminal and updates the UE registration area of the terminal, which is actually to reassign a new TAI list to the terminal. The reason for updating the registration area in this case is the core network in Figure 5.16 contains multiple AMF/UPFs, which are connected to each other in the many-to-many manner with gNB. In this way, when the terminal moves far, the AMF/

UPF where the terminal resides may be different, so the terminal needs to reconfigure the IP address. For this reason, the UE registration area needs to be updated at this time.

(4) The access network assigns a RAN notification area to the terminal according to the RAN area where the terminal is currently located. It contains several cells, at least one cell and at most cells contained in a tracking area. If the terminal moves within the RAN notification area, there is no need to update the RAN notification area of the terminal. If the terminal enters a cell that does not belong to the RAN notification area, the terminal needs to perform RRC RAN notification update, the access network registers the new cell of the terminal and updates the RAN notification area of the terminal.

It should be noted that whenever the terminal performs UE registration area update, it will be accompanied by RRC RAN notification area update.

5.5.3.2 *Mobility in the connected state*

The purpose of mobility management when the terminal is in the connected state is to ensure that there will be no communication interruption or a significant decrease in communication quality during cell handover. To this end, the terminal needs to measure the carrier power of the current cell and neighboring cells. This measurement can be performed on SSB. The terminal reports the measurement result to the network, and the network determines whether the terminal needs to switch to a new cell based on the measurement result. The report is done through RRC signaling.

5.6 5G NR Physical Layer

5.6.1 *5G NR design idea*

The design background of 5G NR and 4G LTE has many differences, mainly:

- *Working frequency*: LTE working frequency band is sub-6 GHz. The licensed frequency band is lower than 3.5 GHz, and the unlicensed frequency band is 5 GHz. In addition to sub-6 GHz, the NR authorized frequency band also needs to support millimeter wave frequency bands. Its working frequency range is 0.45–52.6 GHz, and this frequency range is very wide. The expansion of unlicensed frequency bands is also being planned. Since the characteristics of high-frequency and low-frequency radio wave propagation (such as path loss) and the performance of the radio frequency device itself (such as phase noise) are different, the system design of the high- and low-frequency bands is still very different.

- *Application scenario*: The main application scenario of LTE is MBB, and NR is a variety of application scenarios. In addition to eMBB, mMTC and URLLC must also be supported. The technical requirements are different in different application scenarios, such as data rate, network delay and terminal power consumption and cost.
- *Network deployment*: Due to many NR application scenarios and wide working frequency range, 5G network deployment is more complicated than 4G, and there are more cell types than 4G. There are macrocells, microcells and picocells in NR, and their coverage areas are different. In this way, the cyclic prefix length required for the OFDM waveform is also different.

Due to the different design backgrounds, there are many differences between the design ideas of the 5G NR standard and LTE, mainly:

- *Flexible design*: In order to meet the needs of multiple application scenarios, multiple cell types and a wide frequency range, the standards formulated should be flexible and variable.
- *Minimalist design*: There are some signals in LTE called always-on signals, for example, signals used to detect base station, broadcast signals used for system information and reference signals used for channel estimation. The so-called minimalist design is to minimize the always-on signal. This can reduce signaling load, terminal power consumption and cost and cell interference.
- *Forward compatible*: The R15 version formulated by 3GPP in 2018 is the first version of NR. It is mainly aimed at the eMBB application scenario and also takes into account URLLC to a certain extent. However, the application scenarios of mMTC are still eMTC and NB-IoT based on LTE (see Chapter 4). However, with the continuous development of 5G networks, there will be new technologies in the above three application scenarios, especially mMTC. At this time, new

versions must be developed, and these versions must be backward compatible with R15. This means that when formulating the R15 version, it is necessary to take into account the possibility of new versions in the future, which is called forward compatibility. There are two ways to achieve forward compatibility. One is that the design scheme must be flexible. For example, some time-frequency resources that need to be occupied in R15 should not be fixed, but can be flexibly occupied as needed, that is, occupy when needed and abandon when not needed. The second is to reserve some time-frequency resources for future use.

It can be seen from Figure 5.18 that the physical layer is the bottom layer of the 5G NR protocol stack. Therefore, the design of the physical layer must also be carried out according to the above design principles. It is the core part of the 5G NR design. The 5G NR physical layer technologies mainly include modulation, waveform, coding, multiple access, frame structure, radio resource structure and allocation, reference signal, multi-antenna application, radio frequency unit, NR and LTE interoperability and coexistence.

These technologies are discussed one by one, and we first discuss the modulation methods used in 5G NR.

5.6.2 *Modulation*

Modulation is the mapping in Figure 1.39 in Chapter 1. Regarding the modulation method used in user plane or control plane, it changes differently with 5G network application scenarios, transport channel status and terminal levels, which mainly are as follows:

(i) Both uplink and downlink can use QPSK, 16QAM, 64QAM and 256QAM. They are all high-order modulations, and the modulation order is $M = 2^n$. Therefore, the number of bits of information contained in each symbol output by high-order modulation is $n = \log_2 M$. Thus, the n of the above four modulation methods can be obtained as 2, 4, 6 and 8, respectively. It should be noted that the value of n for QAM is usually an even number, and 1024QAM has been adopted in Wi-Fi (see Chapter 3), and the value of n is 10.

(ii) In the mMTC application scenario, the uplink or terminal will use $\pi/2$-DBPSK modulation. It is called $\pi/2$ differential two-phase phase shift keying. The following first discusses its working principle and then explains the reasons for using it in the mMTC application scenario.

Two-phase phase shift keying BPSK was discussed in Chapter 1. The constellation is shown in Figure 1.4(a), and now it is redrawn in Figure 5.24(a). Figure 5.24(b) shows the $\pi/2$ − DBPSK constellation. It can be seen from the figure that there are two signal points in Figure 5.24(a) and four signal points in Figure 5.24(b). Although the signal points of the two are different, they are still two-phase phase shift keying. The modulation order is $M = 2^n = 2$, which means that each symbol contains only one bit of information. The main difference between the two is the phase modulation method, which is called phase logic, as shown in Table 5.6. It can be seen from the table that BPSK is an absolute phase shift, that is, the initial phase of the carrier is directly controlled by the 1, 0 code input by the modulator. When $a_k = 0$, the carrier initial phase $\phi_k = 0$. When $a_k = 1$, the carrier initial phase $\phi_k = \pi$. The waveform in Figure 5.25(a) is drawn according to this phase logic. It can be seen from the figure that the carrier phase of BPSK always changes 180° at the moment of symbol switching, so there are only two signal points on the constellation in Figure 5.24(a). It can be seen that the instantaneous amplitude of the carrier at the moment of symbol conversion must fall (rise) from the maximum value $+A$ (or $-A$) to zero (pass the coordinate origin) to reach $-A$ (or $+A$). This means that the maximum instantaneous amplitude is A, and the minimum is 0. The maximum instantaneous amplitude change value is also A ($A-0 = A$). The normalized value of the maximum change value to the maximum value, that is, the relative value of the maximum change value is 100% ($A/A = 1$). It shows that the power peak-to-average ratio of BPSK (see Chapter 1 for the definition of power peak-to-average ratio) is relatively large when the phase shift is absolute. When the terminal adopts BPSK, the power efficiency of the power amplifier will be reduced, which is not power saving.

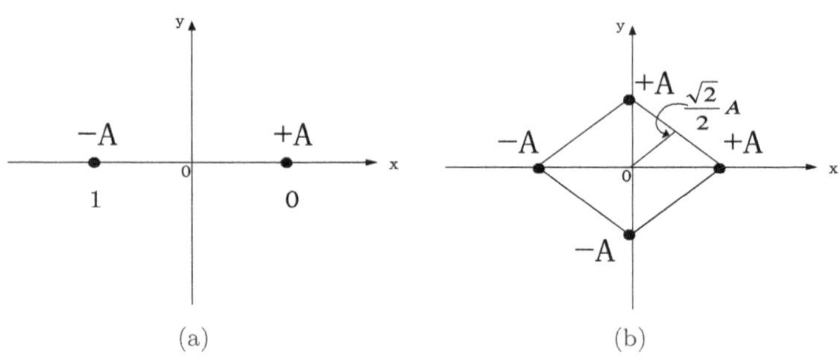

Figure 5.24 Two-phase phase shift keying constellation: (a) BPSK; (b) $\pi/2$-DBPSK.

Table 5.6 Two-phase phase shift keying phase logic.

(a) Absolute phase shift BPSK		
Symbol a_k	0	1
Carrier phase ϕ_k	0	π
(b) Relative phase shift $\pi/2$ − DBPSK		
Symbol a_k	0	1
$\Delta\phi_k = \phi_k - \phi_{k-1}$	$-\pi/2$	$\pi/2$

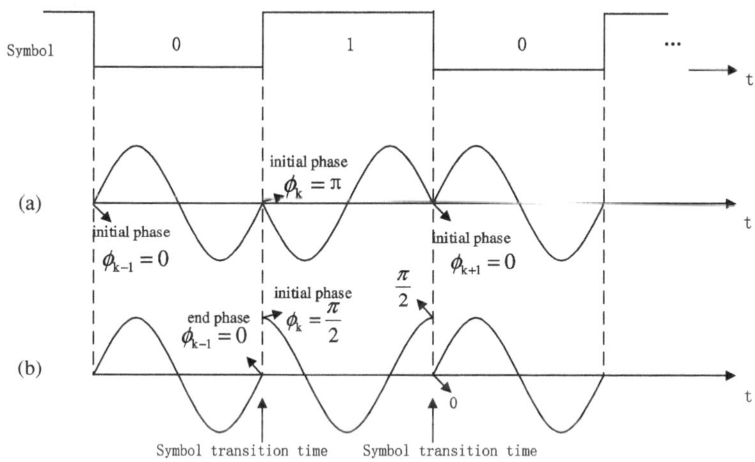

Figure 5.25 Absolute phase shift and relative phase shift: (a) absolute phase shift BPSK; (b) relative phase shift $\pi/2$ − DBPSK.

$\pi/2$ − DBPSK is a relative phase shift. It can be seen from Table 5.6(b) that the 1, 0 code input by the modulator does not directly control the initial phase of the carrier, but the difference $\Delta\phi_k$ of the carrier phase is seen in the two adjacent symbols. When $a_k = 0$, $\Delta\phi_k = -\pi/2$. It means that the carrier in the following symbol is 90° behind the end phase of the carrier in the previous symbol. When $a_k = 1$, $\Delta\phi_k = \pi/2$. It means that the carrier in the following symbol is 90° ahead the end phase of the carrier in the previous symbol. The waveform in Figure 5.25(b) is drawn according to this phase logic. It can be seen from the figure that $\pi/2$ − DBPSK always abruptly changes 90° at the beginning of the carrier at the moment of symbol switching compared to the end phase of the carrier in the previous symbol. Therefore, there are four signal points on the constellation in Figure 5.24(b). It can be seen that the minimum value of the instantaneous amplitude of the carrier at the moment of symbol conversion is $\frac{\sqrt{2}}{2}A$ (it never passes through the origin of the coordinate). The maximum change value is $(1-\frac{\sqrt{2}}{2})A$. The maximum relative change value is $\frac{(1-\frac{\sqrt{2}}{2})A}{A}=1-\frac{\sqrt{2}}{2}=30\%$. It shows that the power peak-to-average ratio during relative phase shift is much smaller than that during absolute phase shift. When the terminal adopts $\pi/2$ − DBPSK, the power efficiency of the power amplifier is much higher, which can reduce the power consumption.

Whether it is BPSK or $\pi/2$ − DBPSK, since only one bit of information is contained in one symbol, it can only be used in low data rate occasions.

When the 5G network is in the mMTC application scenario, the data rate transmitted by the terminal is relatively low, but the terminal requires low power consumption and low cost. In this case, it should be the most suitable terminal to use $\pi/2$ − DBPSK.

Finally, it should be pointed out that which modulation method is adopted in the 5G network is determined by the network side according to the channel transmission characteristics through scheduling adaptively, and the terminal is notified by signaling.

5.6.3 *Waveform*

Like LTE, NR also chose OFDM as the waveform transmitted in the wireless channel. It is the waveform generated by the IFFT in Figure 1.39 in Chapter 1. The OFDM (CP-OFDM) waveform with cyclic prefix has the following two advantages:

(1) Anti-Inter-Symbol Interference (anti-ISI). This feature is very important for systems that transmit high-speed data rates in multipath channels.
(2) It can be used as a two-dimensional radio resource application. One dimension is the number of subcarriers in the frequency domain, and the other dimension is the number of OFDM symbols in the time domain. When it is used for physical channel and physical signal transmission, it can be very flexible to meet the required capacity of the physical channel and physical signal by changing the number of subcarriers and (or) the number of symbols.

5.6.3.1 *NR uplink/downlink transmission waveform*

Currently, in the frequency range of 0.45–52.6 GHz, the following transmission waveforms are selected for NR uplink/downlink:

(i) *Downlink*: Same as LTE, its waveform is OFDM (CP-OFDM). OFDM parameters such as subcarrier spacing and CP length will be discussed in detail in the following.
(ii) *Uplink*: In order to reduce the power peak-to-average ratio in LTE, DFTS-OFDM waveforms are transmitted in the uplink. However, the uplink waveform in NR is OFDM, and the DFTS-OFDM waveform is only an option. The main reasons for this choice are as follows:

- When space division multiplexing (MIMO) is used in the uplink, if the DFTS-OFDM waveform is used in the uplink, the design of the MIMO receiver will be very complicated (because multiple terminals and multiple antennas transmit signals to the base station). Since LTE terminals do not use MIMO, DFTS-OFDM waveforms can be used in the uplink. However, NR terminals support space division multiplexing, so DFTS-OFDM is not suitable for uplink waveforms.
- In the following two situations, it is not advisable to use DFTS-OFDM uplink waveform, and it is convenient and simple to use OFDM waveform:

Case 1: Sidelink, which is the D2D mentioned earlier.
Case 2: Wireless backhaul.

The reasons why the DFTS-OFDM waveform is not used in the above two cases are discussed in the following sections.

5.6.3.2 *Sidelink and wireless backhaul*

(i) *Sidelink*: The schematic diagram of D2D communication is shown in Figure 5.6(b). It can be seen from the figure that when the terminal communicates with the base station, the base station transmits an OFDM waveform. Therefore, the terminal should use an OFDM receiver. If the terminal uses the DFTS-OFDM waveform for uplink, when two terminals perform D2D communication, the terminal transmits the DFTS-OFDM waveform. The OFDM receiver cannot receive this waveform, so D2D communication cannot be realized. Obviously, the terminal uplink

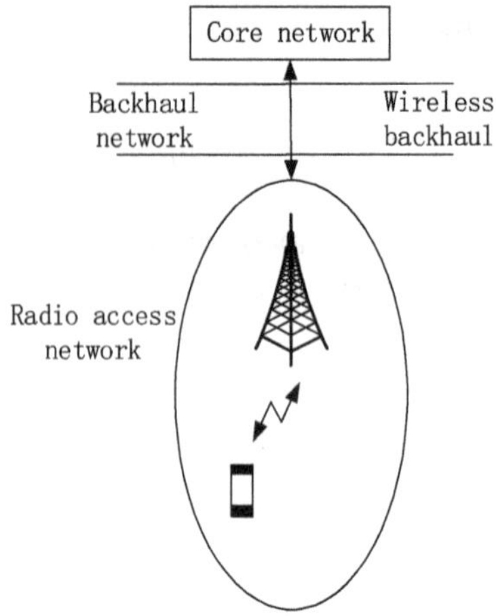

Figure 5.26 Schematic diagram of wireless backhaul.

should be just changed to use OFDM waveform. Then, the communication between the terminal and the base station or the communication between the terminal and the terminal can be realized.

(ii) *Wireless backhaul*: The radio backhaul is shown in Figure 5.26. It refers to the part of the radio access network connected to the core network. Optical fiber is ideal for backhaul networks. However, in environments where optical fiber is difficult to deploy or the deployment cost is too high, the only alternative is to use a wireless solution, that is, wireless backhaul, for example, point-to-point microwave, and millimeter wave backhauls. In the case of wireless backhaul, using the same OFDM waveform for the uplink and downlink will also simplify the system design.

5.6.3.3 *Power spectral density of OFDM waveform*

It can be seen from Figure 1.23 in Chapter 1 that the time-domain envelope of each signal output by the N multipliers is rectangular. The time-domain envelope of the OFDM waveform obtained by adding them is also rectangular. Therefore, the Power Spectrum Density (PSD) of the OFDM waveform is a Sin^2C function. *sinC* is called the singular function, which is equal to $sinx/x$, and x is the variable that is proportional to the frequency f. It can be seen that the PSD of the OFDM waveform is

$(sinx/x)^2$. The shape of the envelope is shown in Figure 5.27. We can understand the following from the picture:

- The Out-of-Band (OOB) of the OFDM waveform is attenuated by f^2. Its decay speed with f is still relatively slow, thus causing larger out-of-band radiation. For frequency division multiplexing, the interference between adjacent spectrums is still relatively large.

The OOB attenuation speed of the OFDM waveform with f is related to the carrier spacing Δf, and the larger Δf is, the slower the attenuation speed is, and vice versa. This feature can be explained as follows. When maintaining the same in-band bandwidth, the number of subcarriers N required for a large Δf is less than that for a small Δf. The larger the number of subcarriers N, the faster the out-of-band attenuation is, and vice versa.

In order to reduce the OOB radiation of the OFDM waveform, the following two measures can be used:

(1) *Frequency domain*: A band-pass filter is connected after the "Add Cyclic Prefix" function block in Figure 1.39 in Chapter 1, to suppress the OOB radiation of the OFDM waveform.
(2) *Time Domain*: A time-domain window should be added as shown in Figure 5.28 after the "Add Cyclic Prefix" function block in Figure 1.39. When (a) and (b) in Figure 1.39 are multiplied in the time

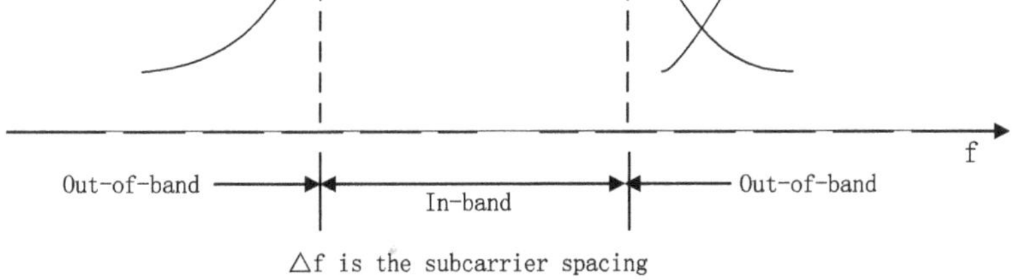

Figure 5.27 Envelope of OFDM waveform PSD.

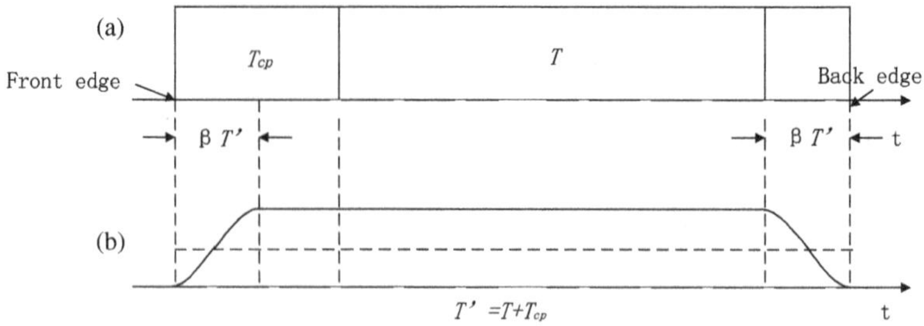

Figure 5.28 Time-domain window and OFDM waveform: (a) OFDM Waveform; (b) time-domain window.

domain, the front and back edges of the resulting OFDM waveform are no longer the rising and falling edges of sudden jumps. It is an OFDM waveform with slowly changing rising and falling edges during the $\beta T'$ period, and the front and rear edges are slowly changing. The decay speed of its power spectral density with f outside the band is much faster than the sudden jump edge. The cost is that it takes up part of the T_{cp} time, which shortens the CP. This is unfavorable against delay expansion.

In practical applications, the implementation complexity of adding time windows is much simpler than adding filters.

5.6.4 *Physical resources*

From the previous discussion, we know that the waveform of 5G wireless channel transmission is CP-OFDM (shortened as OFDM). It is used to carry two types of information, among which the physical channel carries the data information of the physical layer. The other type is data information that carries physical signals, but this type of data information is not data information from higher layers of the physical layer, but data information from various control signaling. We call the carrier used to transmit information the physical resource of the physical layer. So, NR is the same as LTE, its physical resource is OFDM. It is a two-dimensional time-frequency resource, where the time domain corresponds to the number of OFDM symbols, and the frequency domain is the number of subcarriers in an OFDM symbol. This section discusses the structure, parameter set and application of such time-frequency resources.

5.6.4.1 *Physical resource structure*

Like LTE, the resource structure of NR can be represented by a resource grid (refer to Figure 1.46 in Chapter 1) as shown in Figure 5.29, which is explained as follows:

- The resource unit RE has the same definition as LTE. One OFDM symbol and one subcarrier spacing are defined as RE. It is the minimum time-frequency resource. Any physical resource used in transmission is composed of several REs.
- The Resource Block (RB) is defined differently from LTE. In NR, 1 OFDM symbol and 12 subcarriers are defined as RBs, while in LTE, 7 OFDM symbols and 12 subcarriers are defined as RBs.
- A time slot is different from the LTE definition. One slot in NR contains 14 OFDM symbols, and one slot in LTE contains 7 OFDM symbols.

It must be emphasized here that there is only one subcarrier spacing in LTE, that is, $\Delta f = 15KHz$. In NR, there are many kinds of subcarrier

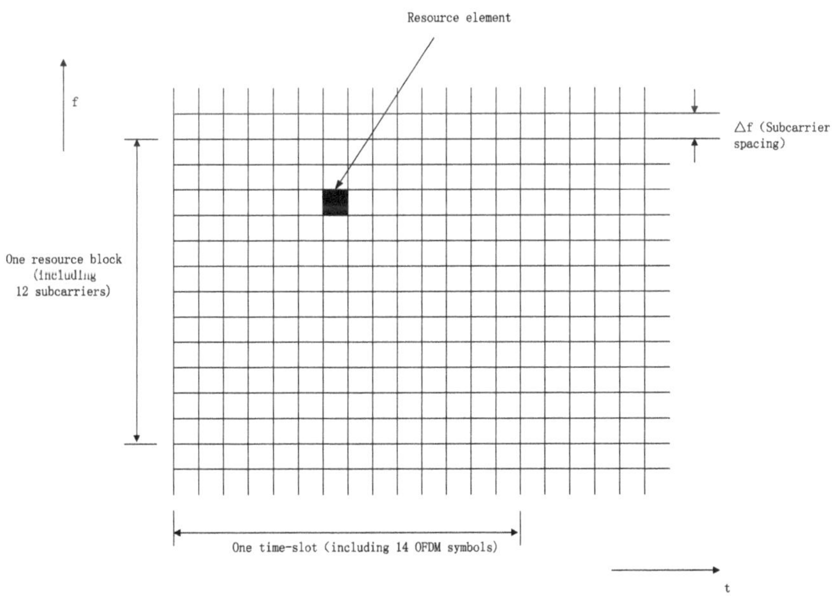

Figure 5.29 NR physical resource grid.

spacing, Δf = 15, 30, 60, 120, 240, 480 *KHz*. We know that the number of subcarriers transmitted in the channel is scheduled with 12 subcarriers as a group. For NR, this group of subcarriers is a RB. For example, several RBs may be required when transmitting certain service data. So now there is a problem, that is, where should these RBs be located in the frequency domain of the resource network? Obviously, in order to clarify the location, it is necessary to number the RBs in the frequency domain. Then, the network side (base station side) arranges the required number of RBs and their numbers according to the transmitted service data and uses signaling to notify the terminal. This process is called network scheduling. It can be seen that the scheduling in the frequency domain is performed with 12 subcarriers as a group.

Since there is only one type of Δf in LTE, the numbering in the frequency domain is simple. However, as there are many kinds of Δf in NR, the numbers in the frequency domain are more complicated at this time. The example in Figure 5.30 is used to illustrate how to number.

In the figure, the subcarrier spacing Δf and $2\Delta f$ are used as examples for discussion. It can be seen from the figure that:

- Although the subcarrier spacing is different, the starting point of the two in the frequency domain is the same, that is, the reference point *A*.
- Taking reference point *A* as a starting point, the 12 subcarriers are divided into a group. Each group is called a Common Resource Block

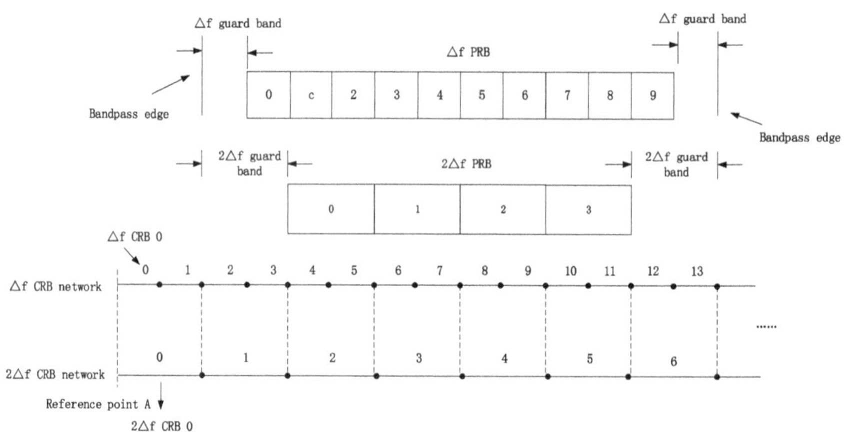

Figure 5.30 Common resource block and physical resource block.

(CRB), and CRB is numbered. It can be seen from the figure that although the subcarrier spacing is different, the starting point of CRB 0 is the same.

The frequency range between the two band-pass edges in the figure is the available bandwidth. The configured Physical Resource Block (PRB) can only be within the available bandwidth. The configuration method for different subcarrier intervals is as follows:

- When using the Δf subcarrier spacing, PRB 0 is actually CRB 3, and PRB 9 is actually CRB 12. There are guard bands on both sides, which are set up to reduce mutual OOB radiation interference during frequency division multiplexing.
- When using $2\Delta f$ subcarrier spacing, PRB 0 is actually CRB 2 and PRB 3 is actually CRB 5. But the guard bands on both sides are wider than Δf. This is because the larger the subcarrier spacing, the larger the out-of-band radiation of the OFDM waveform (see Figure 5.27).

5.6.4.2 *Physical resource parameter set*

The data information transmitted in the wireless channel is carried by physical resources. The location of these physical resources in the resource grid is determined by the network side and implemented through scheduling. In order to facilitate scheduling, the physical resources in the resource grid must be numbered. Since the resource grid is two-dimensional, it is necessary to number them in the frequency domain and time domain separately. The numbering method in the frequency domain has been discussed in the previous section, that is, the common resource block (CRB) is numbered with reference point *A* in Figure 5.30 as the starting point. Then, the CRB is used to determine the position of the actually used physical resource block (PRB) in the frequency domain. The numbering method in the time domain will be discussed in the following. It uses the number of the radio frame to determine the position of the actually used PRB in the time domain. In Chapter 1, we discussed the frame structure of LTE (see Figure 1.45, for example). It can be seen that it is closely related to the subcarrier spacing Δf, and NR has a variety of subcarrier spacing. Each subcarrier spacing constitutes an OFDM parameter set, that is, NR has multiple OFDM parameter sets. For this reason, we must first discuss the parameter set and then the frame structure.

The parameter set (numerology) refers to the two parameters of the symbol length T and the cyclic prefix length T_{cp} selected for OFDM. Among them, T_{cp} changes proportionally with T, and $T = 1/\Delta f$, where Δf is the subcarrier spacing. Therefore, when Δf is determined, the OFDM parameter set is also determined. The choice of Δf is based on the following two principles:

(i) *NR working frequency band*: The working frequency band is high, such as the millimeter wave frequency band, because the power of radio waves decays faster as the propagation distance increases. Therefore, the coverage of the cell is small, and the corresponding delay spread is small. In this case, a shorter T_{cp} can be selected, and a shorter T can be selected. For this reason, a larger Δf should be selected. Conversely, when the operating frequency is low, a smaller Δf should be selected. The R15 version of NR has confirmed two operating frequency bands.

FR1: 0.45–6 GHz; FR2: 24.25–52.6 GHz

In the 6–24.25 GHz range, there is currently no allocated NR spectrum.

(ii) *5G application scenarios*: There are three major application scenarios for 5G, namely eMBB, URLLC and mMTC. The R15 version is mainly aimed at eMBB, but it also takes into account URLLC. Appropriate consideration mainly refers to shortening the time delay as much as possible, that is, adopting some technical measures to ensure the requirements of low delay. For example, the millimeter wave frequency band allows one to work on a larger bandwidth, for which a larger Δf needs to be selected. At this time, a higher data rate is allowed to transmit, so that the transmission delay is reduced.

The OFDM parameter set obtained according to the above principles is shown in Table 5.7. The table is described as follows:

(1) T_s in the table is the LTE sampling interval mentioned in Chapter 1. It refers to the subcarrier spacing in LTE $\Delta f = 15KHz$. The corresponding OFDM symbol length $T = 1/\Delta f = 1/15(ms)$. The number of samples in T is 2048. Therefore, the sampling interval $T_s = 1/(15\times2048) = 1/30720$ ms. However, the sampling interval T_c is sometimes used in R15, which has the following relationship with T_s:

Table 5.7 OFDM parameter set.

OFDM parameter set	Δf = 15 KHz	2 × 15 = 30 KHz	2^2 × 15 = 60 KHz	2^3 × 15 = 120 KHz
Frequency band (GHz)	0.45–6	0.45–6	0.45–6 24.25–52.6	24.25–52.6
OFDM symbol length T (μs)	66.67 (2048 T_S)	33.33 (1024 T_S)	16.76 (512 T_S)	8.33 (256 T_S)
Cyclic prefix length T_{cp} (μs)	4.69 (144 T_S)	2.34 (72 T_S)	1.17 (36 T_S)	0.59 (18 T_S)
$T+T_{cp}$ (μs)	71.35	35.68	17.84	8.91
Time slot length (μs)	1000	500	250	125
Number of OFDM symbols per slot	14	14	14	14
Maximum bandwidth (MHz)	50	100	200	400

Note: $T_s = \frac{1}{15 \times 2048}$ ms.

$$T_c = T_s/64 = 1/(64 \times 15 \times 2048) = (1/480)/4096 \text{ ms}.$$

It can be seen from the above equation that when the subcarrier spacing Δf = 480 *KHz*, the corresponding OFDM symbol length T = 1/480 ms. When the number of samples in T is 4096, the sampling interval is T_c.

(2) The cyclic prefixes in the table are all the standard cyclic prefixes ($144T_s$) mentioned in Chapter 1. At this time, the overhead of CP and LTE is about 7%(144/2048 = 7%). For Δf = 60 KHz, the extended cyclic prefix (512 T_s) can also be configured. Its CP overhead is 25%(512/2048 = 25%).

(3) The four subcarrier intervals given in the table are all used for regular data transmission. For the terminal, NR only requires it to support one of the subcarrier intervals. It should be noted that the subcarrier spacing configured for the Synchronization Signal Block (SSB) is 24×15 = 240 KHz. But it cannot be used for regular data transmission.

(4) The interval of all subcarriers in the parameter set is equal to the nth power of 15×2 KHz, where n = 0,1,2,3,4. The benefits of this design will be discussed in the following.

(5) The maximum bandwidth in the table refers to the 273 maximum physical resource blocks that NR allows for downlink use. Each resource block contains 12 subcarriers. In this way, the maximum

number of subcarriers allowed to be activated is 3276 (273×12). The bandwidth it occupies (3277 × subcarrier spacing) plus the guard bands on both sides is the maximum bandwidth in the table.

(6) Why choose the power of 15×2 KHz for the subcarrier spacing?

The interval of all subcarriers in NR is equal to the power of 15×2 KHz. Therefore, 15 KHz is the baseline of the NR parameter set, that is, the LTE parameter set is selected as the baseline of the NR parameter set. All subcarrier intervals are generated from this baseline by a factor of 2. There are two reasons for this design.

- *Narrowband Internet of Things (NB-IoT)*: As discussed in Chapter 4, when using in-band deployment as shown in Figure 4.3, NB-IoT must use the LTE parameter set.
- The second reason will be discussed in the following section.

5.6.4.3 *NR frame structure*

Like LTE, the NR frame structure is used to represent the structure of the OFDM symbol in the time domain. The frame structure is periodic. Each transmission cycle contains 1024 radio frames. The length of the radio frame is 10 ms. Each radio frame is numbered, which is called System Frame Number (SFN). The number of SFN is #0–#1023. Therefore, the SFN number repeats itself after 1024 radio frames or 10.24 seconds (10 × 1024 ms). Each radio frame is divided into 10 sub-frames, and the length of the sub-frame is 1 ms. Each sub-frame is divided into time slots, and each time slot contains 14 OFDM symbols. The length of the time slot is related to the subcarrier spacing. The greater the subcarrier spacing, the smaller the slot length. The relationship among radio frames, sub-frames and time slots is shown in Figure 5.31.

The following description is given for Figure 5.31:

(i) Since radio frames have SFN numbers and radio frames, sub-frames and time slots have a fixed relationship with each other. Therefore, the position of the actual physical resource in the time domain can be determined. The scheduling on the network side is based on time slots.

(ii) When the subcarrier spacing $\Delta f = 15$ KHz, its sub-frame structure is the same as LTE. But the definition of the two time slots is different.

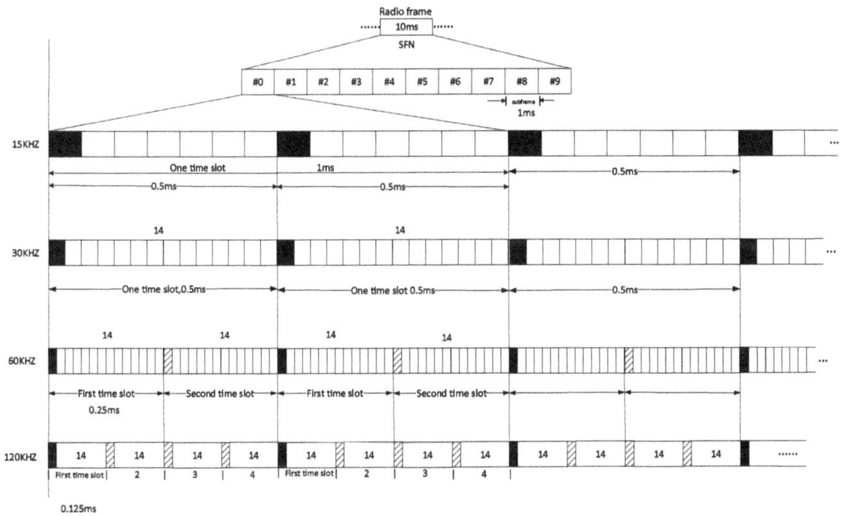

Figure 5.31 NR frame structure.

In LTE, a slot is 0.5 ms and contains 7 OFDM symbols. But in NR, one slot is 1 ms, so it contains 14 OFDM symbols.

The subcarrier spacing in NR is a power of 15×2 KHz. Therefore, the duration of the OFDM symbol decreases by the power of 2 ($\because T = 1/\Delta f$, T_{cp} is proportional to T) because the number of OFDM included in the length of a slot is still 14. Therefore, the length of a time slot is also reduced by the power of 2. That is, when the subcarrier spacing is 15, 30, 60 and 120 KHz, the corresponding time slot lengths are 1, 0.5, 0.25 and 0.125 ms, respectively.

(iii) When the subcarrier spacing $\Delta f = 15$ KHz, its frame structure is the same as LTE. As shown in Figure 1.45 in Chapter 1, it is now redrawn in Figure 5.32(a). The figure requires that under the standard cyclic prefix condition, the occupancy time of 7 OFDM symbols is 0.5 ms. In order to meet this requirement, the number of samples in the first OFDM symbol (OFDM#0) is 16 more than the other 6 OFDM symbols (OFDM#1–#6). In this way, the duration of 7 OFDM symbols is $[16 + (144 + 2048) \times 7]T_s = 15360T_s = 15360/(15\times2048) = 0.5$ ms.

When the subcarrier spacing $\Delta f = 30,60,120$ *KHz*, under the standard cyclic prefix condition, the durations of 14, 28 and 56 OFDM symbols are required to be 0.5 ms, respectively. To meet this requirement, their first

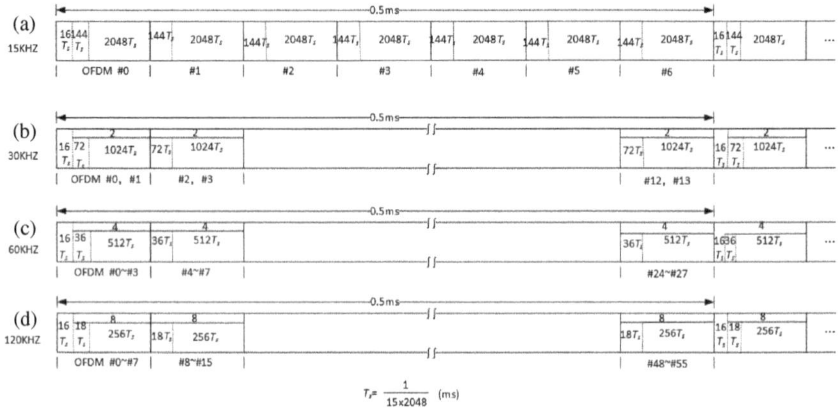

Figure 5.32 Number of samples contained in OFDM symbol.

OFDM symbol (OFDM#0) has 16 more samples than other OFDM symbols (#1–#13, #1–#27 and #1–#55). As shown in Figures 5.32(b)–(d), the time occupied by them can be obtained.

When $\Delta f = 30$ KHz, $[16 + (72 + 1024) \times 14]T_s = [16 + (72 + 1024) \times 2 \times 7]T_s = 15360T_s = 0.5$ ms.
When $\Delta f = 60$ KHz, $[16 + (36 + 512) \times 28]T_s = [16 + (36 + 512) \times 4 \times 7]T_s = 15360T_s = 0.5$ ms.
When $\Delta f = 120$ KHz, $[16 + (18 + 256) \times 56]T_s = [16 + (18 + 256) \times 8 \times 7]T_s = 15360T_s = 0.5$ ms.

It can be seen that no matter which parameter set is used, they have the following two characteristics:

(1) In each 0.5 ms interval, the first OFDM symbol is 16 T_s (= 0.52μs) longer than the other subsequent OFDM symbols. For this reason, the number of time slots and the length of time slots within 0.5 ms in Figure 5.31 are as follows:

When $\Delta f = 30$ KHz, there is one time slot. Time slot length = $[16 + (72 + 1024) \times 14]\ T_s = 15360T_s = 0.5$ ms.
When $\Delta f = 60$ *KHz*, there are two time slots.

The length of the first time slot is $[16 + (36 + 512) \times 14]\ T_s = 7688\ T_s = 0.2503$ ms.

The length of the second time slot is $[(36 + 512) \times 14]\ T_s = 7672\ T_s =$ 0.2497 ms.

It can be seen that the lengths of the two time slots are not equal, and the first is 16 T_s longer than the second.

When $\Delta f = 120$ KHz, there are four time slots.

The length of the first time slot is $[16 + (18 + 256) \times 14]\ T_s = 3852\ T_s = 0.1254$ ms.
The length of the other three time slots is $[(18 + 256) \times 14]\ T_s = 3836\ T_s = 0.1246$ ms.

It can be seen that the length of the first time slot is 16 T_s longer than the other three.

(2) It can be seen from Figure 5.31 that regardless of the frame structure of the parameter set, the time edges of all parameter set slots are aligned every 0.5 ms. That is, one of the multiple time slots in the large subcarrier spacing must be aligned with one of the small subcarrier intervals.

This feature has the following uses. NR can be deployed in the same frequency band as LTE. If there are adjacent LTE-TDD subcarriers, then NR must adopt the same uplink and downlink conversion mode as LTE-TDD. It can be seen from Figure 1.51 in the first chapter that the up/down conversion of LTE is completed within a 1 ms special sub-frame. Due to the above-mentioned characteristics in NR, no matter what subcarrier spacing is used, the sub-frames every 1 ms (2×0.5 ms) must be aligned with the 1 ms special sub-frames in LTE. In this way, NR and LTE can be designed as the same uplink and downlink conversion mode. This is the second reason why 15×2 KHz power is selected for the NR subcarrier spacing mentioned above.

5.6.4.4 *NR mini-slot* [11]

In LTE, user data must be transmitted from the edge of the time slot, as shown in Figure 5.33(a). Although user data has arrived in the previous time slot, it cannot be transmitted, and it must be transmitted when the next time slot arrives. Therefore, a larger time delay is caused. In the R15 version, in order to achieve the low latency requirement of URLLC, a mini-slot transmission method is proposed. It does not require user data

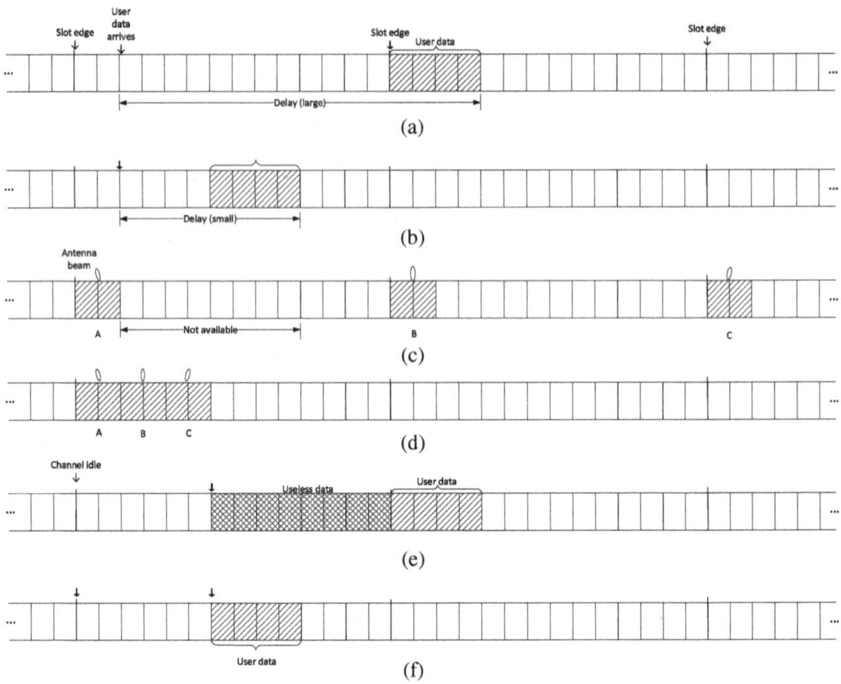

Figure 5.33 Application of mini-slot.

to be transmitted from the edge of the slot, but allows user data to be transmitted from any OFDM symbol edge in the slot, as shown in Figure 5.33(b). Obviously, the delay is reduced. This method uses part of the time slot to transmit data, so it is called mini-slot. In order to reach distinction, the LTE transmission method can be referred to as a full slot transmission method.

When the antenna adopts analog beamforming, the analog beamforming can only emit one beam at a time. Therefore, the network side can only use the time division multiplexing method to serve different terminals. If we want to serve three users *A*, *B* and *C*, the transmission conditions of full time slot and mini-time slot are shown in Figures 5.33(c) and (d), respectively. It can be seen that the time delay of using mini-slots is greatly reduced. However, there is a problem that needs to be pointed out. Since the base station and the terminal can only communicate when their antenna beam directions are aligned, the base station can only communicate with one terminal at a time. In this way, the two can occupy the maximum channel bandwidth, that is, use the highest data rate for communication. Therefore, all data of a user terminal can be transmitted with only a few OFDM symbols. Only under such conditions can the fast service speed shown in Figure 5.33(d) be fully realized.

When working in an unlicensed spectrum, users use the communication method of competition and occupation, that is, the so-called "Listen Before Talk" (LBT) technology. This means that the user receiver monitors whether the wireless channel in the air is free. Once free, the channel needs to be occupied immediately. Otherwise, the channel may be taken up by other users. When using full time slot transmission, in order to occupy the channel, only some useless data can be transmitted first, and user data cannot be transmitted until the time slot comes, as shown in Figure 5.33(e). The useless data transmitted reduces the transmission efficiency of the system, and the delay is also increased. When using mini-slot transmission as shown in Figure 5.33(f), there is no need to transmit useless data.

5.6.5 *Application of NR parameter set*

NR has multiple parameter sets. Many factors are related to the way they are applied, such as system synchronization errors, 5G application scenarios, 5G working frequency bands and cell coverage. All of them are discussed separately in the following sections.

5.6.5.1 *Synchronization error*

From Figure 1.39, it can be seen that the synchronization problem during the transmission of OFDM waveforms between the transmitter and receiver can be represented by Figure 5.34. In order to explain how the synchronization error is generated, the working process of Figure 5.34 is analyzed in the following.

(i) *High-order modulation and demodulation*: High-order modulation (QAM or PSK) is to transform a sequence of binary bits into a m-ary symbol sequence. The mapping relationship between high-order modulation output and input can be represented by a constellation diagram. As an example, Figure 5.35 shows a 16QAM constellation diagram. It can be seen from the figure that the 16 signal points are distributed on 3 circles,

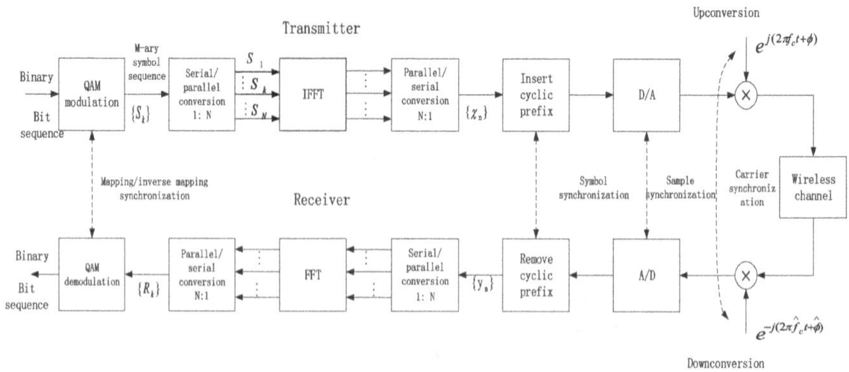

Figure 5.34 OFDM system synchronization.

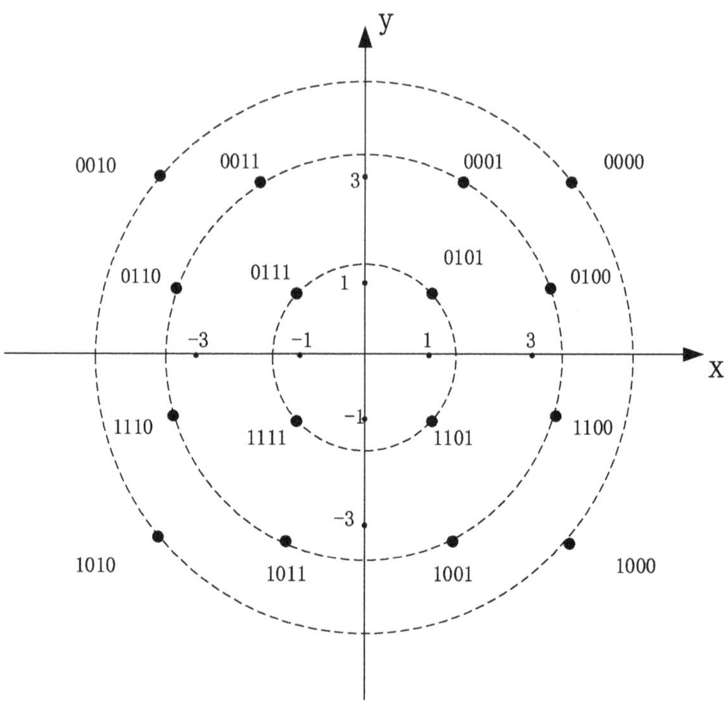

Figure 5.35 16QAM constellation diagram.

which is a standard ideal constellation diagram. When receiving QAM demodulation without any distortion, the ideal constellation diagram should be exactly the same as Figure 5.35. However, when the signal points on the receiving and transmitting constellation diagrams have the following differences, the constellation diagram at the receiver will be distorted.

(1) *There is noise at the receiver*:
At this time, the signal points on the constellation diagram diverge. The greater the noise power, the more scattered the actual signal point positions, otherwise the divergence is smaller (the focus is better), as shown in Figure 5.36(a). For the sake of simplicity, only one signal point is drawn in the figure (the following figures (b)–(d) also have one signal point).

(2) *There is a fixed phase error $\Delta\Phi$ at both ends of the transceiver:*
At this time, the 16 signal points will move $\Delta\Phi$ phase along the three circles. When $\Delta\Phi > 0$, it moves counterclockwise, and when $\Delta\Phi < 0$, it moves clockwise, which makes the actual signal point position away from the ideal signal point position, as shown in Figure 5.36(b).

(3) *Noise* $+\Delta\Phi$:
When noise and phase difference exist at the same time, the actual signal point situation is shown in Figure 5.36(c).

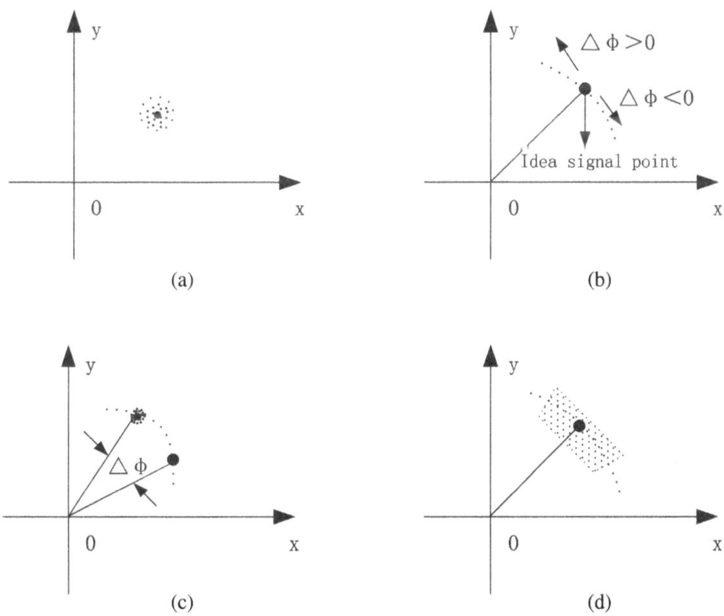

Figure 5.36 The influence of several errors: (a) noise; (b) $\Delta\Phi \neq 0$; (c) noise $+\Delta\Phi$; (d) noise $+ \Delta f_c$.

(4) *There is a fixed frequency error Δf_c at both ends of the transceiver*: The instantaneous phase difference produced by Δf_c is $(2\pi\Delta f_c)t$. It makes all signal points on the constellation diagram rotate periodically along three circles. When $\Delta f_c > 0$, it rotates counterclockwise, when $\Delta f_c < 0$, it rotates clockwise.

(5) *Noise* $+\Delta f_c$:
When noise and frequency difference Δf_c exist at the same time, all signal points on the constellation diagram still rotate periodically along three circles. However, due to noise, the points on the three rings are divergent. The rotation of a signal point on the constellation diagram is shown in Figure 5.36(d).

(ii) *IFFT and FFT*: The discrete fast Fourier transform and inverse transform are shown in Figure 5.37.

- IFFT transforms frequency-domain data into time-domain data. FFT transforms time-domain data into frequency-domain data. The N frequency-domain data input by IFFT comes from the output of the QAM modulator, which is the signal point on the constellation diagram shown in Figure 5.35. The N frequency-domain data output by the FFT are the signal points on the constellation diagram corresponding to the QAM demodulator, as shown in Figure 5.36.
- In implementation, the value of N is an integer power of 2. The specific value is determined this way. According to the required channel bandwidth and subcarrier spacing, the required number of subcarriers can be

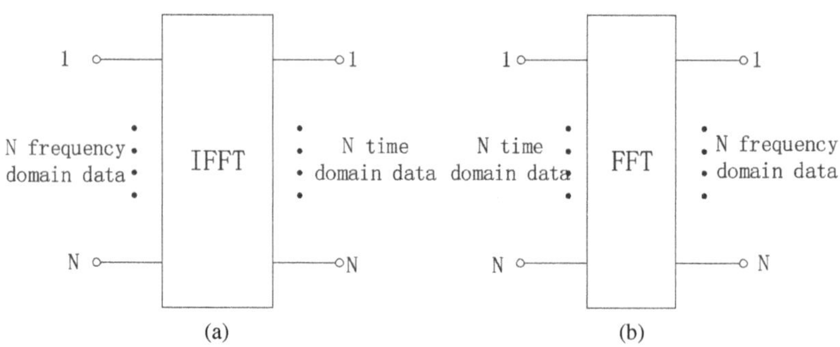

Figure 5.37 IFFT and FFT.

obtained. Since the basic unit of frequency-domain scheduling is a physical resource block, the required number of physical resource blocks N_{RB} can be obtained. In addition, some reference signals need to be inserted into the serial/parallel conversion input terminal in Figure 5.34. Suppose it needs to occupy Nr subcarriers. In this way, the number of frequency-domain data output by serial/parallel conversion is $(12N_{RB} + N_r)$. Therefore, a value of N slightly larger than $(12N_{RB} + N_r)$ should be selected, for example, $N = 2048$. The frequency-domain data values of $[2048–(12N_{RB} + N_r)]$ are all zero. This means that the number of samples in an OFDM symbol length T is 2048, and the number of samples in the cyclic prefix length T_{cp} is 144 (160 for the first OFDM symbol), as shown in Figure 5.38. As mentioned earlier, the maximum activatable subcarrier wave number in NR is 3300 ($N_{RB} = 3300/12 = 275$), and at this time, N should be 4096.

- It can be seen from Figure 5.34 that the receiver must first remove the cyclic prefix and then perform FFT transformation on 2048 samples. However, an error ΔT_s will be generated in the process of removing the cyclic prefix. It can be seen from Figure 5.38 that the best symbol starting point is at time B, where the timing error ΔT_s is 0 at this moment. If the symbol start point is at time A, the timing error ΔT_s is larger than 0 at this time. However, because point A falls within the cyclic prefix, it will not cause distortion of the OFDM signal, but only phase shift the

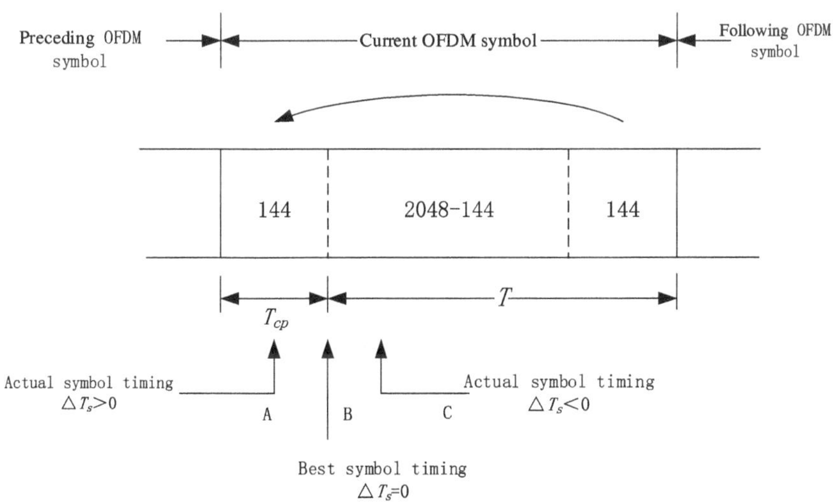

Figure 5.38 Symbol timing error.

original OFDM signal. As shown in Figure 5.36(b), this phase shift can be eliminated by the frequency-domain equalizer in Figure 1.39 in Chapter 1. If the symbol start point is at time C, the timing error $\Delta T_s < 0$ at this time. In this case, the 2048 samples taken out include both the current sample and the sample of the following OFDM symbol. Therefore, Inter-Symbol Interference (ISI) and interference (ICI) between subcarriers are bound to occur (∵ the orthogonality between 2048 subcarriers suffers to destroy), which greatly increases the bit error rate after the system is demodulated. This is not allowed. To solve this problem, the "Symbol Timing and Frequency Synchronization" function block shown in Figure 1.39 in Chapter 1 is used to eliminate this situation. The specific algorithm is not discussed here, as the preceding symbols will still interfere with the following symbols.

(iii) *Up/down conversion*: The up/down conversion and wireless channels in Figure 5.34 produce the following synchronization errors:

- Fixed phase error $\Delta\Phi = \hat{\Phi} - \Phi$
  $\Delta\Phi$ is generated by the channel delay. The impact is shown in Figure 5.36(b). It can be eliminated by frequency-domain equalization.
- Phase noise
  The up/down converter is a component in the radio frequency chip. The RF chip is shown in Figure 1.41 in Chapter 1. It can be seen from the figure that the local oscillator frequency f_{LO} of the up/down converter is provided by the frequency combiner shown in Figure 1.42 in Chapter 1. The frequency combiner is a Phase-Locked Loop (PLL), and the frequency f_s output by the PLL has the same frequency stability as the reference frequency f_r input by the PLL. f_r is generated by a crystal oscillator, and its frequency will drift randomly. The smaller the drift, the more

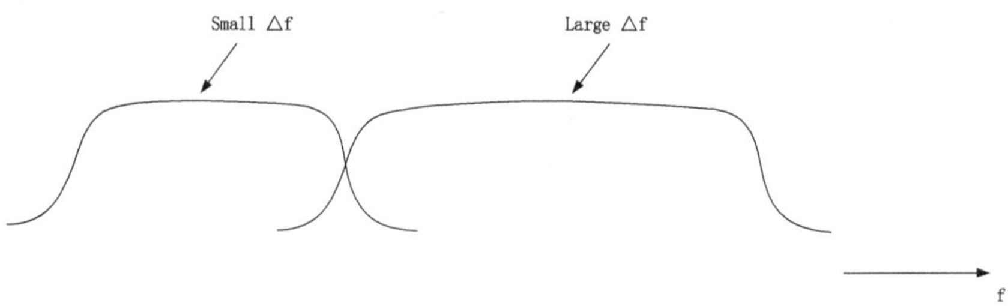

Figure 5.39 Frequency-domain multiplexing of different services.

stable the generated frequency. In order to illustrate the degree of frequency stability of f_r (or f_s), the index of frequency stability is defined:

$$\text{Frequency stability } = \frac{\Delta f_0}{f_0} \tag{5.1}$$

In the equation, f_0 is the nominal frequency, and Δf_0 is the frequency deviation from the nominal frequency.

When the crystal oscillator is at room temperature, its frequency stability is several PPM ($n \times 10^{-6}$).

Phase noise is another way of expressing frequency stability. It is defined as the dBc/ZH value at a given frequency offset Δf_0. dBc is the difference between the power at that frequency and the power at the nominal frequency in dB. For example, −125 dBc/Hz@100 Hz means that the power which is 100 Hz ($f_0 \pm 100$ Hz) away from the nominal frequency f_0 is 125 dB smaller than the power at f_0. It can be seen from equation (5.1) that Δf_0 increases as f_0 rises. Therefore, when 5G works in a high-frequency band, such as a millimeter wave frequency band, the phase noise of the local oscillator frequency f_{LO} of the up/down converter is relatively large. Its influence on carrier synchronization can be analyzed as follows.

Since the local oscillator frequency of the up/down converter in Figure 5.34 has a frequency drift, it results in the carrier frequency difference $\Delta f_c = \hat{f}_c - f_c$ of the transmission and reception, and Δf_c is a function of time. The phase noise $\Phi(t)$ produced by Δf_c can usually be regarded as a generalized stationary random process. In this way, the nth sample in an OFDM symbol at the receiver can be obtained as

$$\gamma_n = x_n e^{j\Phi(n)} \tag{5.2}$$

where x_n is the nth sample value (time domain) in the transmitting OFDM symbol, γ_n is the nth sample value (time domain) in the receiving OFDM symbol, and $\Phi(n)$ is the nth sample value (time domain) in the phase noise $\Phi(t)$.

Since $\Phi(n) \ll 1$, (5.2) can be written as

$$r_n = x_n e^{j\phi(n)} \approx x_n[1 + j\varphi(n)], \text{ In the formula, } e^x \approx 1 + x,\ x \ll 1 \tag{5.3}$$

$$R_k = \frac{1}{N}\sum_{n=0}^{N-1} x_n\left[1 + j\varphi(n)\right] e^{-j2\pi\frac{K}{N}n} \tag{5.4}$$

where R_k is the frequency-domain data carried on the kth subcarrier in an OFDM symbol output by FFT. After analyzing (5.4), we can finally get

$$R_k = AS_k e^{j\Phi} + N_k^{ICI} \tag{5.5}$$

where S_k is the frequency-domain data carried on the kth subcarrier in the transmitting OFDM symbol.

It can be seen from Equation (5.5) that the frequency-domain data carried on the kth subcarrier and obtained by FFT demodulation at the receiver consists of two items. In the first item, A and ϕ are constants related to $\sum_{n=0}^{N-1}\phi(n)$ and produce the same amplitude weight and phase shift for all subcarriers in an OFDM symbol, as shown in Figure 5.36(b). When A and ϕ are eliminated by the frequency-domain equalizer, the receiver constellation will not be distorted. The second item is the additive noise term N_k^{ICI}, and its value varies with different subcarriers. In this way, it destroys the orthogonality between the N subcarriers, thereby generating Inter-Channel Interference (ICI). Therefore, two functional blocks, "Symbol Timing and Frequency Synchronization" and "Frequency Compensation" are set up in Figure 1.39 in the first chapter to reduce this impact. The specific algorithm is not discussed here.

The phase error caused by the first item above is called Common Phase Error (CPE). For fast-changing channels, CPE can be compensated as a part of the channel estimation process. As shown in Figure 1.39 in Chapter 1, the channel estimation function block is estimated and eliminated by frequency-domain equalizer. However, for slowly varying channels, it is necessary to track and compensate CPE more frequently. 3GPP introduces phase tracking reference signal (PT-RS) into NR for CPE compensation. In addition, the demodulation reference signal (DM-RS) can also be used for CPE compensation. PT-RS and DM-RS will be discussed later in physical signals.

It should also be pointed out that by modeling and simulating the phase noise of a phase-locked loop (PLL), the following two points can be obtained:

(1) The higher the operating frequency of the PLL is, the greater the power spectral density (PSD) of the phase noise is and vice versa. For this reason, when realizing millimeter wave local oscillators, a relatively low-frequency phase-locked loop is generally used and then

multiplied to achieve the required local oscillator frequency (multiplication does not change the frequency stability), which can reduce the phase noise.

(2) When the OFDM waveform is used, the signal (OFDM signal power) to interference (phase noise power) ratio (SIR) caused by phase noise will increase with the increase of the OFDM subcarrier spacing Δf. That is to say, the larger the subcarrier spacing is, the stronger its anti-phase noise ability is, otherwise, it is weak. It can be seen that when 5G works in millimeter waves, a larger subcarrier spacing should be selected in order to reduce the influence of phase noise. For example, $\Delta f = 240$ or 480 KHz.

The sample timing error in Figure 5.34 has little effect on the demodulation performance, so it is not discussed here.

5.6.5.2 *Service reuse*

5G can work in a variety of application scenarios, and the corresponding business requirements are different. Their required subcarrier spacing may also be different. For example, in the eMBB scenario, the subcarrier spacing is small, but in the URLCC scenario, a larger subcarrier spacing is required. These two service spectrums with different subcarrier spacing can be frequency-domain multiplexed on the baseband, as shown in Figure 5.39. Then, the multiplexed baseband signal can be modulated on the same radio frequency carrier. This way is called service reuse with different parameter sets.

It can be seen from the figure that in this case, the interference outside the two spectrum bands needs to be reduced to a minimum. To this end, the following two measures can be adopted:

(1) Add a protective band between these two.
(2) Add frequency-domain filters or time-domain windows to OFDM symbols (see Figure 5.28).

5.6.5.3 *Cell size, working frequency band and mobility*

This section discusses the relationship between cell size, working frequency band, mobility and parameter sets.

(i) *Cell size and working frequency band*: The parameter set used by both is shown in Figure 5.40.

It can be seen from the figure that:

- When the working frequency band is higher, the cell range used is smaller. At the same time, a larger subcarrier spacing should be used. This can reduce the time delay and reduce the influence of phase noise. But the cyclic prefix T_{cp} is short.
- When the cell range is larger, the working frequency band used should be lower. At the same time, a smaller subcarrier spacing should be used, so that the cyclic prefix T_{cp} is long. But the delay increases.

(ii) *Mobility*: In Chapter 1, we talked about the Doppler shift f_d when the terminal is in motion. f_d is a function of time. It can be seen from

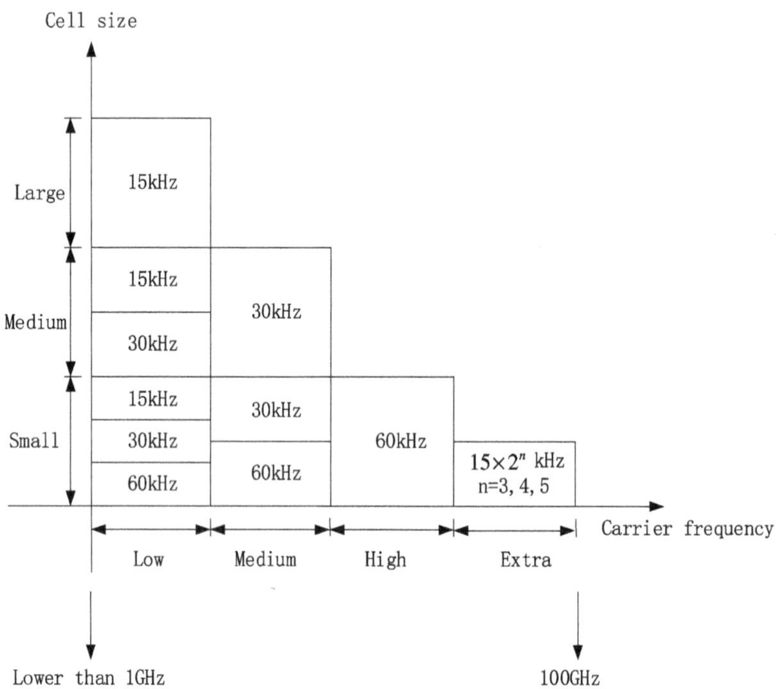

Figure 5.40 The relationship between NR parameter set and cell size and working frequency band.

Figure 5.34 that f_d is the frequency difference between the transmitting and receiving carrier frequencies. The frequency difference will also destroy the orthogonality between all N subcarriers, resulting in ICI. From the equation (1.10) in Chapter 1, we can see that f_d is proportional to the operating frequency f_c. Therefore, only when the operating frequency band is high, such as when 5G is deployed in the millimeter wave frequency band, the influence of f_d is considered. In this case, the influence also depends on the moving speed of terminals. Since the range of cells working in high-frequency bands is small, such as indoors or densely populated urban areas, the radial velocity of the terminal relative to the base station is relatively low. At this time, f_d is still small, so its influence can be ignored. However, when the terminal is in a high-speed motion scene, such as a V2V (vehicle-to-vehicle) communication situation, the radial velocity of the terminal relative to the base station or the terminal relative to the terminal is relatively high. The influence of f_d must be considered at this time. In order to reduce the influence of f_d, it is necessary to select a larger subcarrier spacing, which means that the Doppler effect limits the lower limit of the selection of subcarrier spacing Δf. This is because when Δf is large, the proportion of f_d relative to Δf is reduced, so that the degree of damage to the orthogonality between each subcarrier due to the existence of f_d is also reduced.

5.6.6 *RF channel bandwidth*

5.6.6.1 *Working frequency band*

The R15 version formulated by 3GPP for NR specifies the 5G radio frequency operating frequency bands as shown in Tables 5.2 and 5.3. FR1 contains 27 frequency bands, and FR2 contains four frequency bands. The following points are added here:

- The number of NR bands ranges from *n*1 to *n*512. 4G LTE band numbers use only Arabic numerals without *n* in front.
- The frequency band that has been allocated to LTE is reallocated to NR in R15. This frequency band is called the LTE reused band.
- With the development of 5G, new frequency bands will be used in the future. For this reason, it is stipulated that *n*65–*n*256 are reserved for FR1 new frequency band, and *n*257–*n*512 are reserved for FR2 new frequency band.

- NR is mostly deployed in the 3.3–5 GHz frequency band, which is called the mid-band.

5.6.6.2 *Electromagnetic wave radiation limit*

The regulatory rules for electromagnetic wave radiation levels are based on 6 GHz. When it is lower than 6 GHz (sub-6 GHz), the SAR index is adopted. When higher than 6 GHz, the EIRP index is adopted. We will discuss their meanings separately in the following.

(i) *SAR*: The full name of SAR is Specific Absorption Rate, which is called electromagnetic wave absorption ratio. It is defined as an electromagnetic field induced in the human body under the action of an external electromagnetic field. Since various organs of the human body are lossy media, the electromagnetic field in the body will generate electric current, which leads to the absorption and dissipation of electromagnetic energy. SAR is commonly used in biological dosimetry to characterize this physical process. The meaning of SAR is the electromagnetic power absorbed or consumed per unit mass of human tissue, and the unit is W/Kg (watts/kilogram). At present, there are two mobile phone radiation standards in common use in the world. The SAR standard specified by the United States is 1.6 W/Kg, the European standard is 2.0 W/Kg, and China adopts the European standard.

(ii) *EIRP/TRP*: The full name of EIRP is Effective Isotropic Radiated Power, which is also called equivalent isotropic radiated power:

$$\mathrm{EIRP} = P_T \cdot G_T \tag{5.6}$$

where P_T is the transmitting power of the transmitter, and G_T is the antenna gain of the transmitting antenna.

It can be seen that EIRP means the transmit power in the direction of the maximum antenna gain obtained by the transmitter compared to the omnidirectional antenna ($G_T = 1$). The absolute value of its power is expressed in dBm (1 mw is 0 dBm). According to the terminal type (handheld terminal, fixed terminal, etc.), its EIRP restrictions are different, so EIRP restrictions will lead to the limitation of the transmission power.

TRP, known as Total Radiated Power, can be obtained by calculating the average of surface integral for the emission power of the entire radiating sphere. It reflects the transmitting power of the terminal and is related to the transmitting power of the terminal under conduction and the antenna radiation performance.

5.6.6.3 *Base station channel bandwidth*

The base station channel bandwidth BW_{CH} of FR1 and FR2 specified in the R15 version is shown in Tables 5.8 and 5.9. Under different parameter set conditions, FR1 has 13 channel bandwidths in total, and FR2 has four channel bandwidths. Those with a checkmark in the table indicate the channel bandwidth which is specified.

Table 5.8 FR1 channel bandwidth.

NR frequency band	Subcarrier spacing (KHz)	Channel bandwidth (MHz)												
		5	10	15	20	25	30	40	50	60	70	80	90	100
*n*1	15	√	√	√	√									
	30		√	√	√									
	60													
*n*2	15	√	√	√	√									
	30		√	√	√	√	√							
	60		√	√	√	√	√							
*n*3	15	√	√	√	√	√	√							
	30		√	√	√	√	√							
	60		√	√	√	√	√							
n5	15	√	√	√	√									
	30		√	√	√									
	60													
n7	15	√	√	√	√									
	30		√	√	√									
	60		√	√	√									
*n*8	15	√	√	√	√									
	30		√	√	√									
	60													

(Continued)

Table 5.8 (*Continued*)

NR frequency band	Subcarrier spacing (KHz)	Channel bandwidth (MHz)												
		5	10	15	20	25	30	40	50	60	70	80	90	100
*n*20	15	√	√	√	√									
	30		√	√	√									
	60													
*n*28	15	√	√	√	√									
	30		√	√	√									
	60													
*n*38	15	√	√	√	√									
	30		√	√	√									
	60		√	√	√									
*n*41	15		√	√	√			√	√					
	30		√	√	√			√	√	√	√	√	√	√
	60		√	√	√			√	√	√	√	√	√	√
*n*51	15	√												
	30													
	60													
*n*66	15	√	√	√	√			√						
	30		√	√	√									
	60		√	√	√									
*n*70	15	√	√	√	√	√								
	30		√	√	√	√								
	60		√	√	√	√								
*n*71	15	√	√	√	√									
	30		√	√	√									
	60													
*n*75	15	√	√	√	√									
	30		√	√	√									
	60		√	√	√									

*n*76	15	√												
	30													
	60													
*n*77	15		√	√	√		√	√	√					
	30		√	√	√		√	√	√	√	√	√	√	√
	60		√	√	√		√	√	√	√	√	√	√	√
*n*78	15		√	√	√		√	√	√					
	30		√	√	√		√	√	√	√	√	√	√	√
	60		√	√	√		√	√	√	√	√	√	√	√
*n*79	15							√	√					
	30							√	√	√		√		√
	60							√	√	√		√		√
*n*80	15	√	√	√	√	√	√							
	30		√	√	√	√	√							
	60		√	√	√	√	√							
*n*81	15	√	√	√	√									
	30		√	√	√									
	60													
*n*82	15	√	√	√	√									
	30		√	√	√									
	60													
*n*83	15	√	√	√	√									
	30		√	√	√									
	60													
*n*84	15	√	√	√	√									
	30		√	√	√									
	60		√	√	√									

Table 5.9 FR2 channel bandwidth.

NR band	Subcarrier spacing (KHz)	Channel bandwidth (MHz)			
		50	100	200	400
*n*257	60	√	√	√	
	120	√	√	√	√
*n*258	60	√	√	√	
	120	√	√	√	√
*n*260	60	√	√	√	
	120	√	√	√	√
*n*261	60	√	√	√	
	120	√	√	√	√

5.6.6.4 *Channel frequency band utilization*

The base station channel bandwidth is used to transmit downlink service data. The size of frequency-domain resources occupied by service data is measured by physical resource blocks N_{RB}. One physical resource block occupies 12 subcarriers, and N_{RB} physical resources occupy a total of 12 N_{RB} subcarriers. The bandwidth occupied by them is

$$BW_{RB} = (12N_{RB} + 1)\Delta f \tag{5.7}$$

In the equation, Δf is the subcarrier spacing, and we call BW_{RB} the configured transmission bandwidth. The physical meaning of equation (5.7) can be illustrated by the example in Figure 5.41. Assuming that there are three subcarriers in the figure, the bandwidth occupied by the signal composed of these three subcarriers is $3\Delta f + \Delta f$. From this example, it can be seen that equation (5.7) is correct.

Obviously, the configured transmission bandwidth must be slightly smaller than the channel bandwidth. In this way, protect band B can be left on both sides of the transmission bandwidth. Since the spectrum of the OFDM signal still exists outside the band, it increases (or decreases) as Δf increases (or decreases) (see Figure 5.27). Therefore, a protect band must be left to reduce out-of-band interference. The relationship between BW_{RB}, BW_{CH} and B is shown in Figure 5.42.

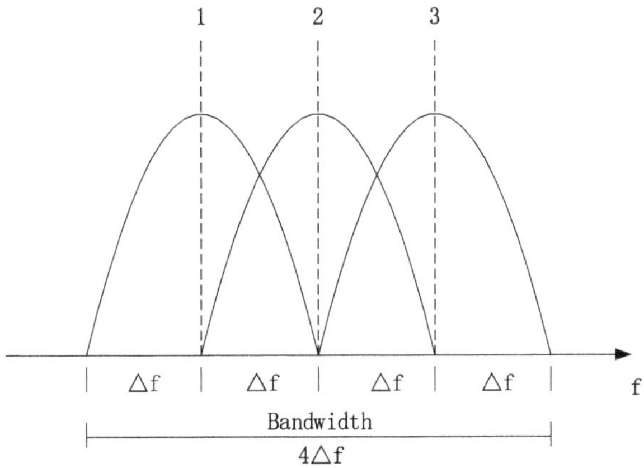

Figure 5.41 Example of configured transmission bandwidth.

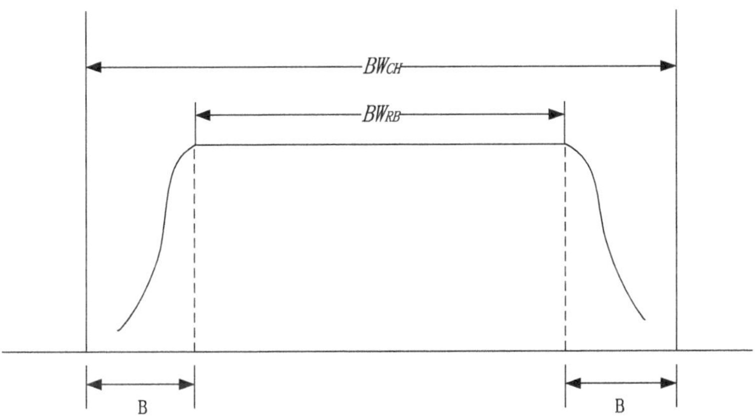

Figure 5.42 The relationship between BW_{CH}, BW_{RB} and B

Thus, we can get

$$\text{Base station channel frequency band utilization} = \frac{BW_{RB}}{BW_{CH}} = \frac{(12N_{RB}+1)\,\Delta f}{BW_{CH}} \quad (5.8)$$

The minimum guard band of each channel of the base station

$$B = \frac{BW_{CH} - BW_{RB}}{2} = \frac{BW_{CH} - (12N_{RB}+1)\,\Delta f}{2} \quad (5.9)$$

Tables 5.10 and 5.11 give the BW_{CH}, N_{RB} and B of FR1 and FR2, respectively. We can use (5.9) to verify the value of B in the table. For example, BW_{CH} = 5 MHz, Δf = 15 KHz, N_{RB} = 25, and we can get $B = \frac{5000-(12\times 25+1)\times 15}{2} = 242.5(\text{KHz})$

We can also use (5.8) to calculate the channel frequency band utilization. Obviously, the larger (or smaller) the N_{RB} is, the higher (or lower) the utilization rate is. For example,

$$BW_{CH} = 100\text{ MHz},\ N_{RB} = 273,\ \Delta f = 30\text{ KHz}$$

$$\text{Channel Band Utilization} = \frac{(12N_{RB}+1)\,\Delta f}{BW_{CH}} = \frac{(12\times 273+1)\times 30}{100000} \approx 98.3\%$$

$$BW_{CH} = 5\text{ MHz},\ N_{RB} = 11,\ \Delta f = 30\text{ KHz}$$

$$\text{Channel Band Utilization} = \frac{(12\times 11+1)\times 30}{5000} \approx 79.8\%$$

Most of the channel frequency band utilization in the table is about 90%.

Table 5.10 FR1 configuration transmission bandwidth (N_{RB}).

		Channel bandwidth (MHz)													
		5		10		15		20		25		30		40	
		N_{RB}	B	N_{RB}	B	N_{RB}	B	N_{RB}	B	N_{RB}	B	N_{RB}	B	N_{RB}	B
Subcarrier spacing (KHz)	15	25	242.5	52	312.5	79	382.5	106	452.5	133	522.5	160	592.5	216	552.5
	30	11	505	24	665	38	645	51	805	65	785	78	945	106	905
	60			11	1010	18	990	24	1330	31	1310	38	1290	51	1610

		50		60		70		80		90		100	
		N_{RB}	B	N_{RB}	B	N_{RB}	B	N_{RB}	B	N_{RB}	B	N_{RB}	B
Subcarrier spacing (KHz)		270	692.5										
		133	1045	162	825	189	965	217	925	245	885	273	845
		65	1570	79	1530	93	1490	107	1450	121	1410	135	1370

Note: The unit of B in the table is KHz.

Table 5.11 FR2 configuration transmission bandwidth (N_{RB}).

		Channel bandwidth (MHz)							
		50		100		200		400	
		N_{RB}	B	N_{RB}	B	N_{RB}	B	N_{RB}	B
Subcarrier spacing (KHz)	60	66	1210	132	2450	264	4930		
	120	32	1910	66	2420	132	4900	264	9860

Note: The unit of B in the table is KHz.

Table 5.12 SS/PBCH configuration transmission bandwidth (N_{RB}).

	Channel bandwidth (MHz)					
	100		200		400	
Subcarrier spacing (240 KHz)	N_{RB}	B	N_{RB}	B	N_{RB}	B
	32	3800	64	7720	128	15560

Note: The unit of B in the table is KHz.

Table 5.12 is specially set for the Synchronization Signal (SS) and the physical broadcast channel (PBCH). It cannot be used to transmit business data.

5.6.6.5 *NR carrier aggregation*

LTE downlink can support up to five component carriers for aggregation, while NR downlink can support up to 16 component carriers for aggregation. In this way, a service bandwidth of 16×400 = 6.4 GHz can be aggregated for the terminal. With regard to downlink carrier aggregation, the following points need to be pointed out:

- If the component carrier is cross-band, then each component carrier is in TDD working state. At a specific moment, NR allows the transmission and reception directions of each component carrier to be different. Therefore, a TDD terminal with carrier aggregation capability must add a duplex filter at the front end of the receiver. For terminals without carrier aggregation capabilities, there is no need to add it.
- A terminal that supports carrier aggregation can transmit and receive data from multiple cells. Only one of these aggregated cells is the

primary cell, which is the cell used for terminal access. Other cells are called secondary cells, which are configured by the network after entering the connected state. The network can quickly activate or deactivate secondary cells to meet changes in service requirements. Different terminals can configure different cells as primary cells.

- The scheduling authorization can be transmitted on the same carrier as the transmission data. This situation is called self-scheduling. The scheduling authorization and the transmission data can also be transmitted on different carriers. This situation is called cross-carrier scheduling. These two scheduling authorization methods are shown in Figure 5.43. In most cases, the network uses self-scheduling.
- The uplink also needs to transmit control signaling during downlink carrier aggregation, such as HARQ confirmation information fed back by uplink. There are two ways to transmit all uplink feedback information to the cell. One is to transmit all feedback information to the primary cell through the physical uplink control channel (PUCCH) group. The other is to divide all the feedback information into two groups and transmit them to the primary cell and a secondary cell through the PUCCH 1 and PUCCH 2 groups, respectively.

The following discusses the base station channel bandwidth during downlink carrier aggregation in the frequency band, as shown in Figure 5.44.

It can be seen from the picture that

$$BW_{CH,CA} = f_{e,h} - f_{e,L} = (f_{c,h} + f_{o,h}) - (f_{c,L} - f_{o,L}) \tag{5.10}$$

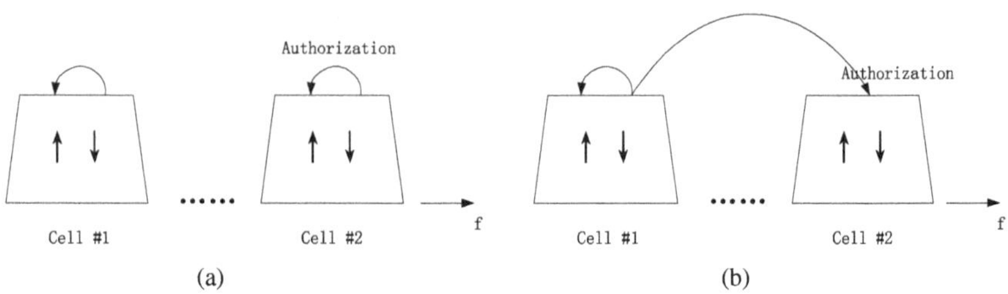

Figure 5.43 Scheduling authorization mode (a) self-dispatch; (b) cross-carrier scheduling.

Note: ...... indicates that the working frequency bands of the two cells are different.

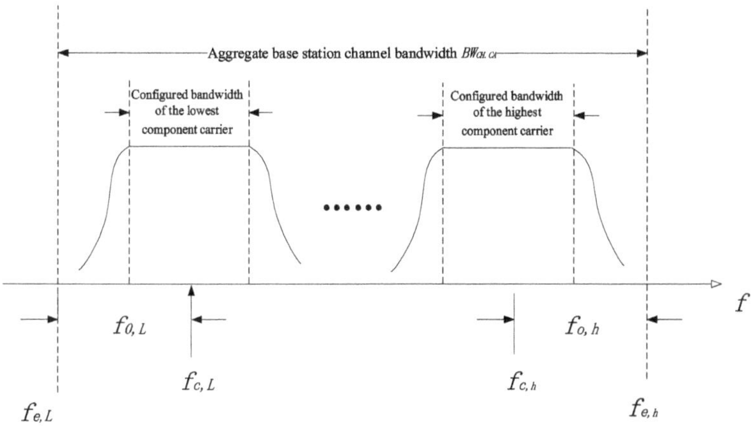

Figure 5.44 The calculation of aggregate base station channel bandwidth.

where $f_{c,h}$ is the carrier frequency of the highest component,

$$f_{o,h} = \text{minimum guard band} + \frac{(12N_{RB}+1)\,\Delta f}{2} \tag{5.11}$$

In equation (5.11), N_{RB} and Δf refer to the minimum protect band, physical resource block and subcarrier spacing used by the highest component carrier, respectively.

$f_{c,L}$ is the carrier frequency of the lowest component carrier:

$$f_{o,L} = \text{minimum guard band} + \frac{(12N_{RB}+1)\,\Delta f}{2} \tag{5.12}$$

In equation (5.12), N_{RB} and Δf refer to the physical resource block and subcarrier spacing used by the lowest component carrier, respectively.

5.6.6.6 *Partial bandwidth of NR terminal*

As can be noted from the first chapter (see Table 1.11), the LTE base station has a total of six channel bandwidths. The maximum channel bandwidth is only 20 MHz. Therefore, it is feasible to require all terminals to support six channel bandwidths and to arrange the carrier frequencies of all terminal channels on the OFDM DC subcarrier of the base station (see Figure 1.55). Since no information is carried on the DC subcarrier, the DC drift caused by self-mixing will not interfere the information. It can be seen from Tables 5.8 and 5.9 that when the NR base station works in the FR1 and FR2 frequency bands, its maximum channel bandwidth is 100 MHz and 400 MHz, respectively. These two kinds of bandwidth are relatively wide. In this bandwidth situation, it is not feasible to require the terminal to be designed for maximum bandwidth in terms of cost and power saving, and it is actually unnecessary (see the following discussion). For this reason, the design of the terminal channel bandwidth is performed independently. The channel bandwidth it supports may also be different from that of the base station and is usually smaller than the base station channel bandwidth. Therefore, in the NR standard design, a receiving bandwidth adaptive technology is introduced, that is, the terminal only monitors the downlink control channel on a smaller bandwidth to receive a small amount of downlink data, and only when the terminal receives a large amount of data, the entire bandwidth is opened for receiving. It can be seen that the channel bandwidth of the terminal is only a part of the channel bandwidth of the base station, which is called Bandwidth Part (BWP). BWP refers to a group of continuous resource blocks under a certain parameter set (subcarrier spacing and cyclic prefix). Since part of the bandwidth actually occupied by different terminals will be in different positions in the frequency domain, their respective DC subcarriers may not all be located on the base station OFDM DC subcarriers, but may be located on some non-DC subcarriers. These non-DC subcarriers carry information, so that the DC drift generated by self-mixing will interfere with the information. In this case, the problem of DC drift needs to be solved, and the specific algorithm is not discussed here.

From the above discussion, it is known that BWP is to allocate part of the bandwidth within the maximum bandwidth of a single carrier for terminals to access and transmit data. The terminal only needs to operate in this part of the bandwidth configured by the system, so as to achieve energy-saving management. The terminal works in part of the bandwidth under the following conditions:

- Transmit VoIP: It is only required that the terminal works in a relatively small bandwidth.
- The terminal is at the edge of the cell. Due to the limited power of the terminal, the terminal is only allowed to transmit a lower data rate. At this time, the terminal works in a smaller bandwidth.

- Since the system is rate-adaptive, only low speed data can be transmitted in certain periods, and the terminal works in a smaller bandwidth at this time.
- Different services require different physical parameters, such as different subcarrier spacing Δf. At this time, the transmitted data rate is different, so the bandwidth of the terminal is also different.
- In order to save power, the terminal is intended to work within a part of the bandwidth.

In order to adapt to the above-mentioned situation, the terminal must realize the bandwidth adaptation adjustment through BWP.

5.6.7 *LTE/NR dual connection base station deployment plan*

(i) *What is dual connection?* The dual connection has been mentioned before. LTE/NR dual connection has the following characteristics:

- The terminal can be connected to both LTE base station (eNB) and NR base station (gNB) at the same time.
- One of these two base stations is the master node, which is responsible for the control plane (various signaling) of radio access and handles all RRC-based configurations of the terminal.
- The other is a secondary node, which provides additional user plane links (various business data) to the terminal.

(ii) *Typical application of dual connection*: 5G non-independent networking (NSA) must apply dual connectivity technology. The series 3, series 7 and series 4 NSA networking schemes have been discussed earlier (see Figures 5.11 and 5.12). Their characteristics are shown in Table 5.13.

(iii) *Base station deployment plan*: Following are the three deployment scenarios:

(1) *No co-site*:
As shown in Figure 5.45, eNB and gNB sites are separated geographically. This deployment has the following characteristics:

- NB is the micro-cell layer, while LTE is the macro-cell layer with overlapping coverage.

Table 5.13 Features of NSA solution.

Scheme	Master node	Secondary node	Core network
Series 3	eNB	en-gNB	EPC
Series 7	ng-eNB	gNB	5GC
Series 4	gNB	ng-eNB	5GC

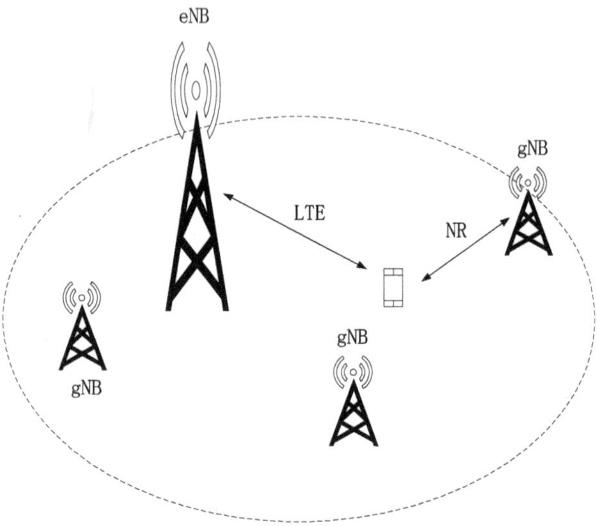

Figure 5.45 Dual connectivity in LTE/NR multi-layer scenarios.

- LTE is the master node and works in a low-frequency band to ensure unimpeded signaling in the entire cell.
- NR is a secondary node that works in high-frequency bands such as millimeter wave frequency bands. It has a large bandwidth to transmit high-speed data, and multiple auxiliary nodes are deployed in different geographical locations of the cell. This enables high-speed data transmission in the entire cell.

It can be seen that in this deployment, the eNB is used to ensure the unblocked signaling of the entire cell, and multiple gNB are used to achieve high-speed data transmission in the entire cell. This improves the capacity of the cell (more users can be accessed in the cell), but this method is expensive.

(2) *Co-site (LTE and NR work in low-frequency band)*:
As shown in Figure 5.46(a), the eNB and gNB are located together, saving the cost of building gNB. Since NR works in a low-frequency band, the throughput of NR and LTE carriers is aggregated through dual connection, so that the terminal can get a higher data rate in the entire cell. But this is not carrier aggregation because the component carriers of carrier aggregation are in different cells.

(3) *Co-site (NR is high-frequency band such as millimeter wave frequency band)*:
As shown in Figure 5.46(b), since NR is millimeter wave, the data transmission rate can be increased. However, NR has poor coverage at the cell edge, and LTE can be used to make up for it at this time. The operating frequency bands of LTE and NR in this deployment are the same as in Figure 5.45. But it mainly relies on one gNB to cover the entire cell, so the coverage effect is worse than that in Figure 5.45. Therefore, the transmission data rate is lower, but the cost of building a station is small.

Finally, it must be emphasized that in the case of LTE and NR dual connection, the terminal needs to transmit two different carrier frequencies in the uplink (see Figure 5.11). One is the LTE carrier frequency, and the other is the NR carrier frequency. When these two carriers are transmitted at the same time, due to the nonlinear distortion of the terminal transmitter power amplifier, intermodulation products will be generated. Some of the

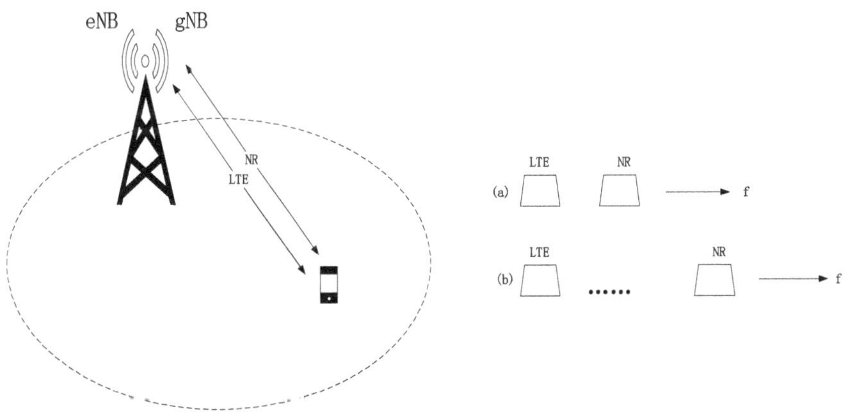

Figure 5.46 LTE/NR co-site dual connection.

intermodulation frequencies may fall in the downlink frequency band, thereby causing interference to the terminal receiver. In order to solve this problem, it is necessary to adopt a single transmission method, that is, LTE and NR uplink carriers need time division multiplexing. This makes the two uplink carriers unable to work continuously. However, whether the intermodulation component will fall in the downlink frequency band depends on the values of the two uplink carrier frequencies, LTE and NR. Among them, only a few frequency combinations will cause low-order intermodulation interference to fall in the downlink frequency band, and most frequency combinations will not appear with such distortions.

5.6.8 *LTE/NR spectrum coexistence*

(i) *NR and LTE spectrum coexist in the following two cases*:

(1) LTE/NR dual connection co-sites, and both work in the low-frequency band (see Figure 5.46(a)).
(2) In Figure 5.9, both the upper (or lower) supplement frequency band of NR works in low bands.

In the above two cases, the working frequency band of NR is low band, so their frequency spectrum and LTE frequency spectrum will coexist in this low-frequency band.

(ii) *Spectrum coexistence method*: Since LTE already occupies most of the low-band spectrum, the low-band spectrum required by NR can only be deployed on the spectrum used by LTE. In this case, the question of how the two spectrums coexist arises. Currently, two coexistence methods are used.

(1) *Static coexistence*:
As shown in Figure 5.47, this method actually gives up part of the LTE spectrum to NR. Obviously, this method reduces the bandwidth of the two terminals, causing the transmission data rate to drop.
(2) *Dynamic coexistence*:
In order to overcome the shortcomings of static coexistence, a dynamic coexistence method as shown in Figure 5.48 is proposed. Through scheduling, LTE and NR work on time-sharing, so that the bandwidths of both will not be reduced. Since NR has the same parameter set as LTE, that is,

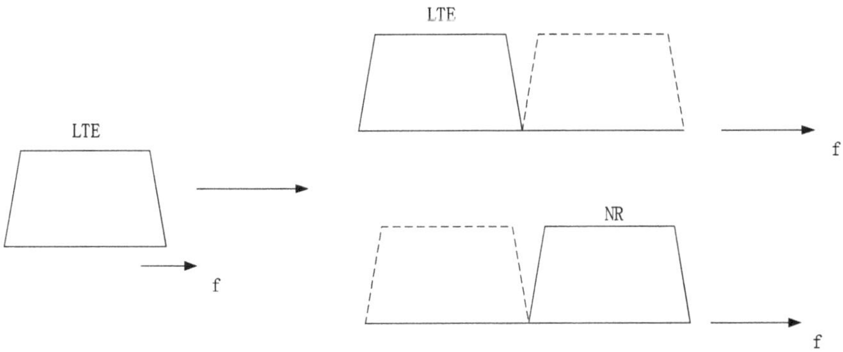

Figure 5.47 Static coexistence.

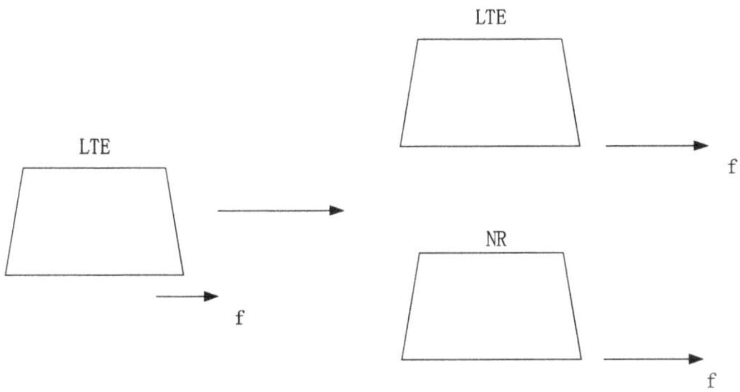

Figure 5.48 Dynamic coexistence.

a parameter set with a subcarrier spacing of 15 KHz, therefore, it is feasible to achieve dynamic coexistence (because the subcarrier spacing is the same, so NR and LTE spectrum bandwidths are the same).

By the way, the supplement uplink carrier has another purpose. The coexistence of the supplement uplink carrier and the LTE uplink carrier is shown in Figure 5.49. If the system is working in the TDD state, the uplink data that needs to be transmitted cannot be transmitted until the TDD state is converted to the uplink time slot, which is disadvantageous for data transmission that requires low delay. But at this time, if the original uplink carrier is not used, the SUL carrier is used to transmit data. Since the SUL carrier is not restricted by the TDD state transition, and data can be transmitted at any time, the time delay can then be reduced.

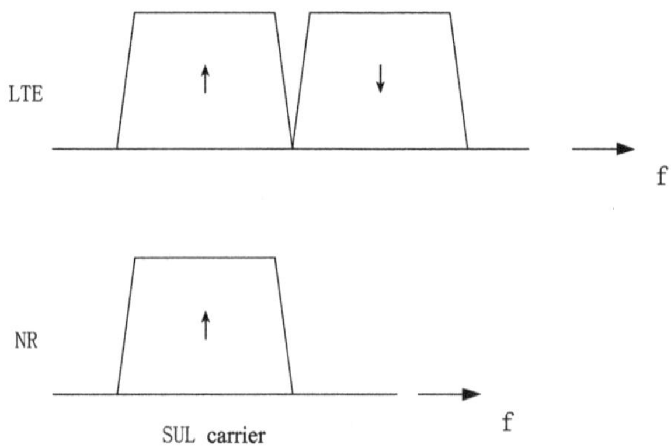

Figure 5.49 Coexistence of supplement uplink carrier and LTE uplink carrier.

5.6.9 *NR duplex mechanism*

Like LTE, NR also supports FDD and TDD duplex mechanisms. Three issues are discussed here, focusing on the difference between NR and LTE:

(i) It can be seen from Tables 5.2 and 5.3 that most low-frequency bands are symmetrical spectrum, so FDD is used. In Chapter 4, we have said that FDD can be divided into full-duplex FDD and half-duplex FDD. The half-duplex FDD transceiver cannot work at the same time, so the duplex filter can be omitted. It is widely used in certain IoT terminals, which have low power consumption and low cost. In the high-frequency band, it is an asymmetrical spectrum, so TDD should be adopted. TDD itself is half-duplex.
(ii) The so-called full-duplex and half-duplex are both for the terminal, and the base station is always full-duplex, that is, the base station needs to receive the uplink signal of one terminal while also transmitting signals to another terminal.
(iii) The differences between NR and LTE are as follows:

- LTE has two sets of frame structures. Figure 1.45 in Chapter 1 is the frame structure for FDD, and Figure 1.51 is the frame structure for TDD. However, NR has only one frame structure shown in Figure 5.31. This set of frame structures can be used for both FDD and TDD.

- When LTE is in TDD, its uplink/downlink sub-frames are configured semi-statically within 10 ms, as shown in Table 1.6 in Chapter 1. The so-called semi-static refers to several pre-defined configuration methods, and the switching cycle is 5 ms and 10 ms. When NR is in TDD, the uplink/downlink configuration is related to network deployment and can be divided into two situations.

(1) *Macro station coverage*: There are many users in the macro coverage, so static (only one configuration) or semi-static configuration is adopted. In the semi-static configuration, the uplink/downlink switching cycle is related to the OFDM parameter set, as shown in Table 5.14. It can be seen from the table that the switching period of the large subcarrier spacing is always compatible with the switching period of the small subcarrier, so this configuration can avoid the problem of mutual interference between adjacent base stations.

(2) *Dense deployment*: In the case of dense deployment, each base station only needs to serve a few terminals. For this reason, the scheduler can dynamically configure a time slot or part of the time slot as uplink or downlink at this time. Through this dynamic configuration, the system can quickly adapt to changes in uplink/downlink service requirements. But dynamic configuration needs to consider the problem of mutual interference between neighboring base stations, and when necessary, the decision of up/down configuration can be made through coordination between base stations.

5.6.10 *NR physical signal*

(i) *What is a physical signal?* Like LTE, the physical signal is a predetermined signal, or the physical signal is a known signal for both the

Table 5.14 Semi-static TDD configuration.

OFDM parameter set	Up/Down switching cycle
15 KHz	0.5,1,2,5,10
30 KHz	0.5,1,2,2.5,5,10
60 KHz	0.5,1,1.25,2,2.5,5,10
120 KHz	0.5,0.625,1,1.25,2,2.5,5,10

transmitter and the receiver, and this type of signal is called the reference signal. It is set according to the different purposes of the network side and the terminal side, such as synchronization, demodulation, channel estimation and acquisition of channel state information. Therefore, both uplink and downlink have reference signals for different purposes. The reference signal is generated and used by the physical layer, and some time-frequency resource units are also required for transmission. But it cannot carry high-level (layers above the physical layer) information. This is because these reference signals are known signals. The so-called known signal means that the value of the OFDM symbol in the reference signal and its position in the resource grid is known in advance by both parties. It is for reasons known in advance (called prior knowledge) that some algorithms can be used to obtain the required information at the receiver. The principle is similar to equation (1.18) in Section 1.2.2.4 of Chapter 1 using known d_{kL} symbols to estimate the channel transfer function.

(ii) *The difference with LTE*: The downlink demodulation of LTE completely relies on the **always-on cell**-specific reference signal (CRS). This kind of **always-on** signal is expected to be avoided as much as possible in NR minimalist design. For this reason, the NR standard uses different downlink reference signals for different purposes, and different reference signals are only transmitted when needed. It is transmitted together with service data in the same time slot. This self-contained transmission mode does not require the reference signal to be a normally open signal.

(iii) *NR reference signal*: The following types of reference signals are used for uplink/downlink:

- *A demodulation reference signal (DM-RS) is available for both uplink and downlink*: DM-RS is used to help the terminal (downlink) or the base station (uplink) to demodulate the physical channel. It is located at the front end of the time slot, and the DM-RS signal configured in the frequency domain of the resource grid is denser. This is because all subcarriers in an OFDM symbol need to be demodulated. But the configuration in the time domain is sparse, as shown in the example in Figure 5.50.

  The function of DM-RS can be seen in the "Symbol Timing and Frequency Synchronization" function block in Figure 1.39 in Chapter 1.

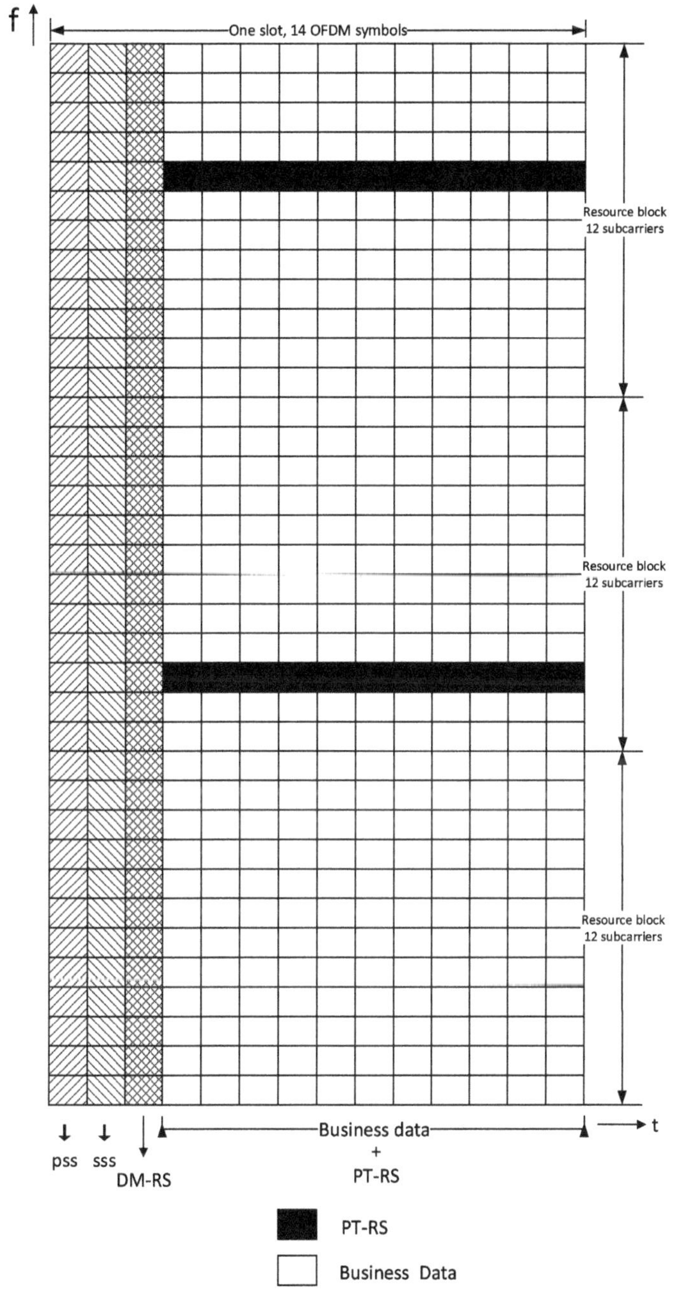

Figure 5.50 Example layout of downlink reference signal in resource grid.

- *The phase tracking reference signal (PT-RS) is available for both uplink and downlink*: The purpose of introducing the PT-RS reference signal is to compensate for the Common Phase Error (CPE) produced by the phase noise of the frequency combiner. The previous analysis shows that the CPE generated by all subcarriers in an OFDM symbol is the same. Therefore, the PT-RS signal configured in the resource grid frequency domain is sparse. But since it is completely random in the time domain, its correlation is very low. Therefore, the PT-RS signals configured in the resource grid time domain are denser, as shown in the example in Figure 5.50. PT-RS always appears at the same time as DM-RS. In the time domain, the two are adjacent, with DM-RS in front and PT-RS in back.
  The role of PT-RS can be seen in the two functional blocks of "Channel Estimation" and "Frequency-Domain Equalization" in Figure 1.39 in Chapter 1.

- *Downlink channel state information reference signal (CSI-RS)*: CSI-RS is used to help the terminal obtain downlink channel state information, which can be used for link adaptation (such as determining the modulation and coding scheme (MCS)), antenna beam management, uplink power control and time/frequency tracking.
- *Downlink tracking reference signal (TRS)*: TRS is a sparse downlink reference signal used to help the terminal achieve fine time/frequency synchronization and estimation of Doppler shift and delay spread.
- *Downlink primary synchronization signal (PSS) and downlink secondary synchronization signal (SSS)*: PSS and SSS are used to help the terminal realize its timing and base station timing synchronization. The former is coarse synchronization, and the latter is fine synchronization. Therefore, the two are adjacent to each other in the time domain, with PSS before SSS, as shown in Figure 5.50.
- *Uplink Sounding Reference Signal (SRS)*: The SRS transmitted on the uplink is used for the network side to measure the status of the uplink wireless channel, and the information obtained by the measurement can be used for scheduling and link adaptation. Based on the reciprocity of the channel state, SRS can also be used in massive MIMO precoder design and uplink beam management in NR (see later discussion).

Finally, a brief description is provided about Figure 5.50. It is a resource grid composed of a time slot and several resource blocks, and the front end of the time slot is configured with PSS and SSS. The purpose is to synchronize the time domain between the terminal and the base station, and then the DM-RS can be found and the information required for demodulation can be obtained. Then, PT-RS can be used to demodulate service data. It should be pointed out that the timing error, frequency error and phase error of the receiver have been eliminated after these processes. At this time, the receiver constellation diagram (see the "Reverse Mapping" function block in Chapter 1.39) will not be distorted, but the signal points in the constellation diagram are divergent. This is caused by thermal noise and ICI interference. The bit error rate after decoding is determined by the size of E_b/N_o.

5.6.11 *NR multi-antenna technology*

In the previous chapters, some of the problems of excessive antenna transmission were discussed in pieces. As it is a key technology in NR, it is necessary to gather details together for a more comprehensive discussion. For the convenience of readers, the detailed process is covered in this section.

Multiple antennas can be used for diversity, spatial multiplexing and beamforming, which are discussed separately in the following sections.

5.6.11.1 *Diversity*

Considering that it is not realistic to configure multiple antennas for terminals, especially mobile phones, here we only discuss the solution of configuring one transmitting antenna for the terminal and multiple receiving antennas for the base station. As shown in Figure 5.51, this configuration method is called receiving diversity.

(i) *Requirements for antenna configuration*: The purpose of diversity is to resist multipath fading of the channel. Therefore, it is hoped that the cross-correlation between the multiple signals received by the receiving antenna should be small, that is, it is hoped that the received NR signals will not fade at the same time. It is hoped that some signals are strong and some are weak, so that a strong signal can always be obtained after processing by the receiver. If the cross-correlation is large, the received NR signals will fade at the same time. For example, if they are all weak signals, then no matter how the receiver processes them, it is impossible to get strong signals.

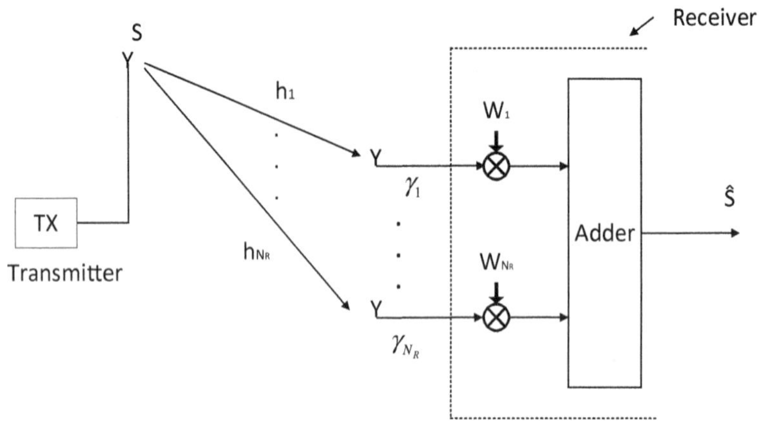

Figure 5.51 Receiving diversity.

The size of the cross-correlation between NR signals is determined by their respective propagation paths. It can be seen from the figure that the greater the propagation path difference, the smaller the cross-correlation and vice versa. Obviously, their propagation path is determined by the distance between the receiving antennas. If the distance between adjacent antennas is greater, the propagation path difference is greater and vice versa. In order to obtain low cross-correlation, the following requirements are needed for the antenna spacing in the case of different cell deployments:

- *The base station antenna is configured in the macrocell*: At this time, the coverage area of the cell is large, and the antenna is mounted at a relatively high position. In this case, the base station antenna spacing should be approximately 10 times the wavelength. By the way, if the terminal has multiple antennas, then the distance between the antennas is about half a wavelength (0.5λ).
- *The base station antenna is configured in the microcell*: At this time, the coverage area of the cell is small, and the antenna height is lower than the roof or configured indoors. In this case, a smaller antenna spacing is sufficient.

In addition to the above methods, when multiple antennas adopt different polarization directions, the cross-correlation can also be reduced.

(ii) *NR received signal combination method*: The received NR signals must be combined into one signal before being transmitted to the receiver, such as coherent demodulation. The method of combining is discussed in the following, and the signals received by NR antennas can be obtained from Figure 5.51:

$$\begin{aligned} r_1 &= h_1 S + n_1 \\ &\vdots \\ r_i &= h_i S + n_i \\ &\vdots \\ r_{N_R} &= h_{N_R} S + n_{N_R} \end{aligned} \tag{5.13}$$

where h_i is the transmission coefficient of the ith transport channel, which is a complex number, that is,

$$h_i = |h_i| e^{j\phi_i} \; i = 1, 2, \ldots, N_R \tag{5.14}$$

where n_i is the equivalent noise of the ith receiving antenna port.

After the above N_R signals are processed by the front end of the receiver, the copy signal output by the adder is

$$\hat{S} = w_1 h_1 s + \cdots + w_i h_i s + \cdots + w_{N_R} h_{N_R} s + w_1 n_1 + \cdots + w_i n_i + \cdots + w_{N_R} n_{N_R} \tag{5.15}$$

The equation (5.15) can be expressed by the following matrix:

$$\begin{aligned} \hat{S} &= \left[w_1 \ldots w_i \ldots w_{N_R} \right] \begin{bmatrix} h_1 \\ \vdots \\ h_i \\ \vdots \\ h_{N_R} \end{bmatrix} S + \left[w_1 \ldots w_i \ldots w_{N_R} \right] \begin{bmatrix} n_1 \\ \vdots \\ n_i \\ \vdots \\ n_{N_R} \end{bmatrix} \\ \hat{S} &= W^T \bullet H \bullet S + W^T \bullet N \end{aligned} \tag{5.16}$$

In the above equation,

$$H = \begin{bmatrix} h_1 \\ \vdots \\ h_i \\ \vdots \\ h_{N_R} \end{bmatrix}$$ is the channel matrix, where h_i is the channel transmission coefficient of the *i*th channel.

$$W = \begin{bmatrix} w_1 \\ \vdots \\ w_i \\ \vdots \\ w_{N_R} \end{bmatrix}$$ is the weighting matrix, where W_i is the *i*th weighting coefficient.

$W^T = [w_1 ... w_i ... w_{N_R}]$ is the transposed matrix of W.

$$N = \begin{bmatrix} n_1 \\ \vdots \\ n_i \\ \vdots \\ n_{N_R} \end{bmatrix}$$ is the noise matrix.

It can be seen from equation (5.15) and (5.16) that the copy signal $\hat{S}$ output by the adder varies with the weighting coefficient. There are two weighting methods.

- *Maximum Ratio Combining (MRC)*:

When the receiver has obtained all the channel transmission coefficients through channel estimation, that is, under the condition that the channel matrix H is known, then the weighting coefficient is

$$w_i = h_i^* = \left(|h_i| e^{j\phi_i}\right)^* = |h_i| e^{-j\phi_i}, i = 1, 2, \ldots, N_R \tag{5.17}$$

Substituting the above equation into (5.15), we can get

$$\hat{S} = \left(|h_1|^2 + \cdots + |h_i|^2 + \cdots + |h_{N_R}|^2\right) S \tag{5.18a}$$

$$+$$

$$\left(h_1^* n_1 + \cdots + h_i^* n_i + \cdots + h_{NR}^* n_{N_R}\right) \tag{5.18b}$$

It can be seen from (5.18a) that this combination method eliminates the phases on the NR channels through weighting coefficients, so the N_R signals are added in phase. Since the amplitude is the square of $|h_i|$, this means that the signal with the original amplitude has a larger amplitude after weighting. It shows that the amplitude weight of each signal is different, so it is called maximum ratio combining. It can be seen from (5.18b) that the N_R road noise will also be weighted, but in general, the combination method of MRC is better.

Knowing from the matrix knowledge, the MRC merging method actually makes (5.19):

$$W^T = H^H = \left[h_1^* \cdots h_i^* \cdots h_{N_R}{}^*\right] \tag{5.19}$$

where H^H is the conjugate transpose of matrix H.

- *Equal Gain Combining (EGC)*:

When the channel matrix is not known, the equal gain combining method can be used, that is, the value of the weighting coefficient is

$$W_i = K, \quad K \text{ is a constant}, \ i = 1,2,\ldots,N_R \tag{5.20}$$

It can be seen from the equation (5.15) that since the phase of the NR signal is not eliminated in this combination method, they are still vector addition, while MRC is in-phase addition, so the signal-to-noise ratio of the combined signal Ŝ after MRC is better than EGC.

5.6.11.2 *Space reuse*

(i) *Space reuse design ideas*: NR defines two frequency bands, that is, FR1 and FR2. The frequency range of FR1 is 0.45–6 GHz, and the frequency range of FR2 is 24.25–52.6 GHz. When discussing multi-antenna technology, we use the frequency band which is lower than 6 GHz as the low-frequency band, and the frequency band larger than 24.25 GHz as the high-frequency band. The high-frequency band is called the millimeter wave band.

The main problem when working in a low-frequency band is that the bandwidth is limited and the allowed frequency band is relatively narrow. It can be seen from the Shannon formula of Chapter 2 (2.1) that when the bandwidth is limited, the data rate that the system can transmit is also limited. In order to meet the requirements for higher data rates, it is necessary to increase the channel frequency band utilization.

As mentioned in Chapter 1, high-order modulation is a way to improve the utilization of the channel frequency band. In addition, the spatial multiplexing discussed in the following is another way to improve the utilization of channel frequency band. Its basic idea is to use space resources, that is, to use multiple antennas configured by transmitter and receiver to establish multiple independent parallel data transport channels in wireless space. Each channel is a point-to-point channel defined by Shannon. Moreover, the time-frequency resources occupied by each channel are the same, that is, the occupied bandwidth and duration are the same. At this time, the data rate transmitted by the system is the sum of the data rates of each channel. Obviously, it is much higher than a single channel, and the occupied frequency band is the same as that of a single channel, so the channel frequency band utilization rate is improved.

We define the number of independent parallel channels established as N_L. The larger the N_L is, the larger the channel band utilization rate is and vice versa. Obviously, the size of N_L is related to the number of transmitting antennas N_T and the number of receiving antennas N_R. When the antenna distance between the transmitter and receiver is relatively small, it is not difficult to obtain

$$N_L = \min \{N_T, N_R\} \tag{5.21}$$

It can be seen that N_L is the minimum of N_T and N_R. This result is not difficult to understand.

When $N_R > N_T$, there are only N_T data channels that the transmitter can transmit at most.

When $N_R < N_T$, there are only N_R data channels that the receiver can transmit at most.

It should be noted that N_L in (5.21) is the maximum value. It refers to the maximum number of independent parallel channels that can be established, which may be smaller in actual application.

(1) *Diagram of space reuse*: From the above analysis, the spatial reuse diagram can be obtained, as shown in Figure 5.52. The various parts are discussed in the following.

(a) *Multi-layer data flow*: Due to spatial multiplexing in space, N_L independent parallel transport channels can be formed. Therefore, the data input by the transmitter should also be a data stream with N_L layers, and these layers are independently coded and modulated. Its modulation and coding scheme (MCS) can be different according to requirement needs (will be discussed later).
(b) *Precoding (precoding matrix)*: Transmitter precoding is essential, and it has two functions.

1. Due to the number of input data stream layers, N_L is not necessarily equal to the number of transmitting antennas N_T. Therefore, N_L layers of the data stream must be mapped to N_T antennas for transmission through precoding (precoding matrix). So, precoding has the function of mapping (transformation).
2. The transmission characteristics of precoding must match the transmission characteristics of the spatial channel. The purpose is to increase the isolation of the N_L transport channels and reduce their mutual

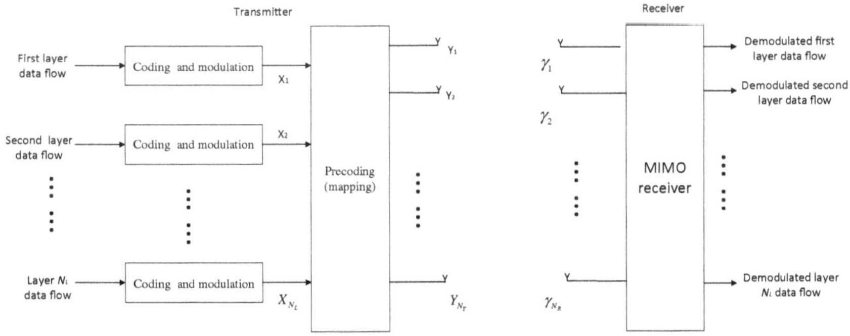

Figure 5.52 Diagram of space reuse.

interference to a minimum, so that they can become independent parallel channels as much as possible. Only in this way can the receiver be able to separate the N_L layers data stream under the condition of the maximum signal-to-interference ratio. Therefore, the precoding must be designed after the channel transmission characteristics are known. The specific design method will be discussed in the following.

- *Description of the transmission process*:

The entire transmission process of the signal from transmission to reception is discussed here. To facilitate understanding, we will first discuss one example and then generalize the results to general expressions.

Example: Suppose $N_T = 3$ and $N_R = 3$. Obviously, $N_L = \min\{3,2\} = 2$. Its transmission process can be shown in Figure 5.53. From the figure, we can obtain that (consider no noise at the receiver).

$$Y_1 = V_{11}X_1 + V_{12}X_2$$
$$Y_2 = V_{21}X_1 + V_{22}X_2$$
$$Y_3 = V_{31}X_1 + V_{32}X_2$$

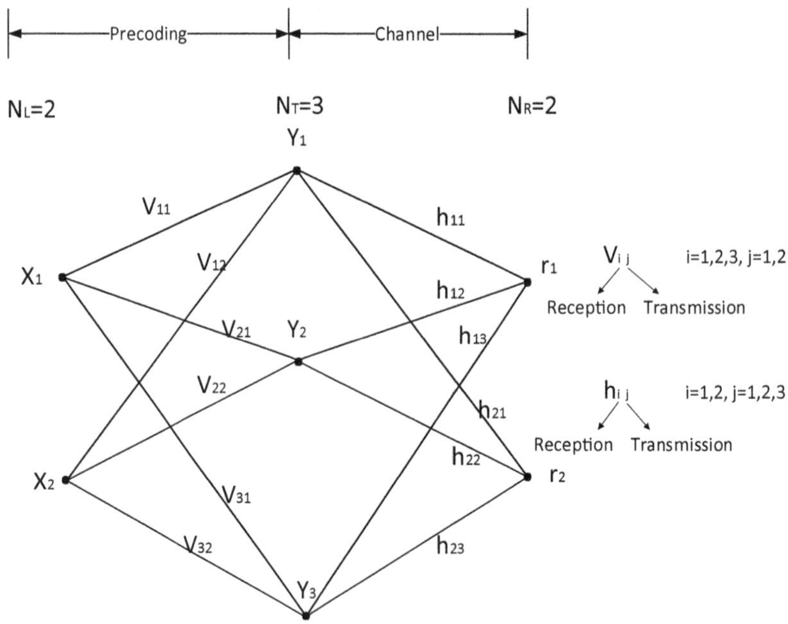

Figure 5.53 Transmission process.

The above equations can be represented by a matrix:

$$\begin{bmatrix} Y_1 \\ Y_2 \\ Y_3 \end{bmatrix}_{3\times1} = \begin{bmatrix} V_{11} & V_{12} \\ V_{21} & V_{22} \\ V_{31} & V_{32} \end{bmatrix}_{3\times2} \bullet \begin{bmatrix} X_1 \\ X_2 \end{bmatrix}_{2\times1} \tag{5.22}$$

$$r_1 = h_{11}Y_1 + h_{12}Y_2 + h_{13}Y_3$$

$$r_2 = h_{21}Y_1 + h_{22}Y_2 + h_{23}Y_3$$

The above equations can be represented by a matrix:

$$\begin{bmatrix} r_1 \\ r_2 \end{bmatrix}_{2\times1} = \begin{bmatrix} h_{11} & h_{12} & h_{13} \\ h_{21} & h_{22} & h_{23} \end{bmatrix}_{2\times3} \bullet \begin{bmatrix} Y_1 \\ Y_2 \\ Y_3 \end{bmatrix}_{3\times1} \tag{5.23}$$

Substituting (5.22) into (5.23), we get

$$\begin{bmatrix} r_1 \\ r_2 \end{bmatrix}_{2\times1} = \begin{bmatrix} h_{11} & h_{12} & h_{13} \\ h_{21} & h_{22} & h_{23} \end{bmatrix}_{2\times3} \bullet \begin{bmatrix} V_{11} & V_{12} \\ V_{21} & V_{22} \\ V_{31} & V_{32} \end{bmatrix}_{3\times2} \bullet \begin{bmatrix} X_1 \\ X_2 \end{bmatrix}_{2\times1} \tag{5.24}$$

Generalized expressions can be obtained by equation (5.24):

$$\begin{bmatrix} r_1 \\ r_2 \\ \vdots \\ r_{N_R} \end{bmatrix}_{N_R\times1} = \begin{bmatrix} h_{11} & h_{12} & \cdots & h_{1N_T} \\ h_{21} & h_{22} & \cdots & h_{2N_T} \\ \vdots & \vdots & & \vdots \\ h_{N_R1} & h_{N_R2} & \cdots & h_{N_RN_T} \end{bmatrix}_{N_R\times N_T} \bullet \begin{bmatrix} v_{11} & v_{12} & \cdots & v_{1N_L} \\ v_{21} & v_{22} & \cdots & v_{2N_L} \\ \vdots & \vdots & & \vdots \\ v_{N_T1} & v_{N_T2} & \cdots & v_{N_TN_L} \end{bmatrix}_{N_T\times N_L} \bullet \begin{bmatrix} x_1 \\ x_2 \\ \vdots \\ x_{N_L} \end{bmatrix}_{N_L\times1}$$

The above equation can be expressed as

$$R_{N_R\times1} = H_{N_R\times N_T} \bullet V_{N_T\times N_L} \bullet X_{N_L\times1} \tag{5.25}$$

$$= H'_{N_R\times N_L} \bullet X_{N_L\times1} \tag{5.26}$$

where

$$H'_{N_R \times N_L} = H_{N_R \times N_T} \bullet V_{N_T \times N_L} \tag{5.27}$$

$$R_{N_R \times 1} = \begin{bmatrix} r_1 \\ r_2 \\ \vdots \\ r_{N_R} \end{bmatrix}_{N_R \times 1} \text{ is the received signal matrix}$$

$$H_{N_R \times N_T} = \begin{bmatrix} h_{11} & h_{12} & \cdots & h_{1N_T} \\ h_{21} & h_{22} & \cdots & h_{2N_T} \\ \vdots & \vdots & & \vdots \\ h_{N_R 1} & h_{N_R 2} & \cdots & h_{N_R N_T} \end{bmatrix}_{N_R \times N_T} \text{ is the channel matrix} \tag{5.28}$$

$$V_{N_T \times N_L} = \begin{bmatrix} V_{11} & V_{12} & \cdots & V_{1N_T} \\ V_{21} & V_{22} & \cdots & V_{2N_T} \\ \vdots & \vdots & & \vdots \\ V_{N_T 1} & V_{N_T 2} & \cdots & V_{N_T N_L} \end{bmatrix}_{N_T \times N_L} \text{ is the precoding matrix}$$

$$X_{N_L \times 1} = \begin{bmatrix} x_1 \\ x_2 \\ \vdots \\ x_{N_L} \end{bmatrix}_{N_L \times 1} \text{ is the transmit signal matrix}$$

$H'_{N_R \times N_L}$ is composed of the channel matrix H and the precoding matrix V, so it is called the equivalent channel matrix.

In equation (5.28), the two numbers before and after each matrix subscript, respectively, represent the number of rows and columns of the matrix.

- *MIMO receiver*:

There are two types of MIMO receivers: linear receivers and nonlinear receivers. The following mainly discusses nonlinear receivers:

o *Linear receiver*:

The entire transmission process of spatial multiplexing can be represented in Figure 5.54. The matrix W^H is a linear receiver.

The demodulated signal matrix can be obtained from equation (5.25) and Figure 5.54:

$$
\begin{aligned}
\hat{R}_{N_L\times 1} &= W^H_{N_L\times N_R}\cdot R_{N_R\times 1} \\
&= W^H_{N_L\times N_R}\cdot H_{N_R\times N_T}\cdot V_{N_T\times N_L}\cdot X_{N_L\times 1}
\end{aligned}
\tag{5.29}
$$

Here, we need to explain how the matrices V and W in Figure 5.52 are obtained. After the channel matrix is known, it will be decomposed into three matrices and multiplied by singular value decomposition. One of the matrices is $V_{N_T\times N_L}$, which is the precoding matrix. The other matrix is $W_{N_R\times N_L}$, and $W^H{}_{N_L\times N_R}$ is the conjugate transpose matrix of $W_{N_R\times N_L}$. This means that as long as the channel matrix H is known, the precoding matrix of the transmitter and the demodulation matrix of the receiver can be

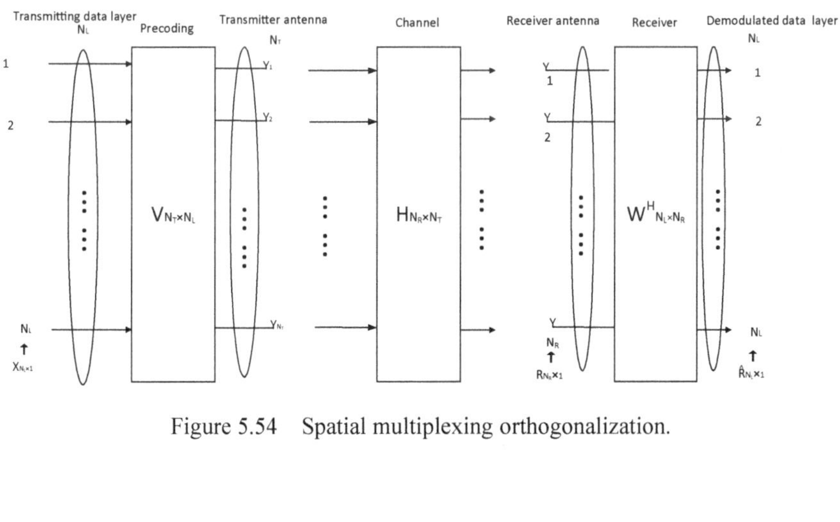

Figure 5.54 Spatial multiplexing orthogonalization.

obtained by decomposition. The specific decomposition process is quite complicated, so we won't discuss it here. Knowing that there exists such a decomposition method is enough.

o *Nonlinear receiver*:
The transmitter of the space division multiplexing is shown on the left side of Figure 5.52. It independently modulates and encodes the N_L layer data stream, so it is a multi-codeword (N_L codeword) transmission. Under such a condition, spatial multiplexing nonlinear receiver is shown as Figure 5.55.

The demodulation method in Figure 5.55 is called the successive interference cancellation (SIC). The basic principle of SIC is this: first demodulate and decode a spatially multiplexed signal (the first layer signal in the figure). If the decoding is correct, use the decoded data to regenerate the signal of the layer (the first layer signal in the figure) and eliminate the signal of this layer in the total input signal (the first layer signal in the figure). Then, the receiver demodulates and decodes the second layer signal. Under ideal circumstances, there will be no interference from the first layer signal, which improves the signal-to-interference ratio. Through such layer-by-layer demodulation and decoding, all layer signal demodulation and decoding can be completed. Obviously, such a layer-by-layer iterative algorithm has the problem of error propagation, that is,

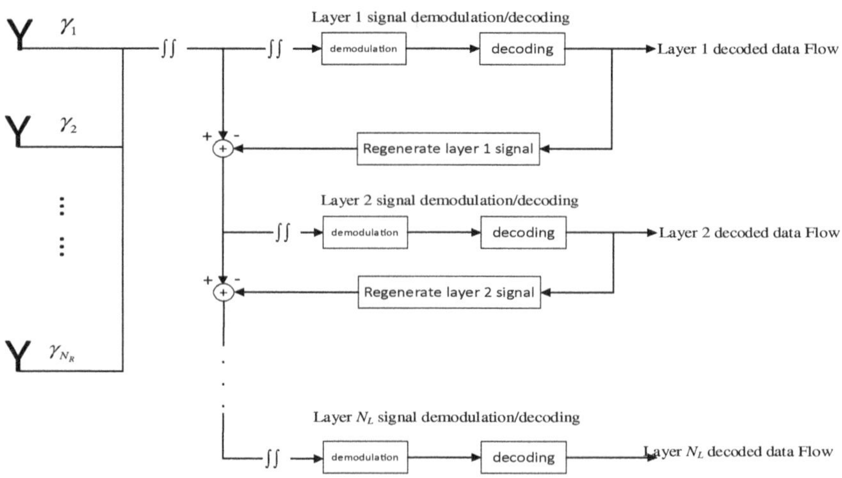

Figure 5.55 SIC nonlinear receiver.

when a layer is not decoded correctly, the signal recovered by that layer is incorrect. Therefore, the signal of this layer cannot be completely deleted from the total input signal. The remaining signal of this layer can interfere with the correct demodulation and decoding of the next layer, and this interference can continue to propagate.

It can be seen from the principle of SIC demodulation that when the first signal is demodulated, the interference is the largest, with a total of (N_L−1) interference signals. When the second signal is demodulated, the number of interference signals is reduced to (N_L−2), and the number of interference signals is gradually reduced, when it comes to the last signal there will be no interference. It can be seen that in order to prevent the above-mentioned error propagation, the robustness of each layer should be different. The first level should have the strongest robustness, followed by the second level and so on. Different robustness can be achieved by the following methods:

- The lower the modulation order, the greater the energy per bit E_b.
- The lower the coding rate and the lower the information rate, the greater the energy per bit E_b.
- The transmit power of each layer is different, the greater the transmit power, the greater the energy per bit E_b.
- Other

The SIC method can be extended to general SIC receivers. The principle is that when the terminal receives the signals superimposed by N_L users, it demodulates gradually in descending order of the signal-to-interference ratio. After a signal is decoded, it needs to be deleted from the total superimposed signal, until its own signal is decoded.

5.6.11.3 *Beamforming*

(i) *Design ideas of beamforming*: As discussed in the previous section, when 5G works in a low-frequency band, the main contradiction is that the frequency band is limited. In order to achieve the demand for high-speed data transmission, the utilization rate of the system frequency band must be improved. In this case, it is necessary to make multiple antennas into a spatial multiplexing method through precoding. However, the main contradiction changes when 5G works in high-frequency bands. It has a large amount of available spectrum, and the available bandwidth of a single carrier is up to 400 MHz, so spectrum efficiency is not so important. However, after the frequency increases, the transmission path loss increases and the radio wave penetration ability decreases, making cell coverage a major contradiction. In order to compensate for the increased transmission loss with increasing frequency, the system gain needs to be increased in this case. Therefore, multiple antennas should be beamformed through precoding to be equivalent to a directional (directional) antenna, thereby increasing the antenna gain.

Beamforming can be implemented in the analog domain or the digital domain. Due to hardware limitations, analog beamforming is currently mainly used in millimeter waves. It puts multiple antenna units on one antenna panel. Figure 5.56 [11] is a photo of a real beamforming device. 64 dual-polarized antenna units are integrated on the antenna panel.

(ii) *Schematic diagram of simulated beamforming*: The principle of simulated beamforming is shown in Figure 5.57. The figure is explained as follows:

- The single-layer data stream *X* comes from the baseband chip. It needs to be converted into an analog signal by a digital-to-analog converter and then entered into the precoding process.

Figure 5.56 Antenna panel.

Note: AAA battery (used to compare the size of the antenna panel).

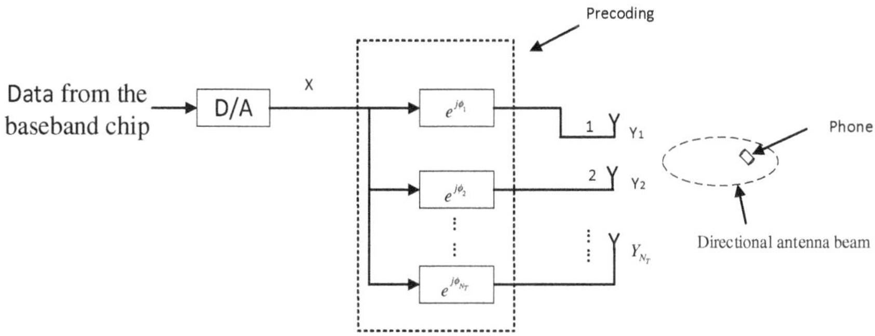

Figure 5.57 Transmit antenna beamforming.

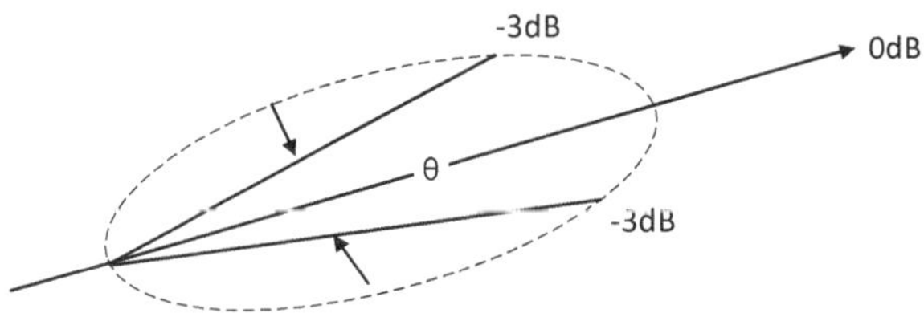

Figure 5.58 Half power angle θ.

- The precoding of analog beamforming only produces different phase shifts for multiple antennas without changing the signal amplitude. When the phase is changed, the direction of the directional antenna will change. In order to communicate with the terminal, the directional antenna must point to the terminal. When the terminal moves, the direction of the antenna must also move. This means that the phase shift value in precoding is determined by the transmission characteristics of the spatial channel. Therefore, the precoding can be designed only after the channel characteristics are known. This concept is the same as spatial reuse, and the specific design method will be discussed in the following.
- The half power angle of the directional antenna is shown in Figure 5.58.

The physical meaning of the directional antenna half power angle θ is very clear. As $-3\ dB$ power is half of $0\ dB$ power, it is called half power angle. Besides,

$$\theta \propto \lambda^2/S \tag{5.30}$$

where S is the physical antenna area and λ is the working wavelength.

It can be seen that when S is larger and/or the frequency is higher, while θ is smaller. At this time, the antenna gain is larger and vice versa.

(iii) *System diagram*: The transmission block diagram from transmitting to receiving is shown in Figure 5.59. When the noise at the receiver is not considered,

$$R_{N_R\times 1} = H_{N_R\times N_T} \bullet V_{N_T\times 1} \bullet x = H'_{N_R\times 1} \times x \tag{5.31}$$

where $H'_{N_R\times 1} = H_{N_R\times N_T} \bullet V_{N_T\times 1}$ is the equivalent channel matrix.

$$V_{N_T\times 1} = \begin{bmatrix} e^{j\phi_1} \\ e^{j\phi_2} \\ \vdots \\ e^{j\phi_{N_T}} \end{bmatrix}_{N_T\times 1} \quad \text{is the precoding matrix}$$

H, *H'* and *R* are the channel matrix, equivalent channel matrix and received signal matrix, respectively.

The demodulation of all downlink physical channels in NR depends on the corresponding demodulation reference signal DM-RS, and DM-RS and data are pre-coded the same, that is, as shown in Figure 5.59. The layout of the downlink signal in the resource grid is shown in Figure 5.50. At this time, the data X in a time slot can be demodulated by the following method.

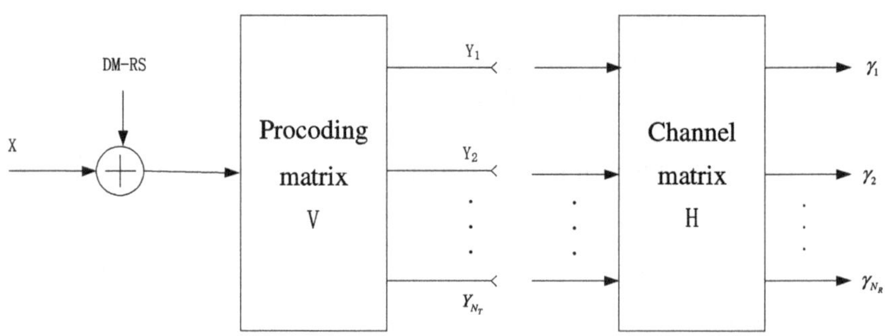

Figure 5.59 Transmission process.

First, put the data in a time slot in the memory, and then process the data in the memory. Since the data of DM-RS is known, it can be processed as follows using equation (5.31):

$$R_{N_R\times1}\bullet\left(\frac{x^*}{|x|^2}\right)=H'_{N_R\times1}\bullet x\bullet\left(\frac{x^*}{|x|^2}\right)=H'_{N_R\times1}\bullet\frac{|x|^2}{|x|^2}=H'_{N_R\times1} \tag{5.32}$$

In the above equation, x^* and $|x|$ are the conjugate value and modulus value of the DM-RS data, which are known.

The $H'_{N_R\times1}$ in equation (5.32) is the equivalent channel matrix when the DM-RS data arrives, and consider that the equivalent channel matrix of all data in a time slot is equal to $H'_{N_R\times1}$ in equation (5.32). In this way, the already obtained $H'_{N_R\times1}$ can be used, and all other data can be solved by doing the following operations on equation (5.31):

$$(H')^{-1}_{1\times N_R}\bullet R_{N_R\times1}=(H')^{-1}_{1\times N_R}\bullet H'^{-1}_{N_R\times1}\bullet x=x \tag{5.33}$$

In the above equation, $(H')^{-1}$ is the inverse matrix of H'.

The above calculation method is actually divided into two steps.

First, use the known DM-RS data to compute the equivalent channel matrix H' according to equation (5.32).

Second, use the known H' to obtain data x according to equation (5.33).

(iv) *Up/down multi-antenna precoding*: Both beamforming and spatial multiplexing require precoding, which must match the channel transmission characteristics. Therefore, it is necessary to know the channel transmission characteristics first, and only after this can the precoding be designed. The following discuss how the base station and terminal obtain channel information.

(1) *Downlink (base station) multi-antenna precoding*:

- When using FFD duplex mode, because the uplink/downlink frequency range is different, the uplink/downlink channel characteristics do not have reciprocity. In this case, the base station first transmits the downlink status information reference signal CSI-RS, and the terminal measures it. According to the measurement result, a precoding codebook (several precoding sets available for downlink) is selected and reported to the base station. The base station may or may not use it, which means that the base station can determine the downlink precoding by itself.

The base station decides to carry out its downlink precoding on its own in the case of multi-user MIMO (MU-MIMO) because in this case, terminals will report multiple precoding schemes to the base station.

The base station needs to comprehensively consider the problem of interference to terminals and then choose an optimal precoding scheme for each terminal.

- When using TDD duplex mode, because the uplink/downlink frequency range is the same, the uplink/downlink channel characteristics have reciprocity. In this case, the base station can determine the downlink precoding scheme by measuring the uplink, which can save signaling overhead.

(2) *Uplink (terminal) multi-antenna precoding*:
The method for determining the precoding scheme on the terminal side is similar to that on the base station side, but it cannot determine the precoding scheme by itself.

(v) *Beam management*: When both the base station side and the terminal side use beamforming, in order to achieve communication between the two parts, a beam pairing problem should be solved first, as shown in Figure 5.60. The so-called beam management is used to establish and maintain such a beam pair. The specific method is not discussed here.

5.6.12 *NR non-orthogonal multiple access*

Currently, NR mainly uses orthogonal multiple access. With the development of 5G, non-orthogonal multiple access will be introduced in subsequent versions of R15, so here is a brief introduction.

5.6.12.1 *Overview of multiple access* [23]

(i) *What is multiple access?* Multiple access is a technical method for multiple users in a communication network to realize communicating

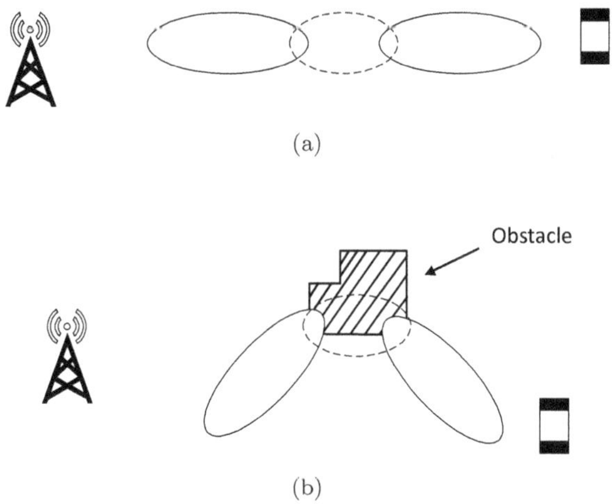

Figure 5.60 Beam pair: (a) direct; (b) reflection.

simultaneously without interfering with each other. As the name implies, it means that the network needs to assign an address to the users participating in this communication, and then the users communicate according to the assigned address. The address used by the user actually occupies the radio resources in the communication network. These radio resources refer to frequency (frequency domain), time (time domain), space (spatial domain), code (code domain) and power (power domain). In this way, there are two methods for allocating addresses. One is to allocate multiple orthogonal addresses (orthogonal divided radio resources) to users. At this time, multiple users participating in communication are called orthogonal multiple access (that is, users occupy multiple orthogonal addresses to access the network). The other is to allocate multiple non-orthogonal addresses (non-orthogonal divided radio resources) to users. At this time, multiple users participating in communication are called non-orthogonal access (that is, users occupy multiple non-orthogonal addresses to access the network).

(ii) *1–4G multiple access mode*: 1–4G uses Orthogonal Multiple Access (OMA), and the divided radio resources and multiple access methods are shown in Table 5.15.

(iii) *The design idea of non-orthogonal multiple access in 5G*: There are multiple application scenarios in 5G, such as eMBB, mMTC and

Table 5.15 1–4G orthogonal multiple access mode.

		Divided radio resources				
Generation	Multiple access name	Frequency domain	Time domain	Spatial domain	Code domain	Power domain
1G	FDMA	√				
2G	TDMA		√			
3G	CDMA			√	√	
4G	OFDMA	√	√	√		

URLLC. Compared to 4G, it has higher requirements for multiple access:

- Further improves channel frequency band utilization (b/s/Hz) (eMBB application scenarios).
- Address scheduling requires low latency (URLLC application scenarios).
- Address scheduling requires low power consumption (mMTC application scenarios).
- There are many addresses (large connections) (mMTC application scenarios).

Among them, the first and fourth requirements can be realized by further increasing the upper limit of the Shannon channel capacity under the same frequency/time/space conditions as 4G. The second and third requirements can be realized as long as the address is free of scheduling.

The following briefly explains how to further improve the Shannon channel capacity. The OFDMA orthogonal multiple access method used in 4G is shown in Figure 5.61. In this figure, it is assumed that two users A and B occupy different 12 subcarrier frequency bands as two orthogonal addresses. This means that only one user's data can be transmitted in a bandwidth of 12 Δf. In 5G, these 12 Δf bandwidths are still occupied, but a power dimension is added to the same frequency band and then the power is divided in the power domain by allocating unequal power to different users. Thus, different powers are used as addresses for different users, so that more users can transmit data in the same frequency band. Thus, the number of addresses increases, and the frequency band utilization

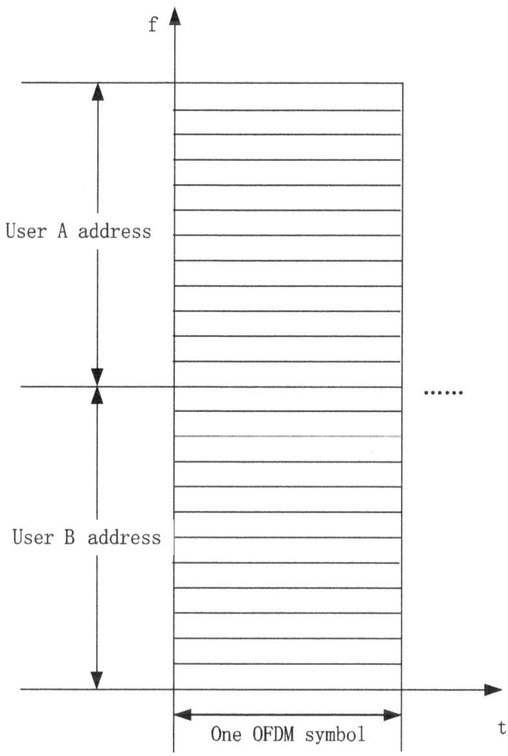

Figure 5.61 OFDMA.

rate is also improved. Obviously, this division is not orthogonal. Therefore, the receiver cannot use the orthogonal method to distinguish between different addresses, but can only use other multi-user detection methods to distinguish between different addresses. For example, the SIC mentioned in the previous section is a commonly used method. In addition, the dimensions added in the same frequency band may be other dimensions such as power dimension, code dimension or a combination of several dimensions. But they are all non-orthogonal divided, so this multiple access method is called non-orthogonal multiple access (NOMA, Non-OMA).

In LTE, user uplink addresses are all scheduled and allocated by the uplink scheduler at the base station side through signaling. Therefore, the signaling overhead and power consumption are increased, and the time delay is also increased. If the user's uplink address can be free of scheduling, then the scheduling signaling is reduced. Therefore, the terminal power consumption and time delay are reduced. There are two ways to solve the user address problem in some IoT scenarios with free address scheduling. One is to predefine the user address, and the other is that the user can choose from a pre-defined codebook (see SCMA discussed later).

(iv) *Uplink/downlink multiple access model*: The uplink/downlink multiple access communication system model is shown in Figure 5.62. We can understand the following from the picture:

- The uplink communication system is multi-point transmission, single-point reception, and single user power is limited. At the same time, the greater the number of users transmitted, the greater the total transmission power. It is difficult for the transmitter to process jointly, but the receiver can process jointly. The corresponding model is called Multiple Access Channel (MAC).
- The downlink communication system is single-point transmission, multi-point reception, and the total transmission power is limited. At the same time, the greater the number of received users, the smaller the power allocated to a single user. The transmitter can process jointly, but it is difficult for the receiver to process jointly. The corresponding model is called the Broadcast Channel (BC).

Due to different uplink/downlink communication models, the upper/lower Shannon channel capacity upper limit is also different.

5.6.12.2 *Comparison of channel capacity of two multiple access methods*

In the AWGN channel, point-to-point communication is the upper limit of its channel capacity:

$$C = W \log_2\left(1 + \frac{s}{n_0 w}\right)(\text{bps}) \tag{5.34}$$

where W is the channel bandwidth.

After analysis, it is known that the difference between uplink and downlink channel capacity is mainly due to the difference in the base in the logarithm, but the logarithm changes very slowly with the base.

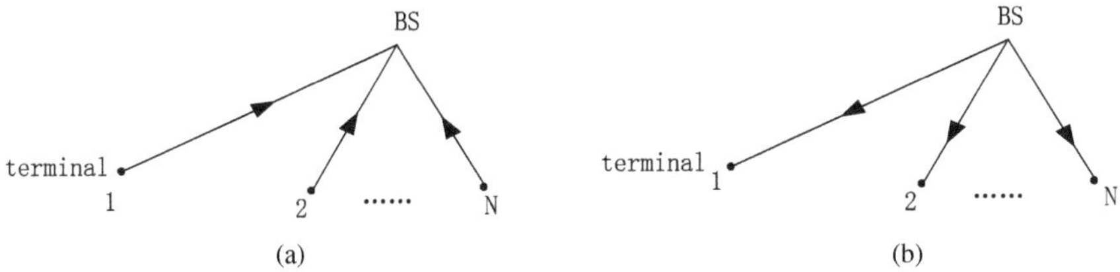

Figure 5.62 Uplink/downlink communication.

For this reason, in the qualitative analysis, the influence of the logarithmic part may not be considered, but the relationship between the channel capacity C and the channel bandwidth is mainly considered:

$$C \propto \text{The channel bandwidth} \tag{5.35}$$

Without loss of generality, we assume that there are two users 1 and 2 in the model of Figure 5.62. It is also assumed that in the two multiple access situations, the two users occupy the total radio resources, that is, the channel bandwidth W (frequency domain), the number of OFDM symbols (time domain) and the number of spatially independent parallel channels (spatial domain) are the same.

(i) *Non-orthogonal multiple access*: Due to the characteristic of non-orthogonal multiple access, the user address may not occupy frequency-domain resources, but occupy other dimensional resources, such as power and other resources. In this way, the channel bandwidth used by the two users to transmit data is still W. It can be obtained by (5.35) that

$$\begin{aligned}&\textit{User 1 maximum rate } R_1 \propto W\\&\textit{User 2 maximum rate } R_2 \propto W\end{aligned} \tag{5.36}$$

(ii) *Orthogonal multiple access*: Due to the characteristic of orthogonal multiple access, the frequency band W should be divided orthogonally as the address. Therefore, the channel frequency band used to transmit data is no longer W. It can be obtained by (5.35) that

$$\begin{aligned}&\textit{User 1 maximum rate } R_1' \propto \alpha W \quad 0 \le \alpha \le 1\\&\textit{User 2 maximum rate } R_2' \propto (1-\alpha)W\end{aligned} \tag{5.37}$$

It can be seen from the equations (5.36) and (5.37) that when both users are connected,

$$R_1 + R_2 > R_1' + R_2' \tag{5.38}$$

It can be seen that the upper limit of channel capacity of the non-orthogonal multiple access is greater than that of the orthogonal multiple access method.

5.6.12.3 *Several non-orthogonal multiple access methods*

(i) *Power domain non-orthogonal multiple access*: Power domain non-orthogonal multiple access (PNMA) divides the power and allocates it to users as addresses. The PNMA diagram for downlink is shown in Figure 5.63.

When the base station has already known the channel characteristics of users, it can be seen from the figure that user 2 is farther from the base station than user 1. In order to ensure that user 2 has a better reception effect, the power allocated to user 2 by the base station should be greater than that of user 1. Then the two signals with different powers are

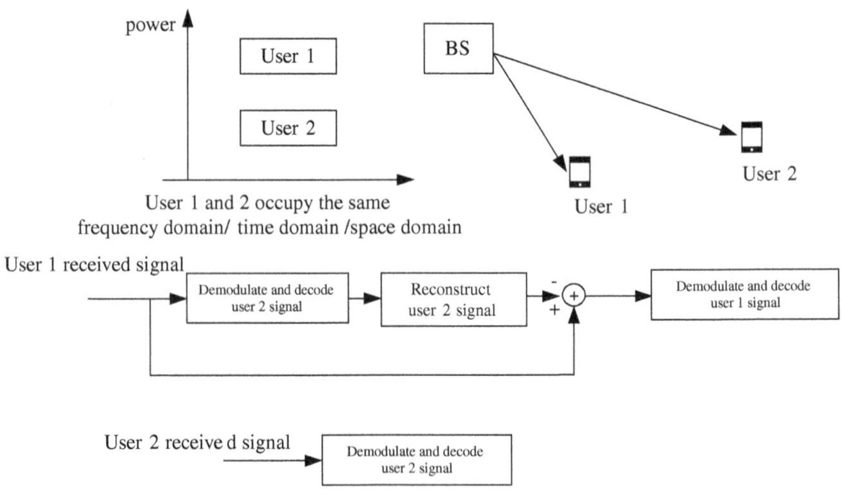

Figure 5.63 PNMA for downlink.

superimposed and transmitted to the user. In order to identify the address expressed in power, users should apply SIC receiver. Since the power of user 2 is greater than the power of user 1, the receiver of user 2 can directly demodulate/decode the signal of user 2. In the demodulation/decoding process, the signal of user 1 is regarded as the signal of user 2 interference. The user 1 receiver first demodulates/decodes the user 2 signal and reconstructs the user 2 signal. Then, the signal of user 2 is deleted from the non-orthogonal superimposed signal received by the user. Then user 1's signal is demodulated/decoded. In this process, there is no interference from the user 2 signal. When PNMA is used in the uplink, its working process is similar.

The principle of PNMA technology is simple, intuitive and easy to implement. But the number of users that can be accessed is still limited.

(ii) *Code domain non-orthogonal multiple access*: The code domain non-orthogonal multiple access technology is to perform non-orthogonal division in the code domain, and there are two schemes currently proposed. One is called Multi-User Shared Access (MUSA), and the other is called Sparse Code Multiple Access (SCMA). Among them, SCMA has better performance than MUSA, and it can be used for both uplink and downlink. According to the analysis, the number of user accesses (user connections) of the SCMA system is 1.5–3 times that of the LTE system (orthogonal multiple access). The following briefly introduces the working process of SCMA for uplink.

The diagram of SCMA used in uplink is shown in Figure 5.64. Assuming there are four users, the main function is discussed in the following.

(1) *User transmitting SCMA code*: The transmitter uses sparse codes for encoding. Figure 5.65 is an example. There are four users in total, and each user has a predefined codebook, and there are four sparse codes in the codebook. In this way, one of the four codes can be selected according to the double-bit output of the channel coding. This code plays a role of spread spectrum. Sparse means that most of the elements in the code are zero, and the positions of the zero elements of all the code in the same codebook are the same. The non-zero elements in the sparse code are all complex numbers. The correlation between each code is low. The advantage of using sparse codes is that

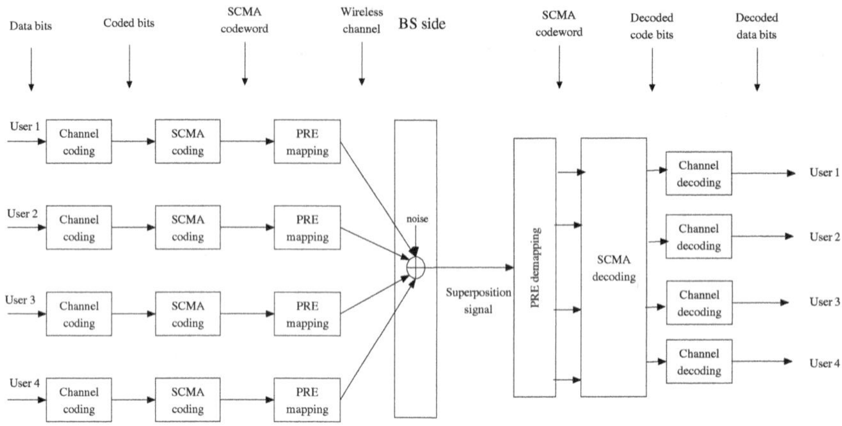

Figure 5.64 SCMA uplink diagram.

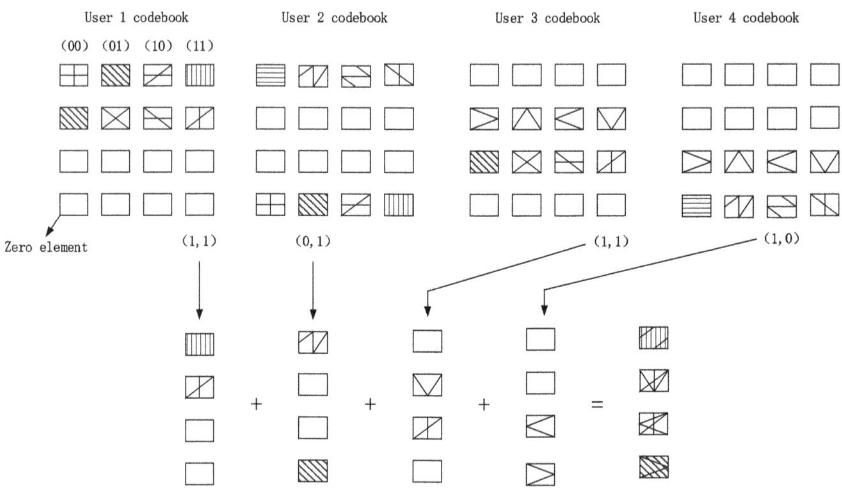

Figure 5.65 SCMA code.

there are more zero elements, which can reduce the complexity of the receiver to separate each code.

(2) *PRE mapping*: PRE mapping is to map the elements of the spreading code output from the SCMA encoding to the time/frequency/space domain. That is, several physical resource units (PRE) in the time-frequency grid are occupied, which are transmitted by the radio

frequency transmitter and the antenna. Its corresponding circuits include all the circuits from the serial/parallel conversion in the baseband chip transmitter shown in Figure 1.39 in Chapter 1, and all the circuits from the RF chip transmitter to the antenna in Figure 1.41 in Chapter 1.

(3) The signals of four users are linearly superimposed at the receiver of the base station, and the subsequent processing is the inverse transformation of the transmitting end. In the case of uplink-free scheduling, the base station does not have to know what information the user wants to transmit in advance. Therefore, the receiver needs to perform active user detection on the channel. The algorithm of SCMA receiver is quite complicated, so it is not discussed here.

There is also a pattern division multiple access technology (PDMA), which is a non-orthogonal multiple access technology that can be divided in all or part of the power domain, code domain, space domain, frequency domain and time domain. In theory, this method can allow the largest number of access users, but its technology is also the most complicated and is not yet mature.

5.6.13 *Summary of key technologies at the physical layer*

According to the above discussion, the key technologies of the 5G NR physical layer can be summarized in Table 5.16.

5.7 NR Terminal Access Process

This section briefly introduces the NR terminal access process, which includes two processes: cell search and random access.

5.7.1 *Cell search*

(i) *Purpose of cell search*: The cell search has the following two purposes:

(1) After the terminal is turned on, a cell search is required to enable the terminal to initially access the network.

Table 5.16 Summary of key technologies at the physical layer.

Key technology	Description
Duplex mode	Supports FDD and TDD mode.
Subcarrier spacing	Below 6 GHz includes 15 KHz, 30 KHz, 60 KHz. Above 6GHz includes 60 KHz, 120 KHz and 240 KHz.
CP	Supports regular CP and extended CP (extended CP is only used for 60 KHz subcarrier spacing).
Frame structure	The frame length is 10 ms. One frame contains 10 sub-frames. Five sub-frames form a half-frame. Supports semi-static and dynamic frame structure configuration.
Basic waveform	Downlink: CP-OFDM; uplink: CP-OFDM, DFT-S-OFDM.
Single carrier support bandwidth	The minimum is 100 MHz below 6 GHz, while the maximum is up to 400 MHz above 6 GHz.
Multiple access	Downlink: Orthogonal multiple access. Uplink: Orthogonal multiple access. Non-orthogonal multiple access (to be determined).
Channel coding	Control channel: Polar code, RM code, repetition code, Simplex code. Data channel: LDPC code.
Modulation	Downlink: QPSK, 16QAM, 64QAM, 256QAM. Uplink: CP-OFDM supports QPSK, 16QAM, 64QAM, 256QAM. DFT-S-OFDM supports $\pi/2$-BPSK, QPSK, 16QAM, 64QAM, 256QAM.
Resource mapping	Supports centralized and distributed resource allocation methods.
Multi-antenna technology	Broadcast, control and data channels adopt integrated multi-antenna design. Downlink data transmission supports transmission schemes such as closed-loop, open-loop, quasi-open-loop and multi-point transmission. Uplink data transmission supports transmission schemes based on codebook and non-codebook. The downlink transmission scheme selection and feedback mode are combined. The uplink transmission scheme is directly configured according to high-level signaling.
Pilot design	Supports DMRS, CSI-RS, PT-RS, uplink SRS and other designs.
Physical layer measurement	Including channel state measurement, channel quality measurement, interference management measurement, and mobility measurement.

(*Continued*)

Table 5.16 (*Continued*)

Key technology	Description
HARQ	Support Chase combination and incremental redundancy (IR) HARQ.
Link adaptation	Adopts a modulation and coding scheme that is adaptively adjusted according to changes in channel status.
BWP	Supports UE initial access bandwidth management and UE working bandwidth adjustment.
Carrier aggregation/dual connectivity	Supports up to 16 NR carrier aggregation or dual connectivity operations. Supports NR to use continuous or discontinuous spectrum over 1 GHz.
Uplink and downlink decoupling	Supports multiple uplink carriers in one NR.

(2) When the terminal moves, the terminal also needs to continuously search for neighboring cell signals to obtain synchronization and evaluate the received signal quality. Then the terminal compares the received signal quality of the neighboring cell with the received signal quality of the current camping cell to determine whether the cell needs to be handed over (when the terminal is in the RRC_CONNECTED state in Figure 5.22, and the handover judgment is made by the network side.) or judge whether the cell needs to be reselected (when the terminal is in the RRC_IDLE or RRC_INACTIVE state in Figure 5.22, and the judgment of whether to reselect is made by the terminal itself).

(ii) *Results of community search*: The following results are obtained through the cell search terminal:

- Obtains frequency synchronization and symbol synchronization with the cell.
- Obtains frame synchronization with the cell, that is, know the start position of the downlink sub-frame.
- Obtains the temporary identification C-RNTI of the radio network of thc ccll.

(iii) *Cell search method*: In order to enable the terminal to search for a cell after being turned on, and to find a new cell when the terminal is moving, each NR cell will periodically transmit a synchronization signal block (SSB) in the downlink. The SSB is composed of the primary synchronization signal (PSS), the auxiliary synchronization signal SSS and the physical broadcast channel (PBCH). Among them, PSS and SSS are the downlink physical signals introduced in Section 5.6.10, and PBCH is the physical channel, which carries the broadcast information of the system. The terminal finds the cell by searching the SSB.

(1) *SSB structure*: The structure of SSB in the OFDM resource grid is shown in Figure 5.66. It can be seen from the figure that SSB occupies four OFDM symbols in the time domain and 240 subcarriers in the frequency domain. Among them, PBCH occupies 576 ($240 \times 2 + 48 \times 2$) time-frequency resource units. The PBCH also self-contains

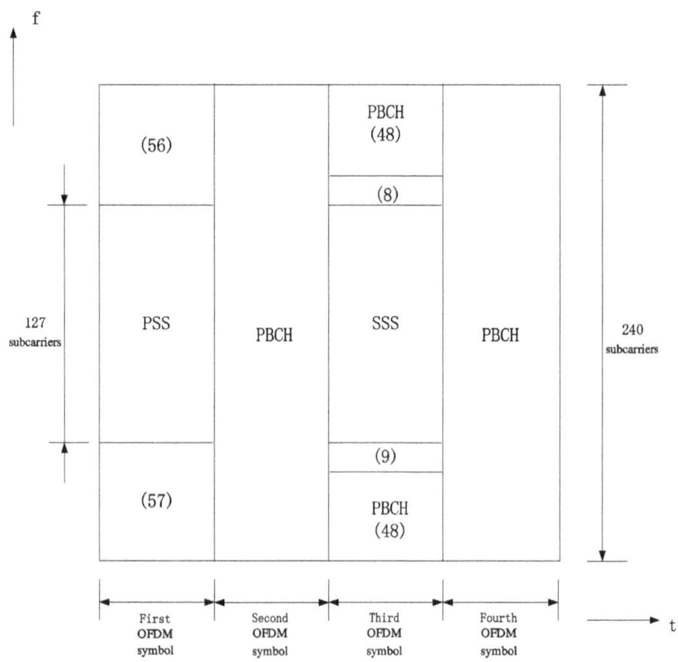

Figure 5.66 SSB time-frequency structure.

Note: The number in ( ) in the figure is the number of subcarriers, of which (56), (57), (8), (9) are not occupied.

Table 5.17 SSB parameter set.

Parameter set (KHz)	SSB bandwidth (MHz)	SSB duration (μs)	Operating frequency range
15	3.6	≈285	FR1(<3 GHz)
30	7.2	≈143	FR1
120	28.8	≈36	FR2
240	57.6	≈18	FR2

the reference signal DM-RS for its coherent demodulation. When using different parameter sets in Section 5.6.4.2, the bandwidth and duration occupied by SSB are different, as shown in Table 5.17. The 60 KHz parameter set cannot be used for SSB.

The value in the first row of the table is calculated like this:

$$\text{SSB bandwidth} = 15 \times 240 = 3.6\ (\text{MHz})$$

$$\text{SSB duration time}=(144+2048)\times 4\times \frac{1}{15\times 2048}(\text{ms}) \approx 285.4(\text{us})$$

The values in the other rows are integer multiples or fractional multiples of 2 in the first row.

(2) *SSB cycle*: SSB is transmitted periodically, and the period can vary from 5 ms to a maximum of 160 ms. The default value is 20 ms.

(3) *SS burst set*: The difference between SSB and LTE is that in high-frequency band base stations will use beam scanning to transmit SSB. That is, different SSBs are transmitted on beams of different directions in a time division multiplexing mode, as shown in Figure 5.67. The SSB set in beam scanning is called the synchronization signal burst set (SS burst set). It should be pointed out that beam scanning can be realized by changing the precoding in Figure 5.57. The default period of the SS burst set is 20 ms. However, an SS burst set that appears within each 20 ms only takes 5 ms. It is the first half-frame or the second half-frame of a radio frame. The maximum number of SSBs in a 5 ms period is related to the NR operating frequency band (see Table 5.17).

- In the frequency band below 3 GHz, an SS burst set has a maximum of 4 SSBs, that is, a maximum of 4 beams can be scanned.

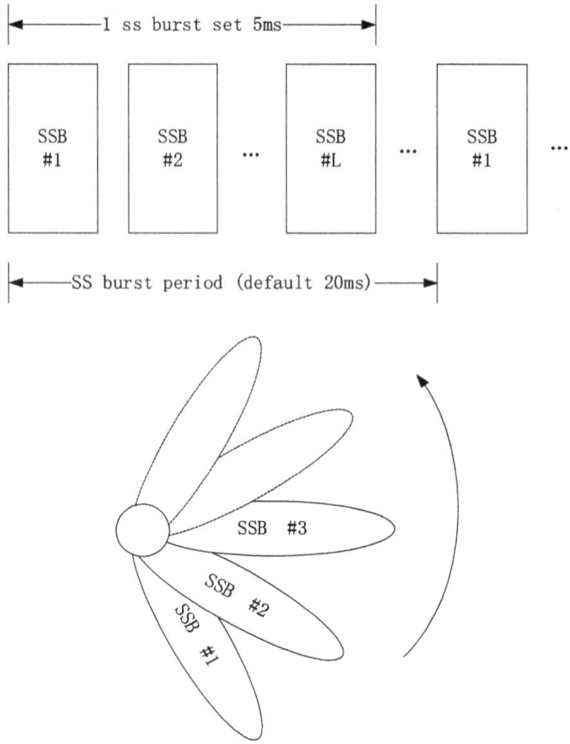

Figure 5.67 Beam scanning and SS burst set.

- In the 3–6 GHz frequency band, an SS burst set has a maximum of 8 SSBs, that is, a maximum of 8 beams can be scanned.
- In the FR2 frequency band, an SS burst set has a maximum of 64 SSBs, that is, a maximum of 64 beams can be scanned.

5.7.2 *Random access*

Random access means that the terminal initiates a request to establish a connection. Similar to LTE, the NR random access process is divided into four steps, as shown in Figure 5.68.

- *Step 1*: The terminal transmits a preamble to the base station on the physical random-access channel (PRACH). Based on this, the base station estimates the transmission delay of the terminal as a basis for realizing uplink synchronization.

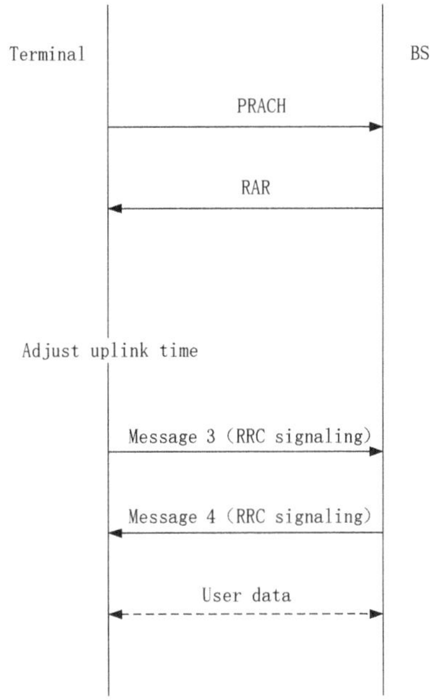

Figure 5.68 Random access process.

- *Step 2*: The base station transmits a random-access response (RAR) to the terminal, indicating that the preamble has been received, and based on the transmission delay estimated in the first step, the transmitting time advance command is used to adjust the terminal's transmitting time to achieve uplink synchronization (see Figure 1.53 in Chapter 1). In addition to establishing uplink synchronization, this step also allocates uplink resources for the terminal to be used in the next step (the third step).
- *Steps 3 and 4*: The terminal and the base station exchange messages (uplink/downlink messages are 3 and 4, respectively), and the purpose is to solve the conflict caused by multiple terminals in the cell that may transmit the same preamble at the same time. If successful, message 4 will move the terminal to the connected state (see RRC_CONNECTED in Figure 5.22). At this time, the terminal and the network can enter the normal communication state.

There is also a paging process corresponding to random access, which is suitable for connection establishment initiated by the network side. In order to save power, the terminal stays in a dormant state for most of the time without any receiving processing and only wakes up briefly within a predetermined time period to monitor the paging message transmitted by the network side. If the paging message contains the identity of the paged terminal, the terminal will establish a connection with the network side, otherwise it will continue to sleep. This process is the same as the process of downlink data transmission.

5.8 Communication Equation and Path Loss

5.8.1 *Communication equation*

The downlink in the cell of the mobile communication network is shown in Figure 5.69. The physical meaning of each level value in the figure is as follows:

- P_t *(dBm)* represents transmitter power.
- L_1 *(dB)* represents the feeder loss of the transmitting antenna.
- G_t *(dB)* represents the gain of the transmitting antenna.
- *PL (dB)* represents the transmission path loss.
- G_r *(dB)* represents the gain of the receiving antenna.
- L_2 *(dB)* represents the feeder loss of the receiving antenna.

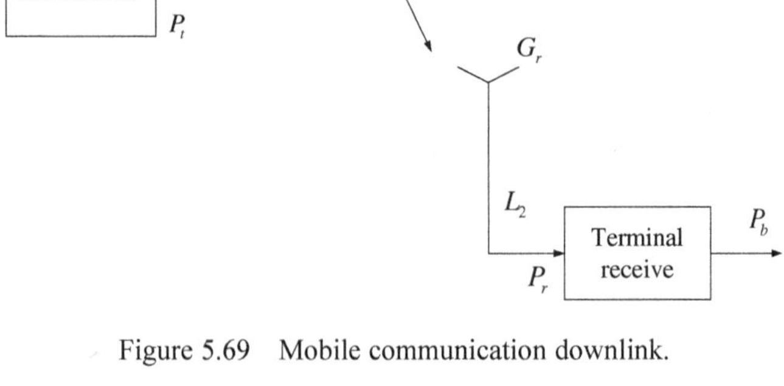

Figure 5.69 Mobile communication downlink.

- P_r *(dBm)* represents the reference sensitivity of the receiver (see Section 1.5 in Chapter 1 for the definition of reference sensitivity).
- *M (dB)* represents the system margin.

It can be seen from the figure that in order to meet the requirements of the receiver bit error rate P_b, the following equation must be established:

$$P_t(dBm) = P_r(dBm) + PL(dB) + L_1(dB) + L_2(dB) + M(dB) - G_t(dB) - G_r(dB) \tag{5.39}$$

Equation (5.39) is the communication equation of point-to-point communication. All parameters in this equation can be obtained. One that needs to be specifically discussed is the Path Loss (PL). There is a similar communication equation for the uplink.

5.8.2 *Estimation of path loss* [4]

5G has two types of path loss in urban areas, namely, Line of Sight (LOS) scenario path loss and Non-line of Sight (NLOS) scenario path loss, which are discussed separately in the following.

(i) $PL_{Los}(dB)$

- When $10 < d \leq d'_{BP}$ 时,

$$PL_{Los}(d)|_{dB} = 10n_1 \log\left(\frac{d}{1m}\right) + 28.0 + 20\log\left(\frac{f_c}{1GHz}\right) + PL_1|_{dB} \tag{5.40}$$

- When $d'_{BP} \leq d < 500m$ 时,

$$PL_{Los}(d)|_{dB} = 10n_2 \log\left(\frac{d}{d'_{BP}}\right) + PL_{Los}(d'_{BP})|_{dB} \tag{5.41}$$

In this equation, f_c is the carrier frequency and 1 GHz is the reference frequency, so the unit of f_c is GHz.

d is the distance between the transmitter and the receiver, 1m is the reference distance, so the unit of *d* is meter.

$n_1 = 2.2$
$n_2 = 4.0$

$PL_1|dB$ is the line-of-sight path loss deviation, and

$$PL_1 \mid dB = -1.38\log\left(\frac{f_c}{1GHz}\right) + 3.34 \tag{5.42}$$

$$d'_{BP} = a_{BP}\frac{4h'_{BS}h'_{MS}}{\lambda} \tag{5.43}$$

$$a_{BP} = 0.87\exp\left[-\frac{\log\left(\frac{f_c}{1GHz}\right)}{0.65}\right] \tag{5.44}$$

λ is the wavelength

$$h'_{BS} = h_{BS} - 1m \tag{5.45}$$

$$h'_{MS} = h_{MS} - 1m \tag{5.46}$$

h_{BS} and h_{MS} are the actual antenna heights of the base station and the terminal (both are greater than 1.5 meters)

$PL_{Los}(d'_{BP})|_{dB}$ is the value obtained by substituting d in (5.40) with d'_{BP}.

(ii) $PL_{NLOS}(dB)$

When $10m < d_2 < 1000\ m$ and $n_j = \max\left[2.8 - 0.0024\left(\frac{d_1}{1ms}\right),\ 1.84\right]$

$$PL_{NLos}\big|_{dB} = PL_{Los}\left(d_1\right)\big|_{dB} + 17.9 - 12.5n_j + 10n_j\log\left(\frac{d_2}{1m}\right) + 3\log\left(\frac{f_c}{1GHz}\right) + PL_2\big|_{dB} \tag{5.47}$$

where d_1 and d_2 are the distances (in meters) from street intersections (crossroads) to base stations and terminals, respectively. The base station and the terminal are located on two intersecting streets.

$PL_{loss}d_1|_{dB}$ is the value obtained by substituting d in (5.40) or (5.41) with d_1.

$PL_2|_{dB}$ is non-line-of-sight path loss deviation. The mean and variance of PL_2 are −9.1 dB and 6.1 dB, respectively, and under shadow fading they are 3.0 dB and 1.3 dB.

5.9 Application of High-Frequency Band in 5G

5.9.1 *The research and development of the high-frequency band is divided into several stages*

Currently, there are fewer and fewer low-frequency spectrums that can be allocated, and the required channel bandwidth is increasing. Due to these two reasons, it will be an inevitable trend for mobile communications to gradually develop toward high-frequency bands. For this reason, 3GPP has determined that the highest carrier frequency supported is 100 GHz at the beginning of the 5G NR research phase. But to achieve this goal requires a process. The process is divided into three stages.

(1) Two version definitions of R15 and R16

- 0.45–6 GHz is the low-frequency band.
- 24.25–52.6 GHz is a high-frequency band.
- 3GPP has standardized the radio access performance of these two frequency bands.

(2) It is expected that the R17 version will study and standardize the performance of the 52.6–71 GHz band. This confirms that the waveform transmitted in the wireless channel of this version is the same as the R15 and R16 versions. In addition to supporting authorized frequency bands, this version also needs to support related radio access technologies for unlicensed frequency bands.

(3) It is expected that the subsequent version of R17 will study and standardize the radio access performance of the 71–100 GHz frequency band.

5.9.2 *High-frequency band performance characteristics*

The radio access network has the following performance characteristics when using high-frequency bands:

(i) *Large bandwidth*: From Tables 5.8 and 5.9, we can see that the maximum bandwidth of a carrier frequency in the low-frequency band in the R15 version is only 100 MHz, but the maximum bandwidth of a carrier frequency in the high-frequency band is 400 MHz. It can be seen that the advantage of high-frequency band compared to low-frequency band is its wide bandwidth. For this reason, high-speed data transmission over the air is allowed.
(ii) *Small coverage*: This is caused by the following two reasons:

(1) From the discussion in Section 5.8.2, it can be seen that the path loss will increase as the operating frequency increases.
(2) From the previous discussion, it is known that the phase noise of RF components increases as the operating frequency increases (see Section 5.6.5.1), and the Doppler frequency shift also increases as the operating frequency increases (see Section 5.6.5.3). The influence of these two on the communication system synchronization is determined by subcarrier spacing Δf. When Δf is larger, the impact on synchronization system is smaller and vice versa. In order to reduce the influence of phase noise and Doppler frequency shift, when 5G NR works in the high-frequency band, a larger subcarrier spacing is selected, such as 120 KHz or 240 KHz. Since the IFFT operation time is $T = 1/\Delta f$ and the prefix of the OFDM waveform is $T_{cp} \propto T$, when Δf is larger, both T and T_{cp} are smaller. Small T_{cp} can only resist small multipath delay extension, which limits the increase in wireless coverage.

Due to the above two characteristics, a high-frequency band is suitable for working in environments that require large bandwidth and small coverage. The next section will discuss its specific application scenarios.

5.9.3 *High-frequency band use cases and deployment methods*

5.9.3.1 *Use cases*

(i) *High-rate eMBB use cases*: The main application of high-frequency band is in eMBB business, such as large stadiums, large conference venues, office buildings, shopping malls, supermarkets, restaurants and other relatively crowded places. The wireless coverage of these places is small, but it is required to support high-speed data.

(ii) *Mobile data offload use cases*: The rapid growth in the number of mobile users and the increase in the transmission data rate (such as ultra-high-definition video) have caused huge pressure on cellular networks. In order to relieve this pressure, D2D data offload technology can be used. This technology allows two mobile users to communicate directly after cooperation, instead of communicating through a base station. Obviously, this offload method will reduce the flow pressure of the base station, thereby reducing network congestion, system energy consumption and delay.

If the mobile users in D2D are close in geographic location, high-frequency bands can be used to achieve high-speed data point-to-point communication. For example, smart home connections, office supplies connections, VR, AR, wearable devices, and wireless USB all can take advantage of high-frequency sharp directivity (antenna beamforming) for short-distance high-speed connections. Besides, in the Internet of Vehicles scenario, short-distance V2V communications, vehicle environmental monitoring and road infrastructure equipment can also use high-frequency resources to communicate.

(iii) *Factory use cases in vertical industries*: Smart factories in vertical industries can also use high-frequency wireless connections, such as remote control, mobile control and robot remote control.

(iv) *LAN use cases*: The high-frequency band can be used in local area networks such as government networks, campus networks, shopping mall networks and home intelligent networks. For low-speed mobile or relatively stationary terminals, high-frequency beamforming technology can also be used to expand coverage and improve service quality.

(v) *Data return use case*:

- Currently, the data backhaul between the radio access network and the core network mainly uses optical fiber backhaul. But when there is no optical fiber or the cost of installing optical fiber is too high, high-frequency wireless backhaul can be used.
- When the base stations are in line of sight, the base stations can use high-frequency bands for data transmission.

(vi) *Smart medical use cases*: Some remote operations require massive transmission of real-time uncompressed images. This requires a high-speed, high-reliability and low-latency transmission system. At this time, a high-frequency band can be used to form such a system.

(vii) *VR/AR use cases*: In various applications of VR/AR services, end-to-end video transmission with high speed, high reliability and low latency is required throughout the entire process. At this time, high-frequency transmission can be used to support these requirements.

5.9.3.2 *Deployment method*

According to the characteristics of high-frequency band, it has the following two deployment methods:

- *Adopt high-frequency and low-frequency joint networking*: In the deployment method shown in Figure 5.45, the low-frequency band is used to cover the entire cell to ensure smooth signaling. The high-frequency band is used for data offloading, thereby increasing the data transmission rate and the cell capacity.
- *Independent deployment of high-frequency band*: In the case of small coverage and low-speed movement of terminals, the high-frequency band can be deployed independently.

5.9.4 *5G NR high-frequency band-related technologies*

The 5G NR design fully considers the characteristics of high-frequency bands and provides the following technologies for high-frequency band applications:

(i) *Supports multiple subcarriers spacing*: In 4G, the OFDM waveform has only one subcarrier spacing, that is, 15 KHz. But there are five subcarrier spacings in 5G, they are 15 KHz, 30 KHz, 60 KHz, 120 KHz and 240 KHz. Such large subcarrier spacing is arranged to reduce the effects of phase noise and Doppler shift in high-frequency bands. Currently, 120 KHz and 240 KHz are used in the high-frequency band.

Using large subcarrier spacing will also result in the following two problems:

(1) *Advantage: Reduces data transmission delay*: This problem needs to be explained.

OFDM is a two-dimensional waveform. Assume that it occupies a bandwidth of B KHz, and the occupied time is 1 ms. It can be proved that the OFDM waveforms occupy the same number of resource units (RE) (see Figure 5.20) in the physical resource grid when different Δf is used. The following uses Δf=15 KHz, 120 KHz and 240 KHz as examples to prove, and the process is shown in Table 5.18. It can be seen from the table that the above conclusion is correct. This means that the number of subcarriers in the frequency domain of the OFDM waveform is reduced by the same factor as the OFDM symbol in the time domain. Thus, under the same bandwidth condition, the number of time slots and the number of OFDM symbols per unit time increase as the subcarrier spacing increases. Therefore, compared to the smaller subcarrier spacing, when the same number of OFDM symbols is transmitted, the required transmission time is shorter.

Table 5.18 Proof process.

Subcarrier spacing Δf (KHz)	15	120	240
Number of subcarriers contained in bandwidth B	$\frac{B}{15}$	$\frac{B}{120}=\frac{B}{15\times 8}$	$\frac{B}{240}=\frac{B}{15\times 16}$
Time slot length (ms)	1	$\frac{1}{8}$	$\frac{1}{16}$
Number of time slots in 1 ms	1	8	16
OFDM number contained within 1 (ms[1])	1 × 14	8 × 14	16 × 14
Number of resource unit([2])	$\frac{B}{15}\times 1\times 14$	$\frac{B}{120}\times 8\times 14=\frac{B}{15}\times 1\times 14$	$\frac{B}{240}\times 16\times 14=\frac{B}{15}\times 1\times 14$

Note: (1) Number of time slots × 14.
(2) Number of subcarriers × number of time slots × 14.

(2) *Disadvantage*: *Poor multipath resistance and limited coverage*: This is because when the subcarrier spacing increases, the cyclic prefix T_{cp} of the OFDM symbol is also shortened.

(ii) *Terminal access design*: It is known from Section 5.6.14.1 that as the carrier frequency increases, the synchronization signal block transmitted within a half-frame of 5 ms also increases, that is, the number of beams scanned increases. Compared with low-frequency band, the increase in the number of scanning beams in high-frequency band is caused by the following two reasons:

(a) The higher the frequency, the smaller the half-power angle of the antenna beam, and the sharper the antenna directivity.
(b) The areas where the high-frequency band works are all areas with high user density.

It can be seen that in the above two cases, in order to enable the terminal to access the network more effectively, the number of beam scanning must be increased.

In the high-frequency band, the subcarrier spacing is 120 KHz or 240 KHz. 64 SSBs need to be transmitted within 5 ms (see 5.67 for the SSB structure). Figures 5.70 and 5.71 show the time-domain positions of 64 SSBs within 5 ms, respectively. The two figures are explained as follows:

(1) Figure 5.70
- 5 ms is divided into four groups. Each group contains five time slot pairs. The first four time slot pairs in each group transmit SSB, while no SSB is transmitted in the remaining time slot pairs.
- Each slot pair transmits four SSBs (each SSB contains four OFDM symbols). In this way, a total of 16 SSBs are transmitted in each group. A total of 64 SSBs are transmitted in four groups.

(2) Figure 5.71
- 5 ms is divided into four groups. Each group contains five time slots. The first four time slot pairs in the first two groups transmit SSB, and the other timeslots do not transmit SSB.
- Each slot pair transmits eight SSBs (each SSB contains four OFDM symbols). Then one group transmits 32 SSBs in total, and two groups transmit 64 SSBs.

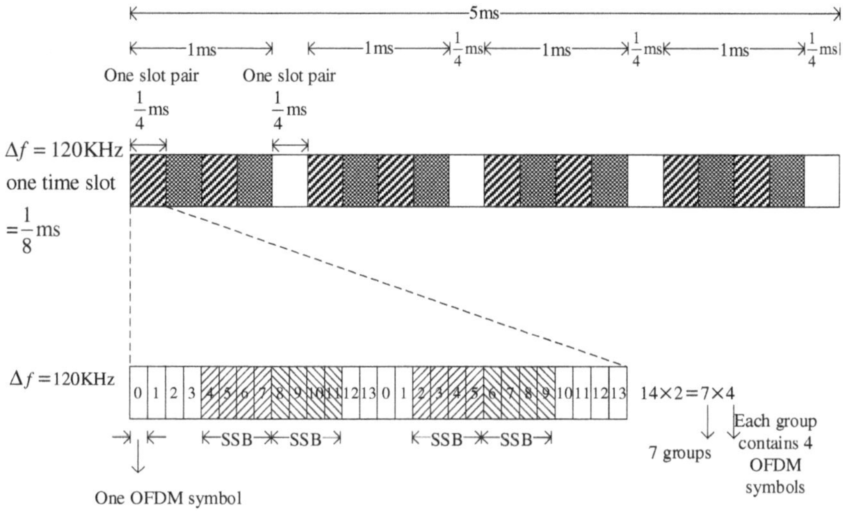

Figure 5.70 Time-domain position of 64 SSBs when spacing is 120 KHz.

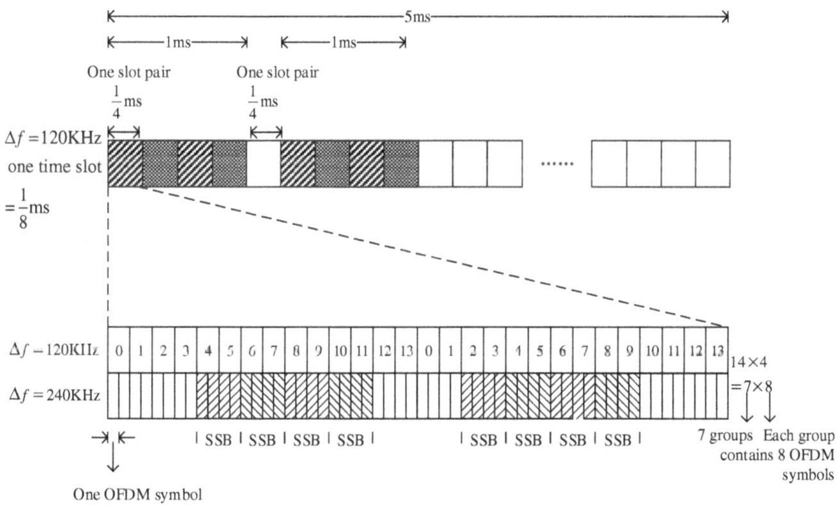

Figure 5.71 Time-domain position of 64 SSBs when spacing is 240 KHz.

Another problem involved in terminal access design is that in order to obtain the downlink carrier frequency, the terminal must perform a frequency-domain search. In the frequency range of 3–24.25 GHz, the frequency search grid is 1440 KHz. In the high-frequency band, the frequency range is 24.25–100 GHz, but the frequency range that needs to be searched becomes wider, so in order to shorten the search time, the frequency search grid must be increased. Its value is 17.28 MHz, which is 12 times that of 1440 KHz.

(iii) *Large-scale antenna design*: Large-scale antenna design is a key technology in high-frequency bands. The following three technologies are supported in base station design:

(1) *Antenna beamforming*:
As the operating frequency increases, the size of the antenna shrinks, and the number of antennas contained in one unit also increases. In this way, multiple antennas can be used for beamforming, so that the antenna has a certain gain, which can compensate the path loss of the high-frequency band to a certain extent.

(2) *Integrated design of terminal access, signaling and data transmission*:
Such integrated design means that access, signaling and data transmission are all performed by narrow beams after beamforming, which is called same coverage of all three. If access and signaling use wide beams for wide coverage, while data is transmitted with narrow beams, it is obviously narrow coverage. The coverage distance of wide coverage is small, while narrow coverage distance is large. The difference in coverage distance between these two greatly limits the application of large-scale antennas in high-frequency bands. When the same coverage of all three is adopted (all use narrow beams), there is no difference in coverage distance.

The method to achieve the same coverage of all three is as follows.

The SSB blocks in Figures 5.70 and 5.71 are transmitted in a group of four OFDM symbols. The direction of the narrow beam transmitted by each group is different, as shown in Figure 5.68. It can be seen that the base station can cover the entire cell with this beam scanning method. It can generally be considered that the geographic location of the terminal when receiving SSB block is basically the same as that when receiving signaling and data, which is called Quasi-Colocation (QCL). In fact, since the terminal is mobile, its geographic location may be different when it receives these three signals, so the colocation of the three is similar, which is the reason that it is called quasi-colocation. When the terminal accesses, it measures the different SSB transmitted by the base station. When the terminal selects a certain SSB, it reports the measurement result of SSB to the base station. Due to the quasi-colocation, the base station can use the reported information to determine the direction of the narrow beam and use this beam to transmit signaling and data. In this way, the requirements of the same coverage of all three are achieved.

By the way, the time required to transmit 64 SSB in Figure 5.71 is shorter than that in Figure 5.70. This is because the former uses 240 KHz subcarrier spacing, which is twice the latter (120 KHz). This shows that its coverage is small and the required scanning time is short.

It should be emphasized that the exact QCL refers to the large-scale parameters of the channel experienced by the OFDM waveform on a certain antenna port (see the following discussion), which can be inferred from the channel experienced by the OFDM waveform on another antenna port. The large-scale parameters can include delay spread, average delay, Doppler spread, Doppler shift, average gain and spatial reception parameters. The so-called two ports are QCL of large-scale parameters, which means that these parameters of the two ports are the same. In other words, as long as some large-scale parameters of the two ports are the same, regardless of whether there is a difference in their actual physical positions or corresponding antenna orientations, the terminal considers that the two ports are transmitted from the same location, that is, quasi co-site.

(3) *Using large-scale antennas, base stations can be designed (virtually) into multiple antenna ports*:
NR supports panel-based large-scale antenna design (see Figure 5.56). As the operating frequency increases, the number of antennas on the panel also increases, which can reach hundreds of antennas at tens of GHz. In NR design, these physical antennas can be arranged in different ways, and virtualized as multiple antenna ports (see Figure 1.13 in Chapter 1 for wireless ports).

(iv) *Bandwidth configuration*: NR supports the following bandwidth configurations:

(a) The maximum bandwidth of sub-6 GHz single carrier is 100 MHz.
(b) The maximum bandwidth of 24.25–52.6 GHz single carrier is 400 MHz.

For the use of single carrier with large bandwidth, considering the power consumption, NR designs a partial bandwidth BWP (see Section

5.6.6.6) technology. That is, within a large carrier bandwidth, multiple BWPs can be configured. According to different services, different BWPs can be activated for data transmission. After the data transmission is over, it will fall back to the default BWP. This can achieve energy saving of the system.

NR supports carrier aggregation and dual connectivity, which can further increase the bandwidth.

5.10 5G Millimeter Wave Technology [21]

The use of millimeter wave technology is a highlight in 5G. Some related technologies have been discussed in the previous sections, and we will gather them together for a further in-depth discussion.

5.10.1 *Introduction to 5G millimeter wave technology*

5.10.1.1 *5G millimeter wave is the inevitable direction of communication technology evolution*

With the continuous increase of business demand for bandwidth, the communication spectrum continues to extend to higher frequency harmonics. 5G millimeter wave has abundant frequency resources, which is the inevitable direction of the evolution of mobile communication technology. The test results of Ookla® SPEEDTESTS® show that the average download rate of existing networks in the 5G millimeter wave band is 4 times that of the sub-6 GHz band (6 GHz and below), and about 20 times that of LTE. In 2020, 5G has already begun large-scale commercial use. The entire industry has focused on the key technologies for deploying the next phase of 5G. Among them, millimeter wave can fully release the full potential of 5G due to its high bandwidth, low latency and other outstanding advantages so as to realize the revolutionary improvement of experience and the digital transformation of thousands of industries and truly realize the vision of "4G changes life and 5G changes society". Therefore, it has received extensive attention from the industry. 5G millimeter wave and medium and low sub-6 GHz wave have their own different advantages. The mutual cooperation and complement between them are the key to achieving a complete and optimal user experience of 5G.

5.10.1.2 *The globally coordinated allocation of 5G millimeter wave spectrum is steadily advancing*

In 2019, the World Radiocommunication Conference (WRC-19) of the International Telecommunication Union (ITU) determined that the millimeter wave frequency band between 24 GHz and 86 GHz will be used for International Mobile Telecommunications (IMT). Among them, the 24.25–27.5 GHz, 37–43.5 GHz and 66–71 GHz frequency bands are globally integrated IMT frequency bands. This marks a solid step taken by the global industry toward the best performance and scale effect of 5G millimeter waves. At present, some countries and regions around the world have successively completed the division and allocation or auction of 5G millimeter wave spectrum. Among them, the United States completed the auction of the 28 GHz and 24 GHz frequency bands in January and May 2019, respectively, and further completed the auction of 37 GHz, 39 GHz and 47 GHz frequency bands in March 2020. The European Commission has passed an implementation decision in May 2019 to unify the radio spectrum of the 26 GHz band, to enable member states to set common technical conditions for the use of frequency bands and open them to use. At present, Italy, Finland and Norway have completed the allocation or auction of part of the spectrum. In Asia, Japan, South Korea, Thailand, Hong Kong and Taiwan have completed the allocation or auction of some of the 26 GHz and 28 GHz spectrum. As of June 2020, 79 operators in 17 countries and regions already have frequency licenses to deploy 5G millimeter waves at 24.25–29.56 GHz.

5.10.1.3 *5G millimeter wave has begun commercial deployment*

As the global deployment of 5G networks advances in depth, commercial deployment of 5G millimeter waves is also gradually unfolding around the world. Several major mainstream communications operators in the United States, including AT&T, T-Mobile and Verizon, have all provided 5G millimeter wave commercial services. Many operators, including NTT DOCOMO of Japan, KDDI of Japan, and SHT of South Korea, have also begun commercial deployment of 5G millimeter wave systems. GSA stated that in June 2020, 123 operators in 42 countries/regions around the world have made 5G investments in the form of trials, licenses,

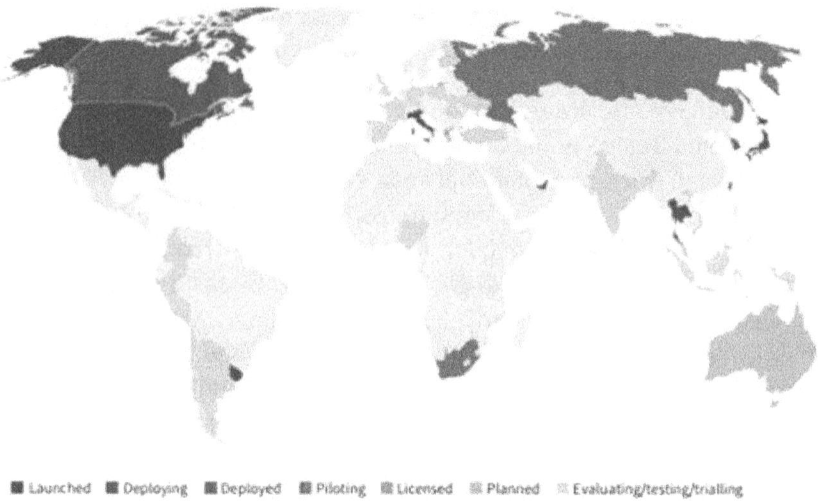

Figure 5.72 Global operators' investment in 5G millimeter wave 24.25–29.5 GHz.

deployments or operating networks between the 24.25 and 29.5 GHz spectrum. The specific distribution is shown in Figure 5.72.

5.10.1.4 *5G millimeter wave is strongly supported by policies in China*

In July 2017, China approved the use of 5G technology in the 5G millimeter wave frequency range of 24.75–27.5 GHz and 37–42.5 GHz to carry out research and development tests. China's Ministry of Industry and Information Technology clearly stated in the "Notice on Promoting the Accelerated Development of 5G" in March 2020 that it would organize millimeter wave equipment and performance tests in conjunction with the national frequency planning schedule, which makes reservations for the commercial use of 5G millimeter wave technology and release the frequency usage plan of 5G millimeter wave. Since 2019, China's IMT2020 (5G) Promotion Group has made overall plans to advance the 5G millimeter wave test work in three stages. In 2019, the focus will be on the verification of key technologies and system characteristics of 5G millimeter wave. In 2020, the focus will be on the verification of functions, performance and interoperability of 5G millimeter wave base stations and terminals. From 2020 to 2021, application verification in typical scenarios will be carried out. Ericsson and One+ have completed the end-to-end test of 5G millimeter wave commercial systems and commercial smartphones. It exhibits stable and excellent performance in various indoor and outdoor environments. Nokia conducts chip-based terminal tests and successfully demonstrates that 5G millimeter wave has 4 Gbps peak performance. In the remote test at 1200 meters, and under test scenarios such as non-line-of-sight scenarios and human occlusion, the downlink rate can still reach hundreds of megabytes to 2 Gbps. OPPO has joined hands with Ericsson to achieve end-to-end testing of 5G millimeter wave commercial systems and commercial CPE, which achieves a downstream rate of 4.06 Gbps and an upstream rate of 210 Mbps. In the remote test, the downlink rate of 200 Mbps is still maintained at 2.3 kilometers. It is foreseeable that 5G millimeter wave will serve the 2022 Beijing Winter Olympics.

5.10.1.5 *Progress of 5G millimeter wave standards*

5G millimeter wave is one of the main components of 5G technology. The standard progress of 5G millimeter wave is an important guarantee for 5G performance indicators. Its standardization work is carried out synchronously with the mid- and low-frequency. At present, 3GPP has completed two versions of the 5G standard, Rel-15 and Rel-16. Both support working in the millimeter wave frequency band. Figure 5.73 is the 3GPP standard timetable.

3GPP New Radio (NR) Rel-15 has completed the first version of the 5G standard including 5G millimeter waves, including two specifications of non-independent networking (NSA) and independent networking (SA). It can provide multiple services for multiple scenarios, including eMBB and URLLC services. Among them, the specifications and ASN.1 corresponding to the 5GNSA mode (Option 3 architecture) were introduced in December 2017 and March 2018, respectively. The specifications and ASN1 corresponding to the 5G SA mode (Option 2/5 architecture) were introduced in June and September 2018, respectively. The delayed version (Option 4 architecture) was introduced in March and June 2019, respectively.

3 GPP NR Rel-16 completed the second version of the 5G standard including 5G millimeter wave. This standard is compatible with NR Rel-15 and focuses on enhancing vertical industry applications and improving overall system performance. NR Rel-16 mainly enhances the

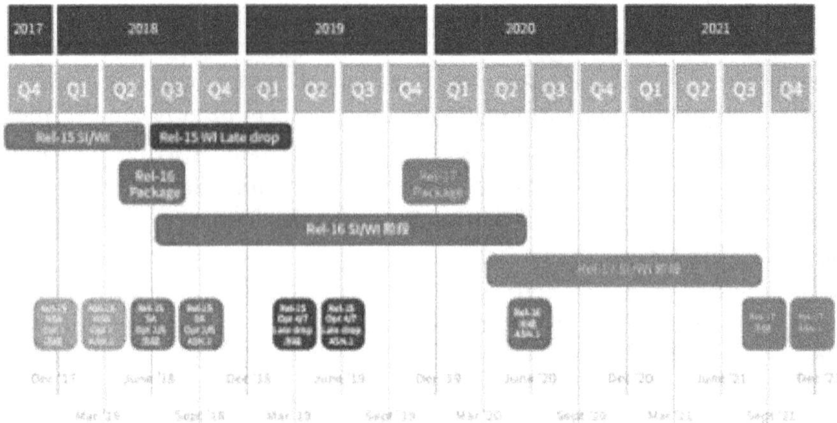

Figure 5.73 3GPP standard timetable.

carrier aggregation function, optimizes beam management function, adds multi-TRP (Transmission and Reception Point) transmission mode and integrated access and backhaul (IAB) functions and introduces the positioning function for the NR network. The NR Rel-16 specification and ASN.1 were introduced in June 2020.

3 GPP NR Rel-17 is the third version of the 5G standard that includes 5G millimeter waves. It will further enhance the 5G millimeter wave function. The upper limit of the 5G millimeter wave frequency band defined in NR Rel-15 is 52.6 GHz (frequency range 2). NR Rel-17 will consider higher frequency bands, that is, research on frequency bands above 52.6 GHz, with an emphasis on 52.6 GHz (frequency range 2). In addition, NR Rel-17 is actively expanding to support future manufacturing and industrial control to make enhancements to IoT and URLLC technologies. Such enhancements include the use of beam scanning technology to improve transmission reliability and the realization of time-sensitive communication between terminals in a wider area. This will enhance MIMO technology, especially enhance the uplink multi-TRP transmission scheme, improve the mobility management function of the low-layer air interface, and provide a more concise beam management mechanism. IAB will also be enhanced to support space division multiplexing of access and backhaul links to improve spectrum utilization and to support higher precision positioning. The corresponding specifications are expected to be introduced at the end of 2021 (*Note*: there is a possibility of extension).

5.10.2 *Advantages of 5G millimeter wave technology* (*Table 5.19*)

Compared with the low- and medium-frequency bands, 5G millimeter wave has many outstanding advantages. These advantages make 5G millimeter wave attract the attention of the industry, so much that it is regarded as the key enabling technology for the revolutionary improvement of user experience and the digital transformation of various industries.

5.10.2.1 *Rich frequency resources and huge bandwidth*

The first, and also the most important, advantage of 5G millimeter wave technology is that it has abundant frequency resources and huge bandwidth. The 5G millimeter wave is valued because it has more abundant spectrum resources than the 5G sub-6 GHz frequency band (FR1) (as shown in Figure 5.74). It is the main way for 5G networks to provide gigabit connectivity, and it is also a strong guarantee for the realization of the original vision of 5G. It is no exaggeration to say that in order to achieve the highest rate of 5G, 5G millimeter waves must be used.

5G millimeter wave networks can easily achieve Gbps-level peak throughput rates. For example, in the 26 GHz frequency band, a continuous 800 MHz spectrum is allocated, and four single-carrier 200 MHz (4×200 MHz) or eight single-carrier 100 MHz (8×100 MHz) are used to implement carrier aggregation transmission. Based on the available channel

Table 5.19 Six advantages of 5G millimeter wave technology.

Advantages	Benefits
Rich spectrum and bandwidth	High peak rate and average rate (Gbps), large capacity
Combine with beamforming	Provides directivity, improves signal strength, enhances performance and reduces interference
Extremely low latency (sub-millisecond)	Industrial IoT, AR/VR, cloud games, real-time computing
Dense cell deployment	Large venues, transportation hubs
High-precision positioning	Industrial Internet, logistics transportation, V2V and indoor rapid high-precision positioning
High equipment integration	Miniaturization and micromation of BS and terminal

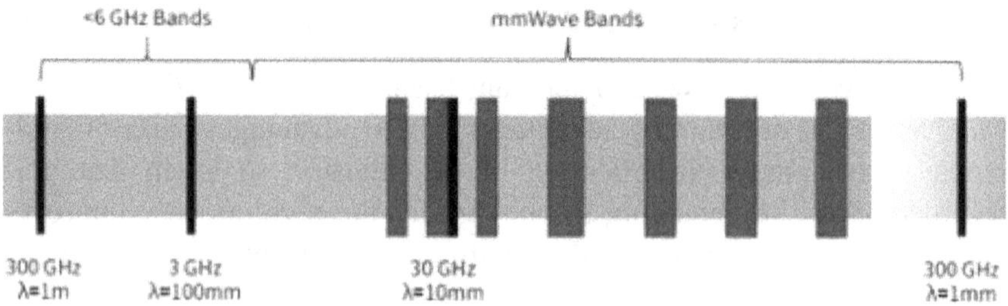

Figure 5.74 Millimeter wave spectrum resources.

width and modulation mode of the 3GPP standard, combined with advanced antenna design and radio frequency processing technology, the 5G millimeter wave system can easily obtain a peak data throughput rate of more than several Gbps. At present, 5G millimeter waves are already commercially available in some countries and regions. Based on the latest measured results of Ookla® SPEEDTESTS®, the average download rate of the 5G sub-6 GHz network is five times faster than that of 4G LTE. The average download rate of 5G millimeter wave networks is four times faster than that of 5G sub-6 GHz. Its average rate is as high as 900 Mbps, and its peak rate exceeds 2 Gbps. This result is achieved in a 400 MHz bandwidth. The future use of 800 MHz bandwidth is expected to get better results. Therefore, compared to 5G sub-6 GHz, 5G millimeter wave technology still has a strong increase in the peak rate of users.

5.10.2.2 *Easy to combine beamforming technology*

The second advantage of 5G millimeter wave technology is that it is easy to combine with beamforming technology. The high-frequency band and short wavelength of 5G millimeter wave give it a space advantage in design and deployment. It is very suitable for combining with beamforming to enhance performance and reduce interference. Beamforming can still provide good signal quality even when the power of a single antenna is very low. Due to the shorter wavelength, the antenna array of the 5G millimeter wave device can place more antenna elements in a limited space. In particular, the number of antenna elements of 5G millimeter wave base stations can reach 256, 512 or even more. Therefore, regardless of the uplink and downlink, the system can obtain a great beamforming gain. In a typical antenna array configuration, it is assumed that the

Table 5.20 Typical beamforming gain of 5G millimeter wave frequency band.

Antenna	Antenna elements	Theoretical BF gain (dB)
BS	256	24 dB
Terminal	8	9 dB

base station has 256 antenna elements and the terminal has eight antennas. At this time, the theoretical beamforming gains that 5G millimeter waves obtained are shown in Table 5.20. Benefiting from beamforming, 5G millimeter wave signal communication distance is greatly enhanced, and it can be used in more scenarios beyond short-distance transmission and indoor hotspot coverage. For example, when 5G millimeter wave base stations are used for outdoor coverage, the coverage radius can reach hundreds of meters or several kilometers. In many cases, it can co-site with existing 4G and 5G low- and medium-frequency small base stations. Figure 5.75 compares and analyzes the downlink coverage when 5G millimeter wave base stations and 4G LTE base stations (working in Band2 and Band4) are deployed together. According to the results in a 0.8 square kilometer area of a large city, the figure compares the deployment of two millimeter wave base stations at 28 GHz and 39 GHz. The baseline deployment density of 4G LTE base stations is 73 stations per square kilometer. It can be seen from the figure that in outdoor coverage, if 5G millimeter wave base stations and 4G LTE base stations are completely co-located and the number is equal, the 5G millimeter wave network downlink coverage rate can reach 77%. If the number of 5G millimeter wave base stations is slightly increased, the downlink coverage rate of the 5G millimeter wave network can reach 95%.

5.10.2.3 *Very low latency can be achieved*

The third advantage of 5G millimeter wave technology is that it can achieve extremely low latency. Generally speaking, 5G networks schedule data in units of time slot pairs. The shorter the time slot length of the air interface, the smaller the delay of the 5G network at the physical layer. As shown in Table 5.21, 5G millimeter wave system air interface time slot length can be as small as 0.125 ms. It is 1/4 of the current mainstream 5G

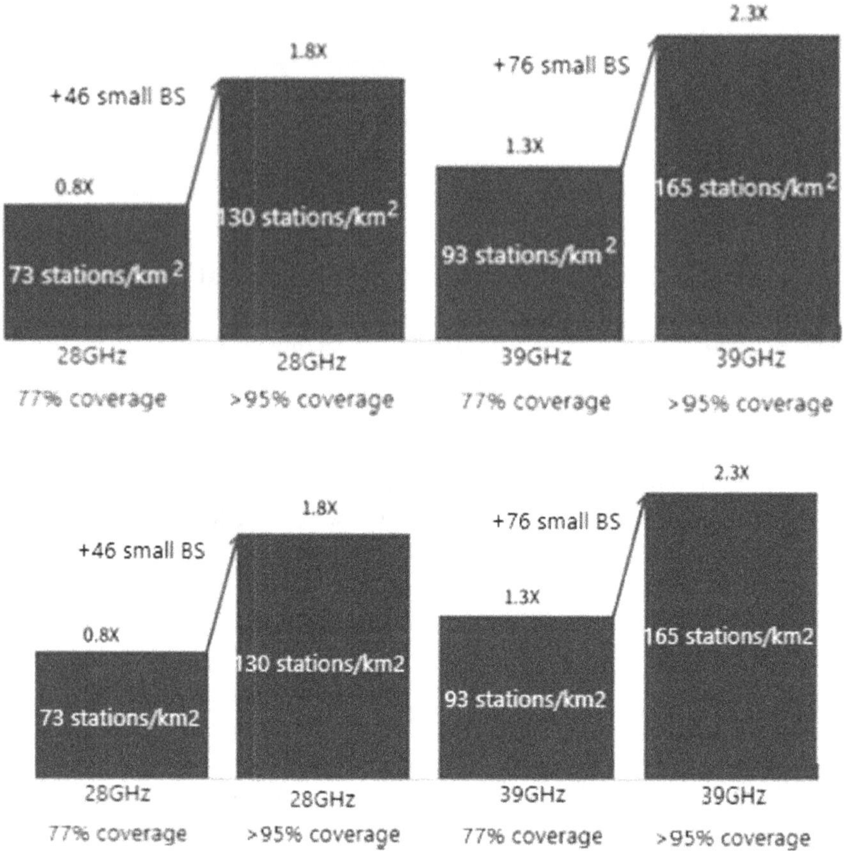

Figure 5.75 Comparison of coverage effects between 5G millimeter wave base stations and 4G LTE base stations.

low-frequency system. If mini-slot scheduling is used, the air interface delay will be smaller. Therefore, the 5G millimeter wave system will significantly reduce the air interface delay compared to the 5G medium- and low-frequency system. This is a strong guarantee that the 5G air interface delay is less than 1 ms. It can ensure the quality of 5G networks for URLLC (high reliability and low latency) services such as industrial Internet, AR/VR, and cloud game real-time cloud computing. For example, AR/VR services require only milliseconds to ensure multi-sensory coordinated experience and interactive capabilities. The Industrial Internet has a very clear demand for low latency. Typical industrial robot networks

Table 5.21 Time slot interval for different frequency bands of 5G.

Frequency band (GHz)	Subcarrier spacing (KHz)	Slot length (ms)
1 GHz	15/30 KHz	1/0.5 ms
1–6 GHz	15/30/60 KHz	1/0.5/0.25 ms
24.25–52.6 GHz	60/120 KHz	0.25/0.125 ms

require millisecond latency. Remote real-time control on the product line also requires millisecond latency guarantees, and industrial vision and other fields which introduce artificial intelligence often require large-scale calculations performed at a certain distance, which also places higher requirements on the air interface delay.

5.10.2.4 *Can support dense cell deployment*

The fourth advantage of 5G millimeter wave technology is that it can support dense cell deployment. Different from 5G medium- and low-frequency systems, through beamforming, 5G millimeter wave systems can not only increase the signal gain of the target object but also focus the signal energy in a specific direction to reduce interference to other non-target objects and ensure the communication quality of proximity link or neighboring cell. Therefore, compared to 5G low-frequency system, with 5G millimeter wave system it is easier to implement dense cell deployment.

This makes 5G millimeter wave systems very suitable for deployment in densely populated areas such as conference rooms, concerts, stadiums and subway stations. For example, AR based on 5G millimeter wave can provide a virtual on-site experience for training and meetings in remote office scenarios. Or when watching football matches in stadium, we can enjoy the game of our favorite sports star through virtual and enhanced methods, and it is the effect of superimposing real-time action images and video clips.

5.10.2.5 *High-precision positioning is possible*

The fifth advantage of 5G millimeter wave technology is that it can perform high-precision positioning. The 5G millimeter wave has a narrow beam, good directivity and extremely high spatial resolution. At the same time, due to the small signal transmission cycle and high time accuracy, 5G millimeter waves are expected to achieve centimeter-level positioning. Compared with GPS, it has advantages in accuracy and speed. Especially in the indoor environment with weak satellite navigation signals, the positioning capability of 5G millimeter wave will play a more important role. It can provide fast and high-precision positioning services in application scenarios such as industrial Internet, logistics and transportation, Internet of Vehicles, transportation hubs, large venues and parks. There are many scenarios in the Industrial Internet that require high-precision positioning, such as the automatic transportation of materials, industrial robots to perform precision mechanical operations such as riveting, welding, assembly, cutting and automatic product detection and packaging.

5.10.2.6 *High device integration*

The sixth advantage of 5G millimeter wave technology is the high level of equipment integration. Compared with sub-6 GHz, the size of 5G millimeter wave components is much smaller. 5G millimeter wave devices are easier to miniaturize. When the 5G millimeter wave is commercialized on a large scale, the cost of related components will be greatly reduced. In the fields of professional equipment, wearable devices and smart components, 5G millimeter wave has the advantage of high integration, which leads to broad application prospects. In addition, 5G millimeter wave base stations also have the advantages of small size, lightweight and easy installation, which can benefit to create a green, efficient and convenient 5G network.

5.10.3 *Challenges and solutions faced by 5G millimeter technology*

5.10.3.1 *5G millimeter wave coverage optimization*

The 5G millimeter wave has high-frequency band, high propagation loss, weak diffraction and inflection capabilities, and relatively limited coverage. This is the biggest challenge for 5G millimeter wave communication systems. The path loss in the propagation process of high-frequency communication is relatively large. The penetration loss from outdoor to indoor is greater, and the impact of buildings, leaves or rainwater is greater.

According to the direct path loss model of 0–100 GHz radio waves in urban areas specified in 3GPP TR38901, it is known that the free space loss is positively correlated with the carrier frequency. The 26 GHz carrier is about 17.42 dB higher than the 3.5 GHz carrier, and the theoretical propagation distance is only about one-sixth of 3.5 GHz. According to the actual measurement results of China Unicom, the penetration loss of 5G millimeter wave is much higher than that of sub-6 GHz (as shown in Table 5.22). At the same time, bad weather such as rain, snow, and fog, also have an adverse effect on the propagation of millimeter waves.

This feature of 5G millimeter wave makes us mistakenly believe that it can only achieve line-of-sight transmission. However, there are many solutions to solve the problem of signal attenuation and blocking. Firstly, 5G millimeter waves increase EIRP (equivalent omnidirectional radiated power) through beamforming technology to improve coverage. It can easily realize the signal transmission of hundreds of meters and alleviate the path loss problem. This technology is not only verified through simulation experiments but also fully verified in field tests and commercial deployments. Secondly, in 5G standardization, 5G millimeter wave beam management has become the focus of 5G millimeter wave standardization. These include beam search, beam tracking and beam switching. The 5G millimeter wave system can quickly capture new beams and dynamically implement beam switching when part of the directional signal is blocked. Finally, the progress of semiconductor technology has promoted the rapid development of 5G millimeter wave technology, which integrates large-scale array antennas and radio frequency links into more cost-effective phased array Radio Frequency Circuits (RFCs), successfully implements smart beamforming, beam search and beam tracking technologies, and provides strong support for the 5G millimeter wave system from the hardware.

These technological advances enable not only 5G millimeter waves to make up for the lack of propagation characteristics through the advantages of beamforming but also enable 5G millimeter wave systems to make

Table 5.22 5G millimeter wave penetration loss test results.

Tree (4 m)	Human body (unilateral/peripheral)	Concrete bearing wall	Wooden door (5 cm)	Glass door	Caravan body
20 dB	11–28 dB	Impenetrable	6 dB	5 dB	17–23 dB

good use of multipath and reflection. Through advanced beam management, 5G millimeter wave system can freely switch between different connections and paths. If the line-of-sight transmission path is blocked, the system can still quickly find and identify non-line-of-sight transmission paths (such as reflection paths) through intelligent beam search and quickly switch to the corresponding transmission path, so as to solve the problem of line-of-sight transmission of 5G millimeter wave signals. In Rel-16, 3GPP also supports the use of IAB (Integrated Access and Backhaul) technology to enhance 5G millimeter wave network coverage. The use of 5G millimeter wave ultra-large bandwidth to provide backhaul links helps to install 5G antennas where it is difficult to deploy optical fibers or where the cost of deploying optical fibers is too high. For example, in high-rise buildings, isolated islands or mountainous where the laying of optical fibers is problematic, the deployment cost can be reduced and the deployment progress can be accelerated to achieve seamless 5G network coverage. When various natural disasters cause traditional backhaul links to fail, IAB can also play an important role. As shown in Figure 5.76, on the one hand, IAB technology can expand the coverage of 5G millimeter wave networks through wireless backhaul. On the other hand, IAB technology can increase capacity by sharing spectrum resources of access and backhaul links. Moreover, IAB supports the functions of multi-hop connection and network topology adaptation. This allows extended 5G millimeter wave network coverage, flexible inter-station connections, and fewer network topology restrictions. This also supports OTA synchronization and node discovery functions, making the synchronization between 5G millimeter wave base stations easier, more intelligent interface, and more convenient deployment. In addition, it also aids in the coordination of access link and backhaul link, realizes the

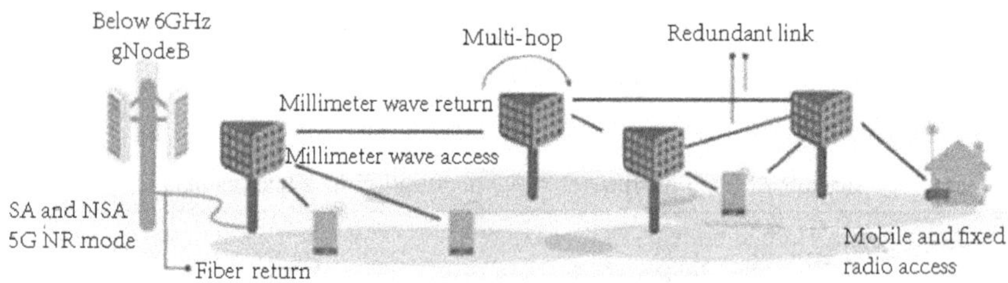

Figure 5.76 IAB can not only expand 5G millimeter wave coverage but also increase capacity.

integration of physical layer resource allocation, beam management, and interference management, fully integrates different link resources and improves overall efficiency. This also improves MAC layer mobility management, fast link recovery, QoS management, load balancing and other aspects to enhance the 5G network which improves the network robustness, so that the 5G millimeter wave network can quickly respond to network failures, dynamically adjust and restore.

Small cell technology is another solution to the 5G millimeter wave coverage problem. In 5G, it is difficult for a single base station to meet all communication requirements and deployment scenarios. Compared with 5G medium- and low-frequency macro stations, 5G millimeter wave small cell base stations have relatively smaller coverage radius and greater deployment density, which guarantees high peak throughput by reducing the communication distance and fully guarantees the coverage effect by increasing the deployment density. The 5G millimeter wave small cell base station has a wide range of deployment scenarios, which can be deployed indoors and outdoors. In order to match different deployment scenarios and usage requirements, 5G millimeter wave small cell base stations can present different 5G network architectures ranging from distributed to centralized. The most typical deployment scenarios for 5G millimeter wave small cell base stations are various hotspots such as conference rooms, large stadiums, concert halls and various transportation hubs such as airports, railway stations, and subway stations. On the one hand, these scenarios usually have a higher demand for connections and data load. If we only deploy 5G mid- and low-frequency macro stations, it is easy to cause base station traffic overload and other problems such as user access failures, which results in a decline in user experience. The high-capacity characteristics of 5G millimeter wave small cell base stations can effectively shunt the load of macro base stations, increase the probability of user access success and ensure user connection stability and consistency of experience. On the other hand, in these scenarios, the high-density deployment of 5G millimeter wave small cell base stations can usually be used to avoid signal penetrating wall losses. There is no need to worry about the influence of rain and blocking. Due to the short coverage distance, the signal quality of the 5G millimeter wave small cell base station can be guaranteed, and the extremely high communication rate can be maintained stably. In addition, compared to macro base stations, 5G millimeter wave small cell base stations can easily achieve low power consumption, low cost and lightweight. At the same time, it is compatible and open and controllable. Therefore, 5G millimeter wave small cell base stations can simplify the deployment of high-density networks, realize plug-and-use, intelligent networking and flexible ultra-dense networks, which reduces the installation requirements of base stations and alleviating the problem of macro base station site selection.

5.10.3.2 *5G millimeter wave mobility management*

The second challenge of 5G millimeter waves is mobility management. Due to the characteristics of high-frequency signal propagation, the coverage radius of 5G millimeter wave cells is usually small. In the mobile state, due to frequent cell switching, data transmit interruption can easily occur in the terminal. To solve this problem, the 3GPP standard provides two solutions to the 5G millimeter wave mobility management in order to ensure a seamless user experience. The first is a variety of flexible and fast cell handover mechanisms. It can be a handover within a base station or a handover between base stations. It can be a handover based on high-level signaling or a handover based on low-level instructions. This fully meets the cell handover requirements of different scenarios. The second is a fast beam recovery mechanism. Once a beam failure occurs, the 5G millimeter wave system can achieve rapid beam recovery in the order of milliseconds without the intervention of the core network and high-level signaling.

In addition, CA and dual connection can be organically combined with low-frequency and 5G millimeter waves. As shown in Figure 5.77, telecom operators can adopt the 5G deployment strategy of "5G Sub-6GHz/4GLTE + millimeter wave". CA or DC technology can be used to take full advantage of the respective advantages of low-, medium- and high-frequency bands. 5G sub-6 GHz or 4G LTE network can be utilized to ensure network coverage and be responsible for functions such as channel acquisition, paging and mobility management. 5G millimeter wave can be utilized to provide multiple Gbps rate and super capacity. It is particularly worth mentioning that in the 3GPP standard, 5G millimeter wave and sub-6 GHz have almost the same air interface protocol (such as a common MAC layer). Therefore, 5G millimeter wave and sub-6 GHz networks can be tightly integrated, and CA can be conducted easily.

In Rel-16, 3GPP has also improved the multi-TRP function. For high-frequency mobile phones that support multi-TRP, they can support

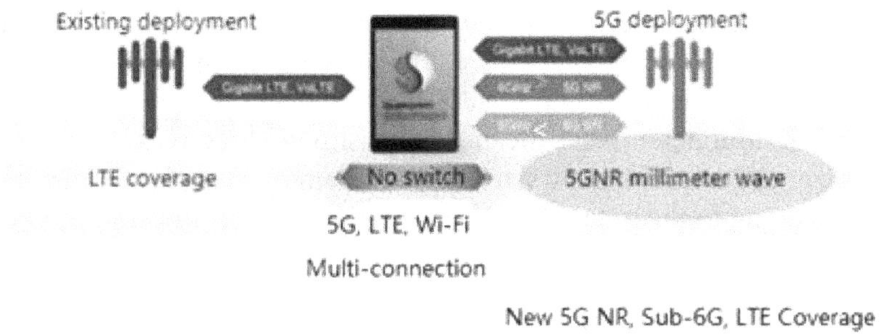

Figure 5.77 The organic combination of 5G millimeter wave network, sub-6 GHz and 4G LTE network that provides the ultimate user experience.

the reception of two beams at the same time. In this way, when the transmission link of one base station is blocked or fast fading, the other transmission link is still unblocked, which can effectively avoid the poor diffraction effect of millimeter waves.

5.10.3.3 *5G millimeter wave product realization*

The third challenge of 5G millimeter waves is high complexity and implementation difficulty. 5G millimeter wave base stations face at least three challenges. First of all, 5G millimeter waves need to install more antennas. The antenna calibration problem becomes more complicated. In order to achieve the precise beamforming necessary for 5G communications, 5G millimeter wave base stations need to be able to accurately control the phase information of each antenna. Limited by the technological level, the actual antenna array may have a large initial phase calibration error. This causes beamforming distortion, which seriously affects the performance of communication. Therefore, 5G millimeter waves need to focus on breaking through the phase calibration problem of large-scale array antennas. Second, 5G millimeter wave requires higher EIRP. At present, CMOS technology can achieve high-frequency radio frequency performance. However, gallium arsenide (GaAs) and gallium nitride (GaN) processes with better performance still have a significant gap in cost and integration. This has led to the relatively high complexity and cost of 5G millimeter wave products. Third, 5G millimeter wave MIMO has the potential to achieve a higher number of streams. This correspondingly puts forward higher requirements for the resolution of radio frequency signals and the ability of baseband digital processing. The complexity and technical difficulty of 5G millimeter wave are also reflected in the terminal. The first is the need to introduce a multi-antenna module. On the one hand, 5G millimeter wave beams have strong directivity and limited coverage. Figure 5.78 shows the signal strength in different directions corresponding to the antenna modules at three different positions. On the other hand, the 5G millimeter wave signal of the handheld terminal may be blocked by the human body. Therefore, the design of 5G millimeter wave antenna modules needs to ensure that the mobile phone can be effectively covered by different antenna modules regardless of the direction and holding position and maintain sufficient link margin during transmission and reception.

The main way to solve this problem is to implement a multi-antenna module in a 5G millimeter wave terminal. This can improve the robustness of 5G millimeter wave communications. For example, in a handheld terminal, individual antenna modules may be blocked by the human body during communication. At this time, the terminal can quickly find and switch to a new transmission path by activating another antenna module on it so as to ensure the stability of the wireless link. Secondly, the design of 5G millimeter wave antennas is more difficult. In the 5G millimeter wave era, the number of mobile phone antennas is increased. However, with the increase in the number of antennas, the space for the antenna is getting smaller and smaller. This is mainly because of the development and design trends of mobile phones. First of all, in order to match the multimedia services provided by the 5G network, 5G mobile phones generally occupy a high proportion of screens and have a design of full screen

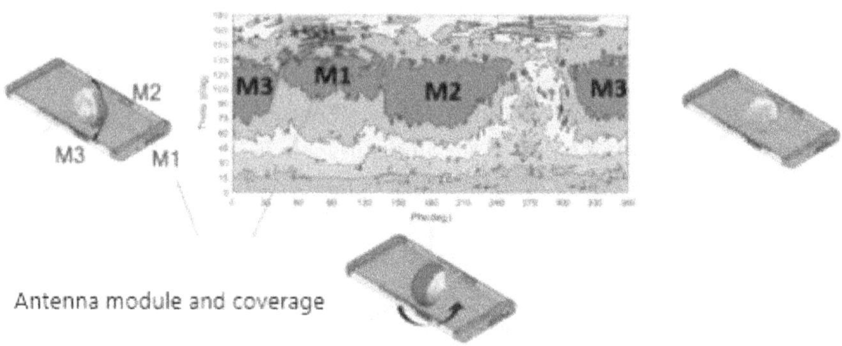

Figure 5.78 Coverage range of 5G millimeter wave antenna.

or even dual screen. Therefore, the place for the antenna is getting smaller and smaller. Secondly, more and more cameras are installed on mobile phones, and the battery capacity is getting larger and larger. Thus, the antenna layout space is further squeezed. In addition, mobile phones generally support multi-mode communication, and 5G millimeter wave antennas also need to coexist or even share space with 4G LTE and 5G low-frequency antennas. At the same time, the high transmission loss of millimeter wave and the high processing precision of the device put forward high requirements on the deployment and production of antennas. These limitations have created design challenges for 5G millimeter wave antennas. In order to meet these challenges, 5G millimeter wave antennas proposed Antenna in Package (AiP) scheme. Based on packaging materials and processes, the antenna, RF transceiver and RF front-end are integrated into the package to achieve system-level wireless functions. AiP conforms to the trend of increasing integration of silicon-based semiconductor processes, while taking into account antenna performance, cost and volume. The third is the radio frequency module integration and low power consumption design of 5G millimeter wave. The key components related to 5G millimeter wave equipment mainly include baseband chips, high-frequency filters, high-frequency power amplifiers, high-frequency antennas and other 5G millimeter wave front-end chips that integrate power amplification functions and amplitude and phase control functions. Compared with the low- and medium-frequency bands, the design and manufacture of 5G millimeter wave antenna need to consider the characteristics of miniaturization, lightweight, broadband, solid state and integration. A series of issues such as large-scale mass production and low cost also need to be considered. This creates higher requirements for process requirements, device chip integration and packaging, that is, to minimize power consumption and maximize mobile phone battery life. In order to expand the coverage of 5G millimeter wave, terminal EIRP needs to be increased as much as possible. However, this brings problems such as rising power consumption and costs, which affects the size and sale price of the terminal and affects the popularity of 5G millimeter wave terminals. At present, some commercial 5G millimeter wave antenna modules have already met the challenges of these terminals. Some 5G commercial chips have integrated 5G new radio transceivers, power management ICs, radio frequency front-end components and phased antenna arrays and can support up to 800 MHz bandwidth on the 24.25–27.5 GHz (*n*258), 26.5–29.5 GHz (*n*257) and the complete 27.5–28.35 GHz (*n*261) and 37–40 GHz (*n*260) millimeter wave frequency bands. Advanced technology can integrate all these functions in a compact package size. This supports the creation of a slimmer, more fashionable mobile terminal with ultra-high rate of 5G millimeter wave new air interface. The 5G millimeter wave network provides fiber-like network speed and low-latency services, helping to open a new generation of networking applications and experiences.

5.10.3.4 *5G millimeter wave test*

The fourth challenge of 5G millimeter wave is testing. Due to the large bandwidth, the small antenna array, the narrow transmitting beam and the large propagation loss of the 5G millimeter wave system cannot be tested by traditional methods. We need to improve the test efficiency and test accuracy and shorten the test time. For example, since the loss of the radio frequency cable is proportional to the frequency, for high-frequency communication, the radio frequency cable connected between the radio frequency transceiver array and the antenna array will have a greater loss, thereby affecting the performance of the equipment. In order to avoid the aforementioned loss, it is necessary to integrate the transceiver array and the antenna array into a whole Active Antenna System (AAS). Due to the high degree of airtightness of 5G millimeter wave base stations, it is impossible to test based on traditional conductive ports. Thus, we need to introduce OTA radio frequency requirements and test methods. After years of exploration, the industry has a deep understanding of the challenges of using OTA method to test AAS indicators. In the Rel-15 stage, the standard formulation of the OTA test method for 5G millimeter wave base station was completed. There are generally three kinds of OTA test site solutions, namely, the far-field test with huge size and very high cost, the near-field test with larger size and higher cost and the compact field test with smaller size. Due to the advantages in construction cost and transmission loss, the industry basically recognizes the compact field test as the mainstream solution for OTA test. The compact field test is a method that uses reflecting surface equipment to convert spherical waves into plane waves at close distances and achieves far-field test performance at close distances. It was originally used to measure the scattering characteristics of radar antennas and radar targets, but it is now used for wireless communication measurements. The OTA test system of 5G millimeter wave base station needs to be equipped with an automated test platform that can control the turntable for global surface radiation power testing and interface data processing automatically, as well as the construction of the test site, instrumentation, turntable and other hardware environments. At present, the 5G millimeter wave OTA test program has made a lot of progress, but there are still some unresolved problems. For example, base station testing needs to test performance indicators under high- and low-temperature conditions, but the tight field test environment requires constant temperature and humidity, so the test conditions contradict the test requirements. In addition, there are problems such as long test time and uncertainty of test. At present, relevant institutions and manufacturers in the industry are exploring, researching and solving the above-mentioned problems. For example, ZTE proposed the ROSE (Rayleigh Optimal Sampling Element) solution, which is Rayleigh optimal sampling and normalized wave vector space sampling and is included in the 3GPP standard. The compact field ROSE test system adopts a small compact field test plan and supports the automatic testing of in-band and out-of-band radio frequency indicators of 5G millimeter wave base stations specified by 3GPP. The test system solves the key problems in the 5G multi-antenna system test and realizes the accurate, fast and low-cost radio frequency index test of the 5G millimeter wave base station, and the efficiency is increased by more than 300 times. Considering the need for increasing the number of antenna elements, minimizing attenuation in the device and reducing costs, 5G millimeter wave terminal equipment also adopts a highly integrated method, which directly integrates the RF transceiver link with the antenna array. Therefore, the terminal no longer has a radio frequency connection port, so that the traditional conduction test cannot be performed, and the OTA test scheme is also required. For terminal equipment, 3GPP first completed the study of 5G millimeter wave terminal test methods in Rel-15. The research has formulated detailed solutions for the testing and verification of radio frequency, demodulation and radio resource management performance of 5G millimeter wave terminals. The measurement methods include direct far-field, compact-field, and near-field conversion. In order to evaluate the performance of millimeter wave MIMO, 3GPP has carried out research work on the 5G millimeter wave MIMO OTA test method in Rel-16. For the static test environment, the standard formulation work of scene definition, channel modeling and test method has been completed. Although the test methods developed by Rel-15 and Rel-16 can verify the basic performance of 5G millimeter wave terminals, there still exist the following problems and challenges. For example, it cannot support test cases that require higher downlink signal power or lower uplink signal power. It cannot support the testing of advanced technologies such as FR2+FR2 inter-band CA. The test equipment of the single-polarization receiver cannot accurately measure the indicators related to the uplink modulation. The static test environment shows that the beam management algorithm was not properly verified. Tests under extreme test conditions are not supported, and the test time is too long. 3GPP will continue to study the above-mentioned problems in Rel-17 to further improve the 5G millimeter wave terminal test solution.

5.10.3.5 *Coexistence of 5G millimeter wave and low frequency*

The fifth challenge of 5G millimeter wave is the coexistence with low frequency. The commercialization of 5G millimeter waves is likely to be a gradual process. For example, the 5G millimeter wave is first applied in industry applications and wireless backhaul scenarios, and then it is gradually applied in dense urban areas and first-tier cities to increase the penetration rate and provide high-speed broadband access in most scenarios. Since 5G millimeter wave deployment finds it difficult to achieve continuous and seamless coverage in the short term, the problems of coexistence with sub-6 GHz, especially coordinated networking, flexible service load sharing and high- and low-frequency switching and interoperability experience are very important. The current mainstream technical route is DC and CA. Relatively speaking, DC is easier to implement, and the terminal support is less difficult. The CA is more efficient and can achieve a deeper level of collaboration. Therefore, if we only need to increase the capacity of the hotspot area, we can use DC to meet the requirements. For long-term evolution, and for certain requirements of continuous downlink coverage, CA can be considered. Another technical challenge for the coexistence of 5G millimeter wave and sub-6 GHz is the synchronization of CA scenario. Due to the hardware difference, the timing of 5G millimeter wave and 5G sub-6 GHz carrier is prone to errors. This leads to problems in the synchronization of the CA system. At present, the 3GPP standard has been optimized for the problems caused by the non-synchronized carrier aggregation system. In addition, in 5G communication systems, 5G mid-/low-frequency and 5G millimeter wave base stations usually coexist on the same site. The terminal also supports both 5G medium- and low-frequency and 5G millimeter wave networks.

The physical limitations of the coexistence of different systems, such as space limitations and heat dissipation, are also one of the technical challenges for the coexistence of millimeter wave and sub-6 GHz systems. For example, 5G terminals need to support millimeter wave signals on 5G networks as well as 5G mid- and low-frequency signals. However, medium- and low-frequency signals are transmitted at longer wavelengths, and 5G millimeter waves cannot be covered by an antenna. Therefore, for the coexistence of 5G low-frequency signal and 5G millimeter wave, the terminal needs to install multiple sets of antennas with different functions. How 5G antennas are placed will make them suitable for mobile devices, balance the limited physical space and ensure the coverage requirements of different signals, without affecting each other's performance, which has become a new design challenge.

5.10.3.6 *Implementation of 5G millimeter wave flexible air interface*

The sixth challenge of 5G millimeter waves is the realization of the flexible air interface. Industry applications in the future will have big differences in business requirements. For example, video surveillance, mobile robots, and remote surgery require a large uplink speed. 4K/8K and VR/AR video require a large downlink rate. However, the current 5G millimeter wave technology has uneven upstream and downstream capabilities of the air interface, which cannot fully meet the needs of flexible deployment. In this regard, China Unicom proposed a flexible frame structure. By adopting three frame structures in the system and flexibly deploying them according to business needs, the differences in business requirements can be met. As shown in Figure 5.79, the DDDSU frame structure is suitable for large downlink services. DSUUU is suitable for large upstream services and can increase the peak rate to more than 25 times compared to DDDSU. DDSUU is suitable for the business of balanced uplink and downlink, and the ratio of uplink and downlink transmission rate is about 1:1.

In addition, as shown in Figure 5.80, the flexible frame structure adaptive adjustment scheme can provide three advantages. The first is to make forecast adjustments based on the long-term business conditions in the coverage area. The second is to make rapid adjustments to the uplink and downlink frame structure based on the emergent situations of 5G industry applications. The third is to meet the application needs of the 5G

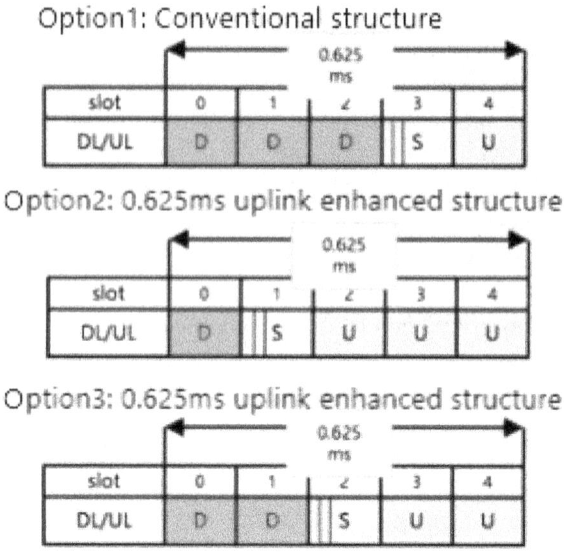

Figure 5.79 5G millimeter wave flexible frame structure meets the differentiated needs of different services.

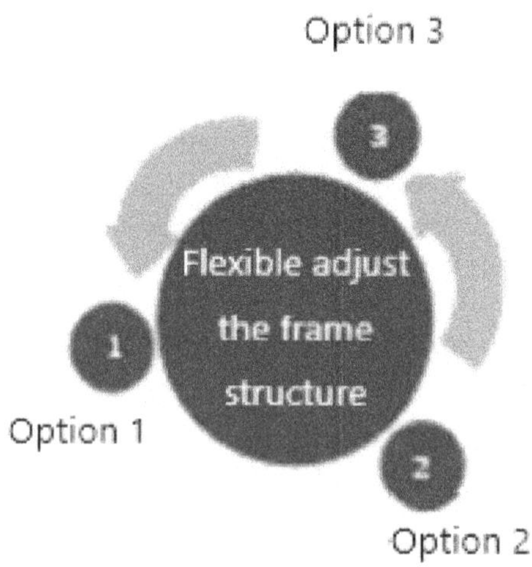

Figure 5.80 Millimeter wave flexible frame structure adjustment scheme.

industry. It can also effectively face public network scenarios where there is a sudden demand for uplink bandwidth, such as concerts and stadiums. Therefore, the 5G millimeter wave adopts a flexible frame structure solution that can cover a variety of scenarios and changes in multiple services and efficiently take advantage of the 5G millimeter wave's large bandwidth.

5.10.4 *Application scenarios and successful cases of 5G millimeter wave*

Integrating the advantages and disadvantages of 5G millimeter wave, its application scenarios can be mainly divided into the three categories described in the following sections.

5.10.4.1 *Indoor and outdoor transportation hubs, venues and other hot spots*

The 5G millimeter wave compensates the additional propagation loss caused by millimeter wave frequency band by using large-scale antenna array technology, beamforming, beam scanning, beam switching and other technologies, which enables 5G millimeter wave base stations to be used for outdoor cellular mobile communications. Typical outdoor hotspots include outdoor squares, stadiums, concert halls and core arterial roads. Typical indoor hotspots include transportation hubs such as airports and railway stations, as well as stadiums and shopping malls. These hotspot areas have the characteristics of a large number of users and a great demand for data traffic. In brand value areas, 5G millimeter waves can be deployed to provide ultra-high-quality services to enhance the brand value. In high-traffic areas, 5G millimeter waves can be deployed to accurately cover and divert traffic to solve hotspot traffic problems. For example, there will be a large number of crowds gathering in outdoor squares, and there will be a lot of communication behaviors such as social media downloads and video and picture uploads. Therefore, the data transmission rate of uplink and downlink is extremely high. Outdoor plazas are generally emptier and there are fewer taller buildings, so that the transmission is basically the line of sight. Therefore, 5G millimeter wave base stations can be deployed around the square for coverage. In the scene of an important transportation hub, due to the large passenger flow, the platform stay time is short. Therefore, the requirements for short-term burst data services and system capacity are both high. In such scenarios, 5G millimeter wave technology can fully exert its advantages of large bandwidth, high speed and large capacity to provide efficient 5G network access for a large number of fast-moving users. Stadiums and gymnasiums have the characteristics of large crowds and concentrated users. For example, when a large-scale concert is held, a large number of people will gather in the stadium center, and there will be a lot of spectators in the auditorium when a sports event is held. Since the number of users is denser and the environment is relatively open, the signal propagation path is mostly line-of-sight propagation (less occlusion, low signal attenuation), which is helpful to avoid the disadvantages of 5G millimeter wave space propagation, so as to fully exert the advantages of large bandwidth, high service rate and large capacity. During large-scale events and sports events, it can provide high-quality 5G network services for a large number of live audiences.

Case 1: 5G millimeter wave coverage simulation of a large subway station in China

In this case, the platform of a subway station in a large metropolis of China is selected. The total area of the station is approximately 1,900 square meters. It is assumed that 5G millimeter wave base stations are deployed at 13 Wi-Fi system access points that have actually been deployed in the station. The base station uses a 128x2 antenna array. Figure 5.81 shows the coverage of 5G millimeter wave signals in the 28 GHz frequency band by simulation. The simulation results show that 96% of the site area can meet the 5G network design downlink coverage

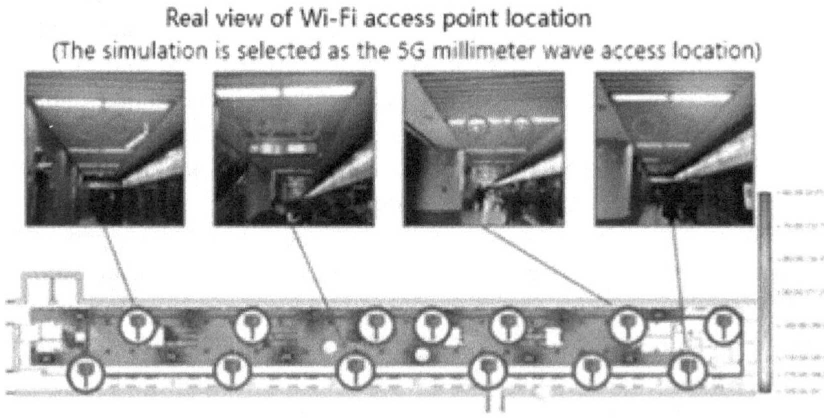

Figure 5.81 5G millimeter wave coverage simulation signal distribution in the platform area of a subway station.

target (0.4 bps/Hz). When the given system bandwidth is 800 MHz and the TDD uplink and downlink time slot ratio is 1:7, the statistical median value of the instantaneous burst data rate of a single user obtained by simulation can be as high as 4.6 Gbps. This simulation result fully shows that for traffic hub scenarios such as subway stations, 5G millimeter wave deployment can achieve excellent 5G signal coverage quality and data service effect by reusing existing Wi-Fi or 4G indoor distribution access point locations without adding new locations.

Case 2: 5G millimeter wave network during the Super Bowl at a stadium in the United States

Eight 5G millimeter wave base stations with a 28 GHz frequency band and a system bandwidth of 400 MHz have been deployed during the Super Bowl in the United States. The placement of the equipment makes the propagation path of the signal to the seating area of the auditorium in the venue all line-of-sight propagation. There are also 100 4GLTE base station equipment deployed in the venue. Figure 5.82 shows the measured signal coverage of each sector of 4G and 5G after the deployment is completed (different colors indicate the signal coverage points of different sectors).

5G millimeter wave signal strength (SS-RSRP) measured results and 4G and 5G millimeter wave download rate comparison test results are shown in Figure 5.83. It can be seen that in the seating area of the stands, it only needs 8 5G millimeter wave base stations to provide good coverage

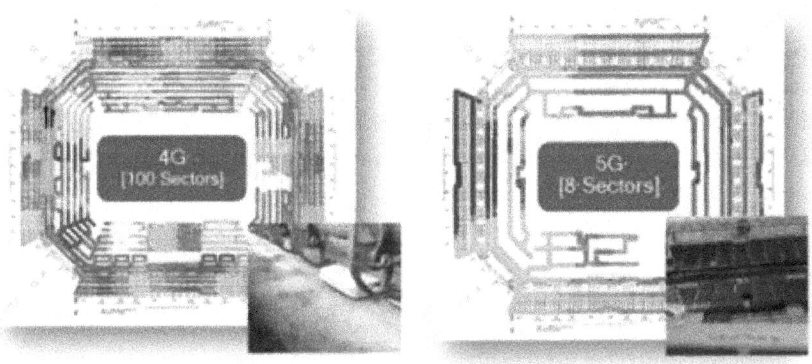

Figure 5.82 Signal situation after 4G and 5G millimeter wave deployment in the grandstand area.

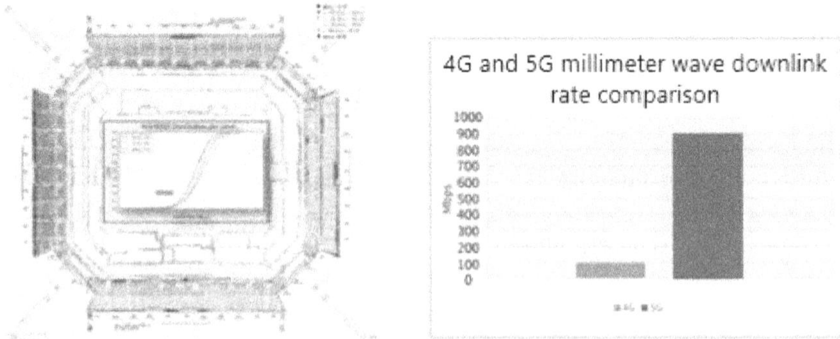

Figure 5.83 Signal measurement results after 5G millimeter wave deployment and comparison test results of 4G and 5G millimeter wave download rates.

of the Super Bowl. The measured download rate of 5G millimeter wave data reached 915 Mbps, which is far more than the 98 Mbps of 4G. (*Note*: In the future, with the increase of 5G terminals and demand for data services, it may be necessary to increase 5G millimeter wave base stations considering the number of users.)

Case 3: Coverage of 5G millimeter wave network in downtown Chicago
As shown in Figure 5.84, a telecom operator and Nokia have commercially deployed 28 GHz millimeter wave applications in a business hotspot in downtown Chicago. Millimeter waves can provide good coverage for blocks, outdoors and open areas. It can reach about 500 meters under the scene of sight distance in dense urban areas. For non-line-of-sight scenes, such as 10 meters away from a street corner, the measured propagation loss is >14 *dB*. 50 meters away from the street corner, the measured propagation loss is >21 *dB*. For lime, brick walls, windows and other types of building materials, the measured propagation loss is 20–50 *dB*. Therefore, the millimeter wave band outdoor coverage indoor scene is basically not considered.

5.10.4.2 *Industry applications (especially the Industrial Internet)*

The industry application of 5G millimeter wave is mainly aimed at large bandwidth and low latency services, plug and play and service access that require mobility, for example, mobile video surveillance, industrial

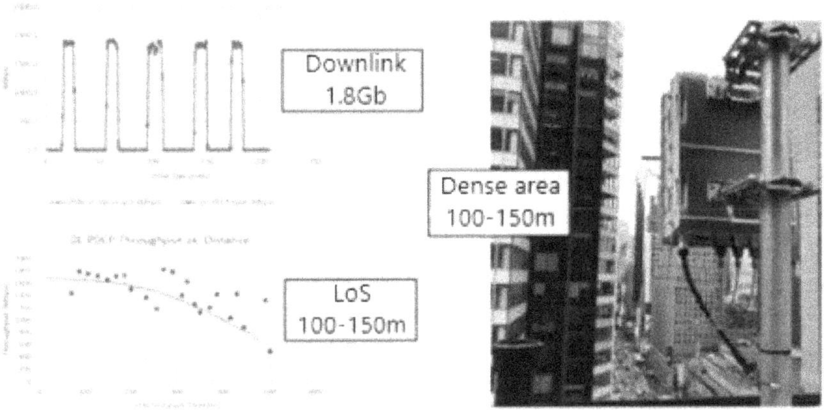

Figure 5.84 Millimeter wave commercial enclosures in downtown Chicago.

vision, AGV, drone control and image transmission, and collaborative robot production line control. These services have very high requirements for bandwidth and delay. Therefore, only 5G millimeter waves can meet the corresponding needs and realize industrial intelligence and network interconnection. 5G millimeter waves can also be combined with industry private networks, enabling operators to provide the industry with flexible bandwidth private network services that are differentiated from public network frequency points. Combining 5G millimeter wave with MEC and AI technologies can provide intelligent solutions with "large capacity and high speed + localization" for the coverage area and provide customized campus private network services for the coverage area.

Case 4: Ericsson and Audi apply 5G millimeter wave to the industrial Internet

With the rapid development of the manufacturing industry, industry leaders such as Audi are looking for ways to improve production automation and assembly process flexibility and reduce personnel safety risks, so that the factory maintains a leading position. 5G URLLC includes the most powerful features of 5G, enabling ultra-high reliability and/or ultra-low latency communication at various data rates. Even in a heavily loaded network, data can be transmitted with the required performance and reliability within the specified delay range to meet the needs of key IoT applications. Typical use cases with a combination of reliability, latency and data rate requirements include AR/VR, autonomous vehicles, mobile robots, real-time human–machine collaboration, cloud robots, haptic feedback, real-time failure prevention and machine coordination and control. Ericsson and Audi have been cooperating to expand the vision of a future factory with 5G technology. In August 2018, it first announced that it would cooperate to create the application of 5G in the automobile manufacturing industry. It also cooperated with Audi and SICK to support the safe operation of AGVs on the 5G network through 5G URLLC communication capabilities based on 5G millimeter waves (as shown in Figure 5.85) and improve factory automation on the radio network to a new level.

Using 5G URLLC to run industry standards and protocols for automated communication such as PROFINET RT and PROFIsafe can provide very low latency and strict latency limits and have the required assurance level to avoid triggering a safe stop in the system. This is of great significance to the industrial automation industry because it opens the door to safe human–computer interaction, which was previously only possible through wired networks. Wireless is the real improvement in the realization of Industry 4.0. This proves the potential of 5G as a future-oriented communication technology that can meet the requirements of automobile production.

Figure 5.85 5G millimeter wave supports AGV operation.

Ericsson and Audi successfully tested the 5G URLLC function based on 5G millimeter wave and actual industrial automation applications in the Ericsson factory laboratory in Kista in January 2020. In the latest demonstration showing the powerful functions of 5G URLLC, Ericsson and Audi jointly constructed a robotic cell. It is the same as other robot units in the current Audi factory, but it is connected via 5G millimeter waves. The robot arm is making a steering wheel (as shown in Figure 5.86, an airbag is being installed), and the laser curtain protects the open side of the robot cell. Due to the ultra-low latency and reliability of 5G URLLC, if a factory worker reaches out, the robot will immediately stop working, so that the worker will not be hurt when operating the machine. This kind of instant response with guaranteed reliability is impossible in traditional Wi-Fi or the previous generation of mobile networks. This frees the automated machines from the wires and greatly improves the flexibility, mobility and efficiency of the production line. This is because the placement of robot cells connected by a wired network on the factory floor is restricted. With 5G URLLC based on 5G millimeter waves, these machines only need one power connection, so they can be used anywhere in the factory. This means that production settings can be easily changed and can be moved on a daily basis to maximize efficiency.

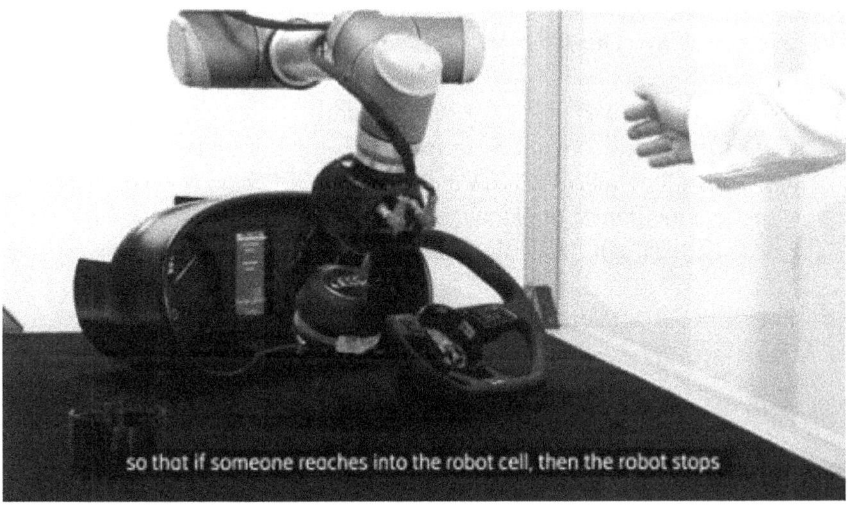

Figure 5.86 5G millimeter wave URLLC capability demonstration.

5.10.4.3 *Wireless broadband access (FWA, Fixed Wireless Access) in homes and office buildings*

The 5G millimeter wave frequency point is higher and the wavelength is shorter, so that more antenna arrays can be deployed in the same area, and the beam energy can be more concentrated, and the 5G millimeter wave system can provide up to 800 MHz bandwidth and a system peak rate of several Gbps, so that the 5G millimeter wave can be used as a wireless backhaul link. This can solve the problem of inability to deploy optical fibers or the high cost of deploying optical fibers in some scenarios, as a supplementary means for conventional access in the last mile. 5G millimeter wave can be used as the backhaul of LTE/5G medium- and low-frequency base stations, or through 5G millimeter wave CPE to provide broadband services for homes or enterprises, as well as radio broadband access for homes and office buildings and achieve good support for high-definition video, AR/VR and other services.

Case 5: Radio broadband access test in corporate office buildings

(i) This case selected an office area of approximately 2,500 square meters in a real office building of an enterprise. 14 of the 20 Wi-Fi system access points actually deployed are selected as the 5G millimeter wave network access point locations. It is assumed that each access point is deployed with a 5G millimeter wave 128 × 2 antenna array. The simulation result of 5G signal cover is shown in Figure 5.87.

When the downlink system bandwidth is set to 800 MHz and the TDD uplink and downlink ratio is 1:7, the simulation results show that under the condition that the area coverage rate reaches 98%, the statistical median of the data download rate in the 5G millimeter wave coverage area can be as high as 4.9 Gbps.

(ii) This case selected the deployment of a 28 GHz band millimeter wave single-sector base station in a real office building of an enterprise. The system bandwidth is 400 MHz. Commercial terminals were used to conduct fixed-point tests in the line-of-sight (near point and midpoint) and non-line-of-sight (midpoint and mid-to-far point) ranges of the coverage area, and it is compared to the 4G data download rate. The case scenario is shown in Figure 5.88.

Figure 5.87 Simulation of 5G millimeter wave access point location and signal coverage strength.

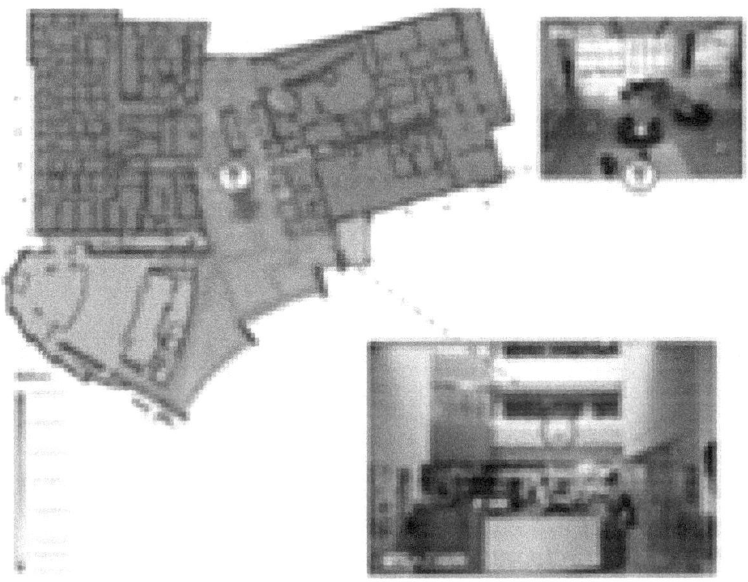

Figure 5.88 Indoor test scenario and path loss distribution.

The location of the fixed-point test in the building is shown in Figure 5.89.

The measured results are shown in Table 5.23. It can be seen that the speed difference of several test points is not significant. Even in non-line-of-sight mid-to-far locations where wireless propagation conditions are poor, 5G millimeter wave terminals can obtain gigabit download rates. This test result provides strong data proof for the good coverage of 5G millimeter wave for indoor office scenes.

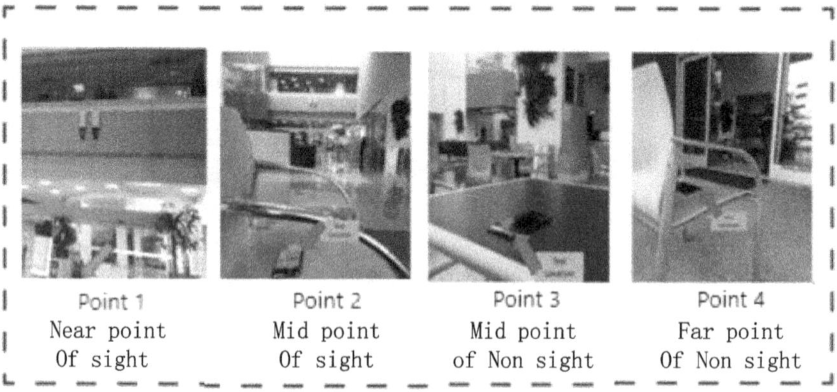

Figure 5.89 Location of selected points for fixed-point test.

Table 5.23 5G millimeter wave indoor fixed-point test results.

No.	Test position	4G	5G mmWave
1	Near LoS	26 Mbps	1.55 Gbps
2	Mid LoS	25 Mbps	1.53 Gbps
3	Non mid LoS	25 Mbps	1.50 Gbps
4	Non far LoS	19 Mbps	1.09 Gbps

Case 6: Rural Fixed Wireless Access Test

Qualcomm, together with Ericsson and U.S. Cellular announced that they have completed the first extended-range 5G millimeter wave new air interface data call. This data call has a length of 5 kilometers and a rate of 100 Mbps, which is the longest connection in a commercial network. This achievement is a key industry milestone, breaking the traditional view that 5G millimeter waves are limited to high-density deployment. The results show that 5G millimeter waves can help narrow the "connection gap" and expand broadband services to cities, suburbs and rural areas. This helps operators to provide optical fiber-speed radio network services to schools, hospitals and cities through 5G millimeter waves in an effective manner. This can solve the "last mile" challenge especially in rural areas. For example, small towns can benefit from networked community spaces such as libraries, stadiums or leisure centers with radio access. Through 5G, it is possible to provide unique performances and immersive experiences to communities that did not have access to high-speed networks before. Another example is local small businesses in remote rural areas. 5G radio access makes it easier for small business entrepreneurs to expand their influence online. Whether it is managing brands on social media, processing payments, product fulfillment, using data analysis to increase website traffic or managing third-party suppliers, a stable high-speed Internet is essential for the growth of home businesses or local stores. Through radio access, people in remote areas can make full use of unique products to meet young consumers' demand for local authentic and rare products. In addition, in rural communities, 5G radio access can transform the animal husbandry industry that requires continuous monitoring for daily breeding and conditions. The introduction of optical fiber in rural areas is too costly for small and medium family farms. But with the help of central base stations, CPEs deployed on barns and other infrastructures can support real-time remote monitoring of livestock and collect and update data through cameras and sensors. The future of small-scale high-tech agriculture will all involve real-time data, and 5G can make these data more accessible.

5.10.5 *5G millimeter wave summary*

5G millimeter wave has six major technical advantages, that is, rich frequency resources, large bandwidth, easy to combine with beamforming technology, can achieve extremely low latency, can support dense deployment, can perform high-precision positioning and high integration. It can fully release the full potential of 5G. It is one of the core technologies to be deployed in the next phase of 5G, and it is also a key enabling technology for 5G to achieve all its original promises and achieve its original vision.

The future 5G millimeter wave technology and product evolution mainly have the following directions:

- Promotes the industrialization of 5G millimeter wave high-frequency devices. The capability index of the high-frequency device determines the equipment capability of the 5G millimeter wave system and the power consumption and energy efficiency of the system.

The industrialization maturity of high-frequency devices determines the equipment cost of 5G millimeter wave systems. This has an important impact on future deployment plans and landing applications, which requires the industry and academia to jointly promote the development of high-frequency simulators and create a mature high-frequency RF device and chip industry chain.

- Improves the 5G millimeter wave test plan. While 5G millimeter wave base station equipment OTA radio frequency index standardization is becoming mature, the feasibility, reliability, accuracy, cost and efficiency of index test schemes are all facing new problems and challenges. Test cost, test efficiency and test accuracy are all OTA test plans that need to be considered and solved. At present, relevant institutions and manufacturers are exploring and researching in this direction. This requires the entire industry to work together from various aspects such as test environment, instrumentation devices and algorithm design to overcome obstacles and promote breakthroughs and progress in 5G millimeter wave base station OTA radio frequency index test technology.
- Supports higher frequency bands. The millimeter wave high-frequency band has abundant spectrum resources, which is the only way to meet the demand for large-capacity communications in the future. The upper limit of FR2 in the millimeter wave frequency band defined in 3GPP NR Rel-15 is 52.6 GHz. At present, 3GPP NR Rel-17 has carried out communication research on the 52.6–71 GHz millimeter wave frequency band. The research gives priority to the development of 5G NR standards for the newly expanded 60 GHz frequency band, and 60 GHz has recently been confirmed by the World Radio Conference as the IMT frequency band in some countries and regions. In 3GPP NR Rel-18 and beyond, it will first consider supporting the millimeter wave frequency band near 100 GHz. For example, the design supports air interface characteristics up to 114.25 GHz. These 5G millimeter wave frequency bands are more abundant than the existing FR2 bands, and the peak throughput provided will be higher. It is expected that the future development direction of 5G millimeter wave will inevitably expand to above 100 GHz and even reach 300 GHz. The higher frequency 5G millimeter wave communication is considered to bring about changes in many aspects, for example, mobile data offloading, short-range machine communication, broadband distribution network,

IAB networking, industrial Internet, AR/VR, intelligent transportation, and data center wireless connection. At the same time, higher frequency 5G millimeter wave communications will also bring complex technical problems, for example, stronger phase noise, more serious path loss, more serious air absorption, and lower power amplifier efficiency. These will be the challenges of 5G millimeter wave in the future.

With the acceleration of 5G construction and the richness of application terminals, the 5G millimeter wave market will get a broad development. All in all, the future of 5G is exciting, more 5G innovations are being bred, and 5G millimeter wave networks will enable continuous technological breakthroughs in the next decade and beyond.

Chapter 6

Main Key Technologies in 5G

So far, South Korea and the United States have announced the commercialization of 5G at almost the same time. China also issued a 5G license in June 2019 and would start commercial use at the end of October 2019. The commercial 5G service mainly meets the needs of eMBB scenarios. For the standards in the other two scenarios (R16, R17), 3GPP would also release R16 in June 2020 and start the research of R17 at the same time. The main technical features and timetable of R17 have been planned to enhance the current version.

Chapter 5 has comprehensively discussed the three major scenarios and eight indicators of 5G, among which the downlink peak rate can reach tens of Gbps, and a whole new set of technologies will be adopted. These technologies are divided into three categories: core network, backhaul and fronthaul network and radio access network (see Figure 6.1) [20]. Some of these technologies have already been discussed in Chapter 5, but for completeness consideration, we will still list them together.

6.1 Core Network

The key technologies of the core network mainly include Network Function Virtualization (NFV), Software-Defined Networking (SDN), Network Slicing and Multi-Access Edge Computing (MEC), Service-Based Architecture (SBA), Control and User Plane Separation (CUPS), etc.

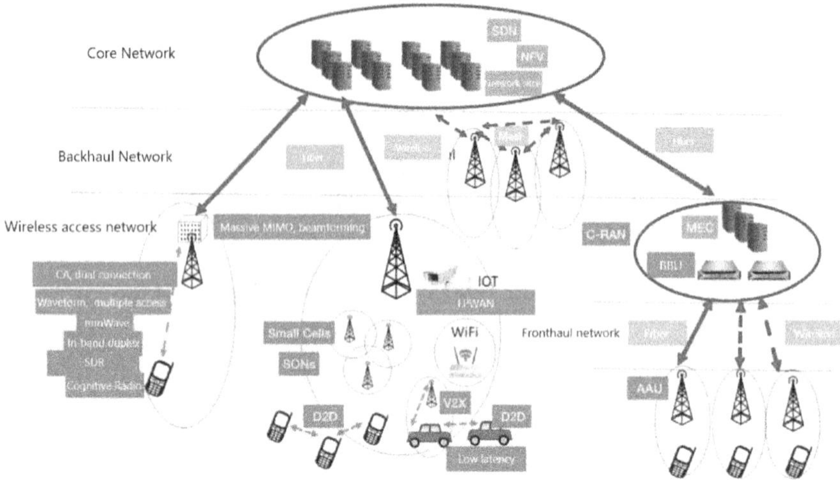

Figure 6.1 Overall architecture of 5G network.

6.1.1 *NFV and SDN*

The core concept of NFV and SDN is to use universal hardware and virtualization technology to carry software processing of many functions, abstract network functions and business processing and control these abstract objects through an external controller, thereby reducing the cost of expensive network equipment. Through software and hardware decoupling and functional abstraction, network equipment functions are no longer dependent on dedicated hardware, resources can be fully and flexibly shared, rapid development and deployment of new services can be realized and conduct automatic deployment, elastic scaling, fault isolation and self-healing based on actual business requirements.

6.1.1.1 *Several benefits of NFV and SDN*

- *Reduced costs*: At present, mobile networks are becoming larger and larger, and more and more hardware devices need to be purchased. These hardware devices are dedicated devices. From the perspective of manufacturers, its research and development costs and production costs are high, and it is bound to be expensive. Introducing NFV and SDN can help make full use of the achievements of computer development, such as Moore's law: performance will double every 18–24 months while the price remains unchanged. If the mobile communication network uses general equipment, the cost is bound to be greatly reduced.

- *Fast update*: Virtualizing network function is equivalent to forming a resource pool, which is also known as the cloud. Thus, it becomes very easy to add new equipment and remove old equipment from the resource pool, which means that operators can decide to invest in updates according to their own networks and can update the latest hardware at any time. As we know that in the past, replacing the original equipment of the mobile communication network is a very huge project, which usually means the replacement of the entire network element. Generally, telecom operators would not make updates unless they had to, and there is not a lot of funds available for these updates.
- *Software openness*: The communication network has one characteristic, high stability. But high stability and high performance are not enough to meet business requirements, while flexibility and agility are more important.

The increasing maturity and successful application of cloud virtualization technology in the IT industry and the open thinking of the Internet have jointly driven major telecom operators to rethink the mobile communication network architecture and service deployment.

The 5G network architecture will also fully adopt SDN and NFV technologies.

6.1.1.2 *NFV function and architecture*

One of the purposes of NFV is also to realize the decoupling of software and hardware functions of specific network communication equipment. NFV uses cloud virtualization-based methods to transform the 4G/5G core network. At present, besides EPC, the most important function of the 4G/5G core network is IMS. After virtualization, they are called vEPC and vIMS, so that a universal server platform can be adopted to replace the original expensive equipment. The cost is much lower than that of telecom equipment, and the cost decreases faster and the renewal cycle is smaller than the telecom equipment, which introduces new IT technology and equipment, as well as maintains the performance of hardware better than the competitors.

NFV is to softwarize network functions through virtualization technology which runs on general hardware in order to replace traditional hardware, as shown in Figure 6.2 [20]. NFV runs network functions on general hardware in the form of virtual machines to achieve flexibility, scalability and mobility of configuration and reduce network Capital Expenditure (CapEx) and Operating Expenditure (OPEX).

The network equipment to be virtualized by NFV mainly includes switches (such as Open vSwitch), routers, Home Location Register (HLR), Serving GPRS Support Node (SGSN), Gateway GPRS Support Node (GGSN), Gateway GSN (CGSN), Wireless Network Controller (RNC), Service Gateway (SGW), Packet Data Network Gateway (PGW), Residential Gateway (RGW), Broadband Remote Access Server (BRAS), Carrier Grade Network Address Translator (CGNAT), Depth Packet Inspection (DPI), PE router, Mobile Management Entity (MME), etc.

Through NFV, the functions of the related network equipment of the existing dedicated 4G core network are virtualized in software, and through cloud computing-related technologies, hardware resources are virtualized into multiple VMs. Using the rapid deployment capability of cloud computing, the capacity configuration adjustment cycle of each EPC software network component is shortened from weeks to minutes, greatly improving the deployment and update agility of EPC network

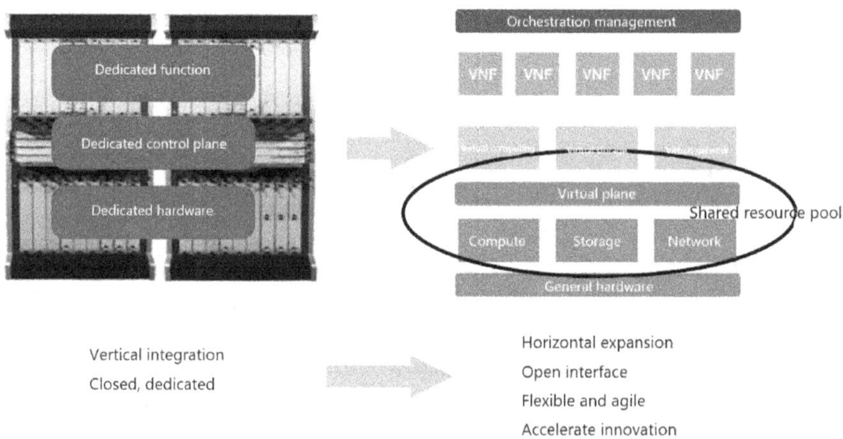

Figure 6.2 NFV functional architecture.

component and the system service level by load balancing mechanism. VMs can be migrated and reinvented and can be backed up locally or remotely to further ensure the high reliability of the network, to realize the dynamic expansion of equipment capacity according to the demands as well as to ensure the maintainability of the system and greatly reduce the deployment, operation and maintenance costs of server hardware infrastructure. As a result, the focus of 4G/5G network telecom operators and equipment vendors can be transferred to service innovation, which further creates higher incomes for telecom operators.

NFV is independent of SDN and can be used alone or in combination with SDN.

6.1.1.3 *SDN function and architecture*

SDN is a network design scheme that separates the network infrastructure layer (also called the data plane) and the control layer (also called the control plane).

The concept of SDN is to allow the software to control the network and fully open the network capabilities. It is a kind of new network architecture and technology which has three major characteristics: control signaling and user data separation (CUPS), centralized control of network functions and open application programming interface (open API).

Since it is a software-defined network, what SDN does is separate the control rights on the network equipment and manage it by a centralized controller, without relying on the underlying network equipment and shielding the differences from the underlying network equipment. The control right is completely open, and users can customize any network routing and transmission rules that they want, making a more flexible and intelligent control. Through SDN, the traditional telecommunication network architecture that is closed and vertically integrated can be transformed into a layered network architecture that is flexible, open, highly integrated, service-oriented and ensures service quality.

The network infrastructure layer and the control layer are connected through standard interfaces, such as OpenFlow.

SDN decouples the network control plane to general hardware devices and centrally controls network resources through software. The control layer is usually implemented by the SDN controller, and the infrastructure layer is usually regarded as a switch. SDN connects the SDN controller and the switch through the southbound API (such as OpenFlow) and connects the SDN controller and the application through the northbound API, as shown in Figure 6.3 [20].

After the introduction of SDN, the new challenges are to reconstruct network functions, to design new interface protocols and to optimize the architecture and end-to-end signaling process based on SDN. On the other hand, a large number of complex control mechanisms are concentrated on the SDN controller, which improves design flexibility. Open-source tools can also be introduced to reduce CapEx and OPEX and stimulate innovation, such as reducing the cost of purchase, management and replacement of SDN switches. At the same time, it also solves the problem of being bound by the special hardware of the network equipment manufacturer.

Since the core concepts of both NFV and SDN technologies are quite similar, and the two have high conditions for complementary integration, in the work of 4G core network virtualization, NFV and SDN are often

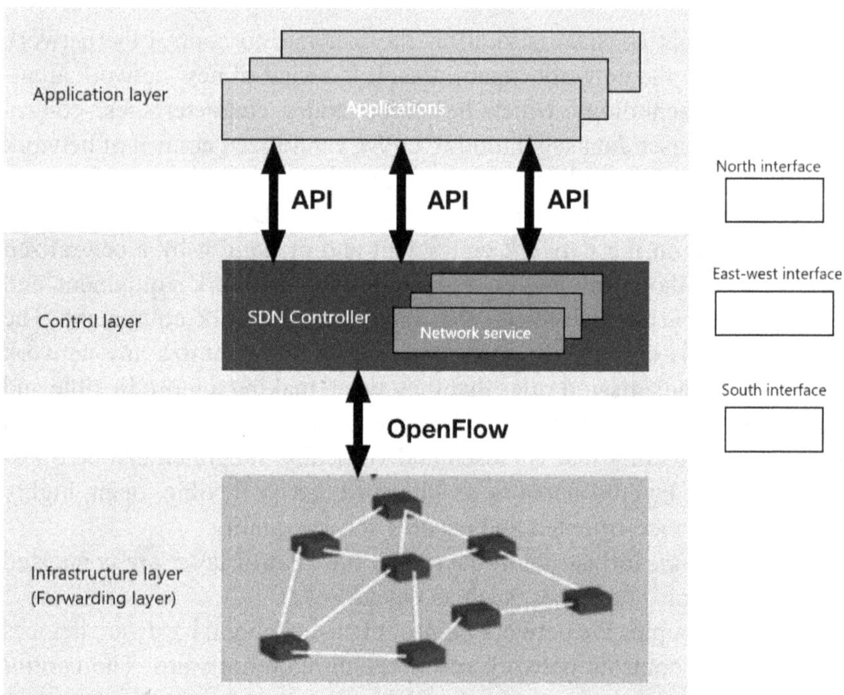

Figure 6.3 SDN architecture.

compared. The coordinated operation model which could be developed in the future is also worth discussing. SDN is responsible for the processing of network infrastructure below Layer 3 and low-level network traffic forwarding, while NFV is responsible for the flexible resource scheduling of network application service facilities above Layer 3. They complement each other and can create an efficient and optimized operator integration service platform in the future.

6.1.2 *Network slicing*

5G networks will face different application scenarios, such as ultra-high-definition video, VR, large-scale IoT, Internet of Vehicles, industrial automation, telemedicine, VR/AR, etc., and different scenarios have different requirements for network mobility, security, latency, reliability and even billing methods. For example, industrial automation requires low latency and high reliability but does not require a high data rate. HD video does not require ultra-low latency but requires ultra-high rate. Some large-scale IoT does not require handover, and mobility management is not its requirement either. Therefore, 5G should divide a physical network into multiple virtual networks, just like a Swiss Army Knife. Each of the virtual networks faces a different application scenario. As shown in Figure 6.4 [20], they are logically independent and do not affect each other but respond to different services.

Of course, such flexible network slicing cannot be undertaken by traditional black box devices. By further refining and modulating the above-mentioned NFV and SDN, business applications can be flexibly assembled.

Only after the realization of NFV and SDN, network slicing can be realized. Different slices are created by NFV and SDN through a shared physical/virtual resource pool, as shown in Figure 6.5 [20]. Network slicing also contains MEC resources and functions.

6.1.3 *Multi-access edge computing*

Multi-access edge computing (MEC) is a cloud-based IT computing and storage environment which is located at the edge of the network, as shown in Figure 6.6 [20]. It enables data storage and computing capabilities to be

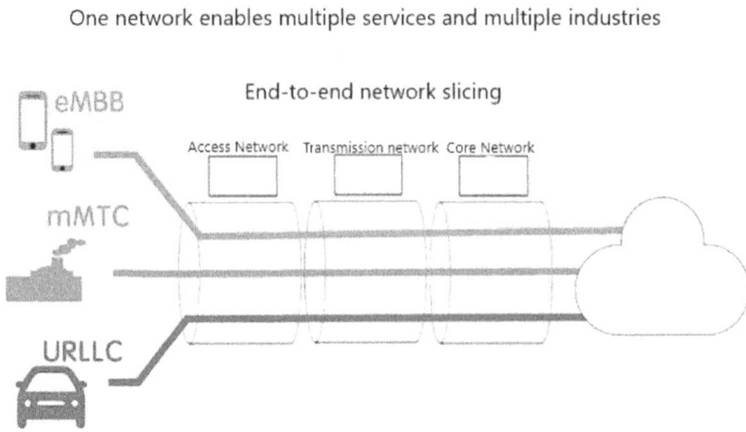

Figure 6.4 5G network end-to-end slicing.

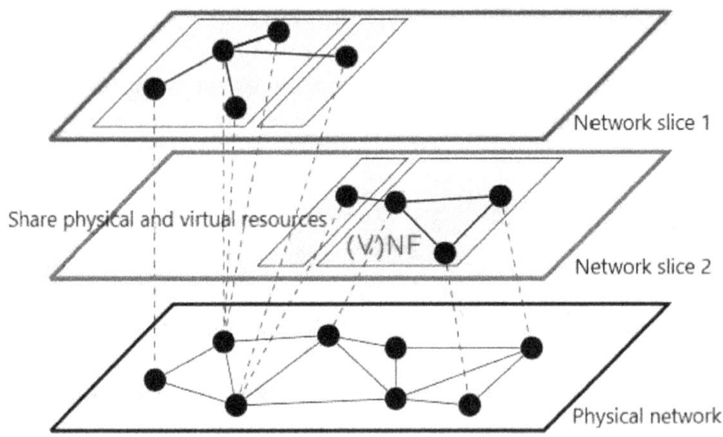

Figure 6.5 5G core network slicing.

deployed closer to the edge of users, thereby reducing network delays, and providing low-latency, high-bandwidth applications.

MEC can introduce new applications through an open ecosystem to help telecom operators provide richer value-added services, such as data analysis, location services, AR and data caching.

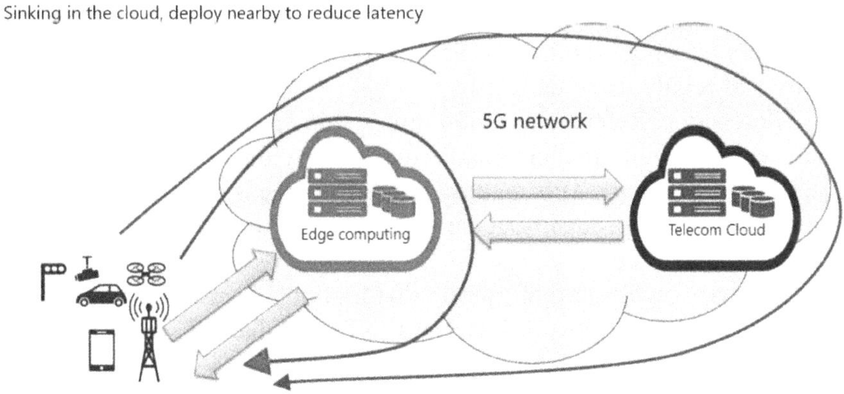

Figure 6.6 Multi-access edge computing.

6.1.4 *Service-based architecture* (*SBA*) [26]

Service-based architecture (SBA) is a framework proposed by 26 companies led by China Mobile. SBA has been confirmed as the 5G core network infrastructure by 3GPP.

However, compared to the traditional core network, this structure has changed so much that people can't help but praise the rapid pace of technological innovation. This change has finally achieved a complete evolution from hardware to software which breaks up the whole into parts (see Figure 6.7).

The SBA is based on the cloud-native architecture design and draws on the "microservice" concept in the IT field.

As we all know, the traditional network element is a tightly coupled black box. NFV decouples the network function software from the black box. But the decoupled software is still a monolithic architecture, which needs to be further broken down into pieces and achieves integration through open API in order to improve the overall agility and flexibility of application development.

To this end, the industry has proposed the design principle of Cloud Native.

The mission of Cloud Native is to change the world on how to build software. It is mainly composed of microservice architecture, DevOps

(Development and Operations) and agile infrastructure represented by containers (see Figure 6.8). The goal is to achieve the resilience, repeatability and reliability of delivery.

Microservice refers to splitting monolithic applications into multiple microservices with smaller granularity (see Figure 6.9). The microservices interact through APIs, and each microservice is deployed, extended and upgraded independently. It can be frequently updated without affecting the customer's use.

It is based on this design concept that traditional network elements are first converted into Network Functions (NF) and then NF is decomposed into multiple "network function services" (see Figure 6.10).

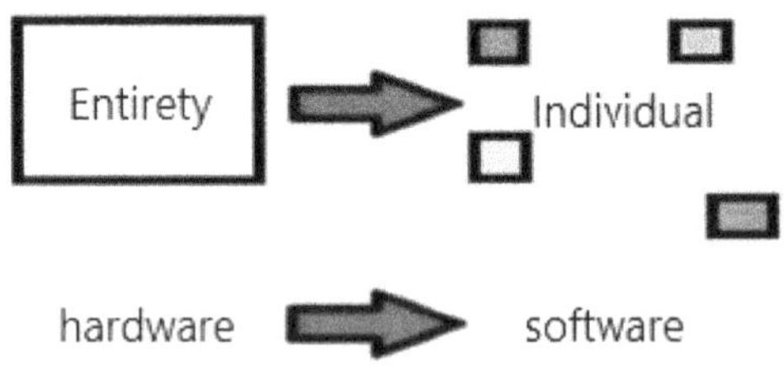

Figure 6.7 SBA breaks up the whole into parts.

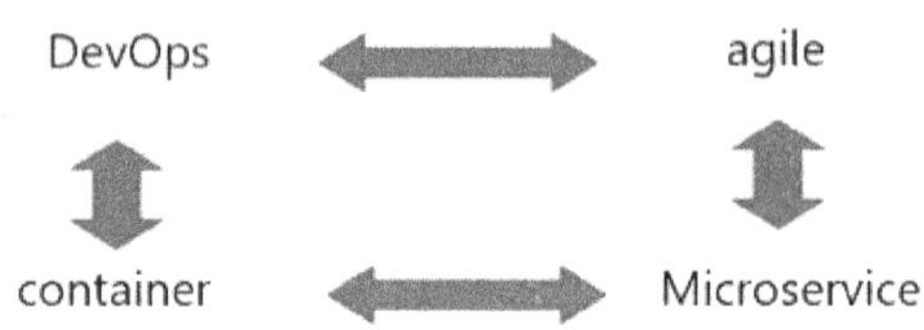

Figure 6.8 Design principle of Cloud Native.

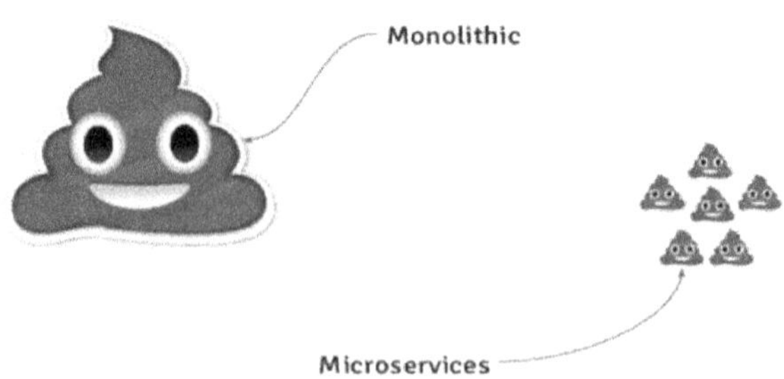

Figure 6.9 Microservices.

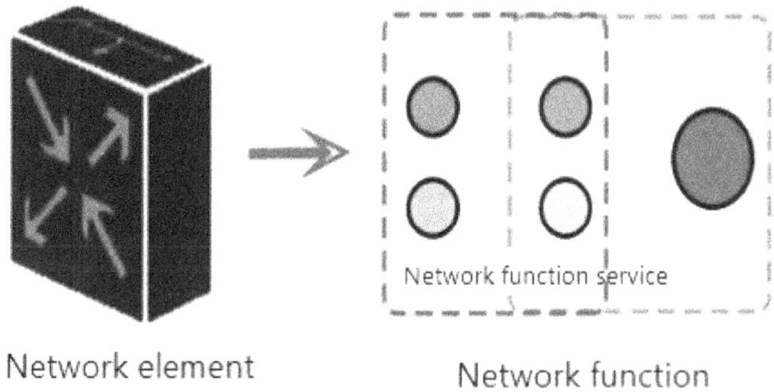

Figure 6.10 Decomposition of the network element.

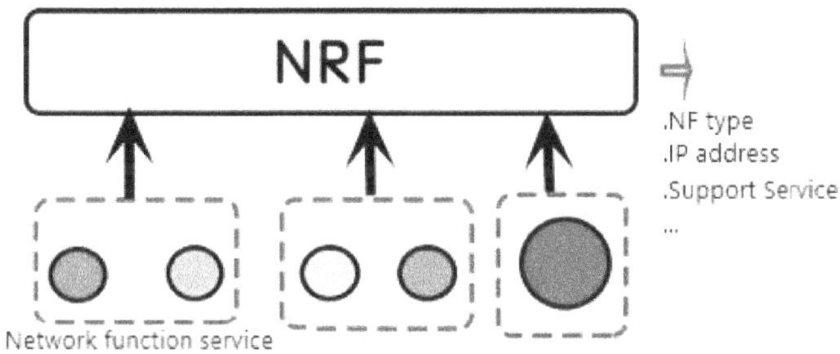

Figure 6.11 Network Repository Function.

SBA is a network function service plus a service-based interface. The network function can be composed of multiple modular network function services and display its functions through the service-based interface, so the network function services can be used flexibly by authorized NFs. Among them, Network Repository Function (NRF) supports network function service registration, status monitoring, etc. (see Figure 6.11), and realizes the automatic management, selection and scalability of network function services.

6.1.5 *Control and User Plane Separation*

The purpose of Control and User Plane Separation (CUPS) is to get rid of the centralization of network user plane functions, so that it can be flexibly deployed in the core network (central data center) or in the access network (edge data center) and finally achieve distributed deployment.

In fact, the core network has been evolving in the direction of separation of the control plane and the user plane (see Figure 6.12). For example, starting from R7, the control plane and the user plane are separated through Direct Tunnel technology, and a direct user plane tunnel is established between 3G RNC and GGSN, and user plane data traffic is directly transmitted between RNC and GGSN, bypassing the SGSN. At R8, pure signaling nodes such as MME appeared.

In the 4.5G and 5G era, this separation trend is more thorough and more necessary. One of the major reasons is to meet the KPI of millisecond-level latency in 5G networks (see Figure 6.13).

The optical fiber propagation speed is 200 km/ms, and the data needs to be transmitted back and forth between the terminal and the core network that are several hundred kilometers apart. Obviously, it cannot meet the 5G millisecond delay. Physical distance is limited, which is an unchangeable drawback.

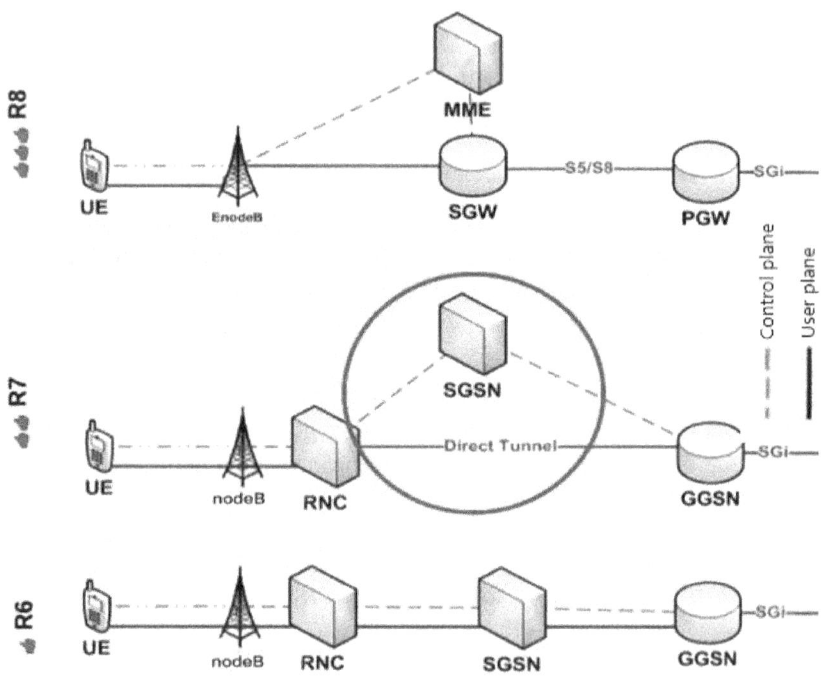

Figure 6.12 2G to 4G network element separation evolution.

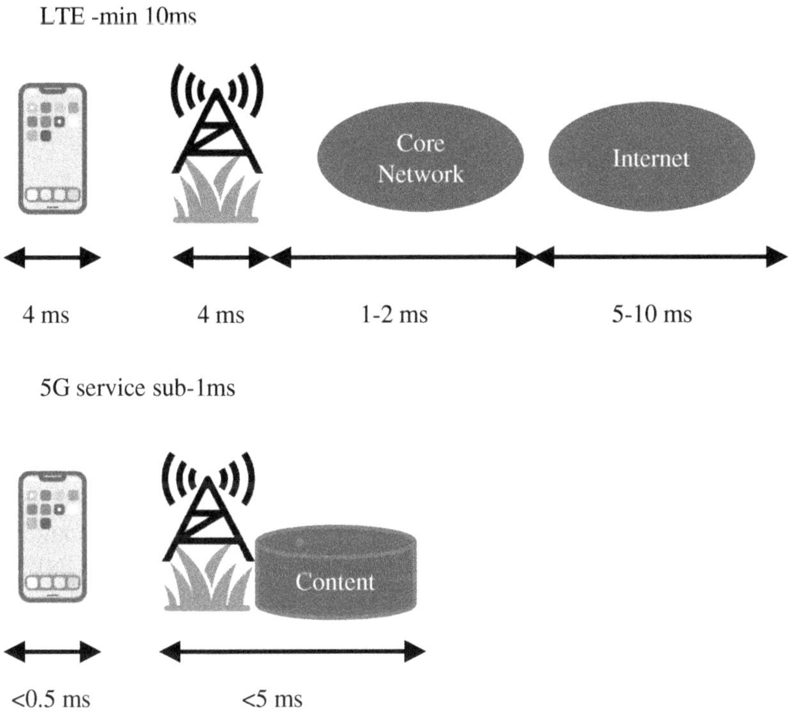

Figure 6.13 4G to 5G network delay evolution.

Therefore, it is necessary to sink and distribute the content on the access network side (edge data center) to make it closer to the user and reduce the delay and network backhaul load.

To better understand CUPS, we will discuss the separation of the control plane and the user plane a little bit more.

6.1.5.1 *Control plane*

The control plane is divided into AMF and SMF. A single AMF is responsible for terminal mobility and access management, while SMF is responsible for dialog management functions, and multiple SMFs can be configured.

Based on flexible microservice architecture, AMF and SMF correspond to different kinds of network slicing (see Figure 6.14).

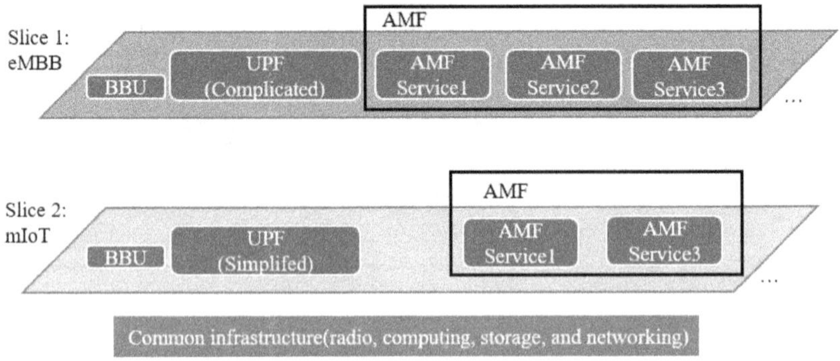

Figure 6.14 Two kinds of network slicing.

AMF and SMF are the two main nodes of the control plane. UDM, AUSF and PCF cooperate with them to perform user data management, authentication and policy control. In addition, two platforms, NEF and NRF, support function nodes to help expose and publish network data and help other nodes discover network services.

6.1.5.2 *User plane*

The user plane of 5G core network is controlled by the User Plane Function (UPF) node. UPF has also replaced the SGW and PGW that perform routing and forwarding functions in the original 4G.

Figure 6.15 shows a simplified 2G and 3G network architecture.

In the 4G era, the core network architecture has changed again. In this time, the Mobile Management Entities (MMEs), Packet Data Gateways (PGWs) and other network elements (see Figure 6.16) are introduced.

In 5G, the MME, SGW and PGW in the 4G core network disappeared. The function of MME in 4G is divided into Access and Mobility Management Function (AMF) and Session Management Function (SMF), and SGW and PGW are replaced by UPF (see Figure 6.17).

From the MSC/HLR in the 2G era to the softswitch, to the introduction of MME and GW in the 4G era, the core network has been evolving in the direction of separation and softwarization in general.

This time, it was more thorough. The traditional black box hardware is decoupled, and the network function software is further decomposed

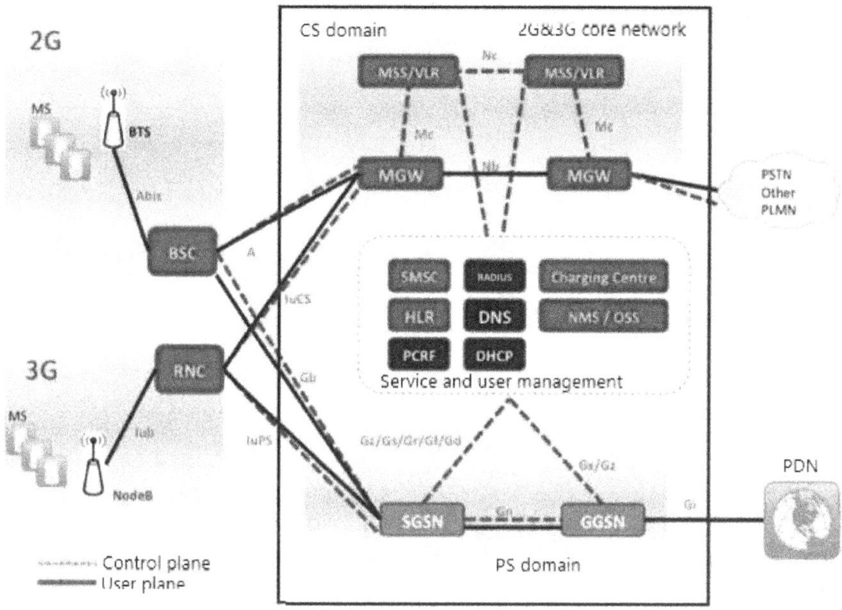

Figure 6.15 Simplified 2G and 3G network architecture.

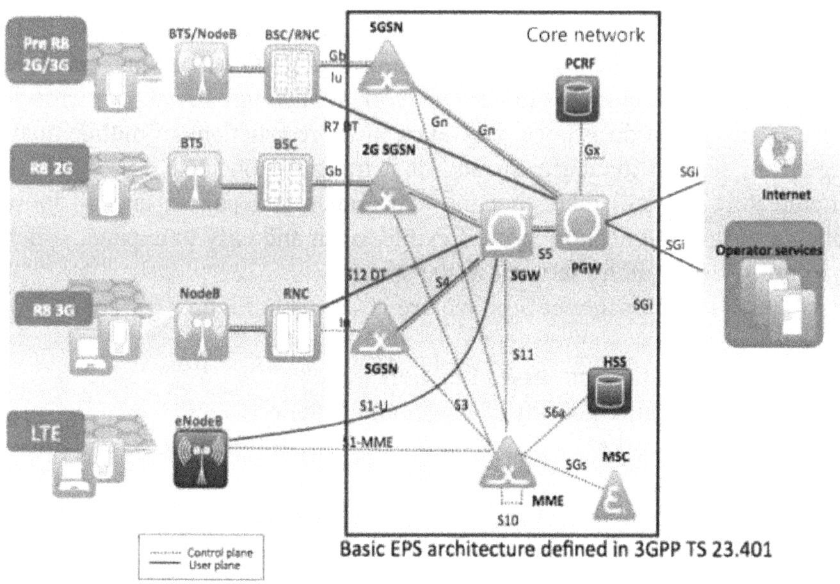

Figure 6.16 General LTE network architecture.

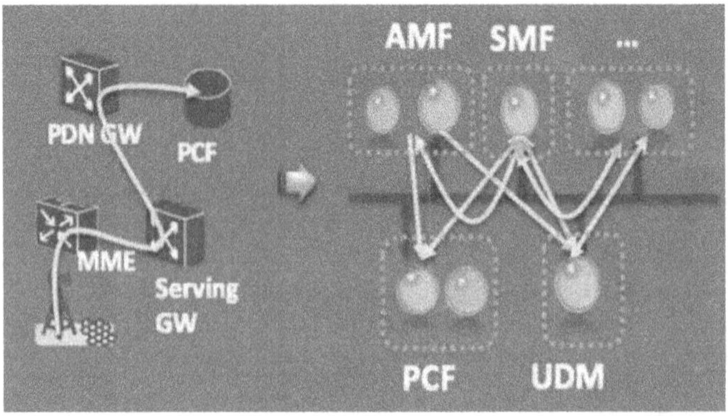

Figure 6.17 Exploded diagram of 4G to 5G network elements.

into microservices to flexibly construct network functions. The network functions run on general COTS servers or migrate to the cloud to realize flexible network slicing. In general, it is a complete evolution from hardware to software.

6.1.6 *Evolution of network architecture*

Figure 6.18 shows the evolution of a traditional 3/4G network architecture based on network elements to a 5G network architecture based on microservices. This evolution is huge. However, the core functions of mobile management, session management and data transmission in the 5G network architecture are still there, but the appearance is repaired, and all these changes make the network more flexible, open and easy to expand, which can bring more quality services to the society.

The 5G system service architecture is shown in Figures 6.19 and 6.20.

- AUSF, authentication server function.
- AMF, access and mobility management functions.
- DN, data networks, such as operator services, Internet access and three-way services.
- SDSF, structured data storage network function.
- UDSF, unstructured data storage network function.
- NEF, network opening function.
- NRF, network repository function.

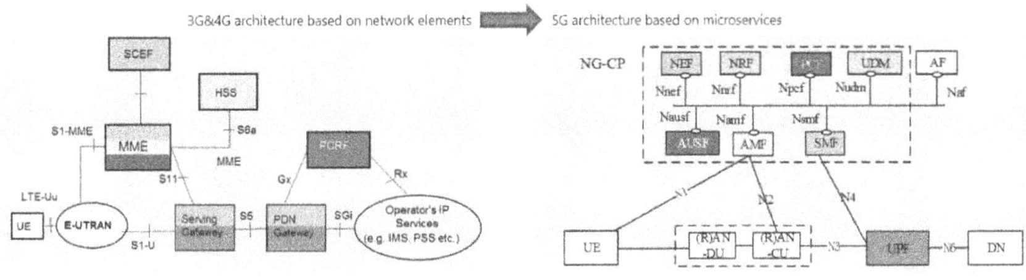

Figure 6.18 3/4G to 5G network architecture evolution.

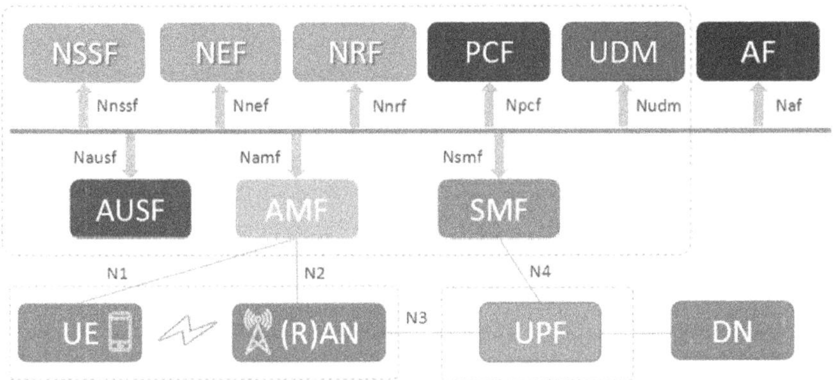

Figure 6.19 5G system service architecture.

- PCF, policy control function.
- SMF, session management function.
- UDM, unified data management.
- UPF, user plane function.
- AF, application function.
- UE, user equipment.
- RAN, radio access network.
- Namf, a service-based interface presented by AMF.
- Nsmf, a service-based interface presented by SMF.
- Nnef, a service-based interface presented by NEF.
- Npcf, the service-based interface presented by PCF.
- Nudm, a service-based interface presented by UDM.
- Naf, a service-based interface presented by AF.
- Nnrf, a service-based interface presented by NRF.
- Nausf, a service-based interface presented by AUSF.

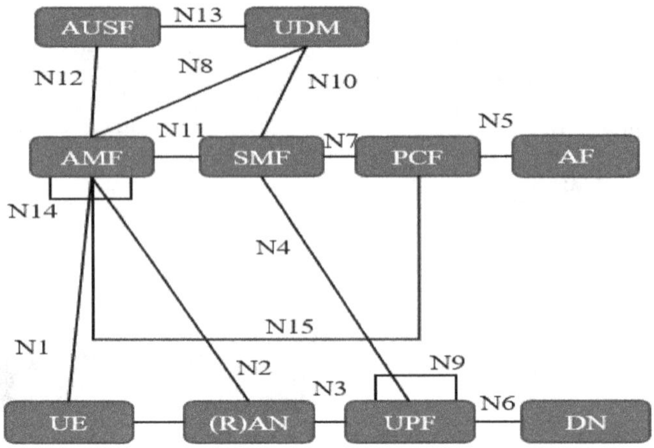

Figure 6.20 Reference point of non-roaming 5G system architecture.

The 5G core network infrastructure is based on the design principles of cloud-native microservice architecture. The 5G core network is constructed in a modular and software-based construction method to efficiently perform network slicing of different service types. Here you can see in the figures (Figures 6.18–6.20) that the network node names are followed by functions. These functions are based on the software to adjust the network dynamically.

6.2 Bearer Network (Fronthaul Network and Backhaul Network)

6.2.1 *Backhaul*

Backhaul refers to the part of the radio access network connected to the core network. Optical fiber is the ideal choice for the backhaul network. However, in environments where optical fiber is difficult to deploy or the deployment cost is too high, wireless backhaul is an alternative, such as point-to-point microwave, millimeter wave backhaul, etc. In addition, a radio mesh network is also an option for 5G backhaul. In R16, 5G wireless itself will be designed as a wireless backhaul technology, that is, IAB (5GNR integrated wireless access and backhaul).

6.2.2 *Fronthaul*

The fronthaul refers to the BBU pool connecting the remote RRU part (or Centralized Unit (CU) and Distributed Unit (DU), DU and RRU), as described in the C-RAN part. The fronthaul link capacity mainly depends on the radio air interface rate and the number of MIMO antennas. The 4G fronthaul link uses the Common Public Radio Interface (CPRI) protocol. However, due to a significant increase in the 5G wireless rate and the doubled number of MIMO antennas, CPRI cannot meet the requirements of capacity and delay of the 5G era. To this end, standards organizations are actively researching and formulating new fronthaul technologies, including sinking some processing power from the BBU to the RRU unit and decomposing the BBU into CU and DU, or DU as described in the following section to reduce delay and fronthaul capacity.

6.3 Radio Access Network

In order to increase capacity, spectrum efficiency, reduce latency and improve energy efficiency to meet key 5G KPIs, 5G radio access network includes key technologies such as C-RAN, new modulation and coding technology, Ultra-dense Heterogeneous Network (ultra-dense HetNet), small cells, Software-Defined Radio (SDR), Cognitive Radio (CR), self-organizing network, D2D communication, massive MIMO, millimeter wave, advanced modulation and access technology, in-band full duplex, carrier aggregation and low-latency, low-power technology.

6.3.1 *Cloud Radio Access Network*

Cloud Radio Access Network (C-RAN) is a green radio access network architecture based on centralized processing, cooperative radio and real-time cloud computing architecture. It softwarizes the network functions of radio access into virtualized functions and deploys them in a standard cloud environment. The concept of C-RAN was developed from the centralized RAN. The goal is to improve design flexibility and computing scalability, improve energy efficiency and reduce integration costs.

The basic idea of C-RAN is to make full use of low-cost high-speed optical transmission networks to directly transmit wireless signals between remote antennas and centralized central nodes to build a radio network access system covering hundreds of base station service areas or even hundreds of square kilometers.

The C-RAN architecture is suitable for the use of collaborative technology, which can reduce interference, reduce power consumption and improve spectrum efficiency. At the same time, it is convenient to realize intelligent networking for dynamic usage. Centralized processing is beneficial to reduce costs, facilitate maintenance and reduce operating expenses. The current research content includes the architecture and functions of C-RAN, such as centralized control, baseband pool RRU interface definition, etc., and closer cooperation based on C-RAN, such as base station clusters, virtual cells, etc.

In fact, C-RAN has a long history, and it is not a new concept in the 5G era. With the continuous construction of large-scale networks from 2G to 3G and then to 4G, C-RAN has always been a topic of exploration in the industry. During the large-scale deployment of 2G and 3G networks, C-RAN involved excessive upgrades to existing network equipment rooms, equipment and transmission, so large-scale development did not be carried out. In the initial stage of 4G network construction, it was mainly technically limited by the huge consumption of fronthaul networks. But through the combination of CPRI (BBU–RRU interface) compression technology and passive Wavelength Division Multiplexing (WDM), the problem of huge consumption of fiber resources in the fronthaul network is resolved to a certain extent.

We say the 5G network architecture is disruptive because the basic processing unit (BBU) of the access network may be reconstructed into two entities: CU and DU. CU equipment handles high-level wireless protocol stack functions, such as RRC layer, PDCP layer, etc., and can even support part of the core network function sinking to the access network. The term is called edge computing network (MEC), which can meet the needs of future communication networks for emerging services. For example, video, online shopping and VR/AR have higher requirements for network delay. It is precisely because of changes in communication network elements and structures that the protocol stack will be adjusted and changed accordingly. Therefore, for naming the next-generation network, 5G New Radio (NR) also vividly illustrates the disruptive changes in many aspects of the 5G network itself. DU equipment mainly deals with physical layer functions and layers two functions with high real-time requirements. Taking into account the transmission resources of RRU and

DU, some physical layer functions of DU can be moved up to RRU. With the miniaturization of RRU, DU can even be combined with RRU.

As can be seen from Figure 6.21, the traditional BBU is divided into two parts. The CU covers the high-level protocol stack of the radio access network and part of the core network functions, while the DU covers the physical layer of the baseband processing and part of the layer two functions. The CU can be centrally deployed, while DU can be deployed according to the actual network environment. For example, it can be deployed in core urban areas with high traffic density, small distance between stations and limited space resources, such as universities and large performance venues. DU can also be centrally deployed, such as in

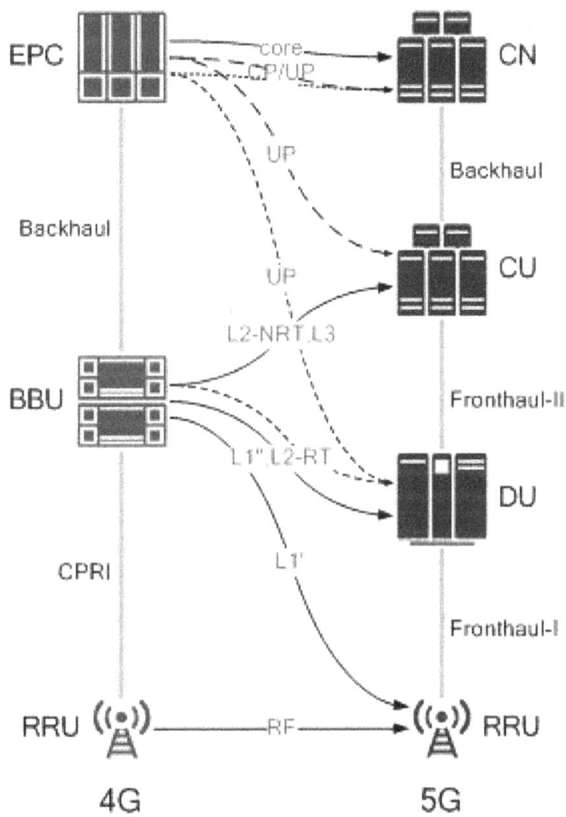

Figure 6.21 BBU decomposition.

areas with sparse traffic and large distance between stations or in a distributed manner. The current 2/3/4G network highlights the functions of the backhaul transmission network due to its star-shaped distributed network architecture, and the local centralized deployment of the access network makes the fronthaul network on its stage. It can be said that the fronthaul network is a series of transmission links from CU to remote RRU.

In the C-RAN architecture, C stands for Cloud. In order to achieve cloudification, adaptability considerations are required from the system design architecture and the implementation of software and hardware. Two concepts of decoupling and slicing are proposed in the 5G system architecture. Decoupling includes two levels. One is the decoupling between service requests and air interface resource allocation. For example, in 4G mobile systems, through QoS, Guaranteed Bit Rate (GBR), Address Interpretation Protocol (ARP) and other network parameter settings and the subscription information of users, the business and resources are relatively loosely coupled. The coupling between services and resources in 2G networks is tighter, for example, each voice user is allocated at least 2M transmission link resources. The other layer of decoupling is the decoupling of software and hardware processing capabilities. The underlying hardware is uniformly deployed with X86 blade server, the middle layer is logically divided by the virtualization layer, and the upper layer is used for software application development. Therefore, the meaning of decoupling these two layers is not only the decoupling of services and resources but also the separation of software and hardware functions of centralized deployed CU elements. The end-to-end slicing is to redivide the end-to-end network functions through new physical and logical network element architecture (CU-DU). For example, the NAS process of the core network is moved down, as the high-level protocol stack is separated from the physical layer and the MAC layer. As traditional network slicing becomes more subdivided, C-RAN cloudification capabilities will continue to mature. Decoupling achieves the freedom of business, and slicing subdivides and standardizes the industry chain. Decoupling can be simply understood as providing a common mobile phone platform, where various OTT (app) providers can provide their services. Slicing defines a standardized interface similar to that of the USB of a laptop. As long as various USB providers design according to the standard protocol, a seamless connection can be achieved.

Relying on the C-RAN network system, not only will revolutions occur in the existing (2, 3, 4) G networking architecture, but there will also be an impact on the existing network equipment providers because IT providers can also provide communication services based on standard protocols on the existing general hardware architecture, which breaks the long-term monopoly of traditional network equipment providers in the communication market. Even telecom operators can themselves program to provide services, which is the foundation of the DevOps model that the telecom operators are currently advocating.

In short, under the C-RAN architecture, the BBU function is virtualized, and its function will be decomposed into CU and DU. CU is centralized and pooled, while DU can be deployed in a centralized manner or distributed with RRU and antenna. When DU and RRU are deployed together, they are connected through the fronthaul network. When deployed separately, they also need fronthaul network for connection. The BBU pool (CU and DU) can share resources and flexibly allocate and process signals from each RRU.

The advantage of C-RAN is that it can improve computing efficiency and energy efficiency, and it is easy to implement than more advanced joint optimization solutions such as Cooperative Multipoint Transmission (CoMP), multi-RAT (Radio Access Technology) and dynamic cell configuration. However, the challenge regarding C-RAN is the complexity of the design and deployment of fronthaul networks.

6.3.2 *New modulation and coding technology*

Modulation and coding technology is the core technology of mobile communication. The new modulation and coding technologies adopted by 5G mainly include 256QAM high-order modulation, LDPC and polar coding and decoding technologies, which will be introduced separately in the following parts.

In 1948, Shannon first proposed a method for reliable communication in noisy channels in his paper *Mathematical Theory in Communication*, and put forward the famous disturbing channel coding theorem, laying the foundation for error correction coding.

At the beginning of the 1950s, Hamming, Slepian, Prange and others designed a series of coding and decoding schemes with excellent performance on the basis of Shannon's theory. Based on this, the Shannon limit of various channels under coding channel conditions is obtained. The Shannon limit is of great significance since it is the performance limit in the communication system. It also drives the research and application of error correction codes that are designed and constructed close to the Shannon limit in the communication field.

Simply put, channel coding is to insert redundant bits into a K-bit data block to form a longer code block. The length of this code block is N bits, $N > K$ and $N–K$ bits which are used for detection and error correction. For redundant bits, the coding rate R is K/N. A good channel coding means that at a certain coding rate, it can infinitely approach the theoretical limit of channel capacity, namely the Shannon limit.

The deciding factors for 3GPP to decide which coding method to use for 5G include decoding throughput, delay, error correction capability, Block Error Rate (BLER), flexibility, complexity, maturity and backward compatibility of software and hardware implementations.

6.3.2.1 *LDPC code*

LDPC code stands for low-density parity-check code. It was first proposed by Dr. Robert G. Gallager of the Massachusetts Institute of Technology in 1963. It is a type of linear block code with a sparse check matrix. Due to its good performance which approaches the Shannon limit, low decoding complexity and flexible structure, LDPC has always been the research hotspot in the field of channel coding.

LDPC is a block code with a very low check matrix density (the number of "1"). The core idea is to use a sparse vector to disperse information into the entire codeword, that is, the number of 1 in the check matrix is required to be much smaller than 0, and the longer the code length, the lower the density.

The density of the ordinary block code check matrix is large. When the maximum likelihood method is used to decode, the error information will be iterated repeatedly and strengthened among the checking nodes, which will cause a decline in performance. On the contrary, the check matrix of LDPC is very sparse, the error information will be scattered throughout the iteration of decoding, and the probability of correct decoding will increase accordingly. To put it simply, the disadvantage of ordinary block codes is that errors are concentrated and diffused, while the advantage of LDPC is that errors are dispersed and corrected.

However, due to the complicated calculation of LDPC decoders, hardware performance drawbacks at that time and the lack of feasible and effective decoding algorithms, LDPC codes were gradually forgotten during the 35 years after its advent.

Until the 1980s, Tanner explained LDPC codes with graph theory and improved the decoding method. In 1993, Berrou *et al.* invented the turbo code. On this basis, David J.C. MacKay of the Cavendish Laboratory of the University of Cambridge once again discovered LDPC, a channel code with excellent performance and proposed a feasible decoding algorithm. The algorithm further discovered the good performance of LDPC codes, which quickly aroused strong response and great attention, and LDPC codes have once again entered the field of academic circles.

After that, the academic circles focused their efforts on LDPC, including research on key technologies such as coding matrix construction and decoding algorithm optimization. Among the key breakthroughs are the multi-edge construction method proposed by Thomas J. Richardson of Qualcomm, which can flexibly obtain LDPC codes at different rates. It is very suitable for the incremental redundancy (IR-HARQ) technology of communication systems. Coupled with the parallel decoding of LDPC, the decoding time and complexity of LDPC were greatly reduced. So far, the obstacles for LDPC have all been removed.

After more than 10 years of research and the rapid development of software and hardware technologies, the related technologies of LDPC have become more mature and have begun to have commercial application results. LDPC has entered related fields such as wireless communications to be used by various communication systems. It has been widely used in deep space communications, optical fiber communications, satellite digital video and audio broadcasting, etc.

- Broadcasting system: Digital Video Broadcasting-Satellite Second Generation (DVB-S2) system, Digital Television Terrestrial Multimedia Broadcasting (DTMB) system and China Mobile Multimedia Broadcasting (CMMB) system.
- Fixed access network: ITU-T high-speed home wired network (G.hn).
- Radio access network: IEEE's 802.11n, 802.11ac, 802.16e (WiMAX).

In addition, LDPC is also used in aerospace communications including Change-2.

So far, only mainstream mobile communication systems dominated by 3GPP have not officially adopted LDPC codes (WiMAX has not been deployed on a large scale by mainstream telecom operators).

The first attempt of LDPC in 3GPP appeared in the LTE R8 discussion in 2006. Due to non-technical factors, LDPC unfortunately lost to the popular turbo code and missed the opportunity to achieve a breakthrough. However, LDPC has not faded out of the public's sight and is still advancing in the field of other communication standards. After 10 years of accumulation in 2016, LDPC, which has been fully verified in the actual communication system, has come to the field of mobile communication and become an alternative to 5G. This time, LDPC has become the consensus of the industry including Chinese companies. After in-depth discussions, driven by mainstream companies in the communications industry (Qualcomm, Samsung, Nokia, etc.), on October 14, 2016, at the 3GPP RAN1 conference held in Lisbon, Portugal, LDPC codes finally defeated Turbo 2.0 and were accepted as the traffic channel data information and long code block coding scheme in the Enhanced Mobile Broadband (eMBB) scenario of 5G system. 53 years after its advent, LDPC has finally been adopted by mainstream mobile communication systems.

6.3.2.2 *Polar code*

At present, the error correction codes that approach Shannon's limit with most research results and more maturity are LDPC and polar codes. Although the performance of the two codewords is excellent, people have always been looking for various coding schemes with better performance, which can reach the Shannon limit very closely or even completely with simple encoding and decoding methods.

Polar code is a new star in the coding world. It was proposed by Professor E. Arikan of Bilkent University in Turkey in 2007 based on the theory of channel polarization. It is a brand new linear channel coding method. This codeword is the only one found so far that can achieve the Shannon limit coding method and has low coding complexity, while when the coding length is N, the complexity is $O(N\log N)$. Since polar code was proposed, it has attracted the interest of many scholars, and it has been a hotspot in the field of information coding in recent years.

The theoretical basis of polar codes is the channel polarization theory. Channel polarization includes two parts: channel combination and channel decomposition. When the number of combined channels tends to infinity, polarization will occur and some channels will tend to be noiseless channels, while the other will tend to be full noise channels. This phenomenon is channel polarization. The transmission rate of the noise-free channel will reach the channel capacity $I(W)$, while the transmission rate of the all-noise channel tends to zero. The encoding strategy of polar code is to apply the characteristics of this phenomenon, using a noiseless channel to transmit useful information for users and an all-noise channel to transmit agreed information or no information.

On November 19, 2016, at the 87th meeting of 3GPP RAN1 just concluded in Reno, Nevada, USA, the International Organization for Mobile Communications Standards 3GPP determined that polar code (polarized code) will be used as the short coding scheme of the control channel in 5G eMBB scenario.

Thus, the channel coding technical solution for 5G eMBB scenario is completely determined, in which the polar code is used as the coding solution for the control channel, that is, the short code block, and the LDPC code is used as the coding solution for the data channel, that is, the long code block.

6.3.2.3 *Performance comparison of the three channel codes*

(1) *LDPC and turbo* [24]:
The following performance comparisons are obtained by simulation, and the simulation conditions are as follows:

- The turbo code is decoded by scaled Max-Log-MAP (Maximum *a posteriori*) with up to eight iterations.
- The quasi-cyclic LDPC codes are decoded by layered Belief Propagation (BP) and iterated up to 25 times.

The simulation results of these are as follows:

(i) *Symbol signal-to-noise ratio*:
It can be seen from Figure 1.60 in Chapter 1 that the data in the physical layer is transmitted by the transmission block, so the parameters related to the transmission block are given in Figure 6.22.

- BLER (Block Error Rate) = 10^{-2}.
- The abscissa is the length of info (information) bits.
- The ordinate is the symbol SNR (dB) required to ensure BLER = 10^{-2}.
- The top-down bit rate of the curve in the figure is gradually reduced.
- AWGN channel.

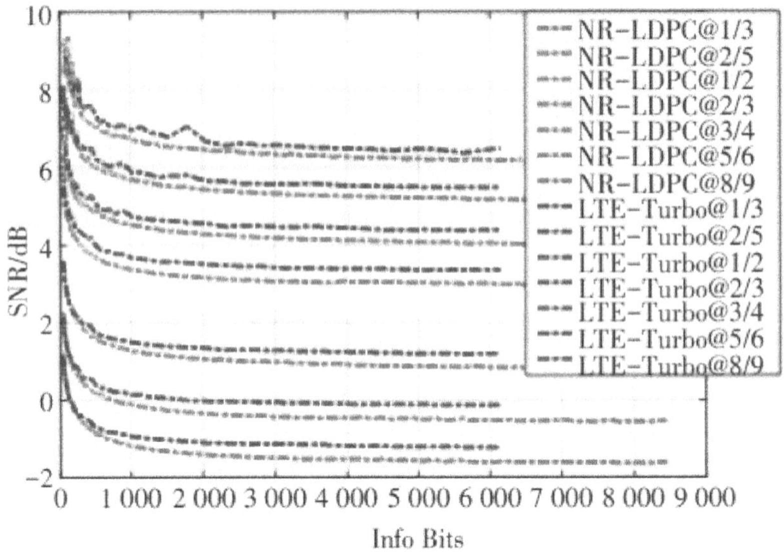

Figure 6.22 Performance comparison of 5G LDPC and 4G turbo codes.

It can be seen from the figure that the LDPC code has 0.3–0.4 dB gain compared to the turbo code.

(ii) *Decoding complexity*:
As shown in Figure 6.23, since turbo codes can only be decoded serially at any code rate, the computational complexity remains unchanged. As quasi-cyclic LDPC codes can intercept the required matrix for decoding, the computational complexity of decoding decreases with the increase of code rate.

(iii) *Decoding delay*:
The decoding of turbo code is based on the serial decoding of mother code, while LDPC can intercept different matrices according to different code rates for decoding, so when the code rate increases, the throughput of LDPC code also increases significantly, while the throughput of turbo code is basically unchanged. The higher the bit rate is, the higher the throughput is required to be in the actual communication system, so LDPC code can better match the requirements of the actual communication system. In particular, quasi-cyclic LDPC codes can use high decoding parallelism, so its throughput potential is huge, which can meet the needs

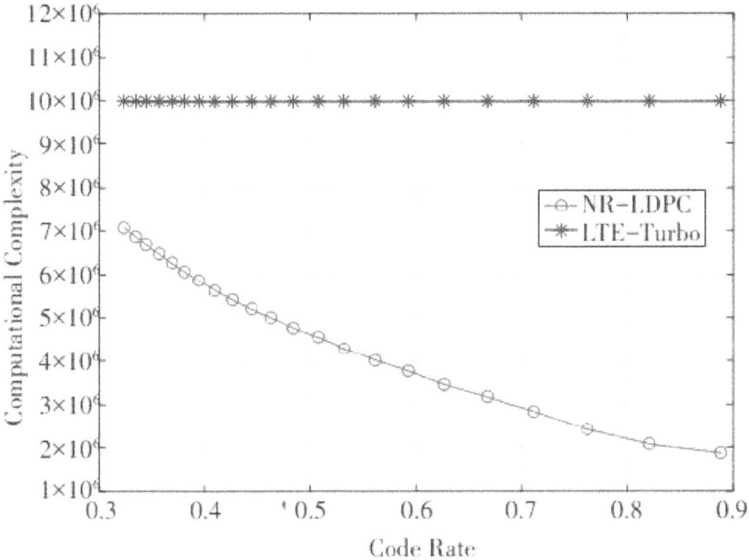

Figure 6.23 Complexity comparison of 5G LDPC and 4G turbo codes.

of different terminals. Some studies show that the throughput of LDPC code is 8.4 times of that of turbo code under the assumption of the same decoding complexity, so the decoding delay of LDPC code is only 1/8 of that of turbo code.

(2) *LDLC and turbo* [25]: On the basis of LDPC codes, Naftali Sommer of Tel Aviv University of Israel and others first proposed a new channel coding technology in 2007, which is called Low-Density Lattice Code (LDLC). It is a practical good code that can achieve the channel capacity of AWGN, and its decoding complexity only increases linearly with the code length. Through simulation, the performance comparison between the LDLC code and the turbo code is shown in Figures 6.25–6.28, which are discussed in the following.

In Figure 6.24, note the following:

- The LDLC code is a binary LDLC code.
- 1/2 turbo code + 16QAM.
- K is the information bit length of turbo code. n is the dimension of the LDLC code, that is, the length of information bits.

- The ordinate is the Bit Error Rate (BER). The abscissa is 1-bit energy signal-to-noise ratio E_b/N_o(dB).
- AWGN channel.

Since turbo + 16QAM with a bit rate of 1/2 is adopted, the spectrum efficiency of the MCS is 4 × 0.5 = 2 (bit/modulation symbol).

It should be pointed out that the spectral efficiency mentioned here is actually another expression of the channel frequency band utilization (also called spectral efficiency) defined by formula (1.5) in Chapter 1 because [(Information bits/Modulation symbol)×(Modulation symbol/Second)]/[Channel bandwidth] = [Data transmission rate]/[Channel bandwidth] (b/s/Hz), which is obviously formula (1.5).

It can be seen from Figure 6.24 that turbo + 16QAM outperforms LDLC codes in the case of spectrum efficiency of 2, when the information bit lengths of the two codes are the same (or similar). For example, when BER = 10^{-5}, turbo codes with 1216 bits of information bit length have a bit error performance gain of about 0.4 dB compared to LDLC codes with

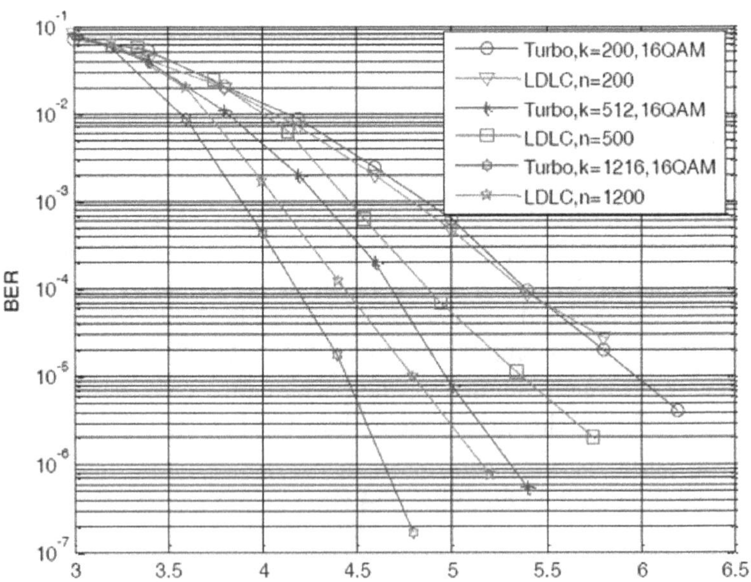

Figure 6.24 Performance of 2-bit/modulation symbol LDLC code and turbo + higher-order modulation.

Note: E_b/N_o(dB).

dimensions of 1200 (similar to 1216). Since the performance curve of turbo codes is steeper, the performance gain of Turbo codes increases with the increase of E_b/N_o (dB).

In Figure 6.25, turbo code with code rate of 4/5 + 16QAM is adopted, so that the spectral efficiency of the MCS is $4 \times 0.8 = 3.2$ (bit/modulation symbol).

As can be seen from Figure 6.25, when the spectral efficiency is 3.2, the performance of the LDLC code is better than that of turbo + 16QAM when the length of information bits of the two codes is the same (or similar). For example, under the condition of BER = 10^{-5}, the bit error performance of LDLC codes with dimension 1200 is about 0.6 dB higher than turbo + 16QAM codes with 1152 bits (similar to 1200).

In Figure 6.26, turbo code with code rate of 4/5 + 32 QAM is adopted, so that the spectral efficiency of the MCS is $5 \times 0.8 = 4$ (bit/modulation symbol).

As can be seen from Figure 6.26, when the spectral efficiency is 4, the performance of the LDLC code is better than that of turbo + 32QAM

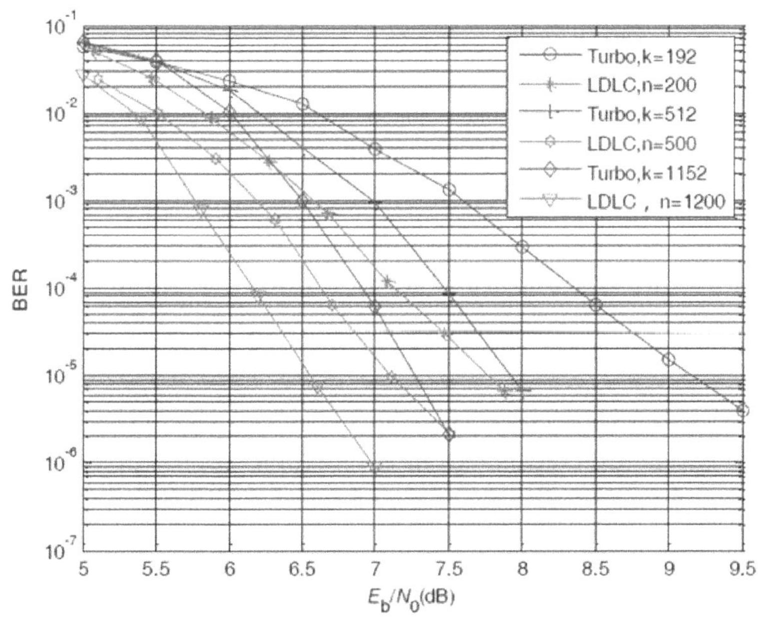

Figure 6.25 Performance of 3.2 bits per modulation symbol LDLC code and turbo + higher-order modulation.

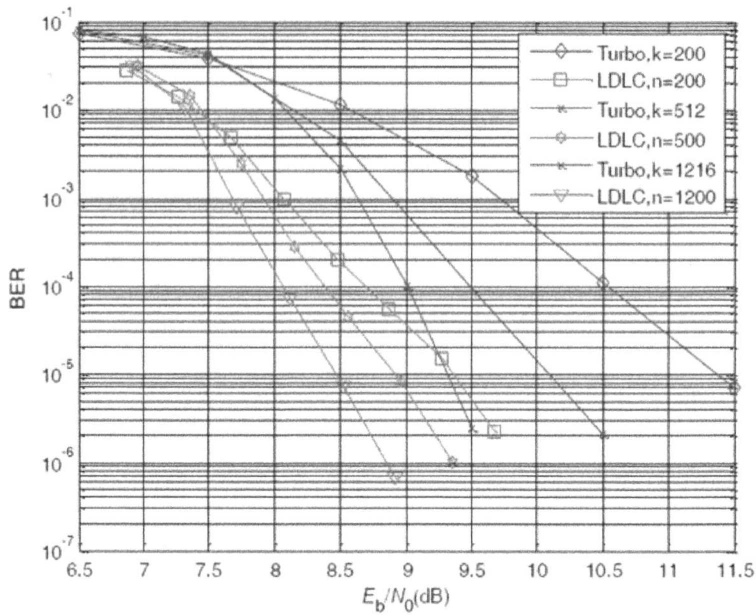

Figure 6.26 Performance of LDLC and turbo + higher-order modulation with 4-bit efficiency per modulation symbol.

when the information lengths of the two codes are the same (or similar). For example, under the condition of BER = 10^{-5}, the bit error performance of the LDLC code with dimension 1200 is about 0.8 dB higher than turbo + 32 QAM with information bit 1216 (similar to 1200).

In Figure 6.27, turbo code with a bit rate of 6/7 + 128QAM is adopted, so that the spectral efficiency of the MCS is $6/7 \times 7 = 6$ (bit/modulation symbol).

As can be seen from Figure 6.27, when the spectral efficiency is 6, the performance of the LDLC code is better than that of turbo + 128QAM when the length of the two information bits is the same (or similar). For example, under the condition of BER = 10^{-5}, the bit error performance of the LDLC code with dimension 1200 is about 1.2 dB higher than that of turbo + 128QAM with 1152 bits (similar to 1200).

As can be seen from the above Figures (6.25–6.28), when the spectral efficiency is high, the performance of the LDLC code is better than that of turbo code, and with the increase of spectral efficiency, the advantage of the LDLC code is more obvious. It can be seen that LDLC code is a channel coding scheme with high-frequency band utilization.

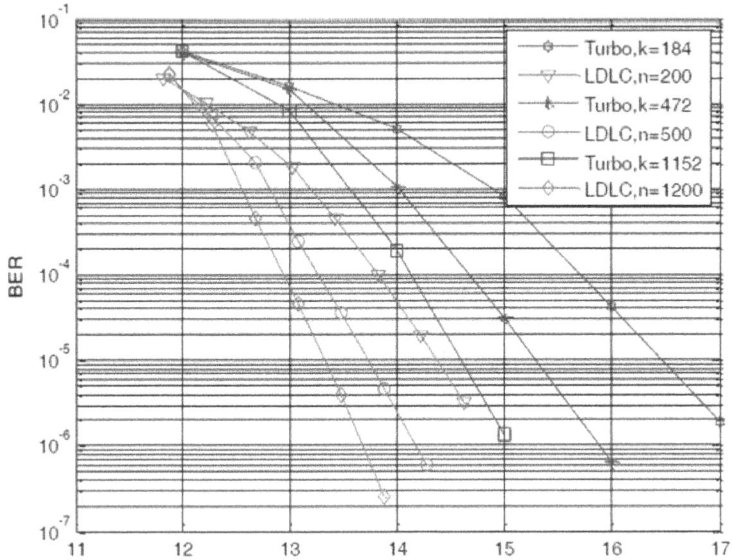

Figure 6.27 Performance of the LDLC codes with spectral efficiency of 6 bits/modulation symbol and turbo + higher-order modulation.

Note: E_b/N_o(dB).

(3) *Polar and turbo codes* [25]:

(i) *Transmit short packets:*

Following are the simulation conditions:

- The length of the information packet is 88 bits, and the length of the coded code group is 256 bits.
- The modulation mode is QPSK.
- The polar code is decoded by SC-List. SC-List can be abbreviated as Successful Cancellation-List (SCL), which is called serial cancellation list algorithm.
- The maximum number of iterations of LTE turbo decoding is eight.
- AWGN channel.

The simulation results are shown in Figure 6.28. For polar code, the list size of different curves in the figure from top to bottom increases gradually. It can be seen from the figure that the performance of LTE turbo is very close to that of polar code with list size = 128; when BLER = 10^{-1},

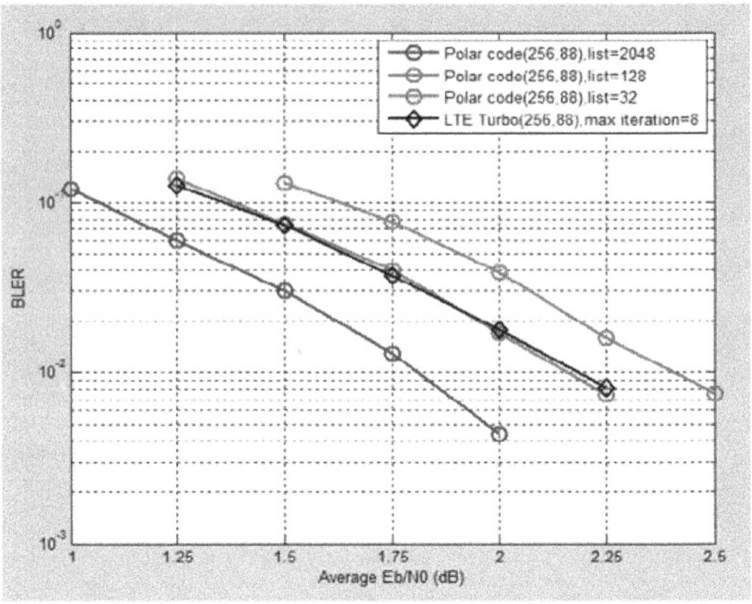

Figure 6.28 Short packet performance comparison.

the BER performance of polar code with list size = 2048 is about 0.3 dB higher than that of LTE turbo code.

(ii) *Transmit long packets*:
Simulation conditions

The length of the information packet is 1024 bits, and the length of the coded code group is 2048.

- The modulation mode is QPSK.
- The polar code is decoded by CA-List. CA-List can be written as CRC-aided SCL, which is called cyclic redundancy check-aided SCL algorithm.
- The maximum number of iterations of LTE turbo decoding is eight.
- AWGN channel.

The simulation results are shown in Figure 6.29. For polar code, the list size of different curves in the figure increases gradually from top to bottom. It can be seen from the figure that for long packets, when

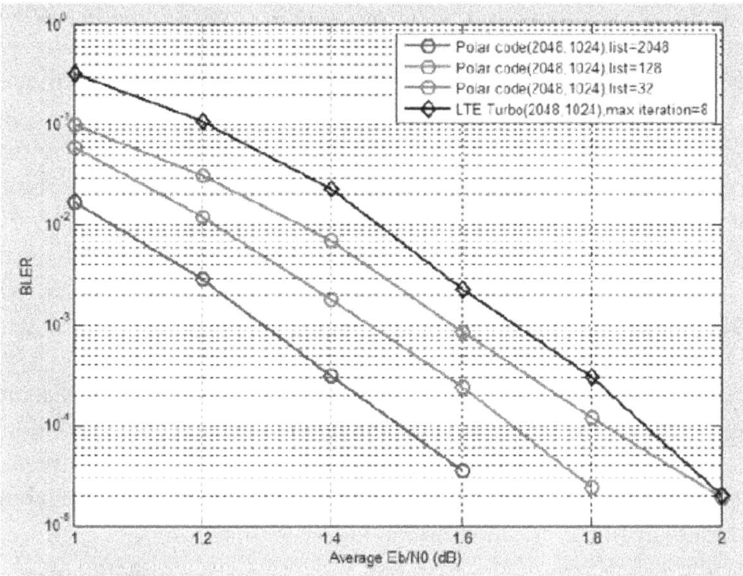

Figure 6.29 Performance comparison of long packets.

BLER = 10^{-1}, the BER performance of all polar codes is at least 0.3 dB higher than that of turbo codes.

6.3.3 *Ultra-dense HetNet and small cell*

In future 5G communication, the wireless communication network is evolving in the direction of network diversification, broadbandization, integration and intelligence. With the popularization of various smart terminals, data traffic will see a blowout growth. In the future, data services will be mainly distributed indoors and as hotspots, which makes ultra-dense HetNet one of the main methods to realize 1000 times the capacity demand of 5G in the future.

The future 5G network will adopt a three-dimensional hierarchical ultra-dense heterogeneous network (HetNet) and deploy a large number of small cells in the macrocell network layer.

Compared to traditional macro base stations, small cells have lower transmit power and smaller coverage, usually covering a range of 10 meters to hundreds of meters. Usually, small cells are based on the size of the coverage area. It is divided into microcell, picocell and femtocell.

The mission of small cells is to continuously supplement the coverage of blind spots and capacity of macro sites and to improve network service quality in a low-cost way. Considering that the 5G wireless frequency band is getting higher and higher and 5G millimeter wave frequency bands will be deployed, the wireless signal frequency band will be higher, the coverage area will be smaller, and the user traffic demand in multiple scenarios will continue to rise in the future. The post-5G era will inevitably deploy a large number of small cells. These small cells will form a super-dense hybrid heterogeneous network with macro stations, which will bring unprecedented complex challenges to network management and frequency interference.

Ultra-dense networks can improve network coverage, greatly increase system capacity and shunt business, with more flexible network deployment and more efficient frequency reuse. In the future, for high-frequency bands and large bandwidth, more dense network solutions will be adopted, and more than 100 small cell sectors will be deployed.

At the same time, the increasingly dense network deployment also makes the network topology more complex. Inter-cell interference has become the main factor restricting the growth of the system capacity, which greatly reduces the network energy efficiency. Interference cancellation, rapid cell discovery, coordination among dense cells, dynamic load balancing and mobility enhancement schemes based on the enhancement of terminal capabilities are all current research hotspots in dense networks.

6.3.4 *Software-defined radio*

Software-defined radio (SDR) can realize a part or all of the physical layer functions defined in software. It should be observed that software-defined radio and software-controlled radio are different; the latter only means that the physical layer functions are controlled by software.

Some traditional physical layer functions such as modulation, demodulation, filtering, channel gain and frequency selection can be realized by SDR. These software calculations can be completed on general-purpose chips, GPU, DSP, FPGA and other dedicated processing chips.

6.3.5 *Cognitive radio*

Cognitive radio (CR) makes decisions in real time by understanding the state of the wireless internal and external environment. SDR is considered to be the enabling technology of CR, but CR includes and enables a variety of technical applications, such as dynamic spectrum access, self-organizing networks, cognitive radio anti-jamming systems, cognitive gateways, cognitive routing, real-time spectrum management and collaborative MIMO.

6.3.6 *Self-organizing network*

A self-organizing network (SON) refers to a network that can automatically coordinate neighboring cells, automatically configure and self-optimize to reduce network interference and improve network operating efficiency.

SON is not a new concept, it was proposed as early as the 3G era, but entering the 5G era, SON will be a crucial technology. As mentioned above, the densification of networks in the 5G era poses unprecedented complex challenges for network interference and management. SON is needed to minimize these challenges. However, even SON may not be able to cope with the super dense 5G network. Therefore, the aforementioned CR technology is needed to help.

6.3.7 *Device-to-device communication*

The traditional networking method of cellular communication system is based on the base station as the center to achieve cell coverage, while the base station and relay station cannot move, and its network structure has certain limitations in flexibility. With the continuous increase of wireless multimedia services, the traditional centralized base station model is unable to meet the business needs of a large number of users in different environments.

Device-to-device communication (D2D) means that data transmission does not pass through the base station, but allows one mobile terminal device to directly communicate with another mobile terminal device. D2D originated in the 4G era and is called LTE Proximity Services (ProSe) technology. It is a short-distance communication technology based on the 3GPP communication system and mainly includes two major functions:

- Direct discovery function, in which the terminal discovers that there are terminals that can be directly connected around.

- Direct communication, which exchanges data with surrounding terminals.

D2D technology can realize direct communication between communication terminals without base stations, which expands network connections and accessing methods. So, D2D communication has the following advantages:

- Due to short-distance direct communication and high channel quality, D2D can achieve higher data rates, fewer delays and low power consumption.
- Through widely distributed terminals, the coverage can be improved and the efficient use of spectrum resources can be realized.
- It supports more flexible network architecture and connection methods to improve link flexibility and network reliability.

At present, D2D adopts broadcast, multicast and unicast technology solutions, and its enhancement technologies will be developed in the future, including D2D-based relay technology, self-organizing network technology, multi-antenna technology and joint coding technology.

Of course, D2D communication technology can only be used as a means of cellular network-assisted communication, but not independent network communication.

In the 4G era, D2D communication is mainly only used in the field of public safety. In the 5G era, due to the emergence of IoT applications such as the Internet of Vehicles, autonomous driving and wearable devices, the application scenarios of D2D communication will be greatly expanded. But it will face challenges such as security and the fairness of resource allocation.

6.3.8 *Massive MIMO*

To increase the radio network speed, one of the main methods is to use multiple antenna technology, that is, multiple antennas are used on the base station and terminal sides to form a MIMO system. The MIMO system is described as $M \times N$, where M is the number of transmitting antennas and N is the number of receiving antennas (such as 4×2 MIMO). MIMO has been widely adopted in 3G and 4G, but at most, it is 8×8 MIMO. MIMO has been introduced many times in the previous chapters.

Multi-antenna technology has undergone the development from passive to active, from two-dimensional (2D) to three-dimensional (3D), from high-order MIMO to massive array, and it is expected to achieve a spectrum efficiency increase of dozens of times or even higher. It is currently one of the important research directions of 5G technology.

Due to the introduction of active antenna arrays and millimeter wave technology, the number of cooperative antennas that can be supported by the same physical space on the base station side will reach 128 or more.

In addition, the original 2D antenna array is expanded into a 3D antenna array to form a novel 3D-MIMO technology, namely stereo multi-dimensional MIMO technology, which supports multi-user beamforming and reduces inter-user interference. Combined with high-frequency millimeter wave technology, 3D-MIMO technology will further improve the wireless signal coverage performance, adding a vertical dimension to the original MIMO technology, so that the beam can be 3D shaped in space, which can avoid mutual interference better. Combined with massive MIMO, multi-directional beamforming can be realized.

Massive MIMO can use a larger number of antennas, and currently, 5G mainly uses 64x64 MIMO. Massive MIMO can greatly increase wireless capacity and coverage, but it faces challenges such as channel estimation accuracy (especially in high-speed mobile scenarios), multi-terminal synchronization, power consumption and computational complexity of signal processing.

6.3.9 *Millimeter wave*

In the past, the traditional working frequency band of mobile communication was mainly concentrated below 3 GHz, which made the spectrum resources very crowded. Millimeter wave (mmWave) refers to radio waves with RF frequencies between 30 GHz and 300 GHz, with wavelengths ranging from 1 mm to 10 mm. The high-frequency band where the millimeter wave is located at is rich in available spectrum resources, which can effectively alleviate the current situation of spectrum resource shortage to realize extremely high-speed short-distance communication and support 5G large-capacity and high-speed requirements. The application of high-frequency bands in mobile communications is the future development trend, and the industry pays great attention to it.

One of the biggest differences between 5G and 2/3/4G is the introduction of millimeter waves. The disadvantages of millimeter waves are large propagation loss and weak penetration. Its advantages are large bandwidth, high speed and the small size of the massive MIMO antenna. So it is suitable for deployment in scenarios such as small cells, indoor, fixed wireless and backhaul.

The following are the main advantages of high-frequency millimeter wave communications:

- The available frequency band of the system is greatly widened.
- Miniaturized antennas and equipment, so user terminals may be equipped with multiple antennas.
- Higher antenna gain.
- Suitable for deploying massive MIMO.

However, high-frequency millimeter wave mobile communications also have shortcomings such as short transmission distance, poor penetration and susceptibility to climate and environment. Problems in radio frequency devices and system design also need to be further studied and resolved.

At present, major research institutions and companies are actively carrying out research on high-frequency band requirements and the selection of potential candidate frequency bands. Although the high-frequency band resources are relatively abundant at present, scientific planning and overall consideration are still needed to optimize the allocation of precious spectrum resources.

Since 5G uses high-frequency millimeter waves, which is a key technical measure to increase the data transmission rate of the system, 5G can easily reach 10 Gbps rate.

Here is an example to illustrate the benefits of increasing the downlink data transmission rate to users, which is the time cost for users to download a 1 GB (gigabyte) HD movie using different mobile communication networks.

3G: $1000 \times 8/14.4 = 555.6$ (seconds), the actual average download rate is lower than the peak rate, so downloading at the average rate requires about $(2–3) \times 555.6$ seconds = (19–28) minutes.

4G (LTE): $1000 \times 8/100 = 80$ (seconds), the time required for downloading at an average rate is about $(2–3) \times 80$ seconds = (2.7–4) minutes.

4G (LTE-A): 1000 × 8/1000 = 8 (seconds), the time required to download at an average rate is about (2–3) × 8 seconds = (16–24) seconds.

5G: 1000 × 8/10000 = 0.8 (seconds), the time required to download at an average rate is about (2–3) × 0.8 seconds = (1.6–2.4) seconds.

6.3.10 *Waveform and multiple access technology*

In the 4G era, OFDM technology is adopted. OFDM has the advantages of reducing inter-cell interference, anti-multipath interference, reducing the implementation complexity of transmitters and receivers and being compatible with multi-antenna MIMO technology. But in the 5G era, due to the definition of the three application scenarios, eMBB, large-scale machine type communication (mMTC) and ultra-reliable low-latency communication (URLLC), these scenarios not only need to consider anti-multipath interference but also MIMO issues such as compatibility and these other issues put forward new requirements for spectrum efficiency, system throughput, delay, reliability, access number, signaling overhead and implementation complexity. To this end, 5G R15 uses the Cyclic Prefix OFDM (CP-OFDM) waveform and can adapt to a flexible and variable parameter set to flexibly support different subcarrier intervals and reuse 5G services of different levels and delays. For the 5G mMTC scenario, because Orthogonal Multiple Access (OMA) may not be able to meet its required connection density, the Non-Orthogonal Multiple Access (NOMA) scheme has become the subject of extensive discussion.

6.3.11 *In-band full duplex*

In-band full duplex (IBFD) may be one of the most promising technologies in the 5G era. Neither FDD nor TDD is full duplex because they cannot transmit and receive signals at the same time under the same frequency channel, while in-band full duplex can achieve simultaneous transmission and reception in the same frequency band, which is the same as half-duplex. Compared to the other schemes, the transmission rate of IBFD can be increased twice.

However, IBFD brings strong self-interference. The key to realizing this technology is to eliminate self-interference. However, it is worth mentioning that self-interference elimination technology is constantly improving. The latest research and experimental results have shown hope, but the biggest challenge is that the complexity and cost are too high.

6.3.12 *Carrier aggregation*

Carrier aggregation (CA), which has been introduced in detail in Chapter 2, Section 2.4 of this book, increases the bandwidth by combining multiple independent carrier channels to increase the data rate and capacity. CA is divided into two types: continuous CA and discontinuous CA. There are also three combinations which are in-band continuous CA, in-band discontinuous CA and inter-band discontinuous CA. The complexity of the implementation increases sequentially.

CA has been widely adopted in 4G LTE and will become one of the key technologies of 5G. The 5G physical layer can support aggregation of 16–32 carriers to achieve high-speed transmission. In addition, the future 5G network will be a converged network, and the CA technology will be greatly expanded to support the following different types of CA between wireless links:

- aggregation of up to 32 carriers in LTE;
- carrier aggregation between systems and 3G-HSPA + wireless links;
- support of FDD and TDD link aggregation, that is, uplink and downlink asymmetric carrier aggregation;
- support of LTE's Licensed Assisted Access (LAA) and LTE's enhanced Licensed Assisted Access (eLAA), that is, supports carrier aggregation with unlicensed spectrum such as Wi-Fi wireless links.

6.3.13 *Dual connection*

DC has also been introduced in Section 5.4 of Chapter 5, where we discussed how the mobile phone can use the radio resources of at least two different base stations (divided into master station and slave station) at the same time when connected. DC introduces the concept of offload bearing, that is, data is offloaded to two base stations at the PDCP layer. The PDCP layer on the user plane of the master station is responsible for PDU numbering, data offloading and aggregation between the master and slave stations. DC is different from CA, which is mainly manifested in the different layers of data offload and aggregation.

When 4G evolved to 5G, the radio access network and core network are disassembled, and 5G radio access network (NR), 5G core network, 4G core network and 4G radio access network (LTE) are mixed and multiple evolution routes of network deployment options are formed. They are as follows:

- Option 1: Independent networking (SA) mode, introducing a 5G core network, and only 5G base stations are connected to the 5G core network.
- Option 2: Non-independent networking (NSA) mode, connected to 4G core network, 4G base station is the primary station, and 5G base station is the secondary station.
- Option 3: Non-independent networking (NSA) mode, introducing a 5G core network, 5G base station as the main station, and 4G base station as the auxiliary station.
- Option 4: Independent networking (SA) mode, introducing a 5G core network, but only 4G base stations are connected to the 5G core network.
- Option 5: Non-independent networking (NSA) mode, introducing a 5G core network, 4G base station as the primary station, and 5G base station as the secondary station.

As can be seen from the above, the so-called non-independent networking is the DC of LTE and NR new radio. Due to the difference an specific implementation, it includes three architectures: EN-DC (option 2), NE-DC (option 3) and NGEN -DC (option 5) architecture.

DC stands for dual connection; E stands for E-UTRA, which is 4G radio access network; N stands for NR, which is 5G new radio air interface; NG stands for next-generation core network, which is 5G core network.

EN-DC refers to the dual connection of 4G radio access network and 5GNR, NE-DC refers to the dual connection of 5GNR and 4G radio access network and NGEN-DC refers to the dual connection of 4G radio access network and 5GNR under the 5G core network.

In the future, 4G and 5G will coexist for a long time. Different dual connection forms such as 4G radio access network and 5GNR dual connection (EN-DC), 5GNR and 4G radio access network dual connection (NE-DC), 4G radio access under 5G core network, the dual connection between the network and 5GNR (NGEN-DC), and the dual connection between 5GNR and 5GNR will exist for a long time in the evolution of 5G networks.

6.3.14 *Low latency technology*

In order to meet 5G URLLC scenarios, such as automatic driving, remote control and other applications, low latency is one of the key technologies. In order to reduce the transmission delay of network data packets, 5G is mainly implemented from two aspects: radio air interface and wired backhaul. For the radio air interface, 5G mainly reduces the air interface delay by shortening the TTI duration and enhancing the scheduling algorithm. For the wired backhaul, MEC deployment is used to bring data and calculation closer to the user side, thereby reducing the physical impact of network backhaul delay.

6.3.15 *Low-power wide-area network technology (LPWA)*

mMTC is a major scenario for 5G. The goal of 5G is to interconnect everything. Considering the exponential growth of the number of IoT devices in the future, LPWA technology is crucial in the 5G era.

Some LPWA technologies are widely deployed (related technologies have been introduced in Chapter 4), such as LTE-M (also known as LTE CAT-M1), NB-IoT (LTE CAT-NB1), LoRa, Sigfox, etc. Low power consumption, wide coverage, low cost and a large number of connections are the common features of these technologies, but these technical features are inherently contradictory. On the one hand, we reduce power consumption, such as making a network enter a dormant state after transmitting data or reducing the coverage area to extend battery life (usually from several years to 10 years). On the other hand, we increase the transmission power per bit and reduce the data rate to enhance coverage. Therefore, weighing the pros and cons according to different application scenarios and seeking the best balance among these contradictions is a long-term issue for LPWA technology.

In the 4G era, two IoT technologies, NB-IoT and LTE-M, have been defined. NB-IoT and LTE-M will continue to evolve from 4G R13 and R14 to 5G R15, R16 and R17. They belong to the future 5G mMTC scenario and is an important part of 5G Internet of Everything.

6.3.16 *Satellite communications*

Satellite communication has been added in the 5G standard. Compared to 2/3/4G networks, 5G is a "network of networks". Satellite communications will be integrated into the 5G framework to achieve a seamless interconnection network composed of satellites, terrestrial wireless and other telecommunication infrastructures. In the future, 5G data will flow dynamically in a seamlessly interconnected network according to bandwidth, delay, network environment and application requirements.

Chapter 7

The Situation and Development of 5G and Future 6G

In this chapter, we mainly discuss the 5G networking situation in China and the United States, since these situations basically represent the global status of 5G, and then present the assumptions and predictions of future 6G.

7.1 Current Situation of 5G in China

7.1.1 *Frequency band of 5G*

China's Ministry of Industry and Information Technology issued the 5G frequency bands (see Table 5.2 in Chapter 5) for testing to the three major telecom corporations in December 2018 according to the frequency bands specified by ITU and then issued the same working frequency bands as the testing bands and commercial licenses on June 6, 2019. Both of them are sub-6 GHz frequency bands, and the specific frequency bands and bandwidths have been discussed in Chapter 5. However, it is worth noting that the China broadcast network also obtained a 5G commercial license on the same day since it controls the 700 MHz frequency band. The 700 MHz frequency band is a low-frequency band, which has the characteristics of wide signal coverage and strong penetration, and the cost of 5G networking is relatively low. At the same time, 4K and 8K ultra-high-definition video is also an important area of commercial 5G.

The 5G license issued by China's Ministry of Industry and Information Technology to the China broadcast network is, in fact, issued to the national cable television and radio industry, so that they can use such frequency band to form a 5G network to transmit ultra-high-definition television. Obviously, it is a vertical Internet of Things network, so that the majority of users can truly experience the smart radio and television services brought by the modern Internet of Things and even the socialized smart city services.

However, it should be noticed that the currently used sub-6 GHz frequency band belongs to the middle- and low-frequency bands, with small path loss and strong penetration, so the cell coverage is wide, which is the advantage of using the sub-6 GHz frequency band. However, due to the low-frequency band, the bandwidth that can be provided is narrow, which means that we can use less frequency resources, which limits the increase in the transmission rate. This is the disadvantage of using sub-6 GHz. It was pointed out in Chapter 5 that some hotspots in 5G networks require a data traffic density of 10 Mbps/M^2, and it is difficult to achieve that with the sub-6 GHz frequency band. Thus, it is necessary to use the millimeter wave frequency band (see Table 5.3 in Chapter 5), and the widely used 28 GHz frequency band in the United States (which is close to the millimeter wave frequency band) is currently a mature solution. Due to high frequency, the millimeter wave frequency band can provide a wider bandwidth and allow higher data transmission rate in the network, so as to achieve the data traffic density requirements mentioned above. However, as the working frequency band increases, its path loss and penetration power decrease, so new technologies such as mMIMO technology, antenna beamforming technology and ultra-dense network (UDN) technology are required, which have already been discussed in Chapter 5.

In summary, China needs to consider the use of millimeter wave frequency bands while gradually improving its sub-6 GHz 5G network.

7.1.2 *5G networking solution*

As mentioned in Chapter 5, there are two schemes for 5G networking: independent networking (SA) and non-independent networking (NSA). SA refers to constructing new base stations, backhaul links and core networks. While SA has introduced new network elements and interfaces, it also adopted new technologies such as network virtualization and software-defined networking on a large scale (see Chapter 6). Although SA is challenging from a technical development perspective, it can basically achieve the three major application scenarios and eight major technical requirements of 5G. NSA refers to the deployment of 5G networks using the existing 4G infrastructure. The 5G carrier based on the NSA architecture only carries user data, but the control signal is still transmitted over the 4G network. Although the NSA network is fast and the temporary investment is small, it needs to use 4G radio air interface when there is no 5G core network (see the third series of the NSA solutions in Chapter 5). While the existing 4G core network architecture and 4G air interface are unable to meet the ultra-low latency and ultra-reliability requirements of 5G. For this reason, NSA cannot support future application scenarios such as the Internet of Things and autonomous driving.

The current (July 2019) 5G networking solutions adopted by China's three major telecom corporations are not exactly the same, which may be related to their reserve funds and technology.

- *China Telecom*: China Telecom has made it clear that its 5G network will give priority to the SA networking solution and realize the synergy of 4G and 5G networks through core network interoperability. Therefore, the possibility of China Telecom adopting the SA solution is 99%.
- *China Unicom*: Although China Unicom has not made a clear statement yet, from the test of 5G experimental networks in various places, most of them are concentrated on NSA, such as Beijing Unicom. Judging from the 5G-related information released by China Unicom, "fast" has been repeatedly emphasized, that is, the rapid establishment of 5G networks. In addition, from a practical point of view, China Unicom's reserve funds are obviously not as much as compared to China Mobile and China Telecom. It can be seen that China Unicom is more than 80% likely to adopt the NSA program.
- *China Mobile*: China Mobile currently has not decided yet. It is testing both SA and NSA solutions and of course it has the strongest resources among the three telecom corporations. It is estimated that China Mobile will focus on NSA in the early stage to solve the rapid deployment of 5G, so that it can quickly start its own 5G brand. However, China Mobile will be soberly aware of the shortcomings of the NSA, so it will also start building SA networking solutions. It can be seen that the possibility of China Mobile first adopting the NSA solution is more than 70%, and at the same time, it will start the SA solution.

There is also a very noteworthy issue that both China Mobile and China Telecom have announced that starting from 2020, 5G phones that only support NSA will not be able to access their 5G networks. This new regulation has two meanings:

(1) It means that China Telecom and China Mobile will soon (by 2020) build the 5G network of the SA solution.
(2) It means that the 5G baseband chip Snapdragon X50 developed by Qualcomm only supports NSA. Therefore, mobile phones made with this chip cannot enter the 5G network built with the SA solution. Some Chinese mobile phone manufacturers, such as Xiaomi MIX3 5G version, OPPO Reno 5G version, Vivo NEX 5G version, and ZTE Axon 10 Pro 5G version, are all "fake 5G phones", because these phones use chips modified with 4G chips and 5G baseband chips (Snapdragon X50). So, these fake mobile phones will soon be unable to connect to the network. Of course, the Qualcomm Snapdragon X55 baseband chip not only supports NSA but also SA.

7.1.3 *5G network functions*

It has been pointed out in Chapter 5 that there are three major application scenarios for 5G networks, and correspondingly there are eight major technical requirements. It can be seen that 5G networks have the following three distinct characteristics.

(1) *Extremely high data rate* (*eMBB*): The basic requirement is 10 Gbps, and the specific scenario requires 20 Gbps.
(2) *Extremely large capacity* (*mMTC*): It is required to reach more than one million users per square kilometer.
(3) *Extremely low latency* (*URLLC*).

Let's take a look at whether the existing 5G network meets the above three requirements (as of July 2019).

- *eMBB*: 3GPP has issued the first version of 5GNR, which mainly refers to the air interface protocol in eMBB application scenarios. However, the current Huawei baseband chip Kirin 980 has a downlink peak data rate of 1.4 Gbps, and the upcoming Kirin 985 or 900 will have a higher data rate. Qualcomm's baseband chip Snapdragon X50 has a downlink peak data rate of 5 Gbps. The downlink peak data rate of the Snapdragon

X55 is expected to be 7 Gbps. It can be seen that the existing 5G network so far only implements part of the functions in the eMBB scenario, and the transmission data rate is far below the required value. There is still a lot of work to be done to truly realize the technical requirements in the eMBB scenario. Of course, the existing 5G data rate is still much higher than 4G.

- *mMTC*: Regarding the air interface protocol of mMTC and URLLC, 3GPP is expected to announce it in the second version of 5GNR in 2020. Therefore, the current 5G network doesn't have these functions, but we can still compare the 5G network capacity requirements with the existing NB-IoT network.

As already known in Chapter 4, the number of users of NB-IoT in the range of 0.87Km^2 is 52K, so that the number of users is 52/0.87 for 60K/km^2, which means that the number of users per square kilometer is 60K. Obviously, it is far from the number of users required by 5G (one million per square kilometer). Therefore, there is still a lot of work to be done to achieve the capacity requirement of 5G network.

- *URLLC*: The various delays produced by 5G networks are shown in Figure 7.1. When 5G is applied in the URLLC scenario, extremely high requirements are put forward on the delay caused by networks, that is, when the reliability of data transmission is 99.999% (five "nine" characters, and the corresponding bit error rate P_b is 1×10^{-5}). Under such a premise, the delay of the network cannot exceed 1 ms. So now what does this "1 ms delay" mean? By searching existing materials, there are three definitions for "1 ms delay". Which one is finally adopted by the ITU can only be known when they release the second version of 5GNR in 2020. The three definitions are discussed in the following.

Definition 1 refers to the one-way wireless transmission delay of the air interface between the mobile phone and the base station in Figure 7.1. Obviously, the 1 ms delay is much smaller than the 4G LTE air interface delay (see Table 2.1 in Chapter 2), so some new technical measures are needed to meet this requirement, but it is still possible from a technical point of view.

Definition 2 refers to the end-to-end delay of the dashed block in Figure 7.1, which refers to the one-way delay from the mobile phone (UE) → base station → network backhaul (optical line) → core network.

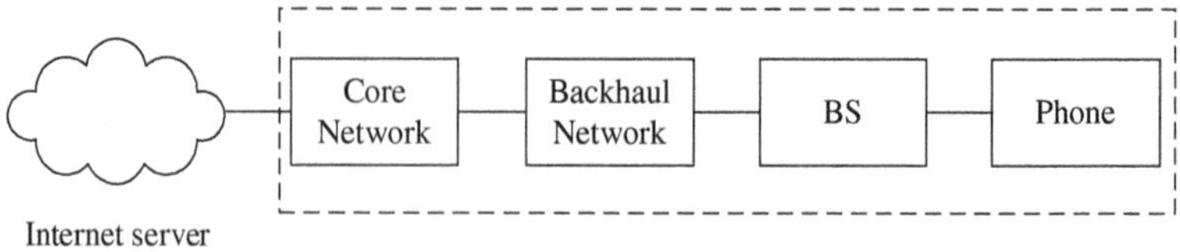

Figure 7.1 Diagram of 5G delay.

Obviously, under this definition, achieving the 1 ms delay requirement is much more difficult than that of definition 1.

Definition 3 refers to the time cost during end-to-end test of ping (ping is a command of Windows, Unix and Linux systems. Ping also belongs to a communication protocol and is part of the TCP/IP protocol. Using the ping command can help us check whether the network is connected or not. It can also help us to analyze and determine the network failure), which is the transmit time of data packet from the mobile phone → base station → network backhaul (optical line) → core network → optical cable transmission → Internet server and then back to the mobile phone. It is the Round-Trip Time (RTT) of the data in Figure 7.1. At this time, the physical distance between the 5G core network and the Internet server will mainly determine the network delay. This is because the Internet servers we want to visit are scattered all over the world, which makes the physical distance between the 5G core network and the Internet servers relatively long.

To meet the 1 ms delay requirement of 5G, we must not only reduce the internal delay of the network but also reduce the physical delay of optical cable transmission. Under this definition, to meet the 1 ms delay requirement, the technical difficulty is much more difficult than that of definition 2. This is because although we can reduce the internal delay of the 5G network by reducing the delay of TTI, frame and HARQ, or even reduce the delay to zero, it is still difficult to meet the requirement of 1 ms, since the propagation speed of optical fiber is 200 km/ms (200,000 km/s), so that the maximum transmission distance of data in 1 ms is only 100 km when the internal delay of the 5G network reaches zero, which is only a distance between two cities. Thus, the scope of the Internet of Vehicles is limited to a city scale, and the distance of remote control is only 100 km, which is far from the true concept of remote control. To solve the physical delay of optical cable transmission, the future 5G technology must sink and split the network, allowing users to access networks nearby (see Chapter 6) to solve the problem of physical delay. Obviously, the 5G network currently built by China cannot meet the above requirements.

In summary, it can be seen that the main focus of China's current 5G network is to solve the problem of data transmission rate in the eMBB scenario. It will take six to eight years to truly realize all functions of the three major application scenarios of 5G. It is estimated that it will take about 10 years for 5G to become mature, and that day might be the day when 6G comes.

China's Ministry of industry and information technology announced that by December 24, 2020, all cities in China have achieved 5G coverage. It is required that the communication industry should continue to promote 5G technology R&D test, accelerate the maturity of network chips, edge computing, chip modules, instrumentation and other technology products, enhance the ability of industrial foundation support, increase the strength of millimeter wave test and promote the maturity of millimeter wave products. They should also accelerate the mature application of R16 version of 5G evolution standard and promote the continuous evolution of R17 and R18 standards.

7.2 Current Situation of 5G in the United States

In 2018, Microsoft conducted a comprehensive survey of broadband access coverage in the United States through its multiple software, such as Office, Windows and Bing. It shows that the current penetration rate of broadband in the United States is about 50%. So, the United States considered the coverage of broadband (optical fiber) when constructing its network. Its design idea is different from China in the following two aspects.

7.2.1 *The US's 5G is built to make up the drawback of low coverage density of 4G and broadband networks*

Since the optical fiber network is not well constructed, the 4G and broadband network cannot cover all areas, resulting in low 4G network coverage density. For this reason, the largest US telecom corporation AT&T and the second largest corporation Verizon want to use 5G technology to make up for the drawback of lacking communication networks in the 4G era and build "fixed 5G wireless services", which is using wireless CPE to convert 5G signals into Wi-Fi (see Chapter 3 for details on wireless CPE) to increase the coverage density of 4G networks. In large cities, Verizon started 5G network services in four US cities in October 2018, and the network name is called "5G Home". It is not difficult to see from the name that the 5G service launched by Verizon this time is mainly for home usage. The converted signal can be connected to a series of smart home appliances such as mobile phones, computers and TVs. It is a fixed network service, which is equivalent to using wireless to replace fiber at home. This means that the United States uses 5G short-distance wireless transmission (connection) to replace the last part of Internet fiber that enters the users' homes. In addition, due to the short wireless transmission distance, it is possible to use millimeter wave frequency band, such as 28 GHz.

The 4G network of China has basically covered all areas of the country, and there is no need to use 5G to make up the difference. Therefore, a SA 5G network can be established according to ITU standards.

7.2.2 *The US favors NSA solution, while China clearly adopts SA solution*

The goal for SA is to split from the existing 4G network and re-establish a new 5G network. Its advantages are ensuring users high-speed Internet access (eMBB) and providing services for IoT and autonomous driving (mMTC and URLLC) to cope with the three major application scenarios of 5G proposed by ITU. As mentioned above (see Section 7.2.1), the first phase of 5G by a major US telecom corporation is to make up for the drawbacks of insufficient 4G network coverage, so it can be inferred that they are using 4G networks to upgrade to 5G networks, which is the NSA networking solution. The NSA cannot support future application scenarios such as IoT and autonomous driving. This problem remains to be resolved.

See Ref. [6] for the development of 5G in the United States.

Currently, China has deployed the largest 5G network and it is serving 80% of the 5G users in the world. The 5G networks operated by the three major operators in China can serve the users with a data rate of up to 1 Gbps, which is three to four times faster than home broadband networks. However, due to the current use of sub-6 GHz frequency band, the data rate is still limited by the bandwidth and far from the requirements in 5G specification. So, it is necessary for us to develop the millimeter wave frequency band. In this respect, Qualcomm has been in a leading position. Its Snapdragon X65 baseband chip (with 4 nm technology) was released in February 2021 with 10 Gbps data rate in the downlink. It has two dual-band wireless access options:

(i) sub-6 GHz FDD band + 28 GHz millimeter wave band;
(ii) sub-6 GHz TDD band + 39 GHz millimeter wave band.

The Snapdragon X65 can be updated through software. This upgradable architecture enables the Snapdragon X65 with great flexibility for network operators, such as home gigabit connections through fixed wireless access, industrial IoT applications and 5G corporate private networks, thus bringing 5G into a new industry.

7.3 R16 Version of 5G

Each generation of mobile communication technology cannot be achieved immediately. It requires a process of gradual improvement. From the evolution of previous generations, it takes about 10 years to replace one generation. During this time, ITU will develop a number of technical standards and its performance is gradually completed and mutually compatible. 5G will also evolve according to this pattern.

As mentioned in Chapter 5, 5G has three major application scenarios as shown in Figure 7.2.

The first layer in the figure is the three major application scenarios of 5G. The second layer is the main technical performance required under the three major application scenarios. The third layer is the specific applications of the three major application scenarios.

7.3.1 *3GPP R15*

In September 2018, 3GPP R15 was released as the first version of 5G, which is applicable in eMBB scenarios and is compatible with the URLLC scenario. Its main feature is a wide bandwidth, which allows transmission with high data rates. For this reason, the four "Methods to Increase Data Transmission Rate" pointed out in Chapter 1, Section 1.1 are adopted in R15 version. For details, see Chapter 5. The transmission signals given in

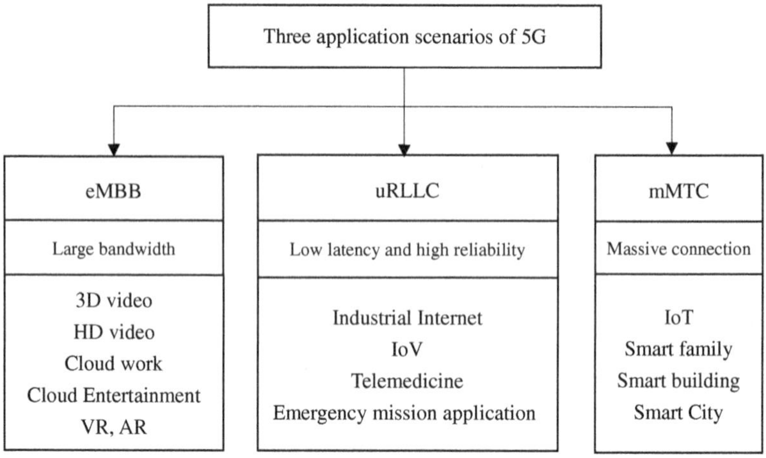

Figure 7.2 Three application scenarios of 5G.

the eMBB scenario in Figure 7.2 are all broadband signals with high data rates. The so-called compatibility with URLLC scenarios mainly refers to reducing the transmission delay as much as possible. According to Chapter 5, the following measures are adopted:

- minimalist design;
- mini-slot;
- various parameter sets;
- use the SUL carrier instead of the uplink carrier (see Figure 5.49).

At present, the commercial 5G network in China is actually designed according to the R15 version. It can be said that it is an upgraded version of 4G and the network speed of experience is faster than 4G. Therefore, the user evaluation of 5G is "usable". We say that 4G changes life and 5G changes society. In fact, the current "usable" 5G is still in the stage of "life-changing" and has not yet entered the realm of "society changing". The true potential and ability of 5G are far from being realized.

The R15 version has completed a series of basic designs in 5G. These basic designs include frame structure design, terminal access design, modulation and coding design, large-scale antenna design, control channel design and multiple access designs. Compared with the 4G system, the 5GNR system design is more flexible and can support more basic parameter configuration, uplink and downlink waveform design, self-contained and flexible frame structure configuration. At the same time, a series of new technologies have been introduced, such as polar codes used for control channels, LDPC codes (low-density parity-check codes) used for data channels and non-orthogonal multiple access.

7.3.2 *3GPP R16*

The R16 version was released in June 2020, and it is the second version of 5G for URLLC scenarios. From the above discussion, it is known that the R15 version has completed a series of basic designs in 5G, so the R16 version only needs to be comprehensively enhanced on the basis of R15 version. The main direction of enhancement is eMBB enhancement and vertical expansion for services and business, including large-scale antenna enhancement, positioning enhancement, integration of access and backhaul, two-step terminal access, dual connectivity/carrier aggregation, 5G Internet of Vehicles, 5G unlicensed frequency band access, ultra-high reliability and low-latency enhancement. Through the comprehensive enhancement of R16 version, 5G will provide stronger network coverage capabilities, lower access and transmission delays, higher transmission rates and lower terminal power consumption, enabling the 5G Internet of Vehicles as a representative of vertical industries put into application and realizing the joint use of authorized and unlicensed spectrum. The main performance indicators of R16 version are as follows:

- It can support synchronization accuracy of 1 μs.
- The air interface delay can be 0.5–1 ms.
- The end-to-end delay within 5 ms can be achieved.

There is no clear quantitative definition for the above three delay indicators in the R15 version, but some technical measures to minimize delay are proposed (see Section 7.5.1).

The reliability of six "nine reliability" can be achieved, that is, the system bit error rate can achieve $P_b = 1\times10^{-6}$. In the R15 version, only five "nine reliability" can be achieved, that is, the system bit error rate can only achieve $P_b=1\times10^{-5}$. It can be seen that the system reliability is improved by an order of magnitude.

The R16 version is also optimized for the high-power consumption of 5G terminals. Through a series of technical measures, it can achieve the same power consumption as 4G terminals while enjoying the high speed and low latency of 5G.

In addition, in the past, the GPS navigation of mobile terminals could only be used for outdoor positioning. After entering a shopping mall or other indoor scenes, it is almost impossible to reach precise positioning. Multiple 5G air interface positioning technologies are introduced in the R16 version, improving the accuracy and reaching the meter-level goal of positioning, which can solve the indoor positioning problem.

Since the R16 version can achieve low latency and high reliability, it can accelerate the realization of the following networks:

- Industrial Internet;
- Internet of Vehicles;
- Support Vehicle to Vehicle (V2V), Vehicle to Infrastructure (V2I), Vehicle to Pedestrian (V2P) and Vehicle to everything (V2X) direct communication;
- Telemedicine.

It can be seen from this that when the above-mentioned network is completed, 5G network will change from the current usable situation into a good situation and from the current "life-changing" situation into "society changing". These brand new networks will show the true ability and power of 5G.

It can be expected that the R17 version will be developed for the application scenarios of mMTC.

7.4 5G R17 Version

In terms of 5GNR, 3GPP has formulated a more powerful radio air interface in the R15 standard version, which can provide a faster and more flexible mobile broadband communication experience.

The 5GNR specification is flexible enough to cope with different device access, frequency spectrum and scenarios. It also greatly reduces latency, improves efficiency and supports connection or aggregation between LTE and NR and between different frequency bands.

The recently released R16 standard version has further extended NR functions to new vertical industries, such as transportation, industrial Internet of things, manufacturing, and vehicles (V2X) industries, while further enhancing capacity and efficiency, including MIMO enhancement, IAB, CLI/RIM, UE energy-saving, URLLC enhancement, NR-U, NR positioning, 2-Step RACH (terminal 2-step access), dual connectivity and carrier aggregation enhancement.

In terms of system architecture, R16 enhances Service-Based Architecture (SBA), improves the flexibility of SMF and UPF deployment, supports the provision of commercial services based on location-based services and enhances UE capability signaling and RAN self-organizing networks.

So, what will the 3GPP R17 version be focused on?

NR-Light

We all know that NB-IoT and eMTC are simplified and lightweight versions of LTE, designed for IoT applications with low power consumption, low cost, low speed, large connections and wide coverage. Entering the era of 5G Internet of Everything, a simplified and lightweight version of 5GNR is also needed, which is NR-Light.

Why do we need NR-Light?

5G defines the three scenarios of eMBB, URLLC and mMTC. eMBB is mainly for 4K/8K, VR/AR and other large bandwidth applications, URLLC is mainly for remote robot control, automatic driving and other ultra-high reliability and ultra-low latency applications, while NB-IoT and eMTC will evolve into mMTC, mainly for low-speed large-scale IoT connections.

To put it simply, URLLC is aimed at "high-end" IoT application scenarios, and mMTC is aimed at "low-end" IoT application scenarios, then comes the problem that there is the blank area between eMBB, mMTC and URLLC of IoT market and this remains to be solved.

Taking 5G intelligent manufacturing as an example, only applications such as robot control and AI quality inspection require ultra-large bandwidth and ultra-low latency network. For surveillance cameras and sensors in factories, ultra-large bandwidth and ultra-low latency could be a waste, while NB-IoT/mMTC cannot meet the demand of latency and bandwidth.

This blank area (see Figure 7.3) is where NR-Light should be.

The performance and cost of NR-Light are between mMTC/NB-IoT and NR eMBB/URLLC. It only occupies 10 MHz or 20 MHz bandwidth.

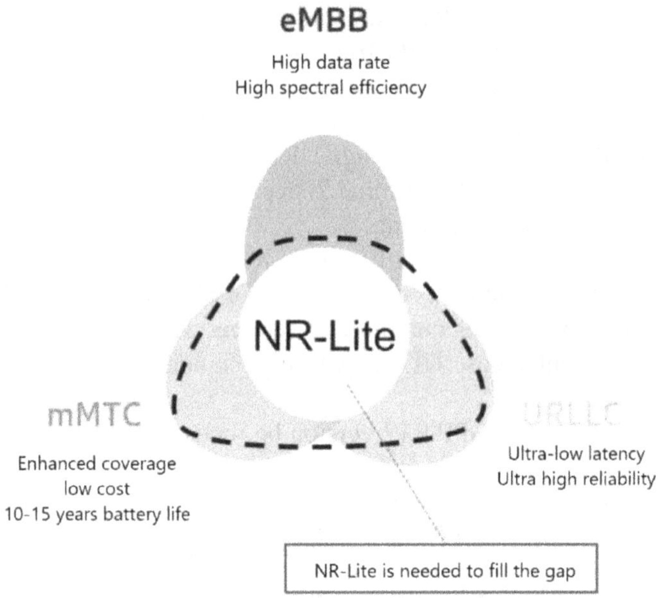

Figure 7.3 Application scenarios of NR-Light.

It supports a downlink rate of 100 Mbps and an uplink rate of 50 Mbps. It is mainly used in industrial IoT sensors, surveillance cameras, wearable devices and other scenes.

The main research directions of NR-Light include the following:

- reduce UE cost and complexity;
- reduce UE uplink and downlink bandwidth;
- reduce UE RX antenna, including 2RX and 1RX;
- reduce baseband complexity;
- reduce UE Tx power level;
- research on technologies to further improve UE energy efficiency;
- research on RRC IDLE/INACTIVE power-saving technology, including idle mode RRM and paging wake-up;
- research on low power consumption technology based on RRC CONNECTED state.

In addition to NR-Light, the R17 version will also conduct research in the following areas:

- NR V2X enhancement;
- NR multicast and broadcast services;
- NR positioning enhancement;
- IAB enhancement;
- XR evaluation;
- NB-IoT/mMTC is integrated with non-terrestrial networks;
- small data packet transmission in inactive state;
- NR coverage enhancement;
- the spectrum range is expanded to 71 GHz;
- others such as Multi-SIM, NR MIMO, NR DSS enhancement, further enhancement of MRDC (Multi-Radio Dual Connectivity), NR UE Power-Saving Enhancement, RAN slicing, SON/Minimization of drive test data collection enhancement, for different 5G business requirements, QoE management and optimization, NB-IoT and LTE-MTC enhancement projects.

7.5 The Future of 6G

6G is the evolution of 5G. Finland first began to study 6G-related technologies in 2018. The current 10 years form the era of 5G, so 6G is expected to be commercialized in 2030.

7.5.1 *What is the main driving force for the evolution from 5G to 6G?*

Mobile communications have a great dream, which is to narrow the digital divide (digital divide refers to the information gap and the trend of further polarization between different countries, regions, industries, companies and communities due to differences in ownership, application and innovation capabilities of information and network technologies during the global digitalization process.) and to achieve global network coverage anywhere and always online. Although 5G is known as an Internet of Everything network, applications such as the Internet of Vehicles and telemedicine require a full-coverage network with almost no blind spots. However, the 5G network is not a true "full-coverage" network. So, in order to achieve "full coverage", the 5G network must be further developed. It can be seen that the main driving force for the evolution of 5G to 6G is the demand and development of the Internet of Things.

7.5.2 *What is 6G?*

So far, there is no unified definition of 6G. For example, some believe that 6G is based on terahertz (THz), some believe that 5G plus AI 2.0 is 6G (AI, Artificial Intelligence), and some believe that 5G plus air networking is 6G. Although the understanding of 6G is different, there are two points in common.

The first is that 6G is the integration of multiple networks, that is, the integration of ground networks and multi-level networks in the air. Specifically, 6G is a combination of satellite networks and terrestrial networks. Satellites are multi-layered. It includes low-orbit satellites, high-orbit satellites and even higher-orbit satellites. In addition to terrestrial communications, terrestrial networks also include maritime communications. Such a three-dimensional network of sea, land and air is possible to achieve the goal of "full coverage", so as to truly realize the intelligent connection of everything on a global scale.

The second is that 6G will use a higher frequency band, such as the terahertz frequency band, which is a sub-millimeter wave frequency band with a frequency range of 100 GHz–10 THz (1THz = 10^3 GHz), while 1G uses the 0.9 GHz frequency band, 4G uses the 1.8 GHz frequency band, and 5G uses the sub-6 GHz and millimeter wave frequency bands (30 GHz–300 GHz). It can be seen that with the evolution from 1G to 6G, its working frequency band is gradually increasing. The purpose of this is that as the frequency band increases, a wider bandwidth will be obtained, more frequency resources can be provided, and the network speed can be improved.

7.5.3 *6G network performance indicators and potential key technologies*

7.5.3.1 *6G network performance indicators*

The 6G network will achieve very large capacity and very small distance communication (VLC and TIC), beyond best effort and high-precision communication (BBE and HPC) and converged multi-type communication (Many-Net). Compared with 5G, 6G's peak rate, user experience rate, time, as well as key indicators such as extension, traffic density, connection density, mobility, spectrum efficiency, positioning capability, spectrum support capability and network energy efficiency have all

Table 7.1 Comparison of key performance indicators between 6G and 5G.

Index	6G	5G	Boost effect
Rate index	Peak rate:100 Gbps–1 Tbps User experience rate:10 Gbps	Peak rate:10 Gbps–20 Gbps User experience rate:0.1 Gbps–1 Gbps	10–100 times
Latency index	0.1 ms, near real-time processing	1 ms	10 times
Flow density	100–10000 $Tbps/km^2$	10 $Tbps/km^2$	10–1000 times
Connection density	Maximum number of connections is 100 $million/km^2$	1 $million/km^2$	100 times
Mobility	Greater than 1000 km/h	500 km/h	2 times
Spectral efficiency	200–300 bps/Hz	100 bps/Hz	2–3 times
Positioning	1 m outdoors, 10 cm indoors	10 cm outdoor, 1 m or less indoors	10 times
Spectrum support	Common carrier bandwidth is 20 GHz, CA can reach 100 GHz	Common carrier bandwidth of sub-6 GHz is 100 MHz, and CA can reach 200 MHz. Common carrier bandwidth of millimeter wave is 400 MHz, and CA can reach 800 MHz	50–100 times
Network energy efficiency	Up to 200 bit/J	Up to 100 bit/J	2 times

been significantly improved. The specific indicators are shown in Table 7.1.

7.5.3.2 *6G potential key technologies*

(1) *Next-generation channel coding and modulation technology*: According to the vision of various countries and related industries, the 6G network will achieve a data rate of 100 Gbps, and by using the

terahertz (THz) frequency band higher than the 275 GHz frequency band, the channel bandwidth will also be in GHz. At the same time, it faces more complex service transmission scenarios such as millimeter wave, space and ocean and poses new challenges to the underlying channel coding and modulation-related technologies.

(i) *New generation of channel coding technology*: As the basic technology of wireless network communication, the new generation of channel coding technology should study and optimize the Tb/s throughput, the GHz channel bandwidth, the THz channel characteristics and the complex transmission model based on the air–space–sea–earth network architecture as well as verify and evaluate the channel coding algorithm and hardware chip implementation scheme. At present, the industry has begun some pre-research, including combining existing turbo, LDPC, polar and other coding mechanisms to develop coding mechanisms and chip solutions for future communication scenarios. Aiming at the complementary research of AI technology and coding theory, research has been carried out on a new channel coding mechanism that breaks through the error correction code technology. At the same time, in view of the multi-user/multi-complex scene information transmission characteristics of the 6G network, the existing multi-user channel coding mechanism is optimized by comprehensively considering the complexity of interference.

(ii) *Design and optimization of polarization multiple access system*: The current industry general view is that non-orthogonal multiple access (NOMA) will become a representative multiple access technology for current 5G and next-generation 6G mobile communications. Introducing the current polarization coding technology into the above-mentioned system, according to the general principle of generalized polarization, optimizes the channel polarization decomposition scheme which is an indispensable part of 5G/6G development. It can be seen that the 6G network will further enable the design and optimization of the polarization multiple access system. It can combine the needs of 6G networks and business scenarios to conduct in-depth research and upgrades on the overall architecture and key technologies of NOMA, build a polarization-coded communication mechanism based on the principle of multi-user (intelligent and ubiquitous connection) and further optimize the related algorithms.

(iii) *Signal processing technology based on deep learning*: Combined with the key parameters of 6G wireless communication, in-depth study and optimization of signal processing technology based on deep learning are required. The industry is currently carrying out related work on channel estimation technology and interference detection and cancellation technology based on deep learning. The channel estimation technology based on deep learning can be modeled by space–time–frequency three-dimensional channel estimation algorithm to independently learn key parameters such as user channels and transmission environment. Prediction of 6G communication system channels mainly involves key technologies such as neural networks and long short-term memory networks. The interference detection and cancellation technology based on deep learning is mainly for autonomous learning and prediction of interference in complex multi-cell scenarios in 6G networks. Optimizing interference detection and cancellation mechanism is mainly based on classic neural network models such as Convolution Neural Network (CNN) and Long Short-Term Memory (LSTM).

(2) *A new generation of antennas and radio frequency technology*: The frequency band of the 6G system can reach terahertz (THz), and the antenna volume is miniaturized. The industry claims that 6G system antennas will be "nano antennas." It brings disruptive changes to traditional antennas and radio frequency, integrated electronics and new materials. It can also empower ultra-large-scale antenna technology, key technologies of integrated radio frequency front-end systems, etc.

(i) *Ultra-large-scale antenna technology*: Ultra-large-scale antenna technology is an important means to better utilize the antenna gain and improve the spectrum efficiency of the communication system. The current 6G terahertz spectrum characteristics research is still in its early stages. Ultra-large-scale antennas face problems such as large-scale cross-frequency bands, full-area coverage theory and technology design of air–space–sea, high power consumption and multiple interference of radio frequency circuits in theory and engineering design. Starting from the above problems, it is necessary to establish a new large-scale array antenna design theory and technology, a highly integrated radio frequency circuit optimization design theory and implementation method and a high-performance large-scale analog beamforming network design technology. In addition, mechanisms such as the development of key technologies for the development of new electronic materials and devices are also needed, and experimental prototypes are developed to support system performance verification.

(ii) *Key technologies of integrated radio frequency front-end system*: In view of the 6G mobile communication technical characteristics such as high integration and large capacity, large-scale antennas and radio frequency front-end technologies within the available frequency bands of 6G networks should be studied. According to the core frequency band technical requirements and circuit modeling theory, the antenna architecture and system integration technology are optimized. High efficiency and easy-to-integrate transceiver front-end key components as well as key technical issues such as radiation and heat dissipation should be explored, and a breakthrough should be achieved in the ultra-large-scale MIMO front-end system technology. At the same time, new device design methods and the integration and packaging technology based on third-generation compound semiconductor chips should also be studied. Methods that can improve circuit performance from the packaging aspect, realize the integration of millimeter-wave chips, packaging and antennas and optimize the overall RF performance of the front-end system should be researched.

(3) *Terahertz wireless communication technology and system*: Terahertz technology has been rated as one of the "Top Ten Technologies to Change the World in the Future" by the industry. A notable feature of 6G is its move toward the terahertz era. At present, research on key technologies of terahertz communication is not yet mature, and many key devices have not been successfully developed which requires a breakthrough. Combining 6G network and business requirements, the main research content of terahertz field is as follows:

- terahertz space and terrestrial communication and channel transmission theory, including channel measurement, modeling and algorithm;
- terahertz signal coding and modulation technology, including high-speed and high-precision capture and tracking mechanism, waveform and channel coding, terahertz direct modulation, terahertz mixing modulation and terahertz photoelectric modulation;

- terahertz antenna and radio frequency system technology, including new material development, new device development, terahertz communication baseband, key antenna technology, high-speed baseband signal processing technology and integrated circuit design methods;
- terahertz communication system experiment, terahertz hardware and equipment development.

(4) *Air–space–sea–earth integrated communication technology*: There is a view in the industry that the 6G network is an effective integration of 5G networks, satellite communication networks and sea networks. The satellite communication network covers various fields such as communication, navigation, remote sensing and telemetry and realizes the global connection of the integration of air, space and sea. The air–space–sea–earth integrated network can optimize land (existing terrestrial cellular, non-cellular network facilities, etc.), sea (sea and subsea communication equipment, marine island network facilities, etc.), air (various types of aircraft and equipment) and sky (various types of satellites, earth stations, and space vehicles) infrastructure to achieve a full element coverage of space, air, land and sea. At present, the inclusion of satellite communications in the 6G network is generally recognized as one of the important subsystems. It is necessary to conduct an in-depth research on key technologies such as network architecture, satellite link selection, space-based information processing and interconnection between satellites. The incorporation of deep sea ocean communication networks into 6G is still in the process of preliminary demonstration.

(5) *Key technologies of software and open-source networks*: The characteristics of 6G network software and open source will be more obvious, so the software and hardware upgrades will be more convenient and efficient. The hardware of the 6G network will be more integrated, modular and white-boxed. The software will be more localized, personalized, flexible and open source. In the future, network infrastructure construction and optimization upgrades will mainly rely on cloud storage resources and software upgrades to fully explore the role of various software and systems in controlling 6G networks. Based on the above development trends, the existing software and key technologies of open-source networks will continue to develop, including big data mining and processing, artificial intelligence AI, software radio (SDR), software-defined networking (SDN),

data cloudification, open-source distributed network software and systems, open-source network security, software and hardware system integration and other technologies.

(6) *AI-based wireless communication technology*: 6G networks inevitably involve general issues such as high-density networks, antenna arrays and data volume. However, a highly autonomous and intelligent ultra-flexible network is one of its most obvious characteristics. 6G intelligence should run through every link of the end-to-end network. AI will perceive learning through multi-dimensional data such as network data, business data and user data and efficiently realize the seamless connections between ground, satellite and airborne devices and can perform real-time high-speed switching. The network's autonomous management and control learning system will continue to be optimized and upgraded. Eventually, an autonomous network like "unmanned driving" will be realized. Key technologies include intelligent core network and intelligent edge network, self-organization and deep learning network technology, channel coding and decoding technology based on deep learning, signal estimation and detection technology based on deep learning, and radio resource allocation technology based on deep learning.

(7) *Blockchain technology*: In order to optimize services, 5G network operators use technologies such as network slicing to control and process traffic and provide differentiated quality services for users. The 6G network will continue to improve customized services for users and adopt more abundant methods. It will provide intelligent and flexible customized services for each user for traffic management and edge computing. The entire network system adopts an automated distributed architecture, and the network tends to be flatter, which makes the blockchain technology highly anticipated. Blockchain is a distributed database, and its distributed information processing technology can be used. The decentralized transmission and storage of data ensure that user information is not stolen by third parties, which steadily improves the collaboration efficiency between network service nodes, improves the collaborative service capabilities of different operators' networks and even changes the way in which radio spectrum resources are used in the future.

(8) *Dynamic spectrum sharing technology*: The terahertz frequency characteristics of 6G make its network density increase sharply. Dynamic

spectrum sharing has become an important means to improve spectrum efficiency and optimize network deployment. Dynamic spectrum sharing adopts an intelligent and distributed spectrum sharing access mechanism. By flexibly expanding the available spectrum and optimizing spectrum usage rules, it can further meet the spectrum resource usage requirements of future 6G systems. In the future, it is an inevitable trend to combine the requirements of 6G large bandwidth, ultra-high transmission rate, air–space–sea–earth scenarios and based on authorized and unlicensed frequency bands continuously optimize spectrum sensing, cognitive radio, spectrum sharing database and efficient spectrum supervision technology. At the same time, it can also promote the synergy of technologies such as blockchain and dynamic spectrum sharing, AI and dynamic spectrum sharing and realize intelligent spectrum sharing and supervision of networks in the 6G era.

7.5.4 *Brief description of 6G white book*

In March 2019, the world's first 6G summit was held in Finland. The organizer of the University of Oulu Summit in Finland invited 70 top communications experts from various countries to hold a closed-door meeting. The main content was to work together to formulate the world's first 6G white book and clarify the basic direction of 6G development.

7.5.4.1 *6G triggers life changes*

The report suggested that by 2030, with the advent of 6G technology, many current dream scenarios will become reality, and human life will undergo tremendous changes.

With the development of technologies such as new display, sensing and imaging equipment and low-power dedicated processors, the current smartphone will be replaced by a lightweight pair of glasses to achieve ultra-high resolution and frame rate through ultra-high network speed, and it can provide the "XR" service that combines virtual reality, augmented reality and mixed reality into one, seamlessly connecting with our senses and movement.

High-resolution sensor imaging, wearable displays and ultra-high-speed radio networks will make remote holography that capture, transmit and render 3D images in real time a reality. For example, in the meeting, each participant is "projected" in real time, and the illusion of perception is created through XR, so that people in different cities feel that they are in the same room. This is widely used in remote education, collaborative design, telemedicine, remote office, advanced 3D simulation and training and national defense.

After 2030, the world will have millions of self-driving vehicles connected to the Internet, and transportation and logistics will be more efficient. These vehicles will include autopilot cars that run between homes, schools and workplaces, as well as autonomous trucks or drones that transport goods. Each car will be equipped with many sensors, including cameras, laser scanners, odometers and terahertz radars. The algorithm must quickly fuse to generate a map of the surrounding environment, including other vehicles, pedestrians, animals and other difficulties it may collide into.

7.5.4.2 *Performance surpasses 5G 100 times*

The white book believes that similar to the previous few mobile communication technology upgrades from 1G to 5G, most of the performance indicators of 6G will be 10–100 times higher than that of 5G (see Figure 7.4).

The white book gives several key indicators to measure 6G technology: peak transmission speed reaches 100 Gbps–1 Tbps, while 5G is only 10 Gbps; indoor positioning accuracy is 10 cm, outdoor 1 m, 10 times higher than 5G; communication delay is 0.1 millisecond, it is one-tenth of 5G; ultra-high reliability, the probability of interruption is less than one in a million; ultra-high density, the density of connected devices reaches more than 100 per cubic meter, and the 5G connection is known from Section 5.2.6 to be only a few per square meter. In addition, 6G will use terahertz frequency band communication, and network capacity will be greatly increased.

In terms of coverage, 6G radio networks are no longer limited to the ground, but will achieve seamless connections between ground, satellite and airborne networks. In terms of positioning accuracy, traditional GPS and cellular multi-point positioning have limited accuracy, making it difficult to accurately deploy indoor objects. 6G is sufficient to achieve high-precision positioning of IoT devices. At the same time, 6G will be deeply integrated with artificial intelligence and machine learning, intelligent sensing, intelligent positioning, intelligent resource allocation, and

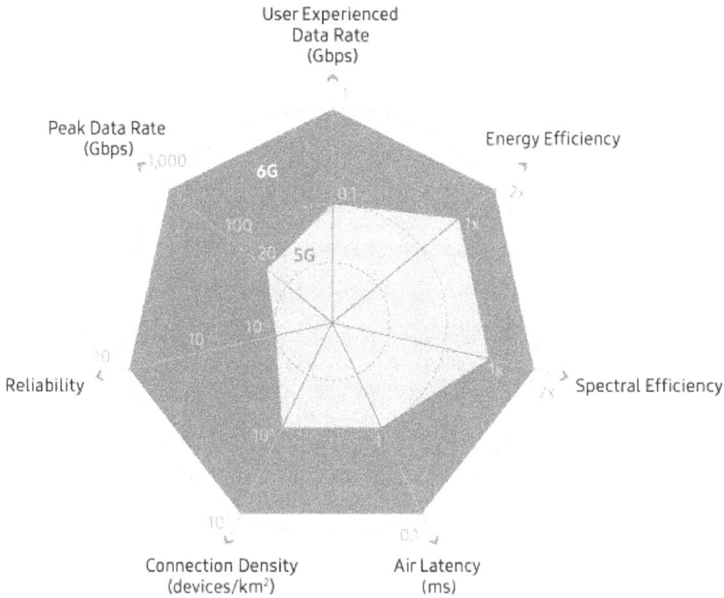

Figure 7.4 6G performance indicators.

intelligent interface switching, which will all become a reality, thus greatly increasing the degree of intelligence.

7.5.4.3 *Technical problems still need to be solved*

The high performance of 6G is very attractive, but there are many technical problems that need to be solved. The first challenge is to overcome the immature terahertz communication technology and realize the ideal communication rate. As the band frequency increases, the antenna volume will become smaller and smaller. When the frequency reaches 250 GHz, 4 square centimeters are enough to install 1,000 antennas. This is a huge challenge for integrated electronics, new materials and other technologies.

The white book believes that by 2030, the digital world will be deeply integrated with the physical world, and people's lives will become more dependent on reliable network operation. This puts higher requirements on the security of communication networks. 6G networks should be able to mitigate and resist cyber-attacks and have the ability to trace the source of the attack.

The arrival of the 6G era will surely bring about the interconnection of everything and generate massive amounts of information. On the one hand, these data are related to personal and corporate privacy, and reliable data protection is a prerequisite for the promotion and application of 6G. On the other hand, real-time processing of these data requires mature edge computing technology, and edge computing also faces limited data access and low computing and storage capacity of devices.

Matti Latva-aho, header of the "6G Flagship Program" at the University of Oulu, stated in a statement of the white book that the root of 6G is data. The way radio networks collect, process, transmit and consume data promotes the development of 6G.

In other words, the research of 6G is to solve the technical problems of data collection and consumption in accordance with 6G standards.

7.6 Starlink

7.6.1 *Starlink project in the United States*

Starlink is a project of SpaceX, the founder of which is Elon Musk. SpaceX plans to build a Starlink network consisting of approximately 12,000 satellites in space between 2019 and 2024, of which 1,584 will be deployed in 550 km low orbit of the earth and start to work in 2020. According to sources, the company is also preparing to add 30,000 satellites, and the total number of satellites in orbit is estimated to be 42,000 by 2027.

In March 2018, the Falcon 9 rocket launched the two test satellites of SpaceX's Starlink program: Ding A and Ding B. A total of eight batches of satellites were launched on May 23, 2019, November 11, 2019, January 6, 2020, January 29, 2020, February 17, 2020, March 18, 2020, April 22, 2020 and June 13, 2020. Each batch uses the Falcon 9 rocket with one-shot multiple satellite technology to launch multiple satellites into space at the same time. In the first seven batches, each batch launched 60 satellites, and the eighth batch launched 61 satellites. The total number of satellites in orbit so far (June 24, 2020) is 483.

Multiple satellites of SpaceX encircle the earth in different orbits. The satellites in the Starlink project are communication satellites, so they form a satellite Internet or space-based Internet. The network is a Low Earth Orbit (LEO), high-speed, low-latency broadband space network.

Figure 7.5 Starlink ground terminal.

The Starlink ground terminal is shown in Figure 7.5. The diameter of the parabolic antenna is 0.48 meters. It can automatically adjust the receiving angle so that the terminal can get the best signal. The terminal can be connected to mobile users through a Wi-Fi router.

Compared with terrestrial 5G mobile communication networks, this network has two advantages:

(1) After the Starlink Space Internet is fully completed, there will be at least three Starlink satellites linked to it at any time anywhere on the earth. Therefore, it can truly achieve full coverage of the earth and no dead ends in the world. This coverage is not restricted and affected by terrain, natural disasters and the degree of ground fiber construction.
(2) The inter-satellite transmission speed in the star chain is faster than that of optical fiber.

The above two points are also the shortcomings of 5G networks.

However, compared with the terrestrial 5G network, the Starlink network has the following three obvious shortcomings:

(1) The one-way delay of the air interface cannot reach the 1 ms required by 5G. This is because when the air interface is transmitted at the speed of light, the delay required for a single-way transmission at a distance of 550 kilometers is 550 km/(300000 km/s) = 1.8 ms>1 ms.

The actual delay of the Starlink network is about 15–20 ms, which is greater than that of the 5G network.

(2) The downlink peak data rate provided for ground terminals is 1 Gbps, which is much lower than the 10 Gbps required by 5G and is basically the current 4G data rate.

(3) Due to the low downlink data rate, high-quality and stable services cannot be provided in areas with dense users (such as areas with dense urban populations).

Through the above analysis, it cannot be predicted that the Starlink network can be used like this.

5G networks can be used in densely populated areas, but Starlink networks can only be used in areas with large areas and sparsely populated areas. This three-dimensional network architecture can greatly enhance the coverage. Through satellite-to-ground interlinking, the goal of eliminating the digital divide can be achieved. From this perspective, it seems to have seen the embryonic form of 6G.

Finally, it must be emphasized that since the Starlink network can cover a wide area with sparsely populated areas, this feature will play a great role in the military or other industry networks. We shall wait and see.

7.6.2 *China's Starlink project*

China's Starlink plan is also in full swing. In 2019, Huawei proposed to launch 10,000 small satellites at the 6G summit to allow 6G to cover the world. The plan is led by China Unicom, Huawei is responsible for R&D and production, and YinHe Aerospace is responsible for satellite network launch. China plans to launch 144 first and then 800, so that a 5G network can be built and a 6G satellite network can be gradually realized on this basis. In addition, Chinese companies have also launched LEO satellite constellation plans such as "Hong Yun", "Hong Yan" and "Sky and Ground Integration".

The first step of the "Hong Yun" plan is to launch the first technology verification star before 2018 to achieve single-satellite key technology verification. The second step is to launch four test satellites to form a small constellation which ensures user experience. The third step is by the end of the "14th Five-Year Plan", all 156 satellites will be connected to complete the construction of the satellite network.

Due to environmental factors such as deserts, mountains and oceans, half of the world's population does not have access to the Internet, and the lack of information limits the development of local economy and culture. "Hong Yun" aims at the deployment of the "One Belt One Road" initiative. With space-based Internet access as the main goal, based on the LEO communication constellation and its ground system, it constructs an information interaction platform for the implementation and industrialization of the "Internet +" strategy.

"Hong Yun" has the characteristics of integration of communication, navigation and remote sensing, global coverage, and system independent and controllable features. With its extremely low communication delay, extremely high-frequency reuse rate and true global coverage, it can meet the needs of Internet access for China and other areas.

The "Hong Yan" plan will be fully completed in 2023. The purpose of this plan is to make sure mobile phones always have a signal and truly achieve no dead ends in the world. It will launch more than 300 base stations into outer space, so that it can solve the signal and network problems in remote areas. The total cost of the project is about 100 billion Yuan. After completion, the Chinese can truly achieve global connectivity. More importantly, it can cooperate with navigation satellites to improve positioning accuracy from sub-meter level to centimeter level.

The LEO satellite orbit adopts the "first-come, first-occupied" rule. When some satellites are damaged or their lifespan ends, new satellites will be launched to make up, without giving up the entire orbital resources, which objectively creates a "first-occupied and forever-obtained" situation. SpaceX announced in October 2019 that it would increase the total number of satellites of Starlink from 12,000 to 42,000 (of which 12,000 have been approved and 30,000 are in application). At present, the number of satellites that have been launched and are still in orbit is about 2,000, and only SpaceX will increase the density of space satellites by 20 times. In the United States, SpaceX needs to submit a satellite frequency application to the Federal Communications Commission (FCC), and the FCC has given a policy tilt. China's current "Hong Yun" plan has 156 satellites, while "Hong Yan" plan has 320 satellites. Adding the number of other companies and judging from the number of satellites which have obtained frequency licenses, the number of satellites is only 5% of SpaceX. The gap between China and SpaceX is huge.

7.6.2.1 *Shanzhi Chen suggested the development of China's LEO satellite communications: Compatible with 5G, integrate with 6G*

From 1G to 5G, after about 40 years of development, mobile communication technology has finally moved from serving people to serving industries, empowering the digital transformation of the economy and society. At the same time, satellite communication has once again become people's focus, especially the "Starlink" proposed by SpaceX, which uses the Falcon 9 recyclable rocket to transmit 12,000 satellites into orbit at a low cost of 60 satellites per rocket, realizing inter-satellite networking, serving mobile broadband communications in space, oceans, forests, deserts and remote areas. Regarding the relationship between LEO satellite communications and 5G/6G, Dr. Shanzhi Chen, who is the Deputy General Manager, Director of the Expert Committee of China Information and Communication Technology Group, Director of the State Key Laboratory of Wireless Mobile Communications and IEEE Fellow, said that judging from multiple dimensions such as demand, application and technology, the current LEO satellite communication and 5G are complementary, while in the 6G era, mobile communication will be moving toward integration of the sky and the ground. The LEO constellation will be organically integrated with the ground mobile communication system to achieve seamless coverage and access for anyone and anything at any time or place. Chen further proposed to accelerate the development of China's LEO satellite communications, suggesting that China's development of LEO satellite communications can reuse the results of 5G technical standards and share the 5G-related industry chain, thereby taking advantage of China's 5G industrial advantages and economies to reduce costs and laying the foundation of sky–ground communications of 6G. China's strategy of LEO satellite communication development is "two strive, two cooperation": to strive for frequency resources and orbital resources and to carry out international and domestic cooperation.

7.6.2.2 *LEO satellite communication is currently complementary to 5G*

Chen analyzed that the construction density of base stations for mobile communications from 1G to 4G is directly proportional to the geographical population density and economic development. The base station construction is prioritized in densely populated and economically developed cities and industrial parks and then expanded to suburbs and rural areas. At present, ground mobile communications only cover about 20% of the land area, which is less than 6% of the earth area. The goal of 5G is to serve thousands of industries and the interconnection of everything, so it is reasonable that the future distribution of 5G base stations will depend on the population distribution and industry application requirements. In industry application requirements, especially air, ocean, forest, desert area and other sparsely populated areas, it is difficult to achieve scenarios, such as airplanes, drones, offshore oil wells, ships, forest fire prevention and wildlife video surveillance, power lines and railway inspections, borderline prevention and control with 5G. Even if such applications on land use 5G networks are possible, the early business model will face great challenges, and the scale of revenue will not match the cost of 5G construction, operation and maintenance.

This brings business opportunities to LEO satellite communication. LEO satellite communication has global coverage, and its cost sensitivity is not directly related to the geographic location of industry applications and the regional density of communication access points, especially for broadband connection and communication in low-density access scenarios. In addition, satellite communication is a quick and effective means of emergency protection in events like disasters. Especially when severe natural disasters such as earthquakes and tsunamis damage the ground base stations, satellite communication becomes the only way of communication for emergency rescue. After major disasters such as the Wenchuan earthquake in 2008 and the Yushu earthquake in 2010, relevant Chinese authorities and telecom corporations urgently deployed various satellite communication equipment to ensure smooth communication in the disaster areas and played an important role in the rescue from the disaster areas. Chen pointed out that due to technical and economic factors, 3G to 4G mobile communications have caused digital divide issues around the world. For example, residents in remote areas do not have mobile communications services and Internet access. This is obvious abroad, especially in developing countries, including Europe and America too. China's policy support for universal services is relatively good. The three major telecom corporations and Administration of Radio and Television have made huge investments in the "Village Access" project. Satellite communications also contributed its own functions. Facing the trend that 5G may further expand the digital division, satellite Internet has become a good tool that can provide universal services and has a relatively low cost. But Chen also pointed out that just like every coin has two sides, LEO satellite communication should be viewed dialectically. It has outstanding advantages and also has limitations. The first is spectrum efficiency. Since the distance, power density and multi-antenna deployment between satellites and ground terminals are restricted, the spectrum efficiency of satellite communication is much lower than that of the cellular mobile communication system in the same period. At present, LEO satellites, including SpaceX and One Web, have an average spectral efficiency of 2.5 bit/s/Hz, which only reaches the 3G level, while the average spectral efficiency of 5G is above 10 bit/s/Hz, which is four times that of satellite communications.

The second is the problem of system capacity. Chen pointed out that in the case of limited frequency resources, to increase the system capacity means to increase the number of satellites, which is the same as the ground cellular communication system through cell splitting (frequency reuse). From the perspective of communication, we can simply view satellite communication as moving ground base stations into outer space. At present, the total number of 4G base stations in the world is about 8.3 million. 5G will build two to three times more compared to the number of 4G base stations, and the Starlink project contains 12,000 stars. If some domestic self-media said that Starlink will subvert 5G and serve the broadband Internet needs of global people, then LEO satellites must achieve a total communication capacity roughly equivalent to 5G, which requires several million satellites densely distributed in the low earth orbit over all the cities of the world. This is completely unimaginable in constellation design and satellite launch. Even a constellation of ten thousand satellites will cause a large amount of space junk and cause environmental problems such as interference with optical observations and a blurred vision of the universe.

Finally, there is the issue of construction, operation and maintenance costs. Although SpaceX has greatly reduced the cost and the threshold of launching satellites, each satellite still faces tens or even hundreds of times the construction, operation and maintenance costs of ground base stations. In addition, satellite terminals are large in size and high in power consumption, and satellite signals cannot cover indoors. The blockage of buildings in cities will also cause user communication interruption, and satellite communication is very sensitive to weather. Therefore, satellite communications can only be used as a useful supplement to ground mobile communications, and its service targets are not the mainstream 5G market, especially for ordinary consumers in cities. Therefore, mobile broadband Internet access in densely populated and economically developed areas such as cities will be based on 5G, supplemented by LEO satellite communications, while communications in space, oceans, forests, deserts and remote areas will use LEO satellite as the main method. The current demand is also very different from the "Iridium" in the 1990s. The range of mobile communication services has begun to shift from traditional interpersonal calls to broadband Internet, Internet of Everything and digital economy era. The market prospects of LEO satellite communications for specific regions, specific user groups and specific applications are still broad.

7.6.2.3 *China's current development of LEO satellite communications is the reuse and compatibility of 5G technical standards*

Chen pointed out that Starlink project is not essentially a contribution to communication technology. Its greatest contribution lies in the innovation of launching multiple satellites with one rocket and rocket recovery. From the perspective of communication technology, maximum reuse and compatibility of sky and ground communication is the correct trend of 5G. Chen said that most of the current LEO satellite communications adopt and improve the third-generation digital satellite television broadcasting standard (DVB-S2/S2X) of the European Union ETSI. The DVB standard initially only supported broadcasting and was later upgraded to support two-way communication (i.e., interactive Internet services). DVB has no mobility management function, no core network and low spectrum utilization efficiency. It can be considered similar to the fixed wireless communication (WiMAX). DVB can support broadcasting service and fixed satellite service, but the ability to support mobile satellite service is poor.

Chen pointed out that with the completion of 5G technical standards and commercial launch, 5G will play an important role in satellite communications from both the perspectives of technology and industry. LEO satellite communication can reuse the key technologies and standards of terrestrial 5G communication to the greatest extent, including large bandwidth, flexible frame structure, signal waveform, channel coding, beamforming and control, mobility management, service architecture and networking methods. On the one hand, it will be very beneficial to the development of satellite communication technology and industry. In addition, the mobile industry chain of China has developed completely from 3G to 5G, including core network systems, base station equipment, mobile terminals, chips, and test instruments.

At present, 5G has officially entered commercial use, and the technology and industrial chain are highly mature. At present, China's LEO satellite communication system is in its infancy. If it can reuse and be compatible with 5G to the greatest extent, industries such as LEO satellite communication gateways and satellite communication terminals can reuse 5G base stations and chips. On the one hand, the most advanced mobile communication technology is introduced while reducing technical risks to a large extent. On the other hand, 5G economies can be used and shared to reduce costs and achieve competitive differentiation. Mature 5G technology and large-scale industries are of great benefit to the development of China's LEO satellite communications industry.

But Chen also pointed out that although LEO satellite communication can reuse and apply 5G key technologies, which is more advanced than DVB technology, due to the many similarities and differences between satellite communication and mobile communication systems, LEO satellite communication cannot directly replicate the current 5G technical standards. The main differences are as follows: (1) The propagation characteristics of satellite channels and terrestrial channels are different, and satellite communications are more susceptible to weather and other factors. (2) The high-speed movement of satellites causes time synchronization tracking performance, frequency synchronization tracking (Doppler effect) and more challenges such as mobility management and modulation performance. (3) The communication propagation distances are different. Satellite communications have long propagation distances, large losses and large transmission delays, which have new impacts on timing relationships and transmission schemes. (4) Satellite communications must adopt a high-point beam to offset the propagation loss, and the attitude of the satellite and the directions of the user antenna can change at any time, resulting in changes to the antenna surface, bringing new challenges to the system design. Therefore, based on these differences and particularities, 5G-based LEO satellite communications must adapt some key technologies of 5G to make adaptive improvements and targeted optimizations. Chen pointed out that since 2016, China Information Technology

(Da tang) began to participate in the 3GPP non-terrestrial network (NTN) standard research work, carried out key technology research, communication system and protocol design based on 5G LEO satellite communication, developed a prototype to participated in key technology verification and cooperated with the domestic industry which resulted in positive progress.

7.6.2.4 *China's strategy for developing satellite communications: Two strive and two cooperation*

In response to China's current 5G-based LEO satellite communications strategy, Chen proposed "two strive and two cooperation", which is to strive for frequency resources and orbital resources, and conduct international and domestic cooperation, respectively. The first is to strive for frequency resources and orbital resources for satellite communications from the International Telecommunication Union (ITU). Frequency resources are as scarce as commercial land in cities. As satellite communications, terrestrial communications and detection services continue to occupy low-frequency spectrum resources, and various applications continue to increase the demand for high-speed mobile communications, the existing Ku-band and Ka-band high-frequency band resources are gradually exhausted, and it will be difficult to meet the huge spectrum demand. It is expected that the higher frequency Q-band, V-band and terahertz band will become the focus of the next-generation communication satellite layout and competition. Therefore, it is necessary to deploy and apply for frequency resources in advance to provide resource guarantee for China's development of satellite communications. In addition, satellite orbits are also limited resources. Starlink's tens of thousands of LEO satellites are planned to be distributed between 340 and 1,300 kilometers, but the satellite orbital position is limited, and satellites cannot be deployed without restrictions. Like frequency resources, orbital resources also require ITU's approval and coordination. In terms of international cooperation, the focus is on participating in the development of international standards such as 3GPP. Chen pointed out that the biggest problem of current satellite mobile communications is that there are too many technical systems and cannot be interconnected. There is no economy scale like terrestrial mobile communications, and it is difficult to interconnect and roam between different satellite operators. Therefore, it is necessary to form a unified LEO satellite communication technology system based on 5G as soon as possible to promote China's satellite communications to the overseas market and participate in international cooperation. At the same time, it can also serve the developing countries of the "One Belt, One Road". Thus, it can not only serve economic cooperation but also solve the problem of digital divide. In terms of domestic cooperation, it should focus on promoting the cross-border cooperation among the commercial aerospace industry, mobile communication industry and integrated circuit industry to jointly promote the advancement of satellite launching, mobile communication and integrated circuit. It can also promote the creation of integrated new industry and realize business model innovation.

7.6.2.5 *Future satellite communications: Integration into 6G*

In the 10-year generation of mobile communications, based on the principle of "Development, commercial and the pre-research of the next generation", it is time to start 6G research when 5G has just been commercialized. At present, China's information technology, Huawei, ZTE and other corporations, three major telecom corporations, Beijing University of Posts, Tsinghua, Southeast University and other communications universities have all begun to conduct 5G- and 6G-related research. Chen believed that 6G will establish an air–space–sea–earth mobile communication network (see Figure 7.6). China's early exploration of satellite communications and 5G compatibility will organically integrate high, medium and low orbit satellite communications and establish the foundation for future development of mobile communications. Future 6G convergence refers to a communication technology standard system that encompasses satellite communications and ground communications, and user terminals adopt uniform identification and access of the entire network and replace the distance between satellite and ground systems with unified management.

Chen said that China Xin ke Group has also established a 6G technology pre-research team, focusing on the pre-research of key technologies such as the integration of ground and satellite communications and promoting the development of coordinated communications around the world. On April 20, 2020, the State-owned Assets Supervision and Administration Commission and the National Development and Reform Commission held a regular economic operation press conference to

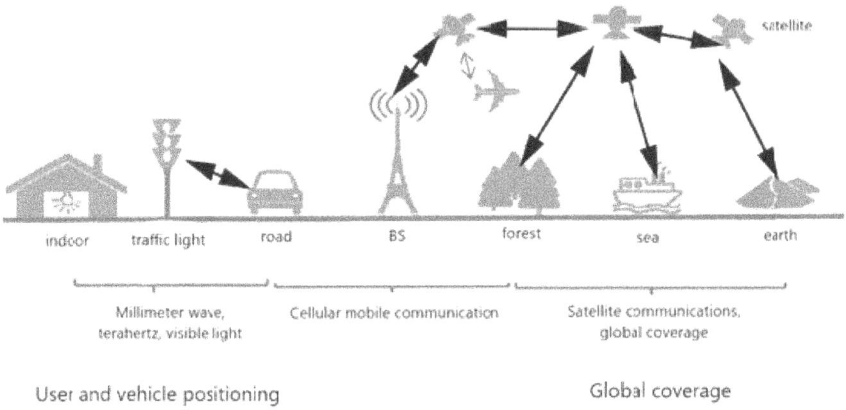

Figure 7.6 6G can organically integrate satellite and ground mobile communications.

clarify the scope of the new infrastructure. The satellite Internet successfully "upgraded" into the new infrastructure strategy. So, 6G opened a new era of converged communications.

7.6.2.6 *The world's first 6G test satellite*

On November 6, 2020, the world's first 6G test satellite "University of Electronic Science and Technology" (Star Times-12/Tianyan 05) successfully lifted off from the Taiyuan Satellite Launch Center by the Long March 6 Yao-san launch vehicle and entered the predetermined track. The "University of Electronic Science and Technology" weighs 70 kilograms and was jointly developed by the University of Electronic Science and Technology of China and National Star Aerospace. The satellite carries a terahertz satellite communication payload designed and developed by the University of Electronic Science and Technology of China and National Star Aerospace. The transceiver link will be established on the satellite platform and the terahertz load test will be carried out. This will also be the world's first technical verification of terahertz communication in space application scenarios.

Abbreviation

3GPP	The 3rd Generation Partnership Project
4G LTE	4G Long Term Evolution
5GC	5G Core
5GNR	5G New Radio
5GS	5G System
8PSK	8 Phase Shift Keying
16/64/256/1024 QAM	16/64/256/1024 Quadrature Amplitude Modulation
AAU	Active Antenna Unit
ACK	Acknowledgement
AD	Active Directory
AES	Advanced Encryption Standard
AF	Application Function
AGV	Automated Guided Vehicle
AI	Artificial Intelligence
AMF	core Access and Mobility Management Function
AMPS	Advanced Mobile Phone System
AR/VR	Augmented Reality/Virtual Reality
ARP	Address Resolution Protocol
ARQ	Automatic Repeat request
AUSF	Authentication Server Function
AWGN	Additive White Gaussian Noise
A/V	Audio/Video

BBU	Baseband Unit
BC	Broadcast Channel
BF	Beam Forming
BLER	Block Error Rate
BRAS	Broadband Remote Access Server
BPSK	Binary Phase Shift Keying
BSC	Base Station Control
BSS	Basic Service Set
BTS	Base Transceiver Station
BWP	Band Width Part
BYOD	Bring Your Own Device
CA	Carrier Aggregation
CAPEX	Capital Expenditure
CC	Component Carrier
CDMA	Code Division Multiple Access
CGNAT	Carrier Grade NAT
CGSN	Gateway GSN
CLI	Cross Link Interference
CMMB	China Mobile Multimedia Broadcasting
CN	Core Network
CoMP	Coordinated Multi-Point
COTS	Connection Oriented Transport Service
CP	Cyclic Prefix
CPE	Customer Premise Equipment
CPE	Common Phase Error
CP-OFDM	Cyclic Prefix OFDM
CS	Circuit Switch
CSFB	CS Fallback
CR	Cognitive Radio
CRB	Common Resource Block
CRC	Cyclic Redundancy Check
CRS	Cell Specific Reference Signal
CPRI	Common Public Radio Interface
CSI-RS	Channel State Information Reference Signal
CSS	Chirp Spread Spectrum

C-RAN	Cloud-Radio Access Network
C-RNTI	Cell-Radio Network Temporary Identifier
C-V2X	Cellular Vehicle to Everything
D2D	Device to Device
DC	Dual Connectivity
DCI	Downlink Control Information
DCI	Data Center Interconnect
DFS	Dynamic Frequency Selection
DFT	Discrete Fourier Transform
DFTS-OFDM	DFT Spread OFDM
DL	Downlink
DMRS	Demodulation Reference Signal
DN	Data Network
DPI	Deep Packet Inspection
DSP	Digital Signal Processing
DTMB	Digital Television Terrestrial Multimedia Broadcasting
DRB	Data Resource Bearer
DRX	Discontinues Reception
DVB	Digital Video Broadcasting
DVR	Digital Video Recorder
DwPTS	Downlink Pilot Time Slot
EAP-TLS	Extensible Authentication Protocol-Transport Layer Security
EDGE	Enhanced Data Rate for GSM Evolution
EGC	Equal Gain Combining
EIRP	Effective Isotropic Radiated Power
eMBB	Enhanced Mobile Broad Band
eDRX	Extended Discontinues Reception
eLAA	Enhanced Licensed Assisted Access
eMTC	Enhanced Machine Type Communication
eNB	Evolved NodeB
EPC	Evolved Packet Core
EPON	Ethernet Passive Optical Network
EPS Fallback	Evolved Packet System Fallback
e LTEeNB	Enhanced LTE Enhanced NodeB
E-UTRA	Evolved Universal Terrestrial Radio Access

E-UTRAN	Evolved UMTS Terrestrial Radio Access Network
EVM	Error Vector Magnitude
FA	Frequency Aliasing
FDD	Frequency Division Duplexing
FDM	Frequency Division Multiplexing
FDMA	Frequency Division Multiple Access
FEC	Forward Error Correction
FFT	Fast Fourier Transformation
FPGA	Field Programmable Gate Array
FSS	Fixed Satellite Service
FWA	Fixed Wireless Access
HyFi	Hybrid Wi-Fi
Gbps	G Bit Per Second
GBPS	G Byte Per Second
GBR	Guaranteed Bit Rate
GGSN	Gateway GPRS Support Node
gNB	(Next)Generation NodeB
GP	Guard Period
GPRS	General Packet Radio Service
GPU	Graphic Processing Unit
GRE	Generic Routing Encapsulation
GSM	Global System for Mobile Communication
GSN	Gateway Service for Network
GTP	GPRS Tunneling Protocol
HARQ	Hybrid Automatic Repeat request
HD	High Dimension
HetNet	Heterogeneous Network
HSDPA/HSUPA	High Speed Downlink/Uplink Packet Access
HSPA	High Speed Packet Access
HSS	Home Subscriber Server
HLR	Home Location Register
IAB	Integrated Access & Backhaul
IBFD	In Band Full Duplex
IBSS	Independent BSS
ICI	Inter Channel Interference

IDFT	Inverse DFT
IFFT	Inverse Fast Fourier Transform
IMS	IP Multimedia Subsystem
IMT-2020	International Mobile Telecommunication-2020
IPv4	Internet Protocol version 4
ISI	Inter Symbol Interference
ISM	Industrial Scientific Medical
ITU-RWP5D	International Telecommunication Union-Radio Communications Sector Working Party 5D
LAA	Licensed Assisted Access
LDAP	Lightweight Directory Access Protocol
LDPC	Low Density Parity Check
LNA	Low Noise Amplifier
LEO	Low Earth Orbit
LOS	Line of Sight
LoRa	Long Rang Radio
LPWAN	Low Power Wide Area Network
LTE	Long Term Evolution
LTE CAT-M1	LTE UE-Category M1(eMTC)
LTE CAT-NB1	LTE UE-Category NB1(NB-IoT)
LTEeNB	LTE Enhanced NodeB
LTE-A	LTE-Advanced
LTE-V2X	LTE Vehicle to Everything
L2TP	Layer-2 Tunneling Protocol
MAC	Medium Access Control
MAC	Multiple Access Channel
MAPL	Maximum Allowable Path Loss
MBB	Mobile Broad Band
Mbps	M bit per second
MCS	Modulation Code Solution
MEC	Mobile Edge Computing
MES	Manufacturing Execution System
MCL	Maximum Coupling Loss
MGW	Media Gateway
MHz	Mega Hertz

MIMO	Massive Input Massive Output
MISO	Multiple Input Single Output
MME	Mobility Management Entity
MQTT	Message Queuing Telemetry Transport
MRC	Maximum Ratio Combining
MSS	Management Support System
MT	Multi Tone
MUSA	Multi User Shared Access
MU-MIMO	Multi User-MIMO
mMIMO	massive MIMO
mMTC	massive Machine Type Communication
mmWave	Millimeter Wave
NACK	Negative ACK
NAS	Non-Access Stratum
NB-IoT	Narrowband Internet of Things
NCDMA	Narrow CDMA
NEF	Network Exposure Function
Ng-eNB	Next Generation eNodeB
NG-RAN	5G Radio Access Network
NFV	Network Function Virtualization
NMT	Nordic Mobile Telephone
NOLS	Non-LOS
NOMA	Non-Orthogonal Multiple Access
NPBCH	Narrowband Physical Broadcast Channel
NPDCCH	Narrowband Physical Downlink Control Channel
NPDSCH	Narrowband Physical Downlink Shared Channel
NPSS	Narrowband Primary Synchronization Signal
NPRACH	Narrowband Physical Random-Access Channel
NPUSCH	Narrowband Physical Uplink Shared Channel
NR	New Radio
NRF	NF Repository Function
NRS	Narrowband Reference Signal
NR-U	5G New Radio in Unlicensed Spectrum
NSA	Non-Stand Alone
NSSS	Narrowband Secondary Synchronization Signal

NVR	Network Video Recorder
OFDM	Orthogonal Frequency Division Multiplexing
OFDMA	Orthogonal Frequency Division Multiple Access
OMA	Orthogonal Multiple Access
OOB	Out of Band
OPEX	Operating Expense
OSI/RM	Open System Interconnection Reference Model
OTA	Over the Air
OTT	Over the Top
OUI	Organizationally Unique Identifier
OVS	Open V-Switch
PAL	Phase Alternating Line
PAR	Peak to Average Power Ratio
PBCH	Physical Broadcast Channel
PCF	Policy Control Function
PCRF	Policy and Charging Rule Function
PDCCH	Physical Downlink Control Channel
PDCP	Packet Data Convergence Protocol
PDMA	Pattern Division Multiple Access
PDSCH	Physical Downlink Shared Channel
PDU	Protocol Data Unit
PGW	PDN Gateway
PE	Provider Edge
PEAP-MSCHAPv2	Protected Extensible Authentication Protocol-Microsoft Challenge Handshake Authentication Protocol version 2
PHY	Physical layer
PIN	Personal Identification Number
PNMA	Power Domain Non-Orthogonal Multiple Access
Polar	Polar Code
POE	Power Over Ethernet
PoS	Point of Sale
PRACH	Physical Random-Access Channel
PRB	Physical Resource Block
PRS	Positioning Reference Signal
ProSe	Proximity Services

PS	Packet Switch
PSD	Power Spectrum Density
PSHO	Packet Service Handover
PSK	Phase Shift Keying
PSK	Pre-Shared Key
PSM	Power Saving Mode
PSS	Primary Synchronization Signal
RTSP	Real Time Streaming Protocol
PT-RS	Phase Tracking Reference Signal
PUCCH	Physical Uplink Control Channel
PUSCH	Physical Uplink Shared Channel
QAM	Quadrature Amplitude Modulation
QCL	Quasi-Co-Location
QoS	Quality of Service
QPSK	Quadrature Phase Shift Keying
RACH	Random Access Channel
Radius	Remote Authentication Dial In User Service
RAN	Radio Access Network
RAT	Radio Access Technology
RAR	Random Access Response
RAU	Route Area Update
RB	Resource Block
RE	Resource Element
RF	Radio Frequency
FRCs	Radio Frequency Carrier shift
RFID	Radio Frequency Identification
RGW	Residential Gateway
RLC	Radio Link Control
RIM	Remote Interference Management
RIT	Radio Interface Technology
RNC	Radio Network Controller
RRC	Radio Resource Control
RRC-IDLE	Radio Resource Control-IDLE
RRC-CONNECTED	Radio Resource Control-CONNECTED
RRC-INACTIVE	Radio Resource Control-INACTIVE

RRM	Radio Resource Management
RRPP	Rapid Ring Protection Protocol
RRU	Radio Remote Unit
RU	Resource Unit
RxD	Receive Data
SA	Stand Alone
SAE	System Architecture Evolution
SAP	Service Access Point
SAR	Specific Absorption Rate
SBA	Service Based Architecture
SCEF	Service Capability Exposure Function
SCH	Shared Channel
SC-FDMA	Single Carrier Frequency Division Multiple Access
SCMA	Sparse Code Multiple Access
SDAP	Service Data Adaptive Protocol
SDM	Space Division Multiplexing
SDL	Supplementary Download
SDN	Software Defined Network
SDR	Software Defined Radio
SDSF	Structured Data Storage network Function
SDU	Service Data Unit
SFN	System Frame Number
SFP	Small Form Pluggable
SGSN	Serving GPRS Support Node
SGW	Serving Gateway
SIC	Successive Interference Cancellation
SISO	Single Input Single Output
SMF	Session Management Function
SMSC	Short Message Service Center
SNR	Signal to Noise Ratio
SON	Self-Organized Network
SUL	Supplementary Upload
SU-MIMO	Single User-MIMO
SS	Spatial Streams
SS	Synchronization Signal

SSB	Synchronization Signal Block
SSID	Service Set Identifier
SSS	Secondary Synchronization Signal
SRS	Sounding Reference Signal
SRVCC	Single Radio Voice Call Continuity
ST	Single Tone
STA	Station
STBC	Space-Time Block Coding
TACS	Total Access Communications System
TAU	Tracking Area Update
TDMA	Time Division Multiple Access
TD-SCDMA	Time Division-Synchronous Code Division Multiple Access
TPC	Transmit Power Control
TRP	Transmission and Reception Point
TRP	Total Radiated Power
TRS	Tracking Reference Signal
TTI	Transmission Time Interval
TWT	Target Wake-up Time
UCI	Uplink Control Information
UE	User Equipment
UDM	Unified Data Management
UDN	Ultra-Dense Network
UDSF	Unstructured Data Storage Network Function
UHD	Ultra-High Definition
UMTS	Universal Mobile Telecommunication System
UL	Uplink
UNB	Ultra-Narrowband
UPF	User Plane Function
UpPTS	Uplink Pilot Time Slot
UWB	Ultra-Wide Band
URLLC	Ultra-Reliable and Low Latency Communication
VLR	Visitor Location Register
VLST	Vertical Layered Space-Time
VoLTE	Voice over LTE
VoIP	Voice over Internet Protocol

VPDN	Virtual Private Dial Network
V2I	Vehicle to Infrastructure
V2P	Vehicle to Pedestrian
V2V	Vehicle to Vehicle
V2X	Vehicle to Everything
xDSL	x (Any Type of) Digital Subscriber Line
WCDMA	Wideband CDMA
WDM	Wavelength Division Multiplexing
WIDS	Wireless Intrusion Detection System
Wi-Fi	Wireless Fidelity
WiMAX	Worldwide interoperability for Microwave Access
WIPS	Wireless Intrusion Prevention System
WRC	World Radio Communications Conference

Appendix

4G/5G frequency band and the four major telecom operator's frequency band allocation.

5G/NR: Frequency Range 1			
Band	Frequency (MHz)	Bandwidth (MHz)	Duplex
n77	3300–4200	10–100	TDD
n78	3300–3800	10–100	TDD
n79	4400–5000	40–100	TDD
n80	1710–1785/N/A	5–30	SUL
n81	880–915/N/A	5–20	SUL
n82	832–862/N/A	5–20	SUL
n83	703–748/N/A	5–20	SUL
n84	1920–1980/N/A	5–20	SUL
n86	1710–1780/N/A	5–40	SUL
n90	2496–2690	10–100	TDD
5G/NR: Frequency Range 2			
Band	Frequency (MHz)	Bandwidth (MHz)	Duplex
n257	26.5–29.5	50–400	TDD
n258	24.25–27.5	50–400	TDD
n259	39.5–43.5	50–400	TDD
n260	37.0–40.0	50–400	TDD
n261	27.5–28.35	50–400	TDD

5G/NR: Frequency Range 1			
Band	Frequency band name (MHz)	Frequency range (MHz)	Bandwidth (MHz)
n1	IMT Core Band	1920–1980/2110–2170	5–20
n2	PCS 1900	1850–1910/1930–1990	5–20
n3	1800	1710–1785/1805–1880	5–30
n5	850	824–849/869–894	5–20
n7	IMT Extension	2500–2570/2620–2690	5–20
n8	900	880–915/925–960	5–20
n12	US 700 Lower A,B,C	699–716/729–746	5–15
n14	US 700 Upper D	788–798/758–768	5–10
n18	850 (Japan #4)	815–830/860–875	5–15
n20	CEPT 800	832–862/791–821	5–20
n25	PCS 1900 G	1850–1915/1930–1995	5–20
n28	APT700	703–748/758–803	5–20
n30	LTE WCS	2305–2315/2350–2360	5–10
n34	TDD 2000 Upper	2010–2025	5–15
n38	IMT Extension Gap	2570–2620	5–20
n39	China TDD 1900	1880–1920	5–40
n40	TDD 2300	2300–2400	5–80
n41	TDD 2600	2496–2690	10–100
n48	US CBRS 3500	3550–3700	5–100
n50	TDDL-band	1432–1517	5–80
n51	TDDL-band, local	1427–1432	5
n65	2GHz LTE	1920–2010/2110–2200	5–20
n66	AWS Extension	1710–1780/2110–2200	5–40
n70	AWS-3/4	1695–1710/1995–2020	5–25
n71	US 600	663–698/617–652	5–20
n74	FDDL-band	1427–1470/1475–1518	5–20
n75	Extended SDL L-band	N/A/1432–1517	5–20
n76	Extended SDL L-band, local	N/A/1427–1432	5

4G/LTE: FDD		
Band	Frequency band name (MHz)	Frequency range (MHz)
1	IMT Core Band	1920–1980/2110–2170
2	PCS 1900	1850–1910/1930–1990
3	1800	1710–1785/1805–1880
4	AWS	1710–1755/2110–2155
5	850	824–849/869–894
7	IMT Extension	2500–2570/2620–2690
8	900	880–915/925–960
9	1700 (Japan #2)	1749.9–1784.9/1844.9–1879.9
10	3GAmericas	1710–1770/2110–2170
11	1500 (Japan #3)	1427.9–1447.9/1475.9–1495.9
12	US 700 Lower A,B,C	699–716/729–746
13	US 700 Upper C	777–787/746–756
14	US 700 Upper D	788–798/758–768
17	US 700 Lower B,C	704–716/734–746
18	850 (Japan #4)	815–830/860–875
19	850 (Japan #5)	830–845/875–890
20	CEPT 800	832–862/791–821
21	1500 (Japan #6)	1447.9–1462.9/1495.9–1510.9
22	FDD 3500	3410–3490/3510–3590
23	US S-Band/AWS-4	2000–2020/2180–2200
24	US L-Band	1626.5–1660.5/1525–1559
25	PCS 1900 G	1850–1915/1930–1995
26	E850 Upper	814–849/859–894
27	E850 Lower LTE	807–824/852–869
28	APT700	703–748/758–803
29	LTE DL FDD	N/A/717–728
30	LTE WCS	2305–2315/2350–2360
31	LTE450 Brazil	452.5–457.5/462.5–467.5
32	SDL L-band	N/A/1452–1496
65	2GHz LTE	1920–2010/2110–2200
66	AWS Extension	1710–1780/2110–2200

4G/LTE: FDD		
Band	Frequency band name (MHz)	Frequency range (MHz)
67	European 700 SDL	N/A/738–758
68	LTE700 Arab Region	698–728/753–783
69	2.6GHz SDL	N/A/2570–2620
70	AWS-3/4	1695–1710/1995–2020
71	US 600	663–698/617–652
72	450 EU BB-PPDR	451–456/461–466
73	450 Region 3	450–455/460–465
74	FDDL-band	1427–1470/1475–1517
75	Extended SDL L-band	N/A/1432–1517
76	Extended SDL L-band, local	N/A/1427–1432
85	Extended band 12	698–716/728–746
87	PPDR PMR/PMAR in EU	410–415/420–425
88	PPDR PMR/PMAR in EU	412–417/422–427

4G/LTE: TDD		
Band	Frequency band name (MHz)	Frequency range (MHz)
33	TDD 2000 Lower	1900–1920
34	TDD 2000 Upper	2010–2025
38	IMT Extension Gap	2570–2620
39	China TDD 1900	1880–1920
40	TDD 2300	2300–2400
41	TDD 2600	2496–2690
42	TDD 3500	3400–3600
43	TDD 3700	3600–3800
44	APT 700 TDD	698–806
45	China TDD 1400	1447–1467
46	LAA	5150–5925
47	V2X	5855–5925
48	US CBRS 3500	3550–3700
49	eLAA 3500	3550–3700
50	TDDL-band	1432–1517
51	TDD L-band, local	1427–1432
52	TDD 3300	3300–3400
53	2400 US	2483.5–2495

LTE-M/Cat-M1&2: FDD (Half/Duplex)		
Band	Frequency band name (MHz)	Frequency range (MHz)
1	IMT Core Band	1920–1980/2110–2170
2	PCS 1900	1850–1910/1930–1990
3	1800	1710–1785/1805–1880
4	AWS	1710–1755/2110–2155
5	850	824–849/869–894
7	IMT Extension	2500–2570/2620–2690
8	900	880–915/925–960
11	1500 (Japan #3)	1427.9–1447.9/1475.9–1495.9
12	US 700 Lower A,B,C	699–716/729–746
13	US 700 Upper C	777–787/746–756
14	US 700 Upper D	788–798/758–768
18	850 (Japan #4)	815–830/860–875
19	850 (Japan #5)	830–845/875–890
20	CEPT 800	832–862/791–821
21	1500 (Japan #6)	1447.9–1462.9/1495.9–1510.9
25	PCS 1900 G	1850–1915/1930–1995
26	E850 Upper	814–849/859–894
27	E850 Lower LTE	807–824/852–869
28	APT700	703–748/758–803
31	LTE450 Brazil	452.5–457.5/462.5–467.5
66	AWS Extension	1710–1780/2110–2200
71	US 600	663–698/617–652
72	450 EU BB-PPDR	451–456/461–466
73	450 Region 3	450–455/460–465
74	FDDL-band	1427–1470/1475–1517
85	Extended band 12	698–716/728–746
87	PPDR PMR/PMAR in EU	410–415/420–425
88	PPDR PMR/PMAR in EU	412–417/422–427

NB-IOT/Cat-NB1&2: TDD		
Band	Frequency band name (MHz)	Frequency range (MHz)
41	TDD 2600	2496–2690

LTE-M/Cat-M1&2: FDD (Half/Duplex)		
Band	Frequency band name (MHz)	Frequency range (MHz)
1	IMT Core Band	1920–1980/2110–2170
2	PCS 1900	1850–1910/1930–1990
3	1800	1710–1785/1805–1880
4	AWS	1710–1755/2110–2155
5	850	824–849/869–894
7	IMT Extension	2500–2570/2620–2690
8	900	880–915/925–960
11	1500 (Japan #3)	1427.9–1447.9/1475.9–1495.9
12	US 700 Lower A,B,C	699–716/729–746
13	US 700 Upper C	777–787/746–756
14	US 700 Upper D	788–798/758–768
18	850 (Japan #4)	815–830/860–875
19	850 (Japan #5)	830–845/875–890
20	CEPT 800	832–862/791–821
21	1500 (Japan #6)	1447.9–1462.9/1495.9–1510.9
25	PCS 1900 G	1850–1915/1930–1995
26	E850 Upper	814–849/859–894
27	E850 Lower LTE	807–824/852–869
28	APT700	703–748/758–803
31	LTE450 Brazil	452.5–457.5/462.5–467.5
66	AWS Extension	1710–1780/2110–2200
71	US 600	663–698/617–652
72	450 EU BB-PPDR	451–456/461–466
73	450 Region 3	450–455/460–465
74	FDDL-band	1427–1470/1475–1517
85	Extended band 12	698–716/728–746
87	PPDR PMR/PMAR in EU	410–415/420–425
88	PPDR PMR/PMAR in EU	412–417/422–427

LTE-M/Cat-0: FDD (Half/Duplex)		
Band	Frequency band name (MHz)	Frequency range (MHz)
2	PCS 1900	1850–1910/1930–1990
3	1800	1710–1785/1805–1880
4	AWS	1710–1755/2110–2155
5	850	824–849/869–894
8	900	880–915/925–960
13	US 700 Upper C	777–787/746–756
20	CEPT 800	832–862/791–821
25	PCS 1900 G	1850–1915/1930–1995
26	E850 Upper	814–849/859–894
28	APT700	703–748/758–803

LTE-M/Cat-1bis: FDD (Duplex)		
Band	Frequency band name (MHz)	Frequency range (MHz)
1	IMT Core Band	1920–1980/2110–2170
2	PCS 1900	1850–1910/1930–1990
3	1800	1710–1785/1805–1880
4	AWS	1710–1755/2110–2155
5	850	824–849/869–894
7	IMT Extension	2500–2570/2620–2690
8	900	880–915/925–960
12	US 700 Lower A,B,C	699–716/729–746
13	US 700 Upper C	777–787/746–756
18	850 (Japan #4)	815–830/860–875
20	CEPT 800	832–862/791–821
26	E850 Upper	814–849/859–894
28	APT700	703–748/758–803
31	LTE450 Brazil	452.5–457.5/462.5–467.5
66	AWS Extension	1710–1780/2110–2200
72	450 EU BB-PPDR	451–456/461–466

LTE-M/Cat-M1&2: TDD		
Band	Frequency band name (MHz)	Frequency range (MHz)
39	China TDD 1900	1880–1920
40	TDD 2300	2300–2400
41	TDD 2600	2496–2690

LTE-M/Cat-0: TDD		
Band	Frequency band name (MHz)	Frequency range (MHz)
39	China TDD 1900	1880–1920
40	TDD 2300	2300–2400
41	TDD 2600	2496–2690

LTE-M/Cat-1bis: TDD		
Band	Frequency band name (MHz)	Frequency range (MHz)
39	China TDD 1900	1880–1920
41	TDD 2600	2496–2690

Frequency band allocation of China's four major operators

5G frequency band				
Telecom operators	Frequency range (MHz)	Bandwidth (MHz)	Band	Note
China Mobile	2515–2675	160	n41	4G/5G spectrum sharing
	4800–4900	100	n79	
China Radio and Television	4900–4960	60	n79	
	703–733/758–788	2×30	n28	
China Telecom, China Unicom, China Radio and Television	3300–3400	100	n78	Indoor coverage sharing
China Telecom	3400–3500	100	n78	Co-construction and sharing
China Unicom	3500–3600	100	n78	

4G frequency band				
Telecom operators	Frequency range (MHz)	Bandwidth (MHz)	Band	Band name
China Mobile	1710–1735/1805–1830	2×25	B3	1800
	1885–1915	30	B39	TDD 1900
	2010–2025	15	B34	TDD 2000
	2320–2370 (Indoor cover)	50	B40	TDD 2300
China Telecom	1765–1785/1860–1880	2×20	B3	1800
	1920–1940/2110–2130	2×20	B1	IMT Core Band
China Unicom	1735–1765/1830–1860	2×30	B3	1800
	1940–1965/2130–2155	2×25	B1	IMT Core Band

2G/3G/NB-IoT frequency band				
Telecom operators	Frequency range (MHz)	Bandwidth (MHz)	Band	Band name
China Mobile	889–904/934–949	2×15	B8	900
China Telecom	824–835/869–880	2×11	B5	850
China Unicom	904–915/949–960	2×11	B8	900

References

[1] G. Yuan, X. Zhang, W. Wang and Y. Yang, "Carrier aggregation for LTE-advanced mobile communication systems" in *IEEE Communications Magazine*, vol. 48, no. 2, pp. 88–93, February 2010, DOI: 10.1109/MCOM.2010.5402669.

[2] M. Iwamura, K. Etemad, M.-H. Fong, R. Nory and R. Love, "Carrier aggregation framework in 3GPP LTE-advanced." *IEEE Communication Magazine*, **48**, 60–67, 2010.

[3] R. Ratasuk, D. Tolli, and A. Ghosh, "Carrier aggregation in LTE-advanced." In *Vehicular Technology Conference* (VTC 2010-Spring), *2010 IEEE* 71st, 2010, pp.1–5.

[4] Translated by M. Chen *et al.*, "5G Mobile Wireless Communication Technology", People's Posts and Telecommunications Press, March 2017.

[5] F. Yang *et al.*, "5G Radio Access Network Architecture and Key Technologies", People's Posts and Telecommunications Press, August 2018.

[6] "The 5G ecosystem: Risks & opportunities for DoD", April 2019 (5G ecosystem risks and opportunities for the US Department of Defense).

[7] S. Wang, Y. Cai, and Y. Xu, "Modern Digital Modulation and Its Application" (Volume 2), Electronic Industry Press, October 2015.

[8] X. Wu, "NB-IoT From Principle to Practice", Electronic Industry Press, September 2018.

[9] X. Xie, "Computer Network" (8th edition), Publishing House of Electronics Industry, 2021.

[10] L. Jiang, "Detailed Explanation of 5G Internet of Things and NB-IoT Technology", Electronic Industry Press, July 2019.

[11] Translated by H. Zhu *et al.*, "5GNR Standard", Mechanical Engineering Press, August 2019.

[12] Architecture, Technology, Implementation, and Operation of 3GPP New Radio, Ahmadi Sassan, June 2019.
[13] "6G Concept and Vision White Paper", China Electronics Information Industry Development Research Institute, March 2020.
[14] Translated by M. Zhu *et al.*, "The Authoritative Guide to 4G Mobile Communication Technology", People's Posts and Telecommunications Press, November 2016.
[15] T. Lu *et al.*, "Narrowband Internet of Things (NB-IoT) Standard Protocol Evolution", People's Posts and Telecommunications Press, July 2020.
[16] https://2ly4hg.smartapps.cn/pages/article/article?articleId=214968099&authorId=99929792&spm=smbd.content.share.0.1613467773744FMhuEXg&_trans_=010005_wxhy_shw&hostname=baiduboxapp&_swebfr=1.
[17] https://post.smzdm.com/p/alpze978/.
[18] https://rf.eefocus.com/article/id-333453?p=2.
[19] "Factory/Workshop Wireless and Internet of Things Construction Solutions", Sundray Network Technology Co. Ltd.
[20] http://www.360doc.com/content/19/0831/23/202378_858366139.shtml.
[21] "2020 5G Millimeter Wave White Paper".
[22] X. Liu *et al.*, "5G Wireless System Design and International Standards", People's Posts and Telecommunications Press, February 2019.
[23] J. Zeng *et al.*, "5G New Multiple Access Mode", People's Posts and Telecommunications Press, June 2017.
[24] Xu Jun, Peng FOCAI and Xu Jin, "Research on 5G Channel Coding", Post and Telecommunications Design Technology, March 2019.
[25] IMT-2020 (5G) Promotion Group "Technical Report of New Modulation Coding Thematic Group".
[26] https://www.sohu.com/a/169764206_120921.

Index

D

E

G

H

S

www.ingramcontent.com/pod-product-compliance
Lightning Source LLC
Chambersburg PA
CBHW061517060426
42492CB00005B/190
* 9 7 8 9 8 1 1 2 4 5 0 5 3 *